Spectroscopy of Pharmaceutical Solids

DRUGS AND THE PHARMACEUTICAL SCIENCES
A Series of Textbooks and Monographs

Executive Editor

James Swarbrick
PharmaceuTech, Inc.
Pinehurst, North Carolina

Advisory Board

Spectroscopy of
Pharmaceutical Solids

edited by
Harry G. Brittain
Center for Pharmaceutical Physics
Milford, New Jersey, U.S.A.

CRC Press
Taylor & Francis Group
Boca Raton London New York

CRC Press is an imprint of the
Taylor & Francis Group, an **informa** business

CRC Press
Taylor & Francis Group
6000 Broken Sound Parkway NW, Suite 300
Boca Raton, FL 33487-2742

First issued in paperback 2019

© 2006 by Taylor & Francis Group, LLC
CRC Press is an imprint of Taylor & Francis Group, an Informa business

No claim to original U.S. Government works

ISBN-13: 978-1-57444-893-1 (hbk)
ISBN-13: 978-0-367-39093-8 (pbk)

Visit the Taylor & Francis Web site at
http://www.taylorandfrancis.com

and the CRC Press Web site at
http://www.crcpress.com

Preface

Spectroscopic techniques have been widely used in the pharmaceutical sciences to obtain both fundamental and applied information. During the development of any given drug candidate, spectroscopy will be used to establish the structure of the compound and understand its interaction with other constituents. It is also often used as a means for evaluating the analytical characteristics of the bulk drug substance and its formulations. It is no surprise, therefore, that solid-state spectroscopic methods have become extremely important to successful modern drug development.

Many scientists believe themselves to be familiar with the principles that govern the interaction of electromagnetic radiation with matter, and yet their knowledge is often based on partial truths. For instance, most would state that for a molecule to absorb ultraviolet light, an electron must be promoted from one energy level to another. While in some cases there is validity to this belief, the true origin of the transition is a change in the orbital angular momentum of the molecule, and the absorption of a quantum of light causes the transition from one molecular state to another. Genuine knowledge as to the origin of spectroscopic phenomena might not change the routine use of a particular technique, but it would provide a basis that could lead to a more advanced application for that technique.

The reasoning just stated has led to the need for the present volume. However great the use of solid-state spectroscopy might be, a greater degree of fundamental understanding is necessary to obtain maximal use out of each technique. In the present work, the underlying principles of each technique will be sufficiently outlined to provide a thorough and proper understanding of the physics involved, and then applications will be used to illustrate what can be learned through the employment of the method under discussion. Whenever possible, the examples will be drawn from the pharmaceutical literature, but this rule will be violated whenever the author feels that an application from another field might inspire analogous work by a pharmaceutical scientist.

In 1995, I edited a volume entitled "Physical Characterization of Pharmaceutical Solids" in which a fairly extensive overview of methods suitable for work at the molecular, particulate, and bulk levels was provided. Since a substantial portion of this earlier book was concerned with the use of spectroscopy for the characterization of solids having pharmaceutical, the present volume may be viewed as being Volume 1 in the second edition of the older book. In the present volume, the use of spectroscopy for the characterization of pharmaceutical solids has been taken much further, and the range of topics has been greatly extended relative to the coverage of the earlier volume.

Harry G. Brittain

Contents

Part IV. Vibrational Level Spectroscopy

Part V.   Nuclear Spin Level Spectroscopy

Contributors

Harry G. Brittain   *Center for Pharmaceutical Physics, Milford, New Jersey, U.S.A.*

David E. Bugay   *SSCI, Inc., West Lafayette, Indiana, U.S.A.*

Robert P. Cogdill   *School of Pharmacy, Duquesne University, Pittsburgh, Pennsylvania, U.S.A.*

James K. Drennen, III   *School of Pharmacy, Duquesne University, Pittsburgh, Pennsylvania, U.S.A.*

Ales Medek   *Pfizer Global R&D, Groton, Connecticut, U.S.A.*

1

Electromagnetic Radiation and Spectroscopy

Harry G. Brittain

Center for Pharmaceutical Physics, Milford, New Jersey, U.S.A.

NATURE OF ELECTROMAGNETIC RADIATION

It had been known since ancient times that a moist atmosphere would split light from the sun into a rainbow of colors. In 1672, Newton used an apparatus similar to that shown in Figure 1 to demonstrate that the same type of splitting could be effected using a glass prism as the active element. In this work, he showed that the refractability of the light increased on passing from red to violet and postulated a corpuscular theory for the nature of light. Unfortunately, Newton's great reputation discouraged others from challenging his theory, and this situation persisted until a new wave theory was presented to the Royal Society by Young in 1801. Needless to say, a fierce scientific and philosophical debate ensued. In 1815, Fresnel developed a mathematical theory to interpret the phenomenon of interference and also explained the polarization of light by assuming that the light vibrations, which pass through a medium, are contained in a plane transverse to the direction of propagation. The great controversy over the wave versus corpuscular nature of light was resolved in 1850 by Foucault, who designed a revolving-mirror apparatus that could measure the velocity of light through different media. He conclusively demonstrated that light travels more slowly in water than it does in air, a finding that was required by the wave theory but incompatible with the corpuscular theory of Newton.

In 1873, Maxwell published his, "Treatise on Electricity and Magnetism," in which he presented the most definitive statement regarding the classical

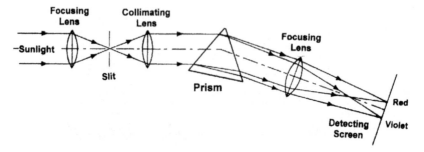

Figure 1 Spectroscopic apparatus of the type used by Newton to observe the splitting of sunlight into visible colors.

theory of electromagnetic radiation. In this theory, electromagnetic radiation is described entirely by a wave theory that associated oscillating electric and magnetic fields with the radiation. Ordinary radiation traveling along the *z*-axis can be treated in terms of electric (**E**) and magnetic (**H**) fields that are mutually perpendicular to each other and to the direction of propagation. Polarized light is then considered to be the electromagnetic radiation whose electric and magnetic vectors are constrained to vibrate in a single plane parallel to the *z*-axis. These relationships are illustrated in Figure 2.

An electric charge has an electric field associated with it, which radiates outward in a uniform manner from its center. If this charge moves with constant velocity or is at rest, then the spherical symmetry of the field is preserved. If the charge accelerates, the field will tend to lag behind, because it will take time to accommodate itself to the changing velocity of the charge. This change causes a disturbance in the direction of the field, which radiates out as a spherical front into the surrounding space. This disturbance, which is propagated with a definite velocity, *c*, constitutes electromagnetic radiation. It has been found that this velocity is constant for all electromagnetic radiation in a vacuum, having a value of $c = 2.998 \times 10^8 \text{ m/sec}$.

By convention, electromagnetic radiation is taken to propagate along the *z*-axis, and the oscillation of the electric and magnetic fields lies in the *xz*- and *yz*-planes. The associated electric field strength vector varies sinusoidally in phase with the oscillating dipole, and the energy spreads outward as a continuous train of electromagnetic radiation, polarized along the direction of oscillation.

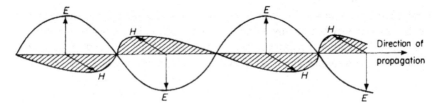

Figure 2 Illustration of the wave nature of electromagnetic radiation.

The value of the electric field strength vector, $\mathbf{E}$, along the z-axis at a given time and position on the z-axis is given by:

$$\mathbf{E} = \mathbf{E}_o \cos 2\pi v(t - z/c) \tag{1}$$

where $\mathbf{E}_o$ is a vector that defines the amplitude of the electric field, and v is the frequency of the electromagnetic radiation.

One often finds it convenient to characterize electromagnetic radiation by the distance between equivalent points along the traveling wave, a parameter known as the wavelength, λ. In any homogenous medium of refractive index, n, the velocity of propagation is now given by c/n. Consequently the frequency and the wavelength are interrelated by:

$$c = nv\lambda \tag{2}$$

A unit that has found extensive use for the description of certain types of electromagnetic radiation is the wavenumber, which is defined as the number of waves contained in a path length of 1 cm. The wavenumber, $\tilde{w}$, is obtained as the reciprocal of the wavelength, so that:

$$\tilde{w} = 1/\lambda = v/c \tag{3}$$

As mentioned before, there is a magnetic field lying in a plane perpendicular to the electric vector, associated with the oscillating electric field, and oscillating in phase with it. The amplitude of the two fields are not independent and are related by:

$$\mathbf{E}_o/\mathbf{H}_o = (\mu/\varepsilon)^{1/2} \tag{4}$$

where $\mathbf{H}_o$ is a vector that defines the amplitude of the magnetic field, μ is the magnetic permeability, and ε is the dielectric constant of the medium in which the radiation is propagating.

The spectrum of electromagnetic radiation is divided into various regions for convenience, and definitions for these are given in Table 1. Most spectroscopists delineate a particular region by its wavelength interval, and it is clear from the table that the preferred SI unit of the meter is only convenient for radiowaves. In the ultraviolet and visible regions, one finds it more convenient instead to specify wavelengths in units of nanometers.

Micrometers are more appropriate for the infrared (IR) region, and centimeters best describe wavelengths in the microwave portion of the spectrum. As will be developed in succeeding chapters, each spectral region is characterized by electromagnetic radiation having a given energy, and the magnitude of this energy determines the nature of its interaction with atoms and molecules.

CLASSICAL DESCRIPTION OF ELECTROMAGNETIC RADIATION

In the classical world, a molecule had to contain either a permanent or transient dipole moment in order to interact with electromagnetic radiation. A bond

Table 1 Spectral Regions of Electromagnetic Radiation

Electromagnetic radiation type	Wavelength (m)	Wavenumber (cm^{-1})	Frequency (Hz)
γ-ray	Less than 1.0×10^{-10}	Greater than 100,000,000	Greater than 3.0×10^{18}
X-Ray	1.0×10^{-10} to 1.0×10^{-8}	1,000,000 to 100,000,000	3.0×10^{16} to 3.0×10^{18}
Ultraviolet			
Far	1.0×10^{-8} to 2.0×10^{-7} (10 to 200 nm)	50,000 to 1,000,000	1.5×10^{15} to 3.0×10^{16}
Near	2.0×10^{-7} to 4.0×10^{-7} (200 to 400 nm)	25,000 to 50,000	7.5×10^{14} to 1.5×10^{15}
Visible	4.0×10^{-7} to 7.5×10^{-7} (400 to 750 nm)	13,350 to 25,000	4.0×10^{14} to 7.5×10^{14}
Infrared			
Near	7.5×10^{-7} to 2.5×10^{-6} (0.75 to 2.5 µm)	4,000 to 13,350	1.2×10^{14} to 4.0×10^{14}
Mid	2.5×10^{-6} to 2.5×10^{-5} (2.5 to 25 µm)	400 to 4000	1.2×10^{13} to 1.2×10^{14}
Far	2.5×10^{-5} to 4.0×10^{-4} (25 to 400 µm)	25 to 400	7.5×10^{11} to 1.2×10^{13}
Microwave	4.0×10^{-4} to 1.0×10^{0} (0.04 to 100 cm)	0.01 to 25	3.0×10^{8} to 7.5×10^{11}
Radiowave	Greater than 1	Less than 0.01	Less than 3.0×10^{8}

formed between atoms having different electronegativity values results in a charge separation and in the formation of a permanent electric dipole. The total electric moment, **P**, of a molecule involves a summation of charge separations over all the electronic and nuclear coordinates:

$$\mathbf{P} = \sum_i q_i r_i \tag{5}$$

where q_i defines the magnitude of charges separated by a distance r_i. To a first approximation, one can assume that the nuclear coordinates are fixed and that inner-shell electrons have a spherical charge distribution. Therefore, the major contribution to the total electric moment arises from the electronic cloud of the bonding and nonbonding valence electrons. The total electric dipole moment of a polyatomic molecule is properly obtained as the resultant from the vectorial addition of the products of the time-averaged electronic and nuclear coordinates, and their corresponding charges, over the entire molecule. Molecules that possess a permanent dipole moment are said to be *polar.*

Neutral atoms or molecules that have a symmetrical charge distribution cannot yield a permanent dipole moment and are termed *nonpolar.* However, when such species are placed in an electric field, both the electronic and nuclear charge distributions become distorted, resulting in an induced dipole moment. For symmetrical atoms or molecules, the magnitude of the induced moment is proportional to the strength of the external field. The proportionality constant in the relationship is the *polarizability* (α) and will consist of contributions from the atomic and electronic polarizabilities.

The classical description as to how atoms and molecules interact with electromagnetic radiation becomes a determination as to the circumstances whereby electric dipole absorption or emission can occur. For example, during the rotation of a polar molecule, the change in its dipole moment will be solely a consequence of the change in orientation. If there is no molecular electric dipole moment, then there can be no interaction with the radiation field. If radiation is to be absorbed or emitted, the distortion associated with a molecular vibration must give rise to a change in the magnitude of the dipole moment. Changes in magnetic dipole moment can also give rise to absorption or emission through an interaction with the magnetic vector in the radiation, but the observed intensity will be many orders of magnitude lower than for the electric dipole case. Such interactions will be seen, however, to be of central importance for nuclear magnetic resonance and electron paramagnetic resonance.

Unfortunately for the workers in the late 1800s, the Maxwell theory was maddeningly vague as to the intimate details of the mechanisms whereby electromagnetic radiation interacted with atoms and molecules. However, the most important property of electromagnetic radiation was that the waves conveyed energy. This energy (w, in units of ergs$/$cm) was distributed in space at the rate of:

$$w = \frac{\varepsilon \mathbf{E}^2 + \mu \mathbf{H}^2}{8\pi} \tag{6}$$

In this description, $\varepsilon \mathbf{E}^2 = \mu \mathbf{H}^2$ (i.e., the amounts of electric and magnetic energies are equal), so:

$$w = \varepsilon \mathbf{E}^2/4\pi \tag{7}$$

This energy density is carried along with the waves but is not really the parameter of importance. What one really needs to know is the amount of energy that flows per second across the unit area of a plane drawn perpendicular to the direction of wave propagation. This quantity is the intensity, $\mathbf{I}$. As mentioned earlier, the wave moves through a medium with a velocity given by c/n, so integration of the motion equation yields:

$$\mathbf{I} = (cw)/n \tag{8}$$

One may take advantage of the fact that w is defined by Equation (7), and that $\varepsilon = n^2$, to obtain:

$$\mathbf{I} = c\mathbf{E}^2/4\pi \tag{9}$$

Since $\mathbf{E}$ is defined by Equation (1), one may determine that the average intensity of the waves is given by:

$$\mathbf{I} = cn\mathbf{E}_o^2/8\pi \tag{10}$$

The classical laws of mechanical and electromagnetic theory permit a discussion of the emission and absorption of electromagnetic radiation by a system of electrically charged particles. According to the classical theory, the rate of emission of radiant energy by an accelerated particle of electric charge, q, is:

$$-\frac{dE}{dt} = \frac{2q^2\tilde{a}^2}{3c^3} \tag{11}$$

in which $-dE/dt$ is the rate at which the energy, E, of the particle is converted into radiant energy, and $\tilde{a}$ is the acceleration of the particle.

First, consider a special system in which a particle of charge q carries out simple harmonic oscillation, with frequency, v, along the x-axis. The equation describing this harmonic motion is given by:

$$x = \mathbf{X}_o \cos(2\pi vt) \tag{12}$$

Differentiating Equation (12), and assuming that the amplitude factor $\mathbf{X}_o$ is independent of time, we obtain an equation describing the acceleration of this particle:

$$\tilde{a} = -4\pi^2 v^2 \mathbf{X}_o \cos(2\pi vt) \tag{13}$$

The average rate of emission of radiant energy by such a system is consequently:

$$-\frac{dE}{dt} = \frac{16\pi^4 q^2 v^4 \mathbf{X}_o^2}{3c^3} \tag{14}$$

As a result of the emission of energy, the amplitude factor X_o of the motion must decrease with time, but if the fractional change in energy during a cycle of the motion is small, Equation (14) retains its validity. The radiation emitted by such a system has the frequency, ν, of the emitting system. It is plane-polarized, with the plane of the electric vector being the plane defined by the x-axis and the direction of propagation of the light.

In case the particle carries out independent harmonic oscillations along all three Cartesian axes, this motion would be characterized by respective frequencies of ν_X, ν_Y, and ν_Z and amplitudes of X_o, Y_o, and Z_o. In that case, the total rate of emission of radiant energy would be given as the sum of three terms similar to the right side of Equation (14). If the motion of the particle cannot be described as simple harmonic, it can be represented by Fourier series or as Fourier integral as a sum or integral of harmonic terms similar to that of Equation (12). Light having a frequency characteristic of each of these terms will then be emitted at a rate given by Equation (14) but with the coefficient of the Fourier term being introduced in place of X_o.

The emission of light by a system composed of several interacting electrically charged particles is handled in an analogous fashion. A Fourier analysis is first made of the motion of the system in a given state to resolve it into harmonic terms. For a given term, corresponding to a particular frequency of motion, ν, the coefficient resulting from the analysis (which is a function of the coordinates of the particles) is expanded as a power series in the quantities x_n/λ, y_n/λ, and z_n/λ. x_n, y_n, and z_n are the coordinates of the particles relative to some origin, and λ is the wavelength of the electromagnetic radiation. The zeroth-degree term in this expansion must equal zero, because the electric charge distribution of the system does not change with time. The first-degree term of the expansion involves (in addition to the harmonic function of time) only a function of the spatial coordinates. The aggregate of these first-degree terms in the coordinates with their associated time factors, summed over all frequency values occurring in the original Fourier analysis, represents a dynamical quantity known as the *electric moment* of the system. This quantity has already been defined in Equation (5).

To the degree that classical theory is applicable, the radiation emitted by a system of several particles can be discussed by performing a Fourier analysis of the electric moment. Corresponding to each frequency term in this representation of **P**, radiation will be emitted having a frequency ν. The rate of this emission is given by an expression similar to that of Equation (14) but with the term (qX_o) being replaced by the Fourier coefficient in the electric moment expansion. The emission of radiation by this mechanism is usually called dipole emission, with the radiation itself sometimes being described as dipole radiation.

However, as shall be discussed in the next section, experimental developments and theoretical explanations arose around the beginning of the twentieth century that amply demonstrated the absolute inadequacy of classical theories to explain the interaction of electromagnetic radiation with atoms and molecules. Hanna, in his treatise *Quantum Mechanics in Chemistry*, has succinctly summarized the three most important statements (and problems) associated with the application of classical mechanics to molecular theory and spectroscopy. First, there could be no limit to the accuracy with which one or more of the dynamical variables of a classical system could be simultaneously measured, except by the limits imposed by the precision of measuring instruments. Second, there was no restriction on the number of variables that could be measured simultaneously. Finally, because the expressions for velocity are continuously varying functions of time, the velocity (and hence the kinetic energy) of a system could be continuously varied so that there were no restrictions on the values permissible to a variable. Although the system of classical mechanics worked well for particles as large as billiard balls, all three assumptions cannot be applied to the mechanics that takes place at the subatomic or atomic scale. For this reason, further discussion of the classical theory will not be pursued, and our focus moves entirely to the use of quantum theory.

Nevertheless, the power of the electromagnetic theory of Maxwell (especially as developed by Lorentz, Hertz, and other workers) proved to be quite successful in explaining many phenomena and in the prediction of new technologies. For instance, the development of radiotechnology at the beginning of the twentieth century represented a direct application of the classical theory for electromagnetic radiation. It also spawned some more dubious research, such as the fruitless search for the *æther*, believed to be the medium by which electromagnetic radiation was propagated. This substance remained an integral part of classical physics until its existence was finally disproved by the experiments of Michelson and Morley.

THE QUANTUM THEORY OF ELECTROMAGNETIC RADIATION

A number of experimental advances were made around the end of the nineteenth century that demonstrated the inadequacy of the carefully constructed and elegant field of physics existing at that moment. One of the most important of these concerned the well-known emission of light by a heated solid. Physicists had known for a long time that objects (such as horseshoes in a blacksmith's fire) would glow when heated to high temperatures and that the color of the emitted light differed with the temperature of the solid.

In 1884, Boltzmann derived a theoretical relationship describing the variation of the emissive power of an ideal heated solid (known as a 'blackbody') as a function of the absolute temperature, T. This relationship arose from a consideration of a Carnot engine operated with radiation as the working substance, and was finally stated as the Stefan-Boltzmann law:

$$W = \sigma T^4 \tag{15}$$

In Equation (15), W is the total amount of energy radiated per square centimeter of surface per second, and σ is the Stefan-Boltzmann constant

(5.6694×10^{-5}erg/sec cm^2). Being a consequence of the second law of thermo-dynamics, this relation was totally compatible with the tenets of the classical theory. Unfortunately, although the Stefan-Boltzmann law provided a reasonably accurate prediction of light emission at long wavelengths, it failed completely at short wavelengths.

Rayleigh and Jeans attempted to address the problem of blackbody radiation using classical statistical mechanics and the properties of wave motion, and derived an alternate formula:

$$\rho(\nu, T)\, d\nu = \frac{8\,\pi\nu^2\,kT}{c^3}\, d\nu \tag{16}$$

where k is the Boltzmann constant, $\rho(\nu,T)\,d\nu$ is the energy density of radiation between frequencies ν and $(\nu + d\nu)$ at absolute temperature T, and is proportional to the intensity of light emitted by the body at this temperature. A plot of Equation (16) for $T = 2000$ K is shown in Figure 3, along with a spectrum of the actual spectral energy output of a real blackbody at that same temperature.

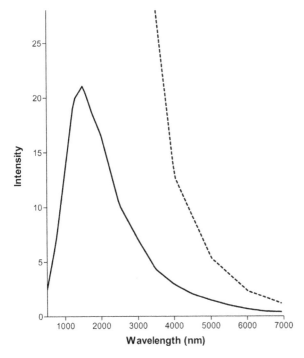

Figure 3 Spectral energy distribution for a blackbody radiator operating at 2000 K (*solid trace*) and the distribution predicted on the basis of the Rayleigh-Jeans equation for a 2000 K radiator (*dashed trace*).

The problem with all of the classical theories advanced to explain the wavelength dependence of blackbody radiation is that although they were fairly successful in predicting the spectrum of energy density at wavelengths longer than the observed maximum, they all predicted an essentially infinite output of energy at shorter wavelengths. This clearly contradicted all experimental observations and was therefore termed the "ultraviolet catastrophe."

Wien also derived an equation based on the second law of thermodynamics and found that the monochromatic energy density in an isothermal enclosure and the monochromatic emissive power of a blackbody when obtained at corresponding wavelengths were both directly proportional to the fifth power of the absolute temperature. From this fact, Wien showed that the wavelength at which the emissive power is a maximum was related to the temperature by the simple relation:

$$\lambda_{max}T = 0.289 \tag{17}$$

This equation came to be known as the Wien displacement law, and the units of the constant on the right-hand side of Equation (17) are cm·deg. Unfortunately, although the Wien law was successful in certain wavelength regions, it failed completely in others.

Classical physics could not be made to yield a solution to the problem of blackbody radiation, but a new theory advanced by Planck was able to explain all of the experimental observations. He began with the assumption made by Rayleigh and Jeans that the radiation on the inside of a heated body arises from the emission of the vibrating constituents of the material. But rather than assuming that these vibrations could emit energy of any value, he proposed that the energy could instead only be emitted in discrete amounts called *quanta*. Planck further hypothesized that these quanta were characterized by an energy given by:

$$E = h\nu \tag{18}$$

where h is the constant now known as Planck's constant (equal to 6.6256×10^{-27} erg sec). The energy of any oscillator would therefore have to be a multiple of $h\nu$. Planck then derived a new equation for the energy density of blackbody radiation, finding:

$$\rho'(\nu, T)\,d\nu = \frac{8\pi h\nu^3}{c^3}(e^{h\nu/kT} - 1)^{-1}\,d\nu \tag{19}$$

The validity of Planck's theory is proven in that Equation (19) was found to yield predictions that were in complete agreement with the experiment. It can be shown that in the limit of small wavelengths, Equation (19) has the form of Wien's law as stated in Equation (17). In addition, for long wavelengths, Equation (19) has the form of the Rayleigh-Jeans law of Equation (16). The most important aspect of the new theory was that one could only obtain agreement with observation if one assumed that the energy of the oscillators could not vary in a continuous fashion. The new theory required that the oscillators could only assume definite energies that were a function of the frequency of oscillation. In addition, energy could not be absorbed or emitted in a continuous manner but only in definite amounts that were equal to multiples of $h\nu$. This required the further assumption that an oscillator had to remain in a state of constant energy (termed a *stationary state*) until a quantum of energy, or its multiple, had been absorbed or emitted.

As one could imagine, the new theory of Plank caused a storm of controversy. The wave theory of electromagnetic radiation had been thoroughly worked out and was able to provide reasonably good explanations for many emission and absorption phenomena. An oscillating electric charge has oscillating electric and magnetic fields, and at distant points these fields constitute the radiation field, or the light wave emitted by the charge. The field carries out energy at a uniform and continuous rate, and the charge loses energy at the same rate, as required by the law of conservation of energy. This situation requires that the system gradually comes to rest, accounting for the emission process. To describe absorption, one assumes a light wave with an alternating electric field that acts on an electric charge that is capable of oscillation. The field exerts its force on the charge, gradually setting it into motion with greater and greater amplitude so that it gradually and continuously absorbs energy. Both the absorption and emission processes are therefore seen to be absolutely continuous according to the classical wave theory. This is to be contrasted with the quantum theory of Planck, which required that energy could only change by finite amounts equal to $h\nu$.

Fortunately for Planck and the new theory, other scientists were conducting investigations in various phenomena that could not be explained by the existing classical theory. For instance, Einstein had been attempting to explain the photoelectric effect, discovered by Hertz in 1887. This effect occurs when light of a suitable frequency illuminates the surface of an electropositive metal, which is maintained at a negative voltage relative to a plate (i.e., acting as a cathode). If the experiment is performed in an evacuated chamber, one observes a flow of electric current from the cathode to the plate as long as the cathode is illuminated. In 1900, Lenard showed that the absorption of light by the metal was accompanied by the emission of cathode rays (shown later to be a stream of electrons) from the metal. Over the course of time it became known that the magnitude of the photocurrent is directly proportional to the intensity of the incident radiation, but paradoxically the maximum kinetic energy of the emitted electrons was found to be independent of the incident radiant intensity. The fact that the kinetic energy of the photoelectrons was rigorously independent of the radiant intensity could not be explained by any application of the classical theory.

In 1905, Einstein published a theoretical explanation for the photoelectric effect that required an assumption extremely similar to that made by Planck. He assumed that the energy of a radiation field could not be considered as being continuously distributed through space (which would be required by classical wave theory), but instead that it was carried by particles. These light quanta (named *photons* by Lewis in 1926) were each of energy equal to $h\nu$. Einstein's hypothesis required that the absorption or emission of light having a frequency of ν must consist of the absorption or emission of a photon having energy of $h\nu$. The simple explanation for the photoelectric effect was that if the energy of a photon was insufficient to remove an electron from the cathode, no current would flow. As the frequency of the radiation was increased, the quanta became characterized by higher and higher energies, and eventually a frequency is reached that is sufficient to eject electrons and cause the flow of current. In addition, because the energy of the photons depends only on the frequency of the irradiating light, its intensity cannot have any effect on the energies of the ejected electrons. But only the absolute number of electrons emitted can be affected by the light intensity.

The philosophical debate that accompanied these new theories revolved around a number of conceptual difficulties. When Planck made his assertion that the energies of an ensemble of oscillators were restricted to certain fixed states, he imposed constraints on a hitherto continuous variable. The postulate by Einstein required that light be transmitted by corpuscular photons, so how did the concepts of frequency and wavelength fit into the picture? Finally, the wave theory had completely explained a large number of phenomena, such as interference fringes, so how could this be if radiation was corpuscular? The only way out of this dilemma would ultimately be through the generation of a quantum wave mechanics theory.

IONIZATION AND EXCITATION OF ATOMS

The final difficulty with wave theory proved to be its failure to explain the structure of atoms, which led to the end of attempts to use classical mechanics as a means to describe microscopic phenomena. As early as 1750, Melvill noticed that when different gases were placed in a flame, light of various colors was emitted. Most importantly, he found that when the emitted radiation was passed through one of Newton's prisms, the light actually was found to consist of bright spots of color separated by completely dark intervals. He also noted that the exact colors of the emitted light were specific and characteristic of the identity of the substance "excited" (the term used by Melvill to describe his observations) by the flame. As an example of typical experimental results of this type, the flame-excited line emission spectrum of hydrogen is shown in Figure 4.

It soon became possible to qualitatively identify elements by the presence of characteristic lines in their emission spectra, and many new elements were discovered for the first time through their spectral emissions. In 1885, Balmer found that the wavelengths associated with the emission spectrum of atomic

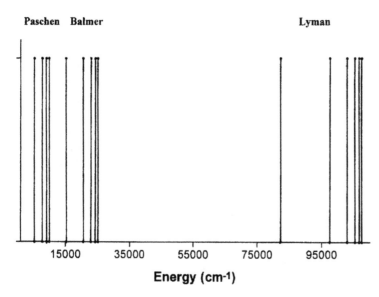

Figure 4 Emission spectrum of the hydrogen atom, indicating the spectral lines associated with transitions within the Lyman, Balmer, and Paschen series. The intensity axis is not drawn to scale.

hydrogen could be empirically fitted by a very simple equation:

$$\lambda = \frac{n^2 K}{(n^2 - 2^2)} \tag{20}$$

where n is one of the integers 3, 4, 5, or 6, and K is a constant (which equals 3.6456×10^{-7} m). The Balmer equation was generalized by Rydberg in 1906 and Ritz in 1908 to yield the energy of atomic hydrogen emission lines in units of wavenumbers:

$$\tilde{w} = \frac{1}{\lambda} = \mathbf{R}\left[\frac{1}{n_1^2} - \frac{1}{n_2^2}\right] \tag{21}$$

where n_1 and n_2 are integers. $\mathbf{R}$ is the factor now known as the Rydberg constant, and has the value of $4/K$, or 109,677.8 cm^{-1}. The success of the Balmer-Rydberg-Ritz equation was in that not only did it accurately fit the energies of all known hydrogen emission lines but that it also accurately predicted the positions of many additional lines that were subsequently discovered. Rydberg also found that the spectra of many other atoms could be predicted fairly well by the general equation:

$$T_n = \frac{R}{(n + \alpha)^2} \tag{22}$$

where T_n represents the differences in energies between two energy states, and α is an empirically determined constant.

The problem for physicists at this time was that although the empirical formula were very successful in interpreting atomic emission spectra, no theoretical explanation could be developed from the existing classical theory that satisfactorily explained the observed phenomena. The best classical theory could only predict integral ratios of frequencies, and the results of computations based on these theories were not even approximately correct. Furthermore, the development of spectral equipment having high degrees of resolution indicated that even the empirical relationships were adequate for the prediction of the fine structure that seemed to be associated with all known spectral lines. All theories advanced prior to 1913 were deemed to be complete failures.

Fortunately, new empirical observations were being made that led to the development of new thinking. Around 1890, Thompson discovered the electron, and shortly thereafter the charge and mass of this elementary particle had been determined. Around 1910, Rutherford bombarded a piece of gold foil with an α-particle beam and studied the angular distribution of the transmitted and reflected particles relative to the incident trajectory. Rutherford used his observations to advance a theory in 1911 that the structure of an atom was basically that of a small positively charged nucleus, about which negatively charged electrons rotated in a type of planetary motion.

Rutherford's model was consistent with experimental evidence but was also totally contradicted by the current electromagnetic theory. According to classical theory, because an electron was subject to the attractive force of the nucleus, it was required to undergo continuous acceleration to maintain a stable orbit. An additional requirement of classical electrodynamics was that this accelerated charge must constantly lose energy through radiation. As energy was lost, the electron would have to spiral closer and closer to the nucleus, eventually completely collapsing with a continuous emission of radiation during the process. The theory indicated that the lifetime of an atom could not be longer than 10^{-8} seconds, a conclusion that obviously disagreed with reality.

In 1913, Bohr proposed a hypothesis to explain the discrepancies between classical theory and experiment, which is now seen to be a logical extension of the theories previously set forth by Planck and Einstein concerning radiation and the energy of stationary states. Noting that the dimensions of Planck's constant were those of angular momentum (energy times time), he proposed that Maxwell's laws were invalid at the subatomic level. Instead, the angular momentum of an orbiting electron in a stable atom had to be quantized in units of $h/2\pi$, a quantity that is ordinarily referred to as $\hbar$. If the orbit of the electron was to be stable, it was further required that the centrifugal force of orbiting electron had to be balanced by the coulombic attraction between the electron and the nucleus. Bohr was then able to obtain a solution to the simultaneous equations for the electron angular momentum, the balance of coulombic attraction and centrifugal force, and Planck's equation, to derive a formula analogous to the Balmer-Rydberg-Ritz equation.

From Bohr's postulates, it is possible to derive the energies of the possible stationary states responsible for the radiation that is absorbed or emitted by an atom consisting of a single electron and nucleus. The specification of these states then permits one to compute the frequency of the associated electromagnetic radiation. To begin, one assumes that the charge on the nucleus to be Z times the fundamental electronic charge, e, and that Coulomb's law provides the attractive force, F, between the nucleus and electron:

$$F = Ze^2/r^2 \qquad (23)$$

where r is the radius of the electron orbit. This force must be exactly balanced by the centrifugal force on the rotating electron:

$$F = mr\omega^2 \qquad (24)$$

where ω is the angular velocity of the electron, and m is its mass. Because Equations (23) and (24) must be equal, we find that:

$$mr\omega^2 = Ze^2/r^2 \qquad (25)$$

The angular momentum, L, of a circulating particle is defined as its linear momentum multiplied by the radius of its circular motion. Bohr's second postulate requires that the angular momentum of a rotating electron be an integral multiple of $\hbar$:

$$L = n\hbar = mr^2\omega \qquad (26)$$

where n is an integer having the values of 1, 2, 3, and so on. Solving Equations (25) and (26) for the orbital radius yields:

$$r = \frac{n^2\hbar^2}{me^2Z} \qquad (27)$$

One now finds that the orbital radius restricted us to certain values, termed orbits, having values equal to $\hbar^2/me^2Z$, $4\hbar^2/me^2Z$, $9\hbar^2/me^2Z$, and so on. For the smallest allowed orbit of a hydrogen atom (defined by $Z = 1$ and $n = 1$), one finds that:

$$r_o \equiv a_o = \hbar^2/me^2 \qquad (28)$$

and that $a_o = 5.29 \times 10^{-11}$ m (0.529 Å).

The Bohr theory also permits a calculation of the total energy, E, of the atom from Hamilton's equation:

$$E = T + V \qquad (29)$$

where the kinetic energy, T, exists by virtue of the rotation motion of the electron:

$$T = \frac{1}{2}mr^2\omega^2 = \frac{1}{2}Ze^2/r \tag{30}$$

and the potential energy, V, exists by virtue of the position of the electron relative to the nucleus:

$$V = -Ze^2/r \tag{31}$$

Substituting Equations (30) and (31) into (29), one finds that the total energy is given by:

$$E = \tfrac{1}{2}Ze^2/r \tag{32}$$

and if the Bohr Equation for r (27) is substituted into Equation (32), one finds:

$$E = -\frac{mZ^2e^4}{2n^2\hbar^2} \tag{33}$$

It is therefore concluded that only certain stationary states are possible and that the state having the lowest total energy is defined by $n = 1$.

A transition between two states requires the absorption or emission of energy equal to the energy difference between the states. By the postulate of Bohr, this difference in energy must be quantized, so that:

$$\Delta E = E_1 - E_2 = h\nu \tag{34}$$

$$= \frac{me^4}{2n^2\hbar^2}\left[\frac{1}{n_1^2} - \frac{1}{n_2^2}\right] \tag{35}$$

Remembering that $\nu = c/\lambda$, one obtains an equation having the form of the Balmer-Rydberg-Ritz formula:

$$\hat{w} = \frac{1}{\lambda} = \frac{2\pi^2 me^4}{h^3 c}\left[\frac{1}{n_1^2} - \frac{1}{n_2^2}\right] \tag{36}$$

with the Rydberg constant, **R**, being found to equal:

$$\mathbf{R} = \frac{2\pi^2 me^4}{h^3 c} \tag{37}$$

substitution of values for m, c, e, and h into Equation (37) yields the exact value as that determined on the basis of the empirical observations ($109{,}737 \text{ cm}^{-1}$).

The power of the Bohr theory was demonstrated in its ability to correlate the series of lines observed in the emission spectrum of atomic hydrogen by Lyman, Balmer, and Paschen, as has been illustrated in Figure 5. These correlations confirmed the postulates of Bohr that energy states were quantized and that absorption or emission spectroscopy could only take place between these

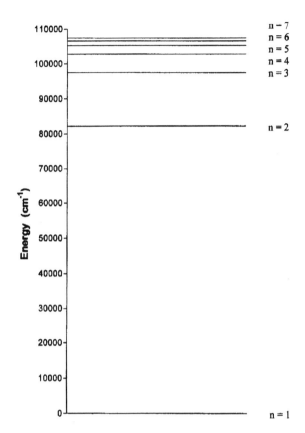

Figure 5 Energy level diagram of the hydrogen atom, illustrating assignments for the observed emission spectral lines. The family of transitions terminating in the $n = 1$ level belong to the Lyman series, those terminating in the $n = 2$ level belong to the Balmer series, and transitions terminating in the $n = 3$ level belong to the Paschen series.

quantized levels. In time, the integers defined as n became known as *quantum numbers*.

Although Bohr's theory quickly became replaced by a wave mechanical quantum theory, one of his enduring conclusions became known as the *correspondence principle*. This represented a recognition that the laws applicable to the subatomic and atomic level must ultimately reduce to the classical laws under the limiting conditions for which the latter were known to be valid.

DEVELOPMENT OF THE QUANTUM THEORY

Although the Bohr theory worked very well for one-electron atoms (H, He$^+$, Li^{2+}, Be^{3+}, etc.), it could not handle the spectrum of any atom containing more than one electron. Furthermore, it could not provide an explanation for the fine structure associated within a given series of emission lines. Refinements were made to the Bohr theory by Sommerfeld to permit the existence of elliptical orbits and to include relativistic effects, but neither of these modifications led to the development of a satisfactory general theory. It was generally agreed among the physicists of the time that a new type of mechanics was needed to treat atomic and molecular behavior.

Prior to 1920, the corpuscular nature of electrons appeared to be well established on the basis of ample experimental evidence. This included the deflection of cathode rays by electric and magnetic fields and the measurement of the charge/mass ratio of an electron. Following this came the oil-drop experiment of Millikan that enabled a direct determination of the charge on an electron. Finally, it had also been observed that electrons would leave cloud tracks in cloud chambers and that the velocities of these electrons could be affected by the nature of the transversed medium.

The field of atomic mechanics was broken wide open by the hypothesis made by de Broglie in 1924 that the dual wave/particulate nature of light was extendable to matter. He reasoned that if light consisted of photons, then the momentum of a photon would be given by its mass times its velocity, and because:

$$E = mc^2 = h\nu \tag{38}$$

then both mc and $h\nu/c$ must each equal h/λ. de Broglie argued that a mass having a momentum equal to p should also have associated with it a matter wave characterized by h/λ. In other words,

$$\lambda = h/p = h/mv \tag{39}$$

where if m and v are the respective mass and velocity of the moving particle, then λ would be the wavelength of its matter wave.

The experimental verification proving the existence of the wave character of matter was provided shortly thereafter by Davisson and Germer, who bombarded the surface of a nickel crystal with a narrow beam of low velocity electrons. They evaluated the intensities of reflected electrons as a function of their angles and found a pattern that was extremely similar to that produced by the diffraction of X-rays off crystalline surfaces. Using a modified Bragg equation, they obtained a value for the wavelength of the electrons that agreed excellently with that calculated by Equation (39).

It was now evident that any valid explanation for atomic or molecular structure would necessarily require a theory that incorporated both the wave and corpuscular properties of matter. Erwin Schrödinger and Werner Heisenberg, who developed theories along two different pathways, solved the problem almost simultaneously. Schrödinger reasoned that electronic motion could be treated as waves and used equations for wave motion and stationary waves available from classical physics to develop a wave mechanical picture. In this approach, techniques of partial differential equations were used to handle the equations for wave motion. Heisenberg used the properties of matrices and developed a matrix mechanics theory based on matrix algebra that was later proved to be equivalent to the wave mechanics of Schrödinger. Dirac and von Neumann developed more general quantum mechanical theories, and both the Schrödinger and Heisenberg approaches were shown to be specific instances of the general theory.

It is beyond the scope of this introductory chapter to fully develop these quantum mechanical theories in detail, and interested readers can consult any of the references cited at the end of this chapter. However, one can look at some of the simplest applications of the quantum theory to gain a degree of understanding as to the defining properties of energy states in atoms and molecules. Knowing that spectroscopy entails transitions among energy states will provide the impetus for this analysis. Because the theory of Schrödinger is more accessible, we shall adopt the usual practice of presenting elements of the wave mechanical theory for our spectroscopic explanations.

In the same way that classical mechanics and thermodynamics were based on a core of assumed postulates, quantum mechanics was formulated upon a series of assumptions that were assumed to be correct as long as they did not lead to the deduction of contradictory predictions. One may gain a considerable degree of understanding into quantum mechanics by considering the fundamental postulates that provides its basis. The first postulate defines the state of a system consisting of N particles by a function, $\psi(q_i, t)$, which is a function of the coordinates of each particle and the time dependence of their motion. Furthermore, this state function must be single-valued at all points, finite, and continuous. The nature of this function is such that its product with its complex conjugate ψ^* (i.e., the function obtained by replacing i by $-i$, where i is the square root of -1) is proportional to the probability of finding the particle in a region between q_i and $(q_i + dq_i)$ at time $= t$.

Among the consequences of this first postulate is the understanding that the wave function describes the system as completely as possible. However, it must be recognized that one cannot specify the exact positions and momenta of all particles contained within a given system and predict their future behavior in the classical manner. Instead, as stated most explicitly by Heisenberg in his *uncertainty principle*, it is possible to specify only the probability that certain events will occur. Because ψ must contain both positional and time dependencies, in some cases it is possible to separate out the time-dependent function and therefore obtain a time-invariant function termed a *stationary state*. Spectroscopists are most interested in the determination and definition of these stationary states, because spectroscopy consists of the characterization of the transitions between such states.

The second postulate of the wave mechanical theory states that there must exist a mathematical operator for every observable property of a system, and the physical properties of the observable can be inferred from the mathematical properties of its associated operator. When one operates on a state function with the operator, the exact experimental values of the observable property are obtained as the *eigenvalues* of the following equation:

$$\hat{a}\psi_i = a_i\psi_i \tag{40}$$

In other words, when the operator, $\hat{a}$, acts on the particular wave function, ψ_i, one obtains a value for the observable property, a_i, multiplied by the unchanged wave function. Examples of operators of interest would be those used to generate atomic coordinates or particle momenta.

Undoubtedly, the most important operator is the one used to obtain the total energy of a system. Given earlier as Equation (29), Hamilton's equation states that the total energy of a system is the sum of the potential and kinetic energies. Therefore, the Hamiltonian operator is defined as:

$$\hat{H} = \hat{T} + \hat{V} \tag{41}$$

where $\hat{T}$ and $\hat{V}$ are the respective operators for the kinetic and potential energy terms. The definition of a Hamiltonian operator is the most important step during an application of the Schrödinger quantum mechanics, so an example is appropriate. In classical mechanics, the kinetic energy of a particle moving in a system defined by Cartesian coordinates is

$$T = \frac{1}{2m}\left[p_X^2 + p_Y^2 + p_Z^2\right] \tag{42}$$

where m is the mass of the particle, and the p_i quantities represent the linear momenta along each Cartesian axis. To deduce a quantum mechanical operator expression, one substitutes the quantum mechanical differential operator for the p_i momenta:

$$p_i \rightarrow i\hbar(\partial/\partial q_i) \tag{43}$$

obtaining:

$$\begin{aligned}
\hat{T} &= \frac{\hbar^2}{2m}\left[\frac{\partial^2}{\partial X^2} + \frac{\partial^2}{\partial Y^2} + \frac{\partial^2}{\partial X^2}\right] \\
&= \frac{\hbar^2\nabla^2}{2m}
\end{aligned} \tag{44}$$

Because $\hat{V}$ is only a function of the particle coordinates, its quantum expression remains equal to its classical expression. In our example, the Hamiltonian operator is then given by:

$$\hat{H} = \frac{\hbar^2\nabla^2}{2m} + V(x, y, z) \tag{45}$$

The third postulate of the wave mechanical theory bridges the gap between the mathematical formalism of quantum mechanical theory and real experimental observation and is used to obtain values of experimental parameters not defined by an eigenvalue equation. When a number of measurements are made on the dynamical variable, A, then the average value of $A(\tilde{A})$ will be given by:

$$\tilde{A} = \int \psi^* \hat{a} \psi \, d\tau \qquad (46)$$

Equation (46) assumes that the wave function, ψ, is normalized (i.e., $\int \psi^* \psi \, d\tau = 1$) and that the integration is carried out over all space permitted to the particle. The average value for the observable, $\tilde{A}$, is also referred to as its *expectation value*, or $\langle A \rangle$. The fourth postulate provides another linkage between theory and experiment. When the operator, $\hat{a}$, and the state function, ψ, are eigenfunctions of each other, then the experimentally observed parameter, a, will be that given by Equation (40).

The fifth, and final, postulate leads to the fundamental equation of the Schrödinger statement of quantum mechanics. The evolution of a state function, $\psi(q_i, t)$, is given by the equation:

$$\hat{H}\psi(q_i, t) = i\hbar \, (\partial\psi[q_i, t]/\partial t) \qquad (47)$$

where the factor $i\hbar(\partial/\partial t)$ is the energy operator for the system, and $\hat{H}$ is the Hamiltonian operator. Equation (47) is often denoted as the time-dependent Schrödinger equation. If neither $\hat{H}$ nor $\psi(q_i, t)$ exhibits time dependence, or if the time dependence in $\psi(q_i, t)$ can be factored out, then Equation (47) reduces to the stationary state Schrödinger equation:

$$\hat{H}\psi_i = E_i\psi_i \qquad (48)$$

where E_i is the energy associated with the wave function ψ_i.

We shall often refer to the state characterized by the lowest energy as the *ground state* of the system, and therefore ψ_0 is wave function defining the ground state. It follows that any other state must necessarily be characterized by a higher energy than the ground state, and such states are referred to as *excited states*. Every excited state is defined by its characteristic excited state wave function that contains a total specification of the state and its properties.

One may now define the two general types of spectroscopic transitions. One of these is *absorption spectroscopy*, where a transition takes place from the unique ground state to one of the excited states. The other is *emission spectroscopy*, where a transition takes place from a state of high energy to a state of lower energy. The terminating state in an emissive transition may be (but is not necessarily) the ground state.

The formalism just described can be brought to life through the exposition of a simple, yet illustrative, example. We will consider the example of a single particle moving in one dimension, known colloquially as the one-dimensional box problem. In this case, the Hamiltonian operator of Equation (45) will involve the second derivative in only one coordinate (such as the x-axis), so

Equation (47) becomes:

$$-\frac{\hbar^2}{2m}\frac{\partial^2 \psi(x,t)}{\partial X^2} + V(x)\psi(x,t) = E\frac{\partial \psi(x,t)}{\partial t} \tag{49}$$

The potential function, $V(x)$, depends on the nature of the interactions that are exerted upon the particle. If the particle is confined to a definite region in space, satisfactory solutions for $\psi(x,t)$ can be found only for certain values of the energy, E, of the particle. The wave functions that satisfy Equation (49) reveal the position of the particle in the system through the probability function, $\psi(x,t)^2$.

A number of assumptions make possible the solution of the Schrödinger equation for this particle in a one-dimensional box. The solutions to this problem will entail wave equations that are time-independent, so the $[\partial \psi(x,t)/\partial t]$ term must equal zero. Outside of the box that is defined by $x = 0$ and $x = a$, the potential energy of the particle must be infinite, so there will be zero probability [i.e., $\psi(x,t)^2 = 0$] of finding the particle outside of the box. This *boundary condition* therefore requires that $\psi(x,t) = 0$ outside of the box as well. To be well behaved, and avoid a discontinuity at $x = 0$ and $x = a$, the wave function must also equal zero at these boundaries. Finally, because the motion of the particle is not constrained in any way when it is inside the box, the potential energy term must equal zero. In that case, the Schrödinger equation becomes:

$$-\frac{\hbar^2}{2m}\frac{\partial^2 \psi}{\partial X^2} = E\psi(x) \tag{50}$$

The form of Equation (50) is well known, and its solution has the general form of:

$$\psi(x) = A\sin(\mathbf{n}\pi x/a) \tag{51}$$

where $\mathbf{n}$ is an integer, having the values 1, 2, 3, 4, and so on, and A is a constant equal to $(2/a)^{1/2}$. The existence of quantum numbers is required by the use of quantum wave mechanics, even for this simple example.

Substitution of Equation (51) back into equation (50) and rearranging yields an equation for the energy of the particle as:

$$E_n = \frac{\mathbf{n}^2 h^2}{8ma^2} \tag{52}$$

The conclusion is obtained that for each wave function, ψ_n, defined by Equation (51) there is a single energy, E_n, defined by Equation (52). In that case, each wave function is defined by a unique quantum number, and the state of lowest energy (the ground state) must be defined by $\mathbf{n} = 1$ (for which $E_1 = h^2/8ma^2$). The state of the next highest energy (the lowest energy excited state) would be defined by $\mathbf{n} = 2$ and would be characterized by $E_2 = h^2/2ma^2$.

One could then envision the simplest absorption spectroscopy possible within an one-dimensional box as consisting of a transition from the ground state defined by ψ_1 to the excited state defined by ψ_2. The energy of this lowest energy transition is given by:

$$\Delta E_{12} = E_2 - E_1 = 3h^2/8ma^2 \tag{53}$$

The energy of the excited state defined by ψ_3 is calculated to be $9h^2/8ma^2$, so the energy pertaining to the $\psi_1 \to \psi_3$ transition is:

$$\Delta E_{13} = E_3 - E_1 = 7h^2/8ma^2 \tag{54}$$

Because it is evident that $\Delta E_{13} > \Delta E_{12}$, then the $\psi_1 \to \psi_3$ absorption transition would be predicted to take place at a higher frequency (shorter wavelength) than would the $\psi_1 \to \psi_2$ absorption. Were one able to empirically measure the energies of these two hypothetical transitions, then a correlation of the experimental ΔE values with the theoretical ΔE values would permit a deduction to be made regarding the quality of the wave functions used in the computation. The correlation of energy states with experimentally observed spectral transitions represents an *assignment* of the empirical data and is the ultimate goal of theoretical spectroscopy.

Additional insight into wave functions, energy states, and transition energies can be obtained by permitting the motion of the particle to take place in three dimensions rather than one. The wave function pertinent to this situation will have the form $\psi(x, y, z)$, and one assumes that the boundaries of the box run from $x = 0$, $y = 0$, and $z = 0$ to $x = a$, $y = b$, and $z = c$.

Applying all of the same assumptions as used previously to the three-dimensional problem, the time-independent Schrödinger equation becomes:

$$-\frac{\hbar^2}{2m}\left[\frac{\partial^2 \psi(x,y,z)}{\partial X^2} + \frac{\partial^2 \psi(x,y,z)}{\partial Y^2} + \frac{\partial^2 \psi(x,y,z)}{\partial Z^2}\right] = E - \psi - (x,y,z) \tag{55}$$

But since $\psi(x, y, z)$ is a simultaneous function of all three x, y, and z Cartesian coordinates, a solution to Equation (55) can only be found if the variables in $\psi(x, y, z)$ are first separated. This can be accomplished by assuming that any motion of the particle along any given direction is not affected by its motion along any other direction. This is equivalent to an assumption that the overall state wave function is the product of three independent wave functions, each of which is in turn a function of only one variable:

$$\psi(x,y,z) = \psi(x)\psi(y)\psi(z) \tag{56}$$

so that the total energy, E, of the $\psi(x, y, z)$ function will be given by:

$$E = E_X + E_Y + E_Z \tag{57}$$

Upon substitution of Equation (56) into Equation (55) and collection of variables, one obtains three equations:

$$-\frac{\hbar^2}{2m}\frac{\partial^2\psi(x)}{\partial X^2} = E\psi(x) \tag{58}$$

$$-\frac{\hbar^2}{2m}\frac{\partial^2\psi(y)}{\partial Y^2} = E - \psi(y) \tag{59}$$

$$-\frac{\hbar^2}{2m}\frac{\partial^2\psi(z)}{\partial Z^2} = E - \psi(z) \tag{60}$$

for which the solutions are:

$$\psi(x) = A\sin(\mathbf{n}\pi x/a) \tag{61}$$

$$\psi(y) = B\sin(\mathbf{p}\pi y/b) \tag{62}$$

$$\psi(z) = C\sin(\mathbf{q}\pi z/c) \tag{63}$$

where $\mathbf{n}$, $\mathbf{p}$, and $\mathbf{q}$ are integers (separately permitted to have values of 1, 2, 3, 4, etc.), and $A = (2/a)^{1/2}$, $B = (2/b)^{1/2}$, and $C = (2/c)^{1/2}$. The complete solution to the particle in a three-dimensional box is obtained by substituting Equations (61–63) into Equation (56) to obtain:

$$\psi(x, y, z) = D\sin(\mathbf{n}\pi x/a)\sin(\mathbf{p}\pi y/b)\sin(\mathbf{q}\pi z/c) \tag{64}$$

where $D = ABC$. One immediately notes from Equation (64) that the wave function of a particle able to move in three dimensions can only be completely specified by the use of three quantum numbers.

The energy of a given $\psi(x, y, z)$ function is given by:

$$E_{\mathbf{npq}} = \frac{h^2}{8m}\left[\frac{\mathbf{n}^2}{a^2} + \frac{\mathbf{p}^2}{b^2} + \frac{\mathbf{q}^2}{c^2}\right] \tag{65}$$

The unique ground state (ψ_{111}) associated with this system is defined by $\mathbf{n} = \mathbf{p} = \mathbf{q} = 1$, and will have an energy of:

$$E_{111} = \frac{h^2}{8m}\left[\frac{1}{a^2} + \frac{1}{b^2} + \frac{1}{c^2}\right] \tag{66}$$

The energies of the lowest energy excited states will be given by:

$$E_{211} = \frac{h^2}{8m}\left[\frac{4}{a^2} + \frac{1}{b^2} + \frac{1}{c^2}\right] \tag{67}$$

$$E_{121} = \frac{h^2}{8m}\left[\frac{1}{a^2} + \frac{4}{b^2} + \frac{1}{c^2}\right] \tag{68}$$

$$E_{112} = \frac{h^2}{8m}\left[\frac{1}{a^2} + \frac{1}{b^2} + \frac{4}{c^2}\right] \tag{69}$$

but without knowing values for a, b, and c, the relative orders of ψ_{211}, ψ_{121}, and ψ_{112} cannot be deduced.

A very interesting situation arises for the case where the three-dimensional box is cubic or where $a = b = c$. In that case, the energy of the $\psi(x, y, z)$ functions is given by:

$$E_{npq} = \frac{h^2}{8ma^2}[\mathbf{n}^2 + \mathbf{p}^2 + \mathbf{q}^2] \tag{70}$$

The unique ground state (ψ_{111}) associated with this cubic system is still defined by $\mathbf{n} = \mathbf{p} = \mathbf{q} = 1$ and has an energy of:

$$E_{111} = \frac{3h^2}{8ma^2} \tag{71}$$

The energies of the lowest energy excited states will then be given by:

$$E_{211} = \frac{3h^2}{4ma^2} \tag{72}$$

$$= E_{121} = E_{112} \tag{73}$$

One finds in this situation the existence of three states, each of which is defined by a unique set of quantum numbers, which are characterized by the same energy. These three states are said to be *degenerate*, and an energy level is termed to have a *degeneracy* corresponding to the number of states that occur at that energy. Figure 6 illustrates the pattern of excited states that is predicted by Equation (70). The three absorption transitions corresponding to $\psi_{111} \rightarrow \psi_{211}$, $\psi_{111} \rightarrow \psi_{121}$, and $\psi_{111} \rightarrow \psi_{112}$ must each take place at identical energies and could not therefore be distinguished in the absence of an additional perturbation on the system.

One can obtain the hypothetical absorption spectrum that would be observed for the particle in a three-dimensional box, by computing the energy differences between the states in Figure 6. The energy difference between the ψ_{111} state and any excited ψ_{npq} state must be equal to the energy required to effect the absorption transition. By Planck's law, this energy difference is pro-portional to the frequency of radiation needed to effect the absorption and inver-sely proportional to the wavelength of the radiation. An absorption spectrum of this type is shown in Figure 7, where the wavelength scale has been scaled so as to appear in reasonable units.

Thus far, we have examined the nature of state functions, the energies of these, and the energy associated with transitions between the levels described by these state functions. The quantum theory also permits one to calculate the intensity of light that is absorbed or emitted during a spectroscopic transition. In quantum mechanics, the intensity associated with a transition is determined by the probability of its occurrence. As discussed earlier, such transitions require the presence of an electric dipole moment within the transitioning

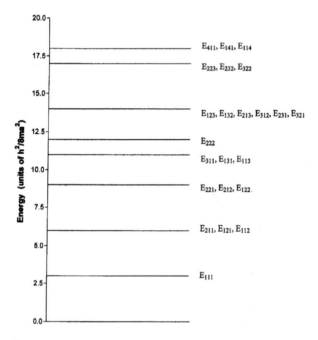

Figure 6   Energy level diagram for the states of a particle in a three-dimensional box.

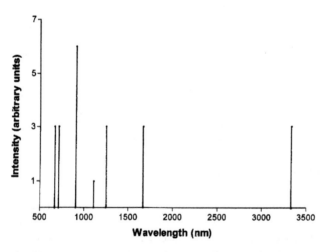

Figure 7   Hypothetical absorption spectrum for a particle in a three-dimensional box. The intensity axis has been plotted to be proportional to the degeneracy of the final state, and the wavelength axis has been arbitrarily fitted to a pseudonanometer scale so as to appear in reasonable units.

species, and wave mechanics indicates that the spectral transition probability, $\mathbf{P}_{ij}$, from a lower state (ψ_i) to an upper state (ψ_j) is given by:

$$\mathbf{P}_{ij} = \int \psi_i \hat{\mu} \psi_j \, d\tau \tag{74}$$

where $\hat{\mu}$ is the total electric dipole moment operator. The probability of the $\psi_i \rightarrow \psi_j$ transition is known as the *transition moment*, and its use replaces the classical Fourier analysis of the electric moment to obtain the radiation intensity associated with the transition. When the value of the transition moment of Equation (74) does not equal zero, the spectral transition is said to be electric dipole allowed.

One can also define the transition moment for a magnetic dipole transition as:

$$\mathbf{M}_{ij} = \int \psi_i \hat{\mathbf{m}} - \psi_j \, d\tau \tag{75}$$

where $\hat{\mathbf{m}}$ is the total magnetic dipole moment operator. In general, pure magnetic dipole allowed transitions are significantly weaker than the corresponding pure electric dipole allowed transitions. However, the simultaneous existence of nonzero transition moments of both types will be required for the observation of chiroptical absorption or emission.

For the example of a particle in a three-dimensional box, the transition moment can be factored into three component integrals after a separation of variables is conducted:

$$\mathbf{P}_{ij}(x) = \int \psi_i \hat{\mu}_X \psi_j \, d\tau \tag{76}$$

$$\mathbf{P}_{ij}(y) = \int \psi_i \hat{\mu}_Y \psi_j \, d\tau \tag{77}$$

$$\mathbf{P}_{ij}(z) = \int \psi_i \hat{\mu}_Z \psi_j \, d\tau \tag{78}$$

where the three $\hat{\mu}_i$ operators represent the components of the electric dipole moment operator along each of the three Cartesian axes. Depending on the nature of ψ_i and ψ_j, any one of the $\mathbf{P}_{ij}(x)$, $\mathbf{P}_{ij}(y)$, or $\mathbf{P}_{ij}(z)$ moments can equal zero (or not). When a given moment has a nonzero value along a particular axis, the transition is said to be *polarized* along that axis. For real molecules, the exact computational task can be formidable, but we shall eventually see that the use of certain symmetry arguments will permit one to make qualitative deductions of great value.

To recapitulate, in classical mechanics all the necessary coordinates and momenta of a system are totally specified, and the interactions among these are totally accounted for. This allows one to calculate the configuration and motion within the system at any future time with complete predictability.

Quantum mechanics, on the other hand, concedes the exact determination of some dynamical variables and provides their computational solutions in terms of probability functions. The inability of a state function to provide exact values for certain variables is often consistent with the inability of some experimental measurements to yield more exact determinations. One concludes that quantum mechanics is consistent with observations for situations where the application of classical mechanics leads to serious inconsistencies and problems.

OVERVIEW OF SPECTROSCOPY

Having developed the basic concepts needed to define the nature of spectroscopy, it now remains to discuss how electromagnetic radiation quanta of various magnitudes may interact with the subatomic and atomic constituents of atoms and molecules. One way to classify the range of spectroscopic possibilities would be the brute force interpretation of each of the spectral energy regions defined in Table 1, but this scheme lacks elegance and impedes clarity in the presentation. Instances exist where the interaction of matter with quanta of different magnitudes (from different spectral regions) leads to the occurrence of the same type of state transition. These processes may proceed through different mechanisms and may take place within different spectral regions, so the requisite cross-referrals unnecessarily complicate matters.

The superior method of spectral classification is to consider the possible types of energy level transitions that might take place in atoms or molecules and to examine the magnitude of the light quanta and associated processes that cause such transitions to take place. This latter approach has the advantage that one might proceed directly from the precepts discussed in this chapter to an area of importance for the practicing pharmaceutical spectroscopist. After that, other types of spectroscopy can be introduced as refinements on this basic model. This approach also permits the introduction of additional theory on an as-needed basis, which will often be spliced into the elements of a preceding section.

Several types of atomic and molecular states are characterized by wave functions that are largely functions of electronic character. Electrons in these states are ordinarily subdivided as to whether they are actively involved in chemical bonding (known as *valence* electrons) or not (known as *core* electrons). Other states are characterized by wave functions that are largely functions of vibrational or rotational nuclear motion. Still other states are defined by the paramagnetic properties and spin states of either the electrons or of the nuclei. Altogether, it is useful to define six broad classes of wave function types and the spectroscopy associated with each grouping.

The first category will concern spectroscopic transitions among states defined as having predominantly core electron properties. The energies separating these states have magnitudes comparable with the energies of X-rays, and so *X-ray absorption spectroscopy* is one of the techniques used to probe coordinative environments of absorbing atoms. Also within this grouping are the various techniques of atomic spectroscopy and photoelectron spectroscopy. These techniques are characterized by the use of light quanta possessing sufficient energy to cause an electron to be completely ejected (i.e., ionized) from an atom or a molecule. One might study the excess kinetic energy associated with the ejected electron when the energy of the excitation photon is larger than the ionization energy. This approach encompasses the various techniques of *photoelectron spectroscopy*, where one allows a beam of photons of known energy to interact with a sample and measures the excess kinetic energy of the ejected electrons. By subtracting the excess kinetic energy from the incident photon energy, one obtains the binding energy of the electron. Electronic energies are then correlated with structural features of the molecule or compared with theoretical calculations. Depending on the choice of experimental circumstances and observing mode of the analyst, the ionization process can be accompanied by the absorption or emission of electromagnetic radiation. This gives rise to the families of spectroscopy conveniently classed as *atomic absorption* or as *atomic emission spectroscopies*.

The second category concerns spectroscopic transitions among states defined as predominantly having valence electron properties. This category contains some of the most familiar types of spectroscopy, namely absorption in the ultraviolet, visible, and near-IR regions of the spectrum. Techniques have been developed that permit such work to be conducted on the basis of either light transmission or of reflected light. A variety of complementary techniques have also been developed that depend on the interaction of polarized light with molecules of certain types, and these have great importance in the study of chiral compounds of pharmaceutical interest. Emission spectroscopies, such as fluorescence and phosphorescence, are also characterized by transitions among valence electron states, and they have found extensive use in pharmaceutically oriented spectroscopy.

The third category concerns spectroscopic transitions among states whose wave functions are independent of electronic character and instead are defined as having predominantly nuclear motion properties. These wave functions describe various states of molecular vibration, and transitions between vibrational states in molecules are caused by the absorption of energy in the IR region of the electromagnetic spectrum. *IR absorption spectroscopy* can be observed in either transmission or reflectance modes. Interestingly, radiation in this region can be produced by hot glowing metal and represents a practical use of blackbody radiation. Transitions among vibrational states can also be measured using Raman spectroscopy, where one observes the frequency of scattered light rather than that of absorbed light. Energy differences between weak scattering lines and the main irradiation line correspond to vibrational and/or rotational transitions in the system being studied.

The fourth category also concerns spectroscopic transitions among states whose wave functions are characterized by nuclear motion properties, but where transitions among these states is associated with pure rotational motion of the molecule. Consequently, transitions among these states constitute *rotational spectroscopy*. Most of these transitions occur in the microwave region, except in the case of very light molecules where the transitions take place in the far-IR. Microwaves are characterized by the fact that they are generated by klystrons and magnetrons and are conducted by waveguides (hollow pieces of metal tubing with either a circular or rectangular cross-section).

The fifth category encompasses the states defined by wave functions specifically related to the properties of unpaired electrons in a molecule. One might think this to be a technique of limited utility, but examples of applicable species include organic free radicals, radiation-induced radicals, paramagnetic transition metal ions, defect centers in solids, and conduction electrons in metals. When a material containing one or more unpaired electrons is placed in an external magnetic field, transitions between different electron spin states can be induced by radiation in the microwave region. Such spectroscopy is properly termed *electron paramagnetic resonance*, but one often encounters the inaccurate term *electron spin resonance* to describe the same phenomenon. Experiments are usually performed at frequencies anywhere in the range of 2000 to 36,000 MHz. A frequency of 10,000 MHz corresponds to a wavelength of 3 cm, and an energy of 0.3 cm^{-1}.

The sixth category encompasses the states defined by wave functions specifically related to the properties of nuclear paramagnetism. Here, transitions among nuclear spin states are induced by absorption of electromagnetic radiation in the radio-frequency region of the spectrum after these have been split by the use of an external magnetic field. *Nuclear magnetic resonance spectroscopy* detects transitions between nuclear spin states in an applied magnetic field, whereas *nuclear quadrupole resonance spectroscopy* detects the splitting in the nuclear spin levels arising from the interaction of an unsymmetrical charge distribution of certain nuclei with an electric field gradient. The size of quanta used in this work is very small (i.e., much less than kT at room temperature), so that these types of spectroscopy require the use of sophisticated techniques to achieve the population differences in nuclear spin states required for transitions to take place.

SUMMARY

It is now firmly established that spectroscopy originates from transitions among states that are properly described by quantum mechanical wave functions. The irradiation of a sample by electromagnetic radiation will only result in its absorption if the energies of the quanta, which constitute that radiation correspond to a difference in energy levels within the substance. Any electromagnetic radiation emitted by a sample under a suitable stimulus condition will be of energy equal to an energy level difference within the emitting substance.

Because the wave functions describing the properties of a substance can ordinarily be factored according to a majority property of the atom or molecule, the absorption or emission of electromagnetic radiation can be considered to affect the electronic, vibrational, or rotational energies of a molecule. Quanta of very low energy will affect the spin states of unpaired electrons and paramagnetic nuclei. Owing to the differing magnitudes of the energy level differences associated with these different states, the type of electromagnetic radiation used to effect the spectroscopic change will generally fall in a characteristic region of the spectrum.

REFERENCES

Classical Mechanics

1. Born M, Wolf E. Principles of Optics. New York: Macmillan, 1964.
2. Corben HC, Stehle P. Classical Mechanics. New York: John Willey & Sons, 1960.
3. Feynman RP, Leighton RB, Sands M. The Feynman Lectures on Physics, Vol. 1. Reading: Addison-Wesley, 1963.
4. Goldstein H. Classical Mechanics. Reading: Addison-Wesley, 1980.
5. Leech JW. Classical Mechanics. London: Methuen, 1958.
6. Kibble TWB. Classical Mechanics. New York: John Wiley & Sons, 1973.
7. Richtmeyer FK, Kennard EH. Chapters 1, 2. Introduction to Modern Physics. New York: McGraw-Hill Book Co., 1942.
8. Rutherford DE. Classical Mechanics. New York: Wiley-Interscience, 1951.

Quantum Mechanics

9. Atkins PW. Molecular Quantum Mechanics. Oxford: Clarendon Press, 1970.
10. Daudel R, Lefebvre R, Moser C. Quantum Chemistry. New York: Wiley-Interscience, 1959.
11. Davis JC. Advanced Physical Chemistry. New York: Ronald Press, 1965.
12. Eyring H, Walter J, Kimball GE. Quantum Chemistry. New York: John Wiley & Sons, 1944.
13. Hanna MW. Quantum Mechanics in Chemistry. Menlo Park: Benjamin/Cummings, 1981.
14. Kauzmann W. Quantum Chemistry. New York: Academic Press, 1957.
15. Levine IN. Quantum Chemistry. Boston: Allyn and Bacon, 1970.
16. Messiah A. Quantum Mechanics. North-Holland, Amsterdam, 1961.
17. Pauling L, Wilson EB. Introduction to Quantum Mechanics. New York: McGraw-Hill, 1935.
18. Pilar FL. Elementary Quantum Chemistry. New York: McGraw-Hill, 1968.
19. Rojansky V. Introductory Quantum Mechanics. Englewood Cliffs: Prentice-Hall, 1938.
20. Slater JC. Quantum Theory of Molecules and Solids. New York: McGraw-Hill, 1963.

General Spectroscopy

21. Barrow GM. Molecular Spectroscopy. New York: McGraw-Hill, 1962.
22. Brittain EFH, George WO, Wells CHJ. Introduction to Molecular Spectroscopy. London: Academic Press, 1970.

23. Guillory WA. Introduction to Molecular Structure and Spectroscopy. Boston: Allyn and Bacon, 1977.
24. King GW. Spectroscopy and Molecular Structure. New York: Holt, Rinehart and Winston, 1964.
25. Reid C. Excited States in Chemistry and Biology. New York: Academic Press, 1957.
26. Richards WG, Scott PR. Structure and Spectra of Molecules. Chichester: John Wiley & Sons, 1985.
27. Sandorfy C. Electronic Spectra and Quantum Chemistry. Englewood Cliffs: Prentice-Hall, 1964.
28. Simons JP. Photochemistry and Spectroscopy. London: Wiley-Interscience, 1971.

--- 2 ---

Core Electron States and X-Ray Absorption Spectroscopy

Harry G. Brittain
Center for Pharmaceutical Physics, Milford, New Jersey, U.S.A.

WAVE MECHANICAL DESCRIPTION OF THE HYDROGEN ATOM

Although the theory of Bohr sufficed to interpret the spectral lines associated with elemental hydrogen, which had been excited by a flame, the extension of this theory to entirely explain the line spectra of heavier atoms proved to be elusive. In addition, an explanation for the doubling of lines in the hydrogen spectrum eluded the simple theory. As indicated in the last chapter, only the wave mechanical theories of Schrödinger, Heisenberg, and Dirac were able to provide the necessary mathematical description of atomic structure that permitted physicists to adequately interpret the empirical data that had become widely available. It became obvious to these workers that an understanding of the spectroscopy of atomic and molecular states derived from core electrons in a molecule required the development of a comprehensive picture for the quantum numbers that define such states. The solution to the wave mechanical equation of the hydrogen atom was found to meet this need, because it could be solved exactly. Furthermore, its solutions would then be used to develop approximate treatments for heavier atoms.

As with all quantum mechanical derivations, one begins with the Schrödinger equation:

$$\hat{H}\psi_i = E_i\psi_i \tag{1}$$

where E_i is the energy associated with the wave function ψ_i. The heart of the derivation is deducing the correct form for the Hamiltonian operator:

$$\hat{H} = \hat{T} + \hat{V} \qquad (2)$$

where $\hat{T}$ and $\hat{V}$ are the respective operators for the kinetic and potential energy terms. The form for the kinetic energy remains as defined in chapter 1:

$$\hat{T} = \frac{\eta^2}{2m}\left[\frac{\partial^2}{\partial X^2} + \frac{\partial^2}{\partial Y^2} + \frac{\partial^2}{\partial X^2}\right] = \eta^2\frac{\nabla^2}{2m} \qquad (3)$$

The potential energy term, $\hat{V}$, for the electron is given by the attractive force between the negatively charged electron and the positively charged nucleus. This expression is available as Coulomb's Law:

$$V = -e^2/r \qquad (4)$$

where r is the distance of the electron from the nucleus, and e is the electron fundamental charge. The Schrödinger equation now takes the form:

$$E\psi = \frac{-\hbar^2}{2m}\nabla^2\psi - \frac{e^2}{r}\psi \qquad (5)$$

The presence of a radial distance as one of the parameters in the potential function precludes one from working in the Cartesian coordinate system. Fortunately, the situation of a single electron moving about a nucleus permits a ready application of the spherical polar coordinate system. In this system, space is defined by a radial distance varying from 0 to ∞, an angle θ varying from 0 to π, and an angle ϕ varying from 0 to 2π. As illustrated in the polar coordinate system illustrated in Figure 1, the position of a point at location (x,y,z) becomes specified by:

$$x = r\sin\theta\cos\phi \qquad (6)$$
$$y = r\sin\theta\sin\phi \qquad (7)$$
$$z = r\cos\theta \qquad (8)$$

By inserting the coordinate transformation relationships of Equations (6) through (8) into Equation (5), one obtains:

$$0 = \frac{1}{r^2}\frac{\partial}{\partial r}\left[r^2\frac{\partial\psi}{\partial r}\right] + \frac{1}{r^2\sin\theta}\frac{\partial}{\partial\theta}\left[\sin\theta\frac{\partial\psi}{\partial\theta}\right] + \frac{1}{r^2\sin^2\theta}\frac{\partial^2\psi}{\partial\phi^2} + \frac{\hbar^2}{2m}\left[E + \frac{e^2}{r}\right] \qquad (9)$$

The advantage of using spherical polar coordinates is that it permits a facile separation of the variables, which is essential to obtain a solution to the equation. One assumes that the motion of the electron along one coordinate direction is independent of its motion along the other two directions (the same assumption that was used to obtain a solution for the particle in the three-dimensional box), so then one can write:

$$\psi(r, \theta, \phi) = R(r)\Theta(\theta)\Phi(\phi) \qquad (10)$$

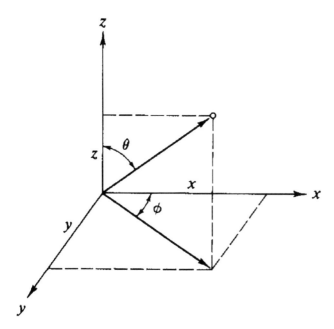

Figure 1   The spherical polar coordinate system used to describe the position of an object
with respect to an origin. The magnitude of the radial distance (r) describes a sphere upon
whose surface the object is found, and its position on the surface of this sphere is defined
by the intersection of the two planes described by the θ and ϕ angles.

In Equation (10), each of the component wave functions is a function of one variable
only. This equation is substituted into Equation (9), whereupon one obtains:

$$0 = \frac{1}{r^2 R}\frac{d}{dr}\left[r^2\frac{dR}{dr}\right] + \frac{1}{r^2(\sin\theta)\Theta}\frac{d}{d\theta}\left[\sin\theta\frac{d\Theta}{d\theta}\right]$$

$$+ \frac{1}{r^2(\sin^2\theta)\Phi}\frac{d^2\Phi}{d\phi^2} + \frac{\hbar^2}{2m}\left[E + \frac{e^2}{r}\right]$$

(11)

In Equation (11), the partial differentiation symbols have been replaced by total
differentiation symbols, because each function is now a function of a single variable.
By multiplying through by $r^2(\sin^2\theta)$ and rearranging, one obtains:

$$\frac{-1}{\Phi}\frac{d^2\Phi}{d\phi^2} = \frac{\sin^2\theta}{R}\frac{d}{dr}\left[r^2\frac{dR}{dr}\right] + \frac{\sin\theta}{\Theta}\frac{d}{d\theta}\left[\sin\theta\frac{d\Theta}{d\theta}\right]$$

$$+ \frac{\hbar^2}{2m}\left[E + \frac{e^2}{r}\right](r^2\sin^2\theta)$$

(12)

Because the left-hand side of Equation (12) contains only r and θ as variables
and because the right-hand side is a function of ϕ only, each side must therefore

be equal to some common constant. For reasons, which will soon become apparent, this constant is denoted as $-m^2$. In that case, the left-hand side of Equation (12) can be written in the form known as the Φ *equation*:

$$\frac{d^2\Phi}{d\phi^2} = -m^2\Phi \tag{13}$$

The right-hand side of Equation (12) must also equal m^2, hence substituting in this value and dividing through by $\sin^2\theta$, one obtains:

$$\frac{1}{R}\frac{d}{dr}\left[r^2\frac{dR}{dr}\right] + \frac{\hbar^2}{2m}\left[E + \frac{e^2}{r}\right]r^2$$
$$= -\frac{m^2}{\sin^2\theta} - \frac{1}{\Theta\sin\theta\,d\theta}\frac{d}{d\theta}\left[\sin\theta - \frac{d\Theta}{d\theta}\right] \tag{14}$$

One has now achieved the separation of all functions in r to the left-hand side of Equation (14) and all functions in θ to the right-hand side. Because these two sets of terms are independent of each other, but their sum is a constant (equal to zero), each set must be equal to a constant. Setting the sum of the terms in θ equal to the constant $-\beta$, we find:

$$\frac{1}{r^2}\frac{d}{dr}\left[r^2\frac{dR}{dr}\right] - \frac{\beta}{r^2}R + \frac{\hbar^2}{2m}\left[E + \frac{e^2}{r}\right]R = 0 \tag{15}$$

and

$$\frac{1}{\sin\theta\,d\theta}\frac{d}{d\theta}\left[\sin\theta\frac{d\Theta}{d\theta}\right] - \frac{m^2\Theta}{\sin^2\theta} + \beta\Theta = 0 \tag{16}$$

Equation (15) is known as the *R equation*, and Equation (16) is known as the Θ *equation*.

Through the use of an appropriate coordinate system, the Schrödinger equation for the hydrogen atom has been reduced to the solution of three separate equations, each of which is constructed in terms of a single variable. It is interesting to note that although the energy of the system depends only on r, a complete description of the wave functions requires that all three equations be solved.

To simplify the course of the present discussion, however, we will not pursue the traditional method of solving each wave equation separately but will instead follow an alternate approach. In our treatment, we will factor the overall wave function into a radial function, $R(r)$, and an angular function, $Y(\theta, \phi)$:

$$\Psi(r, \theta, \phi) = R(r)Y(\theta, \phi) \tag{17}$$

Upon substitution of Equation (17) into Equation (9), one obtains:

$$\frac{1}{R}\frac{d}{dr}\left[r^2\frac{dR}{dr}\right] + \frac{\hbar^2}{2m}\left[E + \frac{e^2}{r}\right]r^2 = \frac{1}{Y(\sin\theta)}\frac{\partial}{\partial\theta}\left[\sin\theta\frac{\partial Y}{\partial\theta}\right] + \frac{1}{Y(\sin^2\theta)}\frac{\partial^2 Y}{\partial\phi^2}$$

(18)

In Equation (18), the partial differentiation symbols for the R equation have been replaced by total differentiation symbols, because that function is a function of a single variable. As soon earlier, both sides of Equation (18) must each be equal to a constant that shall be termed β. One then obtains:

$$\frac{1}{r^2}\frac{d}{dr}\left[r^2\frac{dR}{dr}\right] - \frac{\beta}{r^2}R + \frac{\hbar^2}{2m}\left[E + \frac{e^2}{r}\right]R = 0 \qquad (19)$$

and

$$\frac{1}{\sin\theta}\frac{d}{d\theta}\left[\sin\theta\frac{dY}{d\theta}\right] - \frac{1}{(\sin^2\theta)}\frac{\partial^2 Y}{\partial\phi^2} + \beta Y = 0 \qquad (20)$$

Equation (19) has the same form as the previously derived *R equation*, and Equation (20) is known as the *Angular equation*. A detailed analysis of the mathematics is available in the standard references (1–6), but it shall suffice to note that the constant β is better defined by the term $\ell(\ell + 1)$.

Solution of the Angular Equation

Equation (20) is not a simple differential equation and consequently does not have an obvious solution. However, through appropriate mathematical manipulation one can rearrange the angular equation into a form that is equivalent to that of the *associated Legendre equation*. The solutions to this equation are found by assuming polynomial functions that are single-valued and that exist for only certain combinations of defining constants. The solutions to the Legendre equation are often termed *spherical harmonics*, whose value is determined by two different types of integers (i.e., quantum numbers). The first of these is m, which can only assume the values:

$$m = 0, \ \pm 1, \ \pm 2, \ \pm 3, \ \pm 4, \text{etc.} \qquad (21)$$

The constant m is called the *magnetic quantum number*, because it effectively corresponds to the quantum number of the same number invoked in the Bohr theory to explain the splitting of spectral lines in the presence of a magnetic field. The second constant, ℓ, is called the *azimuthal quantum number* and is restricted to values such that:

$$|m| \leq \ell \qquad (22)$$

Table 1   Selected Angular Wave Functions for the Hydrogen Atom

Azimuthal quantum number (l)	Magnetic quantum number (m)	Angular wave equation
0	0	$(1/4\pi)^{1/2}$
1	0	$(3/4\pi)^{1/2}\cos\theta$
1	± 1	$(3/4\pi)^{1/2}\sin\theta\cos\phi$ and $(3/4\pi)^{1/2}\sin\theta\sin\phi$
2	0	$(5/16\pi)^{1/2}(3\cos^2\theta - 1)$
2	± 1	$(15/4\pi)^{1/2}\sin\theta\cos\theta\cos\phi$ and $(15/4\pi)^{1/2}\sin\theta\cos\theta\sin\phi$
2	± 2	$(15/16\pi)^{1/2}\sin^2\theta\cos 2\phi$ and $(15/16\pi)^{1/2}\sin^2\theta\sin 2\phi$

The solution to Equation (20) is written in the form:

$$Y(\theta,\phi) = P_\ell^{|m|}(\cos\theta)e^{im\phi} \tag{23}$$

where m is permitted to assume the values $0, \pm 1, \pm 2, \pm 3, \ldots, \pm\ell$. An excellent description of the properties of the associated Legendre polynomials, $P_\ell^{|m|}$, is available (7). An examination of associated Legendre functions reveals that the functions equal zero for all cases where the value of $|m|$ exceeds the value of ℓ. Alternatively, the solutions to the angular wave equation can be written in the equivalent real forms:

$$Y(\theta,\phi) = P_\ell^{|m|}(\cos\theta)\sin(m\phi) \tag{24}$$

$$Y(\theta,\phi) = P_\ell^{|m|}(\cos\theta)\cos(m\phi) \tag{25}$$

In Equations (24) and (25), m is a zero or positive integer whose value does not exceed that of ℓ.

A number of angular wave functions that are of particular utility to a discussion of atomic properties are given in Table 1. Considering the restrictions existing for the magnetic and azimuthal quantum numbers, it is indicative that for a given value of ℓ, there are $(2\ell + 1)$ different spherical harmonic functions, each of which is characterized by a different value of m (i.e., $m = -\ell, -\ell + 1, -\ell + 2, \ldots, 0, \ldots, \ell + 2, \ell + 1, \ell$).

Solution of the Radial Equation

Through suitable mathematical manipulation, the radial Equation (19) can be reduced to the form of a standard differential equation whose solutions are in the form of polynomials. This process is facilitated by the introduction of two new constants:

$$\alpha^2 = 2mE/\hbar^2 \qquad (26)$$

and

$$\lambda = me^2/\hbar^2\alpha \qquad (27)$$

The analysis is also aided by the definition of a new variable, ρ:

$$\rho = 2\alpha r \qquad (28)$$

The R equation then has the simpler form:

$$\frac{1}{\rho^2}\frac{d}{d\rho}\left[\rho^2\frac{dR}{d\rho}\right] + \left[\frac{\beta}{\rho^2} + \frac{\lambda}{\rho} - \frac{1}{4}\right]R = 0 \qquad (29)$$

where the radial function now has the form of $R(\rho)$. Because the radial wave function must go to zero as $\rho \to \infty$, the general solution to Equation (29) has the form:

$$R(\rho) = e^{-\rho/2}F(\rho) \qquad (30)$$

where the $F(\rho)$ functions have the form of the *associated Laguerre polynomials* (7). These have the general form:

$$F(\rho) = \rho^\ell \sum a_k \rho^k \qquad (31)$$

where the summation takes place from $k = 0$ to $k = \infty$. If the new expression for $R(\rho)$ is substituted back into Equation (29), one obtains an infinite number of terms in increasing powers of ρ. Because the coefficients of all terms in a given power of ρ must equal zero, it is possible to deduce the recursion formula:

$$a_{k+1} = a_k\left[\frac{k + \ell + 1 - \lambda}{(k + 1)(k + 2\ell + 2)}\right] \qquad (32)$$

It is easily shown that as $\rho \to \infty$, $R(\rho)$ cannot be finite unless the series terminates. To terminate the series defined by the coefficients of Equation (32), it must be that:

$$k + \ell + 1 - \lambda = 0 \qquad (33)$$

which in turn requires that λ be an integer. It is common practice to use the symbol n for λ, so that:

$$n = k + \ell + 1 \qquad (34)$$

It is immediately apparent from Equation (34) that n can have any positive integral value but that it cannot equal zero. This quantum number is known as the *principal quantum number* because it is related to the energy states of the system.

Table 2   Selected Radial Wave Functions for the Hydrogen Atom

Principal quantum number (n)	Azimuthal quantum number (l)	Radial wave equation
1	0	$2\,(me^2/\eta^2)^{3/2}\,e^{-\rho}$
2	0	$(1/2\sqrt{2})\,(me^2/\eta^2)^{3/2}\,(2-\rho)\,e^{-\rho/2}$
2	1	$(1/2\,\sqrt{6})\,(me^2/\eta^2)^{3/2}\,\rho e^{-\rho/2}$
3	0	$(2/81\,\sqrt{3})\,(me^2/\eta^2)^{3/2}$ $(27-18\rho+\rho^2)\,e^{-\rho/3}$
3	1	$(4/81\sqrt{6})\,(me^2/\eta^2)^{3/2}\,(6\rho-\rho^2)\,e^{-\rho/3}$
3	2	$(4/81\,\sqrt{30})\,(me^2/\eta^2)^{3/2}\,\rho^2\,e^{-\rho/3}$

Note: $\rho = rme^2/\eta^2$.

A number of radial wave functions, which are of particular utility to a discussion of atomic properties, are given in Table 2. For a given value of ℓ, n is restricted to the values $\ell + 1$, $\ell + 2$, and so on. But because the value of n determines the energy of the system, it is more convenient to consider n as the fundamental quantum number of the system. In that case, for a given value n. one may have $\ell = 0, 1, 2, \ldots, n - 1$. It is already shown that for a given value of n, one may have $m = -\ell, -\ell + 1, -\ell + 2, \ldots, 0, \ell + 2, \ell + 1, \ell$.

Wave Functions of the Hydrogen Atom and Hydrogen-Like Atoms

Equation (17) defined the wave function of the hydrogen atom, and this description is now extended to include subscripts pertaining to the associated quantum numbers:

$$\psi_{n\ell m}(r, \theta, \phi) = R_{n\ell}(r)Y_{\ell m}(\theta, \phi) \tag{35}$$

Solutions for both the angular wave functions, $Y_{\ell m}(\theta, \phi)$, have been provided in Table 1, and solutions for the radial wave functions, $R_{n\ell}(r)$, are found in Table 2. A particular state of a hydrogen atom is now seen to be defined by a definite set of n, ℓ, and m quantum numbers, and its wave function is obtained by multiplying the appropriate individual wave functions.

We can readily extend the scope of these derivations to include all possible one-electron atoms, which are ordinarily termed as being "hydrogen-like" atoms. Atoms in this series include He^+, Li^{2+}, Be^{3+}, and so forth. The main difference that arises for these hydrogen-like atoms is found in the statement of the potential energy, which is still derived from Coulomb's Law but is now given as:

$$V = -Ze^2/r \tag{36}$$

where Z is the atomic number of the atom in question. The inclusion of higher values for the nuclear charge does not materially affect any of the wave function derivations and is simply carried through during the solution of the radial equation.

As we have stated earlier, for a given value of n, the azimuthal quantum number can have values of $\ell = 0$ to $\ell = n - 1$. Furthermore, a state characterized by a given value of ℓ can, in turn, exist as any one of the substates characterized by different values of the magnetic quantum number, m. Because m can possess all values from $-\ell$, through 0, and up to $+\ell$, it follows that the number of such states is given by the expression $(2\ell + 1)$. Following the nomenclature of Mulliken, we will refer to the $\psi_{n\ell m}(r, \theta, \phi)$ wave function as an *orbital*.

States of the hydrogen atom characterized by $\ell = 0$ are known as s orbitals, those by $\ell = 1$ are p orbitals, those by $\ell = 2$ are d orbitals, and those by $\ell = 3$ are f orbitals. For example a state or orbital characterized by $n = 3$ and $\ell = 1$ is termed a $3p$ state or a $3p$ orbital. There are actually three substates within a p state, and these are characterized by $m = -1$, $m = 0$, and $m = 1$. Because these three p states differ only in their angular properties, we can plot their angular properties using the pertinent aspects of their angular wave functions. The advantage of having obtained the solutions to the angular equation in the real forms that were summarized in Table 1 is that their three-dimensional probability density graphs have recognizable shapes that permit some physical interpretation to be assigned to them. For instance, all p orbitals characterized by the $(\sin \theta \cos \phi)$ term have their major density aligned along the x-axis and are therefore known as p_x orbitals. Similarly, all p orbitals characterized by the $(\sin \theta \sin \phi)$ term have their major density aligned along the y-axis and are therefore known as p_y orbitals. Finally, the p orbitals characterized by the $(\cos \theta)$ term have their major density aligned along the z-axis and are therefore known as p_z orbitals.

Taking advantage of this additional nomenclature, one may write the full wave functions that describe the various orbitals for hydrogen-like atoms, and a number of these are collected in Table 3.

Energy Levels and Spectroscopy of the Hydrogen Atom

Substitution of n for λ back into the defining Equations (26) and (27) yields the result:

$$E_n = \frac{-Z^2 m e^4}{2\eta^2} \frac{1}{n^2} \tag{37}$$

Equation (37) represents exactly the same result that had been obtained from the Bohr theory (Equation 32 of chap. 1). One immediately notes that the total energy of a state is solely determined by the identity of the principal quantum number, n. All of the substates associated with a given value of n (and defined by their ℓ and m quantum numbers) are all therefore of equal energy in the absence of an additional perturbation. These states of the same energy are said to be *degenerate*.

Table 3 Normalized Wave Functions of the Hydrogen Atom

Principal quantum number (n)	Azimuthal quantum number (l)	Magnetic quantum number (m)	Normalized wave equation
1	0	0	$\psi_{1S} = (1/\sqrt{\pi})\,(Zme^2/\hbar^2)^{3/2}\,e^{-\rho}$
2	0	0	$\psi_{2S} = (1/4\sqrt{2\pi})\,(Zme^2/\hbar^2)^{3/2}\,(2-\rho)\,e^{-\rho/2}$
2	1	0	$\psi_{2P_z} = (1/4\sqrt{2\pi})\,(Zme^2/\hbar^2)^{3/2}\,\rho\,e^{-\rho/2}\cos\theta$
2	1	± 1	$\psi_{2P_x} = (1/4\sqrt{2\pi})\,(Zme^2/\hbar^2)^{3/2}\,\rho\,e^{-\rho/2}\sin\theta\cos\phi$
			$\psi_{2P_y} = (1/4\sqrt{2\pi})\,(Zme^2/\hbar^2)^{3/2}\,\rho\,e^{-\rho/2}\sin\theta\sin\phi$
3	0	0	$\psi_{3S} = (1/81\sqrt{3\pi})\,(Zme^2/\hbar^2)^{3/2}\,(27-18\rho+\rho^2)\,e^{-\rho/3}$
3	1	0	$\psi_{3P_z} = (2/81\sqrt{\pi})\,(Zme^2/\hbar^2)^{3/2}\,(6\rho-\rho^2)\,e^{-\rho/3}\cos\theta$
3	1	± 1	$\psi_{3P_x} = (2/81\sqrt{\pi})\,(Zme^2/\hbar^2)^{3/2}\,(6\rho-\rho^2)\,e^{-\rho/3}$ $\sin\theta\cos\phi$
			$\psi_{3P_y} = (2/81\sqrt{\pi})\,(Zme^2/\hbar^2)^{3/2}\,(6\rho-\rho^2)\,e^{-\rho/3}$ $\sin\theta\sin\phi$
3	2	0	$\psi_{3d_{z^2}} = (1/81\sqrt{6\pi})\,(Zme^2/\hbar^2)^{3/2}\,\rho^2\,e^{-\rho/3}$ $(3\cos^2\theta - 1)$
3	2	± 1	$\psi_{3d_{xz}} = (\sqrt{2}/81\sqrt{\pi})\,(Zme^2/\hbar^2)^{3/2}\,\rho^2\,e^{-\rho/3}$ $\sin\theta\cos\theta\cos\phi$
			$\psi_{3d_{yz}} = (\sqrt{2}/81\sqrt{\pi})\,(Zme^2/\hbar^2)^{3/2}\,\rho^2\,e^{-\rho/3}$ $\sin\theta\cos\theta\sin\phi$
3	2	± 2	$\psi_{3d_{x^2-y^2}} = (1/81\sqrt{\pi})\,(Zme^2/\hbar^2)^{3/2}\,\rho^2\,e^{-\rho/3}$ $\sin^2\theta\cos 2\phi$
			$\psi_{3d_{xy}} = (1/81\sqrt{\pi})\,(Zme^2/\hbar^2)??\,\rho^2\,e^{-\rho/3}$ $\sin^2\theta\sin 2\phi$

Note: $\rho = rZme^2/\hbar^2$.

The azimuthal quantum number ℓ is the quantum number associated with the total angular momentum of the electron in the given state (i.e., orbital). In quantum mechanical terminology, the spherical harmonic functions ($P_\ell^{|m|}$) are eigenfunctions of the total orbital angular momentum operator ($\hat{L}$), and have an eigenvalue equal to $\ell\,(\ell+1)\,\hbar^2$, or:

$$\hat{L}^2 P_\ell^{|m|} = \ell(\ell+1)\hbar^2 P_\ell^{|m|} \tag{38}$$

As stated earlier, the quantum number ℓ is restricted to integral values between 0 and $(n-1)$ and gives the number of nodes existing in the angular wave function. For more complex atoms, the energies of the substates defined by differing values of ℓ will not be equal, and these states will not be degenerate.

The magnetic quantum number m is associated with the component of angular momentum along a specific axis in the atom that is usually the z-axis. Because all atoms are spherically symmetric, there is no way to define a particular axis, unless the atom is placed in an electric or magnetic field. Consequently, the energy of a state is not affected by its value of m unless an external field is present. The value of m does determine the degeneracy of a state, however, because there are $2\ell + 1$ values of m for a state defined by the quantum number ℓ. The splitting of states defined by different values for m is called the Zeeman effect. The magnetic quantum is defined explicitly during the solution to the Φ equation, and the solutions to this equation are found to be eigenfunctions of the z-component of the angular momentum operator:

$$\hat{L}_z\Phi(m) = m\hbar\Phi(m) \tag{39}$$

It should be completely evident at this point that the properties of a given electron in an atom are critically determined by its position relative to the nucleus (through the radial wave function) and by its angular momentum state (through the angular wave function).

Having evaluated some of the properties associated with the wave functions of the hydrogen-like atoms, it now becomes possible to discuss the spectroscopy associated with these systems. In the previous chapter, we established that the intensity associated with a transition is determined by the probability of its occurrence, for which the electric dipole spectral transition probability, P_{ij}, from a lower state (ψ_i) to an upper state (ψ_j) was given by:

$$P_{ij} = \int \psi_i \mu \psi_j \, d\tau \tag{40}$$

where μ is the total electric dipole moment operator. For the hydrogen atom, the instantaneous dipole moment is $-e\,\mathbf{r}$, where $\mathbf{r}$ is the vector from the nucleus to the electron.

Substitution of the various wave functions provided in Table 3 into Equation (38) and performance of the necessary integration operations leads to the realization that definite restrictions exist on the number and type of spectral transitions that may actually take place. Recognizing that every wave function is defined by a unique set of n, ℓ, and m quantum numbers, one defines the lower state wave function by ψ_i (n', ℓ', m') and the upper state wave function by ψ_j (n'', ℓ'', m''). In that case, Equation (38) becomes:

$$P_{ij} = -\int \psi_i(n',\ell',m')er\psi_j(n'',\ell'',m'') \, d\tau \tag{41}$$

To evaluate this integral, each component must be evaluated separately. This is accomplished by using the definitions of the spherical polar coordinate system as given in Equations (6) through (8) to define the components of the electric dipole spectral transition probability along each of the Cartesian axes. One obtains these quantities through the solution of the triple integrals:

$$P_{ij}(x) = e \iiint \psi_i(n', \ell', m')(r \sin\theta \cos\phi)\psi_j(n'', \ell'', m'')r^2 \sin\theta \, dr\,\theta \, d\phi \quad (42)$$

$$P_{ij}(y) = c \iiint \psi_i(n', \ell', m')\psi_j(n'', \ell'', m'')r \sin\theta \cos\phi \, \mathbf{r}^2 \sin\theta \, dr \, d\phi \quad (43)$$

$$P_{ij}(z) = e \iiint \psi_i(n', \ell', m')\psi_j(n'', \ell'', m'')r \cos\theta \, \mathbf{r}^2 \sin\theta \, dr \, d\theta \, d\phi \quad (44)$$

where the integration over dr takes place from 0 to ∞, the integration over dθ takes place from 0 to π, and the integration over dϕ takes place from 0 to 2π.

The integrals defined in Equations (42) through (44) can be evaluated from general Legendre polynomials to obtain selection rules that govern the spectroscopy of the hydrogen atom, but these operations are outside the course of this narrative. Interested readers can look up the details of the solution (2), but the outcome of these derivations can be summarized in three relations.

$$n' \rightarrow n'', \quad \text{no restriction} \quad (45)$$

For a hydrogen-like, one-electron atom, Equation (45) signifies that there are no rules imposed on spectral transitions that are related to the relative magnitudes of the principal quantum numbers of either the initial or final states.

$$\ell' \rightarrow \ell'', \qquad \Delta\ell = \pm 1 \quad (46)$$

Equation (46) signifies that transitions may take place from s-orbital wave functions to p-orbital wave functions or that transitions may take place from p-orbital wave functions to d-orbital wave functions, but that the transition from s-orbital wave functions to d-orbital wave functions are strictly forbidden.

$$m' \rightarrow m'', \qquad \Delta m = 0, \pm 1 \quad (47)$$

The selection rule of Equation (47) is restrictive only in those situations where significant degeneracy exists in the wave functions by virtue of their possessing larger values of ℓ.

It is worth noting that the symmetry of the wave functions describing the states of a system has an important bearing on whether integrals such as these will vanish or not. Often, the selection rules can be determined without needing to undertake a detailed evaluation of the integrals. It is a requirement of all atomic wave functions that they have either even or odd parity, or that they must either be symmetric or antisymmetric with respect to inversion of the coordinates of all the particles through the origin. The electric dipole moment operator involved in the matrix element will also be even or odd in parity, because it changes sign upon inversion of the coordinates. Now if both ψ_i and ψ_j are even, then the elements of the integral $\int \psi_i \, \hat{\mu} \, \psi_j \, d\tau$ will also be odd, because the product of two even functions is even and that of an even function with an odd function is odd. But if the integral is odd, it will contribute equal positive and negative amounts to the integration over all values of the coordinates, and therefore the integral must equal zero. Similarly, two odd states and the odd dipole moment will result in an odd integral, which also results in a vanishing integral. Only when one state is even and the other odd will the integral be even and have a finite value.

The effect of wave function symmetry is summarized in the Laporte rule, which states that dipole radiation transitions can take place only from even states to odd states or from odd states to even states. In the hydrogen atom situation, states characterized by $\lambda = 0, 2, 4, 6, \ldots$ are even, and states characterized by $\lambda = 1, 3, 5, 7, \ldots$ are odd. When one deals with complex atoms, the use of the Laporte rule greatly simplifies the evaluation of possible spectroscopic transitions, because it is relatively easy to assign a parity value to even the most complicated wave functions.

In the earlier discussion regarding the Bohr theory and the spectra of the hydrogen atom, it was concluded that spectral transitions could take place between any two energy levels specified by the principal quantum number, n. The quantum mechanical treatment just concluded for this system shows the assumption made by Bohr to be correct, but a restriction was found for which the azimuthal quantum number, λ, could only change by ± 1. This means that because the Lyman series terminates in the $1s$ ground state, its spectral lines must originate from p-states of differing values of $\mathbf{n}$. Similarly, transitions with the Balmer series must originate only from s-states and d-states, because they terminate in the $2p$-state.

Electron Spin

Although the methods of quantum mechanics just outlined were found to yield excellent agreement with the observed spectra of atoms, it was recognized that the theory still failed to explain several features in observed atomic spectra. These consisted of small shifts in the expected frequencies of spectral lines and the resolution of many spectral lines into closely spaced multiplets of several lines. This multiplicity of lines, implying the existence of additional electronic states and transitions, could not be predicted from the theory.

As it happened, an explanation for the multiplicity of spectral lines had been independently advanced by Goudsmit and Uhlenbeck (8) and by Bichowsky and Urey (9) shortly before quantum mechanics had been fully developed. These workers postulated that an electron had an intrinsic angular momentum that they called "spin angular momentum." Some examples of phenomena that had been explained by the concept of spin were the Stern-Gerlach experiments, the degeneracy of the excited states of atoms and molecules, the anomalous Zeeman effect, and the fine structure splittings of atomic spectra. The proper theory of these properties was not obtained until Dirac applied relativity theory to the quantum mechanical formulation (10). The mathematical considerations necessary in Dirac's theory are extremely complex, but there is some value in briefly examining the consequences of his approach.

According to the theory of relativity, the energy of a body depends not only on its kinetic and potential energies but also on its rest mass, or its mass when it is stationary with respect to the frame of reference being used to make measurements. If this energy term, in addition to the momentum and potential energy terms, is inserted into the Hamiltonian operator during the construction of the Schrödinger equation, it is found that the equation is of such a form that it cannot be solved without the introduction of several additional operators, whose nature cannot be discussed here. In addition, the resulting equation cannot be solved as an ordinary differential equation but requires the use of a more complex matrix formulation.

Some understanding of the important findings that result from the relativistic equation that Dirac solved can be acquired through a qualitative discussion of the results. If one considers the simple case of a single particle moving in one dimension and confined in a potential well with infinite walls, with the nonrelativistic Schrödinger equation, one finds that only certain positive energy levels are possible. When the Dirac theory is used, the essential difference between the relativistic and nonrelativistic solutions is that the former includes the relativistic rest-mass energy equivalent mc^2. However, because transitions between electronic states involve differences in energies, the constant mc^2 is not observed experimentally.

In the nonrelativistic solution of the Schrödinger equation, one finds that for each energy level there is a single wave function describing the state of the system. The solution of the relativistic equation gives the surprising result that there are two independent, orthogonal wave functions, each of which are the solutions for each energy level. Unfortunately, it is difficult to describe the significance of these two states with a simple physical picture. An investigation of the angular momentum properties of the wave function solutions reveals that the z-component of the orbital angular momentum of a state in a hydrogen atom does not really remain constant with time, as predicted. One is forced to construct another angular momentum operator for which the relativistic solutions are eigenfunctions. One of the two solutions is an eigenfunction having a positive sign on one of its terms, and the other is an eigenfunction with a negative sign on the analogous term.

It is plausible, then, to advance a physical explanation for the self-contained angular momentum, which an electron apparently possesses in addition to angular momentum arising from its orbital motions. The intrinsic angular momentum arises because of a spinning motion of the electron mass about an axis through its center. Such a rotational motion of mass would obviously give rise to angular momentum, even though the particle might not itself be moving in some path about a central nucleus. Ordinarily, this intrinsic angular momentum of the electron is termed spin angular momentum, or simply electron spin. One should remember, however, that this interpretation is not required by relativistic quantum mechanics and is, in fact, introduced only by a process of analogy (i.e., comparing an operator, which is a constant of motion in the relativistic case with the orbital angular-momentum operator from nonrelativistic quantum mechanics). Be that as it may, the concept of electron spin is useful and will be used henceforth whenever necessary.

Because there is no classical analog of electron spin, one cannot use the normal procedure of stating a classical expression and substituting in the appropriate quantum mechanical operators. The concept of electron spin is therefore introduced into quantum mechanics through a series of three postulates. The first of these is that the operators for spin angular momentum commute and combine in the same way as those for ordinary angular momentum. This yields a series of spin angular momentum operators ($\hat{S}^2, \hat{S}_x, \hat{S}_y$, and $\hat{S}_z$), which are exactly analogous to the orbital angular momentum operators ($\hat{L}^2, \hat{L}_x, \hat{L}_y$, and $\hat{L}_z$) that are discussed earlier. The second postulate states that for a single electron, there are only two simultaneous eigenfunctions of the operators $\hat{S}^2$ and $\hat{S}_z$. By convention, these are called ψ_α and ψ_β and obey the eigenvalue equations:

$$\hat{S}_z \psi_\alpha = \tfrac{1}{2}\hbar\psi_\alpha \tag{48}$$

$$\hat{S}_z \psi_\beta = -\tfrac{1}{2}\hbar\psi_\beta \tag{49}$$

A comparison of Equations (48) and (49) with Equation (39) indicates that spin angular momentum, like orbital angular momentum, can be expressed in multiples of η. The eigenvalue of α is seen to be $\tfrac{1}{2}$, and the eigenvalue of β is $-\tfrac{1}{2}$, leading to the definition of a new quantum number, m_s. The permissible values of m_s are seen to be $\pm\tfrac{1}{2}$.

The third postulate is that the spinning electron acts like a magnet, having a magnetic dipole moment operator equal to:

$$\underline{M} = -g_0\beta_m\hat{S} \tag{48}$$

where g_0 is the spectroscopic splitting factor (equal to 2.0023), and β_m is the Bohr magneton (equal to 9.2732×10^{-21} erg/G). It is interesting to note that the value of g_0 was originally obtained empirically but was later derived a priori from the Dirac theory.

Because the spin operators can only affect coordinates in "spin space," they must commute with all operators that are a function only of spatial coordinates. With this in mind, one is able to partition the total wave function between orbital and spin contributions:

$$\psi_{total} = \psi_{orbital}\psi_{spin} \tag{49}$$

where $\psi_{orbital}$ is the wave function previous denoted as $\psi\,(\mathbf{r}, \theta, \phi)$ in Equation (10), and ψ_{spin} is the spin wave function (ψ_α or ψ_β). Incorporation of the spin operators in the Hamiltonian operator ultimately yields atomic wave functions, which are characterized by four quantum numbers, namely n, λ, m_λ, and m_s. The m_λ quantum number is the one previously termed m in the spin-free (nonrelativistic) Schrödinger system but which is termed m_λ when electron spin is included in the relativistic wave functions. A single electron can therefore be considered as moving in four-dimensional space, which consists of three space coordinates and one spin coordinate.

The Pauli Exclusion and Aufbau Principles, and Multielectron Atoms

It has now been established that a single electron moving in a spherically symmetrical, but non-Coulomb force field can be characterized by two quantum numbers (n and ℓ). The energy differences between states of different ℓ and equal n are ordinarily less than that between states with different n. The possible states of an electron can therefore be divided into principal groups (or levels) that differ from one another in their n values, and into subgroups (or sublevels) of the same n but different ℓ values. In the presence of a magnetic field, states of given n and ℓ, the states characterized by different values of m_λ, will not be degenerate.

In an atom containing several electrons, the motion of each individual electron can be approximated as motion in a centrally symmetric but non-Coulomb field of force. This field results from the overlapping of the Coulomb field of the nucleus and the mean field of the other electrons. In this approximation, a definite value of n and ℓ can also be attributed to each electron in a complicated atom. The approximation will be particularly good when considering a single electron with large n, as is usually the case for most of the higher terms of an atom. The action of the remaining electrons may then be approximated as being due to their mean field. This approximation permits the characterization of electrons as being either *core electrons* or as being *valence electrons*.

But when considering electrons having equal n and ℓ, these must necessarily be roughly equidistant from the nucleus, and thus the assumption of a mean field is relatively poor. The action of the other electrons on a given electron will be strongly dependent on their momentary positions, hence the field in which the given electron moves can no longer be considered as being centrally symmetric. This causes the n and ℓ quantum numbers to lose an exactly definable and physically interpretable meaning. Nevertheless, we shall see that it will still be possible to obtain the number and type of the electronic states, and this will permit major deductions to be made about spectroscopic transitions.

In order to understand the assembly of the periodic system, and the periodicity in the properties of elements and their energy level diagrams, a new assumption must be introduced. This is termed the *Pauli exclusion principle*, and its use prevents the filling of various electron shells with an arbitrary number of electrons. To formulate the principle conveniently, it is easy to imagine an atom brought into a very strong magnetic field, which is so strong that each individual ℓ_i and s_i are space-quantized independently of one another in the direction of the field. In other words, for each single electron, the components of ℓ in the direction of the field will take one of the values $m_\lambda = \ell, \ell - 1, \ell - 2, \ldots,$ $-\ell$, and the components of s can take one of the values $m_s = \pm\frac{1}{2}$. The total number of possible states, of course, will not be altered by the assumption of a strong magnetic field.

The Pauli principle states that in one and the same atom, no two electrons can have the same set of values for the four quantum numbers, n, ℓ, m_λ, and m_s. It follows that only a limited number of electrons can have the same set of values for the n and ℓ quantum numbers. The Pauli principle does not result from the fundamentals of quantum mechanics but is an assumption that fits very well into quantum mechanics. It cannot be properly theoretically justified without going into the Dirac theory. Accepting the validity of the Pauli principle, however, permits the ready classification of electrons in an atom according to a pattern that explains a considerable body of spectroscopic information.

The utility of the Pauli principle has been illustrated in Table 4, which list out all of the possible states of an electron in an atom, and the divisions into groups and subgroups, for shells up to $n = 3$. On the basis of the Pauli principle, only two electrons can be present in each cell defined by a specific set of $(n, \ell,$ and $m_\lambda)$ quantum numbers, and then only when these have antiparallel spin directions (different values of m_s).

The ordering of states in Table 4 is roughly that of the relative energies of the various states, hence according to the Pauli principle electrons would fill the states according to the following order:

$$1s \; 2s \; 2p \; 3s \; 3p \; 3d \; 4s \; 4p \ldots \tag{50}$$

The electronic configuration of an atom is built up by the filling of electrons into the lowest possible sequence of levels that is permitted by the Pauli principle. This concept is commonly referred to as the *Aufbau*, or *Building-Up*, principle. Thus, the single electron of the hydrogen atom is placed into the $1s$ orbital of the K shell (the shell nomenclature being that of Bohr), and one writes its electronic configurations as $1s^1$. Similarly, the Pauli principle permits both electrons of the helium atom to occupy the $1s$ orbital of the K shell, yielding the electronic configuration of $1s^2$. Of course, the two electrons of helium must exist in different spin states so as not to violate the Pauli principle.

Owing to the restrictions on quantum numbers imposed by the Pauli principle, it is not possible to place any more electrons into the K shell. When any shell in a multielectron atom contains the maximum number of electrons permitted by the Pauli principle, it is termed a *closed shell* or a *full shell*, and electrons in a full shell are often denoted as *core electrons*. For example, the $1s^2$ configuration of helium represents the maximal filling of the K shell, and these 1s electrons will be core electrons for any atom containing additional electrons. The outer electrons of an atom are often termed the *valence* electrons, because it will be only these electrons that become involved in chemical bonding.

Table 4   Possible States of an Electron in a Multielectron Atom

Shell	Orbital type	Principal quantum number (n)	Azimuthal quantum number (l)	Magnetic quantum number (m_λ)	Spin quantum number (m_s)
K	1s	1	0	0	+1/2
		1	0	0	−1/2
L	2s	2	0	0	+1/2
		2	0	0	−1/2
	2p	2	1	+1	+1/2
		2	1	+1	−1/2
		2	1	0	+1/2
		2	1	0	−1/2
		2	1	−1	+1/2
		2	1	−1	−1/2
M	3s	3	0	0	+1/2
		3	0	0	−1/2
	3p	3	1	+1	+1/2
		3	1	+1	−1/2
		3	1	0	+1/2
		3	1	0	−1/2
		3	1	−1	+1/2
		3	1	−1	−1/2
	3d	3	2	+2	+1/2
		3	2	+2	−1/2
		3	2	+1	+1/2
		3	2	+1	−1/2
		3	2	0	+1/2
		3	2	0	−1/2
		3	2	−1	+1/2
		3	2	−1	−1/2
		3	2	−2	+1/2
		3	2	−2	−1/2

The *ground state* of an atom is the one in which all electrons are in the lowest possible energy states. On the basis of the Pauli principle, this is the state in which all the lower shells are filled as far as the Pauli principle allows. We shall return to this point later, but excited states of an atom result when one or more of its electrons are raised to any of the higher orbitals.

The configuration of an atom having a nuclear charge of $Z + 1$ is obtained by taking the ground state electronic configuration of the atom characterized by the nuclear charge of Z and adding an additional electron to one of the shells not yet filled. The electron configurations of the atoms belonging to the first four rows of the periodic table are dictated by the pattern of quantum numbers given in

Table 5   Configurations of Multielectron Atoms Making Up the First Four Rows of the Periodic Table

Z	Element	Electron configuration	Z	Element	Electron configuration
1	H	$1s^1$	19	K	[Ar] $4s^1$
2	He	$1s^2 \equiv$ [He]	20	Ca	[Ar] $4s^2$
3	Li	[He] $2s^1$	21	Sc	[Ar] $3d^1\,4s^2$
4	Be	[He] $2s^2$	22	Ti	[Ar] $3d^2\,4s^2$
5	B	[He] $2s^2\,2p^1$	23	V	[Ar] $3d^3\,4s^2$
6	C	[He] $2s^2\,2p^2$	24	Cr	[Ar] $3d^5\,4s^1$
7	N	[He] $2s^2\,2p^3$	25	Mn	[Ar] $3d^5\,4s^2$
8	O	[He] $2s^2\,2p^4$	26	Fe	[Ar] $3d^6\,4s^2$
9	F	[He] $2s^2\,2p^5$	27	Co	[Ar] $3d^7\,4s^2$
10	Ne	[He] $2s^2\,2p^6 \equiv$ [Ne]	28	Ni	[Ar] $3d^8\,4s^2$
11	Na	[Ne] $3s^1$	29	Cu	[Ar] $3d^{10}\,4s^1$
12	Mg	[Ne] $3s^2$	30	Zn	[Ar] $3d^{10}\,4s^2$
13	Al	[Ne] $3s^2\,3p^1$	31	Ga	[Ar] $3d^{10}\,4s^2\,4p^1$
14	Si	[Ne] $3s^2\,3p^2$	32	Ge	[Ar] $3d^{10}\,4s^2\,4p^2$
15	P	[Ne] $3s^2\,3p^3$	33	As	[Ar] $3d^{10}\,4s^2\,4p^3$
16	S	[Ne] $3s^2\,3p^4$	34	Se	[Ar] $3d^{10}\,4s^2\,4p^4$
17	Cl	[Ne] $3s^2\,3p^5$	35	Br	[Ar] $3d^{10}\,4s^2\,4p^5$
18	Ar	[Ne] $3s^2\,3p^6 \equiv$ [Ar]	36	Kr	[Ar] $3d^{10}\,4s^2\,4p^6 \equiv$ [Kr]

Table 4 and are summarized in Table 5. It must be noted that interelectronic interactions can cause the relative energies of the orbitals to vary slightly from atom to atom, and the order can differ from the ideal sequence when two orbitals lie close together in energy. For example, the ground state configuration of chromium is [Ar] $3d^5\,4s^1$ rather than [Ar] $3d^4\,4s^2$ as a result of the favorable situation of an exactly half-filled shell of d-electrons.

One will note that the electron configurations of the ground states show *periodicity*, because after a certain number of electrons have been added, the outermost electron will be of a type previously encountered. For instance, all the alkali metals of Group IA contain a single outer electron and exhibit electron configurations that can be written as [core] ns^1. Similarly, all the alkaline earth metals of Group IIA contain two electrons outside of the core and exhibit electron configurations that can be written as [core] ns^2.

Spectroscopy of Core Electron States

To discuss the possible spectroscopic transitions, which can originate out of core electron states, a reexamination of the energy levels of the hydrogen atom is appropriate. Earlier, Equation (21) of chapter 1 was cited as the relation that fitted the emission spectrum of atomic hydrogen:

$$\tilde{W} = \frac{1}{\lambda} = R\left[\frac{1}{n_1^2} - \frac{1}{n_2^2}\right] \tag{51}$$

where n_1 and n_2 are each integers. Having completed the quantum mechanical treatment of this system, we now see that the integers n_1 and n_2 are in fact the principal quantum numbers of the initial and final states, respectively. This permits calculation of transition energies and are collected in Table 6 for transitions originating out of the $n = 1$ ($1s$) ground state of the hydrogen atom.

An examination of the pattern of transition energies and wavelengths of Table 6 yields three important conclusions. The first of these is that the energy required to effect the promotion of an electron out of the $1s$ level is exclusively associated with the electromagnetic radiation within the X-ray region of the spectrum. Second, when the principal quantum number of the terminal state is fairly low (i.e., less than 7), the spectral transitions are separated by a sufficient energy difference so that they remain discrete and are easily resolved. However, when n exceeds 10, the close spacing of succeeding transitions requires such high

Table 6 Absorption Spectra of the Hydrogen Atom Originating from the $n = 1$ State

n of terminal (state)	Energy (cm^{-1})	Wavelength (Å)	Wavelength (nm)
2	82,258.4	1215.68	121.57
3	97,491.4	1025.73	102.57
4	102,822.9	972.55	97.25
5	105,290.7	949.75	94.98
6	106,631.2	937.81	93.78
7	107,439.5	930.76	93.08
8	107,964.1	926.23	92.62
9	108,323.8	923.16	92.32
10	108,581.0	920.97	92.10
11	108,771.4	919.36	91.94
12	108,916.1	918.14	91.81
13	109,028.8	917.19	91.72
14	109,118.2	916.44	91.64
15	109,190.3	915.83	91.58
16	109,249.4	915.34	91.53
17	109,298.3	914.93	91.49
18	109,339.3	914.58	91.46
19	109,374.0	914.29	91.43
20	109,403.6	914.05	91.40
50	109,633.9	912.13	91.21
100	109,666.8	911.85	91.19

resolution to detect individual spectral transitions that these eventually degrade into a continuum of unresolvable spectroscopy. Finally, no energy levels exist when the transition energy exceeds the value of the Rydberg constant (109,677.8 cm^{-1}, equivalent to a wavelength of 91.18 nm), hence any electron excited at a higher energy must be ionized out of the atom.

These conclusions permit the classification of core electron spectroscopy into two broad divisions. The first entails the excitation of core electrons into states, which are defined by the quantum numbers previously discussed and is termed *X-ray absorption spectroscopy*, or XAS. The excited state produced by XAS can undergo a multitude of secondary transitions, leading to a variety of phenomena, which can be grouped together under the broad category of atomic spectroscopy.

The second type of spectroscopy concerns transitions where the electron is excited above the continuum of energy levels and is completely ionized. This type of spectroscopy is termed *X-ray photoelectron spectroscopy*, where one follows these processes through a measurement of the kinetic energy of the ionized electron. The excited state can be deactivated through the emission of electromagnetic radiation, a process known as *X-ray fluorescence*. When the excited state is deactivated through nonradiative processes, but which results in the ejection of electrons nonetheless, the phenomena constitute *Auger electron spectroscopy*.

X-RAY ABSORPTION SPECTROSCOPY

To understand the specificity of the X-ray absorption spectra, it is first necessary to consider the mechanism of the interaction of X-rays with a material. When an X-ray beam passes through a specimen, the intensity of the rays becomes weaker because of the dual processes of absorption and scattering. The mechanism of X-ray absorption differs from that of optical absorption, in that the absorption of X-ray energy occurs as a result of a single process—ionization of the atom at the expense of inner electrons. The energy of the absorbed radiation is thus transformed into kinetic energy of these displaced electrons (known as photo-electrons), in addition to the potential energy of the excited atom that equals the bonding energy of the displaced electron.

The X-ray radiation characterized by the least energy (i.e., greatest wavelength) causes the displacement of electrons from the outer shells. With increasing energy of the excitation quanta, an ever smaller quantity is required to remove electrons from a given shell, which is characterized by reduced absorption. A continuous lessening of absorption continues for as long as the emission energy is sufficient to force an electron out of the next, deeper-lying, inner shell. This phenomenon gives rise to a sharp increase in absorption that manifests itself in the form of an absorption edge.

Thus, the X-ray absorption spectra of materials are characterized by the existence of three elements. These are a continuously rising absorption coefficient with increasing wavelength, an absorption edge, and the fine structure of the absorption edges.

Most experimental work is conducted using synchrotron radiation, although in principle there is no reason why conventional high-intensity X-ray sources (i.e., rotating anode generators) could not be used. The absorption spectroscopy experiment is conducted by inserting the sample into the X-ray beam and measuring the incident and transmitted beam flux.

Another phenomenon that produces a weakening of X-ray intensity during the passage through the substance is scattering that is known to occur in two types. Coherent (or Thompson) scattering can take place as a result of the collision of the X-ray photon (having energy equal to $h\nu$) by the electrons (which have energies of E_{el}) of the atom. Should energy of the X-ray photons be less than that of the electron bonding ($h\nu < E_{el}$), the photons cannot displace the electron from a given inner shell. Following an elastic collision with fixed electrons, the photons are scattered, hence their energy (and wavelength) remain unchanged. On the other hand, incoherent (or Compton) scattering results if the energy of the X-ray photons is greater than that of the binding electrons ($h\nu > E_{el}$), so that the photons force the electron out of the corresponding inner shell. When the photoelectrons collide with electrons, they partition some of their energy to them. This results in the scattering photons displaying a lesser energy and a greater wavelength. Because the ejection of an electron is the first condition for the emergence of all X-ray and electron spectra, incoherent scattering must accompany the effect. Furthermore, because the atom has at the same time more and less strongly bound core electrons, one can observe two lines in the spectrum of scattered emission, those with unchanged and changed (increased) wavelength.

The intensity of scattering increases with the atomic number of the material, so that the more electrons present in an atom, the greater will be their scattering effect. The X-rays are scattered poorly by light atoms and strongly by heavy ones. The correlation of the coherent and incoherent scattering depends on two factors, namely, the wavelengths of the incident radiation and the atomic number of the absorbing atom. The smaller the wavelength of the incident radiation (i.e., the higher its energy) and the lesser the atomic number (i.e., the smaller the energy of the electrons bonding in these atoms), the greater will be the incoherent scattering of the emission.

The quantitative estimation of the reduced intensity of the X-rays following their passage through the substance is effected by using the attenuation factor μ, which is the sum of the coefficient of pure (photoelectric) absorption (τ) and the scattering (dispersion) factor (σ). The attenuation factor is often referred to as the coefficient of absorption. With wavelengths exceeding 0.5 Å and for elements with atomic numbers exceeding 26, the attenuation is effectively caused entirely by absorption. In that case, one concludes that $\mu = \tau$.

The coefficient of absorption (in units of cm^{-1}) is deduced from the usual statement of Beer's law. This is employed in evaluating the transparency or opacity of the specimen with its given thickness (at a specific wavelength), but because it depends on the state of the substance it is not a constant characterizing the absorption of a given element.

More frequent use is made of a mass absorption coefficient, equal to the absorption coefficient divided by the density of the substance. The mass absorption coefficient is independent of the state of the substance and of the chemical compound to which the given element belongs, and it has a definite significance for the given element at the given wavelength. It is a common practice to indicate the wavelength for which the value of the mass absorption coefficient is cited or to indicate the characteristic X-ray line whose radiation is used in measuring the absorption. Values for the most important absorption coefficients of the elements, measured at commonly used X-ray lines, have been published (11).

When viewed under conditions of low resolution, the X-ray absorption spectrum appears as a continuously rising absorption with increasing wavelength but which is superimposed with abrupt discontinuities. When the spectrum is examined at higher resolution, one finds the existence of a fine structure made up of distinctive elements. One element is the position of the absorption edge itself. Another is the Kossel structure of the initial region of absorption, consisting of additional weaker maxima near the absorption edge and having energies up to 20–30 eV at higher energies. Finally, one observes the Kroning extended fine structure, continuing from the absorption edge up to higher energies as large as 400 eV and being evident in the form of weak absorption fluctuations. Because substances in different phases will exhibit differences in their electronic structure, the nature of the fine structure transitions is a function of the phase under study.

In atoms, the absorption edge corresponds to the first allowed transition to an empty level. In the case of argon, for example, the electron configuration (Table 5) is $1s^2 \, 2s^2 \, 2p^6 \, 3s^2 \, 3p^6$, and the lowest unoccupied levels are the $4s$, $4p$, and $3d$ orbitals. Recalling the selection rule $\Delta \ell = \pm 1$ from Equation (46), one finds that the K-edge of absorption corresponds to the transition to the first vacant level, or $1s \rightarrow 4p$. The next allowed transitions from the $1s$ (K) level will be $1s \rightarrow 5p$, $1s \rightarrow 6p$, $1s \rightarrow 7p$, and so on right out to the limit of the p-level series. Superposition of these lines forms a fine structure of the argon K-edge absorption.

The X-ray absorption spectrum can be divided into two basic regions, each of which provides different information. One of these is near the absorption edge, and is termed the XANES (*X-ray absorption near edge structure*) region. The XANES region is known to be sensitive to the coordination symmetry and the oxidation state of the target atom. The other is the EXAFS (*extended X-ray absorption fine structure*) region, and its spectroscopy can be used to obtain information on particle size and composition, coordination numbers and distances, and interparticulate interactions.

Owing to the favorable absorption coefficients of heavy atoms, XANES and EXAFS studies are usually conducted on systems containing at least one atom having an atomic number exceeding 26. As an example, the K-edge absorption in metallic nickel (for which the binding energy of the Ni *K*-electron is 8332.8 eV) is plotted against the energy of the ejected photoelectron in Figure 2A. The EXAFS spectrum extracted from this spectrum is found in Figure 2B, plotted in the usual manner of EXAFS absorption function as a function of the photoelectron wave vector.

The X-ray absorption spectrophotometric techniques have found extensive use in the characterization of all systems that contain heavy atoms, such as homogeneous and heterogeneous catalysts (12). Not surprisingly, this methodology has proved to be extremely useful in the characterization of small-molecule inorganic compounds and extraordinarily informative in the study of metalloproteins. The utility of this approach will be outlined through the description of appropriate selected illustrative examples.

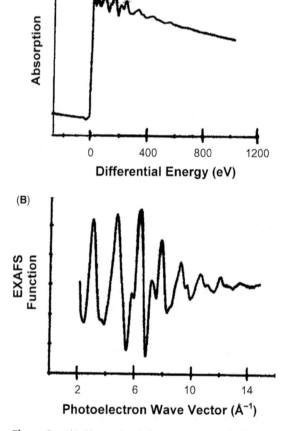

Figure 2 (A) X-ray absorption spectrum and (B) corresponding EXAFS spectrum for metallic nickel. *Abbreviation*: EXAFS, extended X-ray absorption fine structure.

XAS Studies of Inorganic Compounds

Different EXAFS and XANES in the K-edge of Mn(III) bound to dodecaphenyl-porphyrin complexes were used to study the activation mechanism induced upon reaction with ozone (13). It was deduced that the coordination environment of the Mn(III) ion consisted of the four equatorial pyrrolic nitrogen atoms of the porphyrin ligand plus two ozone molecules, these being arranged symmetrically with respect to the porphyrin core. The two Mn–oxygen bonds were found to be nonequivalent, exhibiting bond lengths of 2.09 ± 0.02 Å and 2.49 ± 0.05 Å. Interestingly, only one of these bonds was affected upon this formation of the catalytically active species.

Both K-edge and L-edge absorption spectroscopies were used to study the light-induced excited spin state trapping of *bis*(phenanthroline)-*bis*(thiocyanate)-Fe(II) (14). The spectra obtained for low- and high-spin forms exhibited trend discontinuities that paralleled trends in other structural work, indicating the utility of the spectroscopic study. In another study of this type, XAS was used in conjunction with other techniques to study the spin crossover phenomena associated with other Fe(II) complexes (15). It was possible to obtain values for changes in Fe–N bond distances, which were caused by changes in the spin state of the iron atom in the complexes.

A number of Fe(III) complexes with various saccharide ligands were characterized using K-edge XAS techniques, together with a number of other methodologies (16). A variety of mononuclear, dinuclear, and trinuclear complexes were obtained, and in some cases mixed ligand complexes were obtained. The XANES studies of these complexes exhibited a pre-edge structure possessing $1s$-$3d$ transition structure, which is indicative of the existence of octahedral Fe(III) centers. The EXAFS studies delineated the binding of the saccharide ligands through the pattern of Fe—O bonds and also indicated the existence of weaker interactions between the central Fe(III) ion and carbon atoms in the ligands.

The K-edge XANES spectra were obtained for two polymorphs of *bis*(1,2-dione-dioximato) Ni(II), and the compound dissolved in pyridine (17). In the crystalline materials, the XANES features consisted mainly of the $1s \rightarrow 4p$ transition of Ni(II), with the features around 8335–8350 eV being significantly affected by the interaction between the central Ni(II) ion and the atoms of adjacent molecules in the stacking structures. In pyridine solution, the molecules were found to stack above each other, with slight deviations from the central axis being noted.

The XAS studies were performed to study the Cu(I) complexes of crown ether appended *bis*([2-pyridyl]-ethyl)amines and their oxygen adducts (18). Following a crystallographic study of a model compound, the contribution of pyridine ring atoms to the EXAFS was simulated. Oxygenation of the complexes resulted in a large increase of the intensity of the major peak in the phase-corrected Fourier transform, which was interpreted as evidence for a peroxo coordination mode. This in turn caused a valency change from Cu(I) to Cu(II), as judged from the edge position.

One compound of pharmaceutical interest, which has been studied by XAS is zinc stearate (19). It was concluded in this work that the Zn(II) ion contains four oxygen atoms in its inner coordination sphere and that the Zn—O distances were 1.95 Å. Furthermore, the carboxylate groups were found to bind in a bridging bidentate mode. Zinc stearate was observed to melt at 130°C, but the coordination structure of the Zn(II) ions was found to be maintained in the molten state. Examples of the Zn K-edge EXAFS spectra, and the Fourier transforms of these, at various temperatures are found in Figure 3. The Zn—O bond lengths at 25°C, −203°C, and 170°C were all found to be 1.95 Å, and the coordination number of the central Zn(II) ion was deduced to be 4.1 ± 0.1 at all temperatures.

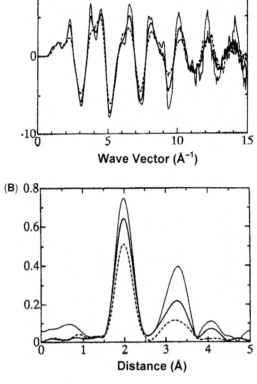

Figure 3 (A) Zn K-edge EXAFS spectra of zinc stearate at 25°C (*thick curve*), −203°C (*thin curve*), and 170°C (*dotted curve*), and (B) the Fourier transforms of these. *Abbreviation*: EXAFS, extended X-ray absorption fine struction. *Source*: Adapted from Ref. 19.

The XAS Studies of Metallo-Proteins

Without a doubt, the most powerful method available to establish the three-dimensional structure of a chemical species is that of single-crystal X-ray diffraction. However, this technique cannot be used when the substance in question cannot be crystallized. This difficulty is often encountered during the study of peptides and protein systems, and as a result a large number of investigations have been conducted where structural information has been obtained from the conduct of the XAS work. The ability of this technique to provide information on the immediate coordinative environments of metal ions has enabled workers to obtain important information about such binding sites in peptides and proteins.

The solution structures of the binuclear Mn(II) centers in arginase, Mn-catalase, and the Mn(II) substituted forms of selected iron enzymes (ribonucleotide reductase and hemerythrin) have been determined using XAS (20). The spectra obtained on model Mn(II) compounds showed an inverse correlation between the XANES peak maximum and the root-mean-square (RMS) deviation in metal–ligand bond lengths. Arginase and Mn-substituted ribonucleotide reductase protein systems were found to have symmetric nearest neighbor environments characterized by a low RMS deviation in bond length, whereas Mn catalase and Mn-substituted hemerythrin appeared to have a larger bond length deviation. The $1s \rightarrow 3d$ areas for arginase and Mn-substituted ribonucleotide reductase were consistent with six coordinate Mn, whereas the $1s \rightarrow 3d$ areas for Mn catalase and Mn-substituted hemerythrin were larger, suggesting that one or both of the Mn(II) ions were five-coordinate in these proteins. The EXAFS spectra for Mn-substituted ribonucleotide reductase and arginase were nearly identical, with symmetric Mn–nearest neighbor environments and outer-shell scattering being consistent with a lower limit of one histidine per Mn(II) core. In contrast, the EXAFS data for Mn catalase and Mn-substituted hemerythrin showed two distinct Mn–nearest neighbor shells and outer-shell carbon scattering consistent with a lower limit of approximately two to three histidine residues per Mn(II) core.

The XAS in the Mn(II) *L*-edge was used to study the metal–ion binding sites of Mn-catalase from *Lactobacillus plantarum* and the coordination chemistry of mixedvalence Mn(II) complexes that served as model compounds (21). Both reduced and superoxidized Mn-catalase were examined by fluorescence-detected soft XAS, and it was learned that the corresponding Mn(II) *L*-edge spectra were dramatically different. The spectrum of reduced Mn(II)-Mn(II)-catalase was interpreted using ligand field atomic multiplet calculations and by comparison to model compound spectra. For the interpretation of mixed valence Mn(II) spectra, an empirical simulation procedure based on the addition of homovalent model compound spectra was developed and tested on a variety of Mn(II) complexes and superoxidized Mn-catalase.

The Oxygen-Evolving Complex (OCE) of photosystem II catalyzes the four-electron oxidation of two water molecules to dioxygen. The importance of this system is that it is the source of most of the molecular oxygen in the atmosphere and provides the reducing equivalents needed for the reduction of carbon dioxide to glucose. The XAS has proven to be of extreme value in the study of this system, because the activity appears to center around the inorganic cofactors required for oxygen evolution (four Mn(II) ions, Ca(II), and chloride). In one study, EXAFS spectroscopy was used to characterize the local structural environment of Mn(II) in the resting state and in two different reduced derivatives of the photosynthetic OEC complex (22). The same group has reported on the use of EXAFS for the characterization of the structural consequences accompanying Ca(II) replacement in the reaction center complex of the OEC complex (23). Others have used XANES and EXAFS to assess the degree of similarity between the OEC in photosystem II and a series of synthetic manganese complexes containing the distorted cubane $\{Mn_4O_3X\}$ core (where $X =$ benzoate, acetate, methoxide, hydroxide, azide, fluoride, chloride, or bromide) (24). Most recently, edge spectroscopy and EXAFS was used to study the structural change of the Mn(II) cluster during the onset of substrate/water oxidation by the system (25). It was proposed in this work that the changes in Mn–Mn distances observed during the transition were the result of ligand or water oxidation, leading to the formation of an oxyl radical intermediate at a bridging or terminal position. The reaction of the oxyl radical with hydroxide ion, water, or an oxo group during the subsequent conversion was proposed to lead to the formation of the O—O bond.

The iron L-edge X-ray absorption spectra of deoxymyoglobin (deoxy-Mb), oxymyoglobin (Mb-O_2), carbonmonoxymyoglobin (Mb-CO), and some photoproducts of these (MB*-CO and Mb*-O_2) were obtained and compared to simulated spectra derived from a ligand field multiplet calculation (26). The analysis indicated that both Mb-CO and Mb-O_2 were low spin contradicting other work that suggested Mb-O_2 to be of intermediate spin. These workers summarized the advantages of L-edge XAS (relative to the more energetic K-edge spectroscopy) as being a direct metal-selective probe for metalloproteins, characterized by intense electric dipole allowed transitions, capable of yielding higher resolution in the structural details, and having energy levels accessible to d-orbital calculations.

The iron K-edge X-ray absorption spectrum of *Rhodococcus* sp. R312 nitrite hydratase (in frozen solutions at pH 7 and 9) has been analyzed to determine details of the iron site coordination (27). The EXAFS analysis permitted the deduction of two or three sulfur ligands per iron, and an overall six-fold coordination sphere implies an N_3S_2O ligation sphere. The bond lengths obtained from the EXAFS analysis support *cis*-coordination of two cysteine ligands and conclusively rule out nitric oxide coordination to the iron. Iron K-edge spectroscopy was used to study the binding of nitric oxide to iron in inactive nitrile hydratase, confirming a 1:1 stoichiometry of nitric oxide bound to low-spin Fe(III) (28). Studies were also performed on four iron complexes of various pentadentate ligands that served as model compounds for the metal-ion binding site in the protein system, enabling a better modeling of the metal-ion coordination environment.

The binding of nitric oxide to ferric nitric oxide reductase from *Fusarium oxysporum* is studied using IR, resonance Raman, and XAS (29). Trends in vibrational frequencies indicated that NO bound to Fe(III) more strongly in the parent enzyme than it did with the *d*-camphor-bound form of *Pseudomonas putida* camphor hydroxylase cytochrome (a typical model of the monooxygenase). Support for this conclusion was provided from an EXAFS study, which yielded a Fe—NO bond distance of 1.66 ± 0.02 Å for the *Fusarium oxysporum* ferric nitric oxide reductase, and 1.76 ± 0.02 Å for the *d*-camphor-bound form of *Pseudomonas putida* camphor hydroxylase cytochrome.

The EXAFS studies of the heme in human cystathionine β-synthase have been performed in order to assign the axial ligands provided by the protein (30). The ray absorption data revealed that iron in ferric cystathionine β-synthase is six-coordinate, and the scattering intensities were judged consistent with the presence of five nitrogen donors and one sulfur ligand. All of the data support this assignment of the axial ligands as cysteine and imidazole.

A key step in the catalytic cycle of coenzyme B_{12}-dependent enzymes is the homolysis of the organometallic bond of the cofactor, leading to the formation of a 5'-deoxyadenosyl radical. For the adenosylcobalamin-dependent enzyme methylmalonyl CoA mutase (MCM), it has been suggested that this step is mediated by a protein-induced lengthening of the cofactor axial cobalt—nitrogen bond. An EXAFS study indicated a normal axial cobalt—nitrogen bond of the cofactor in the related coenzyme B_{12}-dependent enzymes glutamate mutase from *Clostridium cochlearium* and 2-methyleneglutarate mutase from *Clostridium barkeri*. The XANES part of the various spectra were similar to each other but deviated significantly from the corresponding spectra of aquocob(III)a-lamin and for cob(II)alamin. In addition, the spectra contained a pronounced pre-edge peak, indicating the presence of a covalently attached sixth carbon ligand to the cobalt center (31).

Structural information has been obtained from the analysis of Ni(II) *K*-edge absorption spectroscopy of [Ni—Fe] hydrogenases from a number of sources (32). It was reported that the nickel sites of all but the oxidized sample of *Alaligenes eutrophus* hydrogenase were quite similar, with the nickel *K*-edge energies shifting 0.9–1.5 eV to lower energy upon reduction from oxidized to fully reduced forms. Analysis of the XANES features assigned to 1*s* → 3*d* electronic transitions indicates that the shift in energy, which accompanies a reduction of the enzymes, may be attributed at least in part to an increase in the coordination number from five to six. With the exception of the oxidized sample of *Alaligenes eutrophus* hydrogenase, the EXAFS data were dominated by the scattering from S-donor ligands at approximately 2.2 Å. The data obtained from this hydrogenase were unique in that they indicated a significant structural change to occur upon reduction of the enzyme. Data obtained from the oxidized enzyme indicate that the EXAFS is dominated by scattering from 3-4 N,O donor atoms at 2.06(2) Å, with contributions from 2-3 S-donor ligands at 2.35(2) Å. This coordination shell changes upon reduction to a more typical nickel site composed of four S-donor ligands at a Ni–S distance of 2.19(2) Å. The results of the EXAFS analysis was shown to be in general agreement with published crystal structures.

Nickel *L*-edge absorption spectroscopy is used to study the electronic structure of Ni(II) in several Ni—Fe hydrogenases under a variety of conditions (33). The *L*-edge spectra were interpreted by ligand field multiplet simulations and by the comparison with data for Ni(II) model complexes, and evidence was obtained for high-spin Ni(II) in the active enzyme. The heterogeneous Ni(II) sites in carbon monoxide dehydrogenases derived from *Clostridium thermoaceticum* and *Rhodospirillum rubrum* have been characterized using nickel L-edge spectroscopy (34). The data indicated that most of the Ni(II) in the isolated enzymes is low-spin, but upon treatment with CO, a fraction of the Ni(II) was converted either to high-spin NI(II) and/or to Ni(I).

The zinc site in tRNA-guanine transglycosylase derived from *Escherichia coli* has been studied by zinc *K*-edge absorption spectroscopy (35). The EXAFS data were most consistent with a tetracoordinate zinc atom being bound to one nitrogen and three sulfur ligands. It was concluded in this work that careful analysis of EXAFS data could reveal subtle conformational changes in metal binding sites, which were not observable using more conventional spectroscopic methods. The zinc *K*-edge X-ray absorption spectra of the ferric uptake regulation protein isolated from Escherichia coli have been obtained in frozen solution to determine details of the zinc coordination (36). The EXAFS spectra of the apo-protein and of the Co(II)-substituted protein indicated the existence of a tetrahedral environment for the zinc atom, with two sulfur donor ligands (from cysteine residues) located at a distance of 2.3 Å and two N/O donor ligands at 2.0 Å. The XANES of the two forms of the protein were sufficiently different so as to suggest a change of the conformation of the zinc site upon incorporation of cobalt.

Zinc is essential for thousands of proteins in organisms, but despite the undisputed need for information regarding local zinc-site structure, the majority of these environments remain uncharacterized. This is primarily due to the difficulty of studying the diamagnetic Zn(II) ion with conventional spectroscopic probes. In many cases, the coordinative environment of Zn(II) in proteins has been inferred on the basis of sequence similarity with one of the crystallographically characterized Zn proteins. Unfortunately, many zinc sites do not show significant sequence similarity to crystallographically characterized Zn proteins, and thus sequence alignment alone is often not sufficient to give the correct Zn coordination sphere. To address this problem, X-ray absorption fine structure spectra have been measured for a series of structurally characterized zinc model complexes that mimic the zinc sites found in metalloproteins (37). Because Zn—N and Zn—S EXAFS oscillations are nearly out of phase over the accessible energy range, it is difficult to determine the relative number of scattering sulfurs and nitrogens. The authors described a protocol that can be used to obtain an accurate quantitation of the number of low-Z (S) ligands. It was shown that two of the spectroscopic variables (the scale factor and shift in the threshold energy) ordinarily treated as freely adjustable parameters could lead to erroneous results if not carefully controlled.

The molybdenum enzymes comprise a group having a wide range of functionalities, catalyzing a variety of two-electron redox reactions coupled to the transfer of an oxygen atom. All of the structural information obtained on the molybdenum site indicates that the enzymes contain a novel pterin-molybdenum cofactor, in which the molybdenum appears bound by a dithiolene side chain of the pterin ring. A number of molybdenum complexes, which could serve as model compounds for the enzymes, have been characterized by XAS. In one such study, molybdenum K-edge spectra and EXAFS analyses were used to measure shifts of edge energies over the M(IV,V,VI) oxidation states, and features in the second-derivative edge spectra were correlated with the number of bound oxo ligands (38). In another work, the X-ray absorption spectra at the molybdenum and selenium K-edges and the tungsten $L_{2,3}$-edges were acquired for a set of 14 Mo(IV) and W(IV,VI) *bis*(dithiolene) complexes related to the active sites of molybdo- and tungstoenzymes (39). The edge shifts of these complexes were correlated with ligand electronegativities, and terminal ligand binding was clearly distinguished in the presence of four methyl-dithiolene interactions.

Sulfite oxidase is an oxo-transferase, which is responsible for the physiologically vital oxidation of sulfite to sulfate. The XAS at the molybdenum and sulfur K-edges was used to probe the active site of wild-type and cysteine 207—serine mutant human sulfite oxidases (40). The active site structure in the wild-type enzyme consisted of Mo=O ligands at 1.71 Å, and three Mo—S ligands at 2.41 Å. The mutant molybdenum site consisted of a trioxo site characterized by Mo=O bond lengths of 1.74 Å, and two Mo—S ligands at 2.47 Å. The X-ray absorption spectra were measured at the sulfur K-edge, chlorine K-edge, and Mo L_3- and L_2-edges for several model compounds of sulfite oxidase to investigate ligand–metal covalency and its effects on oxo transfer reactivity (41). It was concluded that anisotropic covalency of the Mo–cysteine bond in sulfate oxidase could promote preferential transfer of one oxo group during the performance of its catalytic function.

In another noteworthy study, XAS at the molybdenum K-edge is used to probe the molybdenum coordination of dimethyl sulfoxide reductase obtained from *Rhodobacter sphaeroides* (42). The molybdenum site of the oxidized protein possessed a novel Mo(VI) mono-oxo site (Mo=O at 1.68 Å), with additional coordination by approximately four thiolate ligands at 2.44 Å and either an oxygen or nitrogen ligand at 1.92 Å. The reduced Mo(IV) form of the enzyme is a *des*-oxomolybdenum, coordinated with 3-4 thiolates at 2.33 Å and two different Mo—O/N ligands at 2.16 Å and 1.92 Å. The stable Mo(V) glycerol-inhibited species was found to be a *des*-oxomolybdenum, with approximately four thiolate ligands at 2.40 Å and two similarly coordinated oxygen or nitrogen ligands at 1.96 Å.

The X-ray absorption spectroscopy at the molybdenum and selenium *K*-edges is used to probe the active site structure of formate dehydrogenase H obtained from Escherichia coli (43). The selenium *K*-edge EXAFS not only agreed with the molybdenum *K*-edge data but also indicated the unexpected presence of Se—S ligation at a bond length of 2.19 Å. This finding led to the suggestion that the active site of *Escherichia coli* formate dehydrogenase H contains a novel selenosulfide ligand to molybdenum, where the selenium and sulfur originated from selenocysteine and one of the pterin-cofactor dithiolenes.

Not all of the systems studied by XAS have involved edge studies of metal ions, and some very interesting work have been performed using i-edge XAS of the bound ligands (44). The sulfur *K*-edge EXAFS studies of cadmium-, zinc-, copper-, and silver-rabbit liver metallothioneins are used to obtain the first direct evidence for mixtures of bridging and terminal sulfur ligands (45). The XAS at the sulfur *K*-edge is applied to a series of mononuclear iron-sulfur complexes to determine covalency, and its distribution over *d*-orbitals split by the ligand field (46). It was found that the covalency decreased in the proteins relative to the models, indicating that the thiolate—Fe(III) bond was highly covalent and that a modulation of covalency in the proteins could contribute to the redox properties of the active site.

Investigation of the electronic structure of 2Fe-2S model complexes and the Rieske protein is performed using sulfur *K*-edge XAS (47). The degree of sulfide-iron covalency in the $[Fe_2S_2(SR)_4]^{2-}$ complexes was higher than the degree of thiolate-iron covalency, indicating extensive charge donation of the bridging sulfides. It was determined that the degree of thiolate covalency of the Fe(III) center is the same in both the oxidized and reduced Rieske clusters and similar to that of the $[Fe_2S_2(SR)_4]^{2-}$ model complexes.

The sulfur *K*-edge XAS was also used to investigate the distribution of sulfur types in whole blood cell samples, in selected subcellular blood fractions, and in cell-free plasma from the tunicate *Ascidia ceratodes* (48). The sulfur *K*-edge XAS spectrum of washed blood cell membranes revealed traces of sulfate and low-valence sulfur but no sulfate ester or sulfonate. It was concluded that the sulfur species contained within the blood cells was almost exclusively cytosolic.

REFERENCES

1. Atkins PW. Molecular Quantum Mechanics. Oxford: Clarendon Press, 1970.
2. Eyring H, Walter J, Kimball GE. Quantum Chemistry. New York: John Wiley & Sons, 1944.
3. Hanna MW. Quantum Mechanics in Chemistry. Menlo Park: Benjamin/Cummings, 1981.

4. Kauzmann W. Quantum Chemistry. New York: Academic Press, 1957.
5. Levine IN. Quantum Chemistry. Boston: Allyn and Bacon, 1970.
6. Pauling L, Wilson EB. Introduction to Quantum Mechanics. New York: McGraw-Hill, 1935.
7. Davis JC. Advanced Physical Chemistry. Appendix D New York: Ronald Press, 1965:596–600.
8. Goudsmit GE, Uhlenbeck S. Naturwissenshaften. 1925; 13:953.
9. Bichowsky R, Urey HC. Proc Nat Acad Sci 1926; 12:80.
10. Dirac PAM. Proc Roy Soc 1928; A117: 610; Proc Roy Soc 1928; A118:351.
11. Birks LS X-Ray Spectrochemical Analysis. New York: Interscience, 1969.
12. Evans J. Chem Soc Rev 1997; 26:11.
13. Gotte V, Goulon J, Goulon-Ginet C, Rogalev A, Natoli CR, Perie K, Barge J-M, Guilard R. J Phys Chem B 2000; 104:1927.
14. Lee JJ, Sheu H-S, Lee C-R, Chen J-M, Lee J-F, Wang C-C, Huang C-H, Wang Y. J Am Chem Soc 2003; 122:5742.
15. Real JA, Castro I, Bousseksou A, Verdaguer M, Burriel R, Castro M, Linares J, Varret F. Inorg Chem 1997; 36:455.
16. Rao PP, Geetha K, Raghavan MSS, Sreedhara A, Tokunaga K, Yamaguchi T, Jadhav V, Ganesh KN, Krishnamoorthy T, Ramaiah KVA, Bhattacharyya RK. Inorg Chim Acta 2000; 297:373.
17. Matsuo S, Yamaguchi T, Wakita H. J Phys Chem B 2000; 104:3471.
18. Feiters MC, Gebbink RJMK, Sole VA, Nolting H-F, Karlin KD, Nolte JM. Inorg Chem 1999; 38:6171.
19. Ishioka T, Maeda K, Watanabe I, Kawauchi S, Harada M. Spectrochim Acta 2000; A56:1731.
20. Stemmler TL, Sossong TM, Goldstein JI, Ash DE, Elgren TE, Kurtz DM, Penner-Hahn, JE. Biochem 1997; 36:9847.
21. Grush MM, Chen J, Stemmier TL, George SJ, Ralston CY, Stibrany RT, Gelasco A, Christou G, Gorun SM, Penner-Hahn JE, Cramer SP. J Am Chem Soc 1996; 118:65.
22. Riggs-Gelasco PJ, Mei R, Yocum CF, Penner-Hahn JE. J Am Chem Soc 1996; 118:2387.
23. Riggs-Gelasco PJ, Mei R, Ghanotakis DF, Yocum CF, Penner-Hahn JE. J Am Soc 1996; 118:2400.
24. Cinco, RM, Rompel A, Visser H, Aromi G, Christou G, Sauer K, Klein MP, Yachandra VK. Inorg Chem 1999; 38:5988.
25. Liang W, Roelofs TA, Cinco RM, Rompel A, Latimer MJ, Yu WO, Sauer K, Klein MP, Yachandra VK. J Am Chem Soc 2000; 122:3399.
26. Wang H, Peng G, Miller LM, Scheuring EM, George SJ, Chance MR, Cramer SP. J Am Chem Soc 1997; 119:4921.
27. Scarrow RC, Brennan BA, Cummings JG, Jin H, Duong DJ, Kindt JT, Nelson MJ. Biochem 1996; 35:10078.
28. Scarrow RC, Strickler BS, Ellison JJ, Shoner SC, Kovacs JA, Cummings JG, Nelson MJ. J Am Chem Soc 1998; 120:9237.
29. Obayashi E, Tsukamoto K, Adachi S-I, Takahashi S, Nomura M, Iizuka T, Shoun H, Shiro Y. J Am Chem Soc 1997; 119:7807.
30. Ojha S, Hwang J, Kabil O, Penner-Hahn JE, Banerjee R. Biochem 2000; 39:10542.
31. Champloy F, Jogl G, Reitzer R, Buckel W, Bothe H, Beatrix B, Broeker G, Michalowicz A, Meyer-Klaucke W, Kratky C. J Am Chem Soc 1999; 121:11780.

32. Gu Z, Dong J, Allan CB, Choudhury SB, Franco R, Moura JJG, Moura I, LeGall J, Przybyla AE, Roseboom W, Albracht SPJ, Axley MJ, Scott RA, Maroney MJ. J Am Chem Soc 1996; 118:11155.
33. Wang H, Raiston CY, Patil DS, Jones RM, Gu W, Verhagen M, Adams M, Ge P, Riordan C, Marganian CA, Mascharak P, Kovacs J, Miller CG, Collins TJ, Brooker S, Croucher PD, Wang K, Stiefel EI, Cramer SP. J Am Chem Soc 2000; 122:10544.
34. Ralston CY, Wang H, Ragsdale SW, Kumar M, Spangler NJ, Ludden PW, Gu W, Jones RM, Patil DS, Cramer SP. J Am Chem Soc 2000; 122:10553.
35. Garcia GA, Tiemey DL, Chong S, Clark K, Penner-Hahn JE. Biochemistry 1996; 35:3133.
36. Jacquamet L, Aberdam D, Adrait A, Hazemann J-L, Latour J-M, Michaud-Soret I. Biochem 1998; 37:2564.
37. Clark-Baldwin K, Tierney DL, Govindaswamy N, Gruff ES, Kim C, Berg J, Koch SA, James E. Penner-Hahn. J Am Chem Soc 1998; 120:8401.
38. Musgrave KB, Donahue JP, Lorber C, Holm RH, Hedman B, Hodgson KO. J Am Chem Soc 1999; 121:10297.
39. Musgrave KB, Lim BS, Sung K-M, Holm RH, Hedman B, Hodgson KO. Inorg Chem 2000; 39:5238.
40. George GN, Garrett RM, Prince RC, Rajagopalan KV. J Am Chem Soc 1996; 118:8588.
41. Izumi Y, Glaser T, Rose K, McMaster, J, Basu P, Enemark JH, Hedman B, Hodgson KO, Solomon EI. J Am Chem Soc 1999; 121:10035.
42. George GN, Hilton J, Raiagopalan KV. J Am Chem Soc 1996; 118:1113.
43. George GN, Colangelo CM, Dong J, Scott RA, Khangulov SV, Gladyshev VN, Stadtman TC. J Am Chem Soc 1998; 120:1267.
44. Glaser T, Hedman B, Hodgson KO, Solomon EI. Acc Chem Res 2000; 33:859.
45. Gui Z, Green AR, Kasrai M, Bancroft GM, Stillman MJ. Inorg Chem 1996; 35:6520.
46. Rose K, Shadle SE, Eidsness MK, Kurtz DM, Scott RA, Hedman B, Hodgson KO, Solomon EI. J Am Chem Soc 1998; 120:10743.
47. Rose K, Shadle SE, Glaser T, de Vries, S, Cherepanov A, Canters GW, Hedman B, Hodgson KO, Solomon EI. J Am Chem Soc 1999; 121:2353.
48. Frank P, Hedman B, Hodgson KO. Inorg Chem 1996; 38:260.

3

X-Ray Photoelectron and X-Ray Fluorescence Spectroscopy

Harry G. Brittain

Center for Pharmaceutical Physics, Milford, New Jersey, U.S.A.

X-RAY EMISSION SPECTROSCOPIES

Up to this point, we have considered only the spectroscopy associated with transitions from states characterized by core electron wave functions to other states that are equally well defined by different wave functions. This spectroscopy description is appropriate as long as the quantum numbers of the terminal state are relatively low in magnitude, but different phenomena become important when the quantum numbers become sufficiently large. At high quantum numbers, the energy spacings between higher excited states become successively smaller until the states effectively merge into a continuum. In fact, when an anharmonic potential energy term is used in the Hamiltonian operator, one finds that at sufficiently high quantum numbers dissociation of an excited electron may take place. This information is of great interest, because knowledge of the energy required to dissociate an electron would be directly related to the energy of the state from which it had been removed.

The *ionization potential* of an electron is defined as the energy required for its removal from its orbital of occupancy. Originally, investigation of the magnitude of ion currents obtained as a function of the energy of an impacting species were used to deduce ionization potentials, but these studies generally yielded information only on a few electrons in the valence shells. However, the ionization of electrons from atoms upon irradiation by X-rays is a phenomenon known as the *photoelectric effect*, the process being expressed as:

$$A + h\nu \rightarrow A^+ + e^- \tag{1}$$

where hv is the energy of the impacting photon. One could study the characteristics of the ionized atoms (as is done in the practice of mass spectrometry), or one may study the characteristics of the ejected photoelectrons. For the latter, the Einstein photoelectric law provides a relation that describes the partitioning of energy during the photoionization process:

$$\mathbf{KE} = hv - \mathbf{IP} \tag{2}$$

where **IP** is the ionization potential of the state from which the electron is ejected, and **KE** is the kinetic energy of the electron.

Ejected photoelectrons are separated according to their kinetic energies in an analyzer system. The photoelectron spectrum is therefore a measure of the number of electrons detected, and using Koopman's theorem one associates the kinetic energy of the observed electrons (E_{obs}) with their characteristic binding energies (E_B):

$$E_{obs} = hv - E_B \tag{3}$$

If the system undergoing photoionization consists of molecules, then the observed energies will exhibit fine structure due to the additional possibilities of vibrational or rotational excitation upon ionization. In that case:

$$E_{obs} = hv - E_B - E_{VIB} - E_{ROT} \tag{4}$$

If the resolution of the spectrometer is sufficient, then the individual vibrational and/or rotational fine structure can be resolved and observed. If not, then the observed photoelectron peak will consist of a broader band that encompasses all of the unresolved fine structure.

From its introduction, electron spectroscopy has been attractive for the study of atomic and molecular electronic structure of each band in the spectrum corresponding to ionization from a single atomic or molecular orbital. In addition, each occupied orbital having binding energy less than the excitation energy will yield a single band in the spectrum.

Although there are a multitude of photoionization spectroscopies that have been developed for the study of core electronic states, the present discourse will center on two main techniques that have found significant use in the characterization of pharmaceutical solids. These have been termed *x-ray photoelectron spectroscopy* (XPS) and *x-ray fluorescence* (XRF), and the relevant photophysics associated with these is illustrated in Figure 1. In XPS, irradiation by X-rays of energy hv is used to photoeject core electrons, whose kinetic energies are directly determined. In XRF, one does not observe the photoejected electron but instead monitors the fluorescence that is emitted by the system after one of the valence electrons drops down into a core electron level to fill the vacancy created by the initial photoionization. It is worth noting that the pathway of the XRF process is independent of the primary photoionization step and can be initiated by electron impact and also photon impact.

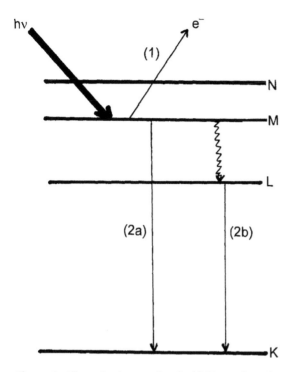

Figure 1 Photophysics associated with X-ray photoelectron spectroscopy and X-ray fluorescence. As illustrated, in the XPS experiment one monitors the energy of the electron ejected from the M-shell upon photoionization (*process 1*). In the XRF experiment, one monitors the fluorescence emitted either from the M-shell after photoionization (*process 2a*) or from the L-shell after photoionization and radiationless decay (*process 2b*).

X-RAY PHOTOELECTRON SPECTROSCOPY

Because the energy of the x-ray source used to effect the photoionization is accurately known, the XPS experiment requires the determination of the kinetic energies of ejected core photoelectrons to deduce their ionization potentials and hence the energy of their original binding energies (1). As is the case for most spectroscopic methods, this process is easy to describe but not necessarily easy to execute. Equation (3) is only appropriate to photoionization studies performed in the gaseous state, and the equation applicable to the solid state is somewhat more complicated:

$$E_{obs} = h\nu - E_B - \mathbf{W} - C \tag{5}$$

$\mathbf{W}$ is a quantity termed the *work function*, and is the energy required to remove the already photoionized electron out of the solid-state matrix in which it is located. One may envision this energy as the overcoming of all barriers that are associated with interaction of the photoelectron with the material.

For many substances, the work function is approximately 5 eV, and this is not insignificant compared with the magnitude of typical binding energies. Unless one chooses to measure the work function, XPS spectra are ordinarily presented in terms of energy shifts with respect to a defined reference and not as absolute binding energies. The *surface charging effect*, *C*, is another correction that reflects the charge distribution in a solid, and the effect of atomic potentials on core electron levels.

Siegbahn developed a model to refine the concept of binding energy, attempting to account for the interaction of electron **i** with all nuclei in the solid and all other electrons (1):

$$E_B = E_0 + k\,q + \Sigma_{i \neq j} q_j / r_{ij} \tag{6}$$

where E_0 is the binding energy of the electron associated with the isolated atom, q is the valence charge, k is an empirical constant, and $\Sigma_{i \neq j}\, q_j/r_{ij}$ is a Madelung-type term reflecting the potential field of the environment. Sets of charge-binding energy relations deduced from quantum mechanical calculations are available (2).

Experimental Details

Only the briefest outline of the experimental methodology will be provided here, because very excellent reviews of instrumentation (3,4) are available. In addition, the ongoing annual reviews in *Analytical Chemistry* (5) provide another source of information regarding the state of the art. The essential components of a typical system consist of the X-ray source, an energy analyzer for the ejected photo-electrons, and a detector for the analyzed electrons, with the entire system being enclosed in a high-vacuum chamber. A schematic diagram of the basic XPS spectrometer system is given in Figure 2. The photoelectrons ejected from the sample upon excitation by the source are drawn into the analyzer, separated on the basis of their kinetic energy, and their intensity as a function of kinetic energy measured by a detector. The sample is mounted on a probe face in such a way that the photoejected electrons can easily pass into the analyzer.

It is clear from equation (5) that the energy of the irradiation energy must exceed the ionization potential of the electronic states to be studied, so any X-ray source having sufficient energy can be used. The two most important sources are the Kα lines of magnesium (1254 eV, line width 0.8 eV) and aluminum (1487 eV, line width 0.9 eV), and for measurement of higher binding energies one may use the Kα lines of titanium (4511 eV, line width 1.4 eV) or copper (8048 eV, line width 2.5 eV). A discussion of the relative merits of different sources is available (6). For very high-resolution work, the line width of the X-ray source can be further reduced through the use of a monochromator.

The determination of the kinetic energy of ejected photoelectrons is the most important aspect of the technique, and hence the analyzers used must be simultaneously selective and sensitive. One analyzer class is the *retarding*

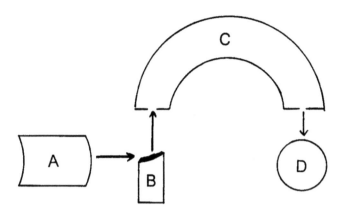

Figure 2 Basic components of the apparatus used for the measurement of X-ray photo-electron spectroscopy. X-rays from the source (**A**) are allowed to impinge on the sample mounted on a probe (**B**), the energies of the ejected photoelectrons discriminated by the electron analyzer (**C**), and finally measured by the electron detector (**D**). All of these components are contained in a high-vacuum chamber.

potential type, where photoelectrons are subjected to a variable retarding voltage and the photoelectron spectrum results from differences in the relative transmission of electrons of different energies at a given retarding field. Examples of retarding potential analyzers are the cylindrical grid and spherical grid analyzers. The other analyzer class is the *dispersive* type, where one obtains a spatial resolution of photoelectrons on the basis of their differing kinetic energies. Examples of dispersive analyzers are the cylindrical condenser and spherical condenser analyzers.

Once the ejected photoelectrons have been discriminated on the basis of their kinetic energies, detection of these is usually effected using electron multipliers. These devices can vary significantly in their design, with the simplest consisting of a plain charge collector (termed a Faraday cup). More typical applications make use of electron multipliers having gains in the range of 10^6, a factor that is necessary to obtain accurately measurable currents in the microampere range.

Assignment of Observed XPS Bands and their Chemical Shifts

Because the binding energies associated with a given electron in a defined core electron state is largely determined by the identity of the atom and the particular set of **n**, **l**, and **m** quantum numbers defining the state, assignment of XPS peaks can be readily achieved through the use of standard tables (1,7,8). Some of the assignments made for core electron binding energies of elements in the free atom state are found in Table 1, and Figure 3 illustrates the regular increase in binding energies of various states with increasing atomic number. Thus,

Table 1   Binding Energies (eV) of Core Electrons in Various Free Atom States

Element	Electron configuration	K-shell ($1s$ orbital)	L_1-shell ($2s$ orbital)	L_2-shell ($2p_{1/2}$ orbital)	L_3-shell ($2p_{3/2}$ orbital)
H	$1s^1$	13.60	—	—	—
He	$1s^2 \equiv$ [He]	24.59	—	—	—
Li	[He] $2s^1$	54.7	*(5.4)*	—	—
Be	[He] $2s^2$	111.5	*(9.3)*	—	—
B	[He] $2s^2\,2p^1$	188	*(12.9)*	*(8.3)*	—
C	[He] $2s^2\,2p^2$	284.2	*(16.6)*	*(11.3)*	—
N	[He] $2s^2\,2p^3$	409.9	*(37.3)*	*(14.5)*	—
O	[He] $2s^2\,2p^4$	543.1	*(41.6)*	*(13.6)*	—
F	[He] $2s^2\,2p^5$	696.7	*(37.9)*	*(17.4)*	—
Ne	[He] $2s^2\,2p^6 \equiv$ [Ne]	870.21	48.47	21.7	21.6
Na	[Ne] $3s^1$	1070.8	63.5	30.4	30.5
Mg	[Ne] $3s^2$	1303.0	88.6	49.6	49.2
Al	[Ne] $3s^2\,3p^1$	1559.0	117.8	72.9	72.5
Si	[Ne] $3s^2\,3p^2$	1839	149.7	99.8	99.2
P	[Ne] $3s^2\,3p^3$	2145.5	189	136	135
S	[Ne] $3s^2\,3p^4$	2472	230.9	163.6	162.5
Cl	[Ne] $3s^2\,3p^5$	2822.0	270	202	200
Ar	[Ne] $3s^2\,3p^6 \equiv$ [Ar]	3205.9	326.3	250.6	248.4

Note: Binding energies of some valence electron states have been listed for comparison purposes.
Source: Adapted From Refs. 4,5.

identification of a particular XPS peak with a state of known **n, l,** and **m** quantum numbers is relatively straight-forward.

Once the peak system is identified, the real chemical information is derived from the fine structure within each system. Because the valence electrons are intimately involved in chemical bonding and because they provide an electrostatic shield for the core electrons, the interaction between the core electrons and the valence electrons causes the energies of the core electrons to be affected by details of their molecular environment. Any change in the valence electron shield will be reflected in a change in the amount of energy required to remove a core electron through that valence shell. For example, the beryllium 1S band in the oxide is shifted by 2.9 eV relative to that of the metallic element, which is attributed to the difference in oxidation state between the beryllium atom in its two states (9). In beryllium fluoride, the 1S band is shifted further by 4.6 eV, which is relative to that of the metallic element, reflecting the additional electronegativity of fluorine relative to that of oxygen.

For the organic compounds of pharmaceutical interest, the chemical shifts of the 1S bands of carbon and oxygen would be of primary interest, and consideration of a few simple molecules will serve to illustrate the energy ranges. Electronegativity is found to exert a strong influence on the carbon 1S binding

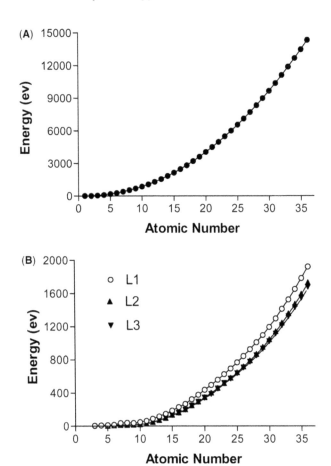

Figure 3 Binding energies of (**A**) K-shell electrons and (**B**) L-shell electrons as a function of atomic number. *Source*: From Ref. 5.

energies, as illustrated by the series CH_4 (290.7 eV), CH_3OH (292.3 eV), CHF_3 (298.8 eV), and CF_4 (301.8 eV) (10). The two carbons in ethanol can be differentiated, with the carbon of the terminal methyl group exhibiting a 1S binding energy of 290.9 eV and the methylene carbon directly bound to the hydroxyl group exhibiting a binding energy of 292.3 eV. The chemical shift differentiation in acetic acid is even more pronounced, where the carbon of the terminal methyl group exhibits a 1S binding energy of 291.4 eV and the carboxylate carbon exhibits a binding energy of 294.5 eV (10).

Applications of XPS to the Characterization of Solids

In its simplest application, XPS can be used as a complementary technique to X-ray diffraction or solid-state nuclear magnetic resonance in the study of the properties of different compounds. For instance, the strong ionic $[N-H\cdots N]^+$ hydrogen bonds in a series of perisubstituted naphthalenes were found to strongly influence the binding energies of the core electrons of the donor and acceptor atoms (11). These authors were able to use the difference in binding energies of the donor and acceptor as a means to evaluate the strength and asymmetry of the strong hydrogen bonds. The properties of the $[N-H\cdots N]^+$ hydrogen bonds developed between 1,8-bis(dimethylamino)naphthalene with various organic and inorganic acids were further studied in other publications (12,13).

XPS has been used to study the core electronic levels of triaminotrinitrobenzene and as a means to understand the processes associated with its decomposition (14). The nitrogen and oxygen core spectra exhibited unusually high-energy satellite structure, with the separation of the satellites from the main core lines decreasing with the successive removal of donor amine groups. The core level spectra of the component nitrogen and oxygen decreased when the compound was subjected to isothermal and photolytic decomposition, indicating the breakage of the $C-NO_2$ bond.

The use of XPS as a means to yield quantitative elemental analysis through the use of detail scans was used to characterize human immunoglobulins (15). In this methodology, a small portion of the binding energy range is studies with a higher degree of resolution to obtain better definition of the fine details in the spectra, and thus minimizing the use of curve resolution techniques. This particular formulation consisted of IgG-dispersed monolithic matrices, prepared using a nonbiodegradable polymer carrier. An initial burst of drug release was noted, and the XPS studies showed that this effect might be associated with the surface concentration of IgG molecules. This and other findings suggested that the kinetics of in vitro drug release were determined by the rate of IgG release through the polymer matrix.

The incident X-ray radiation used as the source in XPS does not penetrate far beyond the upper molecular layers of a solid sample, typically reaching to depths approximately 50Å beneath the surface. Consequently, it is not surprising that the technique has been used to study the surfaces of materials having pharmaceutical interest. For instance, XPS was used to study the surface concentration of PEG 400 distearate contained in poly(D,L)-lactic acid microspheres (16). In this study, the carbon 1S lines were found to contain more useful information than did the oxygen 1S lines. As illustrated in Figure 4, as the amount of PEG 400 distearate increased in the microspheres, the carbon lines were found to exhibit a greater shift to higher energies.

XPS and time of flight secondary ion mass spectrometry were used to show that the polypropylene oxide component of adsorbed poloxamers adhered to the surface of polystyrene particles (17). Prior to the performance of this work, it had been known that the adsorption of poloxamers altered the electrostatic charge, adhesion behavior, and handling properties of such particles, and from the XPS studies the authors were able to deduce that there was a polyethylene oxide-rich outer surface that appeared to influence the charge alteration.

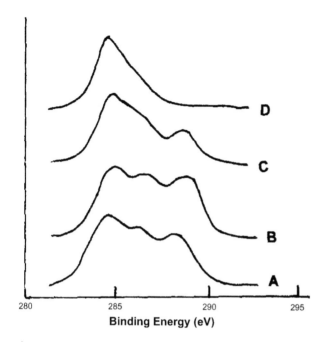

280   285   290   295
Binding Energy (eV)

Figure 4 X-ray photoelectron spectra of poly(D,L)-lactic acid microspheres containing PEG 400 distearate at levels of (**A**) 0%, (**B**) 1%, (**C**) 10%, and (**D**) 100%. Source: From Ref. 16.

Poly(methyl methacrylate) is used to fill the gap between a prosthesis and its surrounding bone in cemented arthroplasties; biocompatibility problems related to bone cement application are known to limit the clinical success of these devices. XPS was used to learn if there was a correlation between surface changes and aging time (18). Evidence for the breaking of old bonds, and the formation of new bonds, with the new bonding patterns are clearly visible in oxygen 1S spectra of the hydroxyl groups. These effects were interpreted as indicating the hydrolysis of PMMA ester groups, and the subsequent formation of hydrogen bonds between the hydroxyl and carbonyl groups of adjacent polymer chains.

The changes in chemical composition after the surface modification and coating of dextran-based ion-exchange microspheres were studied by XPS (19), because with the incorporation of proper standardizations the technique can be used to deduce elemental analyses of surface species. For instance, after introduction of palmitoyl groups to sulfopropylated dextran microspheres, the relative content of hydrocarbon carbon 1S response could be correlated with concentration. Because the relative abundance of carbon associated with carboxylic acid groups increased effectively from 0% to over 6%, this suggested that the palmitoyl groups were linked to the microspheres through ester bonds.

Given the utility of XPS as a technique for surface analysis, it is logical that the method could be used to study the properties of coating films. For instance, XPS has been used to study the surface chemical composition of a range of poly(methacrylate) films that are commonly used as pharmaceutical film coatings (20). Through the use of curve resolution methods, it was determined that the carbon 1S region consisted of four unresolved peaks (Fig. 5— representative example of a carbon 1S spectrum) derived from carbon atoms in the polymer backbone and also in the ester side chains. Knowledge of chemical shifts in model compounds enabled peak assignments, and these were used to elucidate the differences in film character associated with Eudragits S, L, E, and RS. In a similar application, XPS was used to develop a method that could be used to quantitate the interfacial thickness of a film-tablet layer (21).

Another application of XPS that takes advantage of its responsiveness to surface phenomena is in the study of molecular adsorption. In one study, XPD was used to evaluate the absorbed state of glucose that had been sorbed onto alumina (22). The alumina used in this work was a specific type (Mizusawa Kagaku, type Neobead-P) containing a little percentage of sodium, which acts as a basic site in the mutarotation of the adsorbed glucose. However, the XPS data indicated that the aluminum-oxygen position of the surface served as the

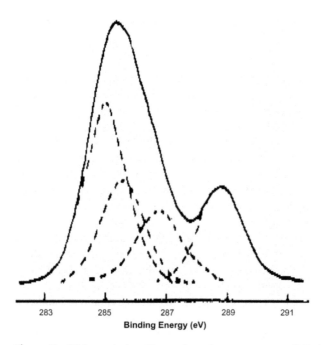

Figure 5 High-resolution X-ray photoelectron spectra of Eudragit L. *Source*: From Ref. 20.

adsorption sites for the glucose and that no preferential adsorption of glucose onto the sodium sites took place.

XPS has been used to identify the functional states of carbon existing on the surfaces of different activated charcoals (23). Because the adsorption on phenobarbital on such materials had been studied in the past, the system represented a good model to evaluate the utility of XPS. A good correlation was found to exist between the apparent areas occupied on the adsorbent surface per phenobarbital molecule and the relative percentages of a C—O functional state, which was determined to be the primary site involved in the binding.

Evaluating the ability of a pharmaceutical powder to be wetted is an important criterion in understanding its dissolution properties, its ability to be granulated, the adhesion of film coatings, and its tendency to aggregate when placed in suspended form. In order to learn whether XPS could be used as a method to study the wet ability of a powder, the binding energies of various barbiturates in powdered and compressed form, and compared with wet ability data obtained by other methods (24). There appeared to be no correlation, however, between binding energies and surface wet ability. Although compaction of the powders would change surface energetics, that process would not change the chemical composition of the surface.

A capillary penetration technique was used to determine the wet ability of morphine sulfate powders, and the surface chemistry of these powders studied by XPS (25). Contact angles were shown to correlate with both the aspect ratio of the component crystals and the atomic nitrogen-to-oxygen ratios at the surface. It was concluded that the relative exposure of different crystal faces (which could be evaluated by XPS) played an important role in controlling the wet ability of the powdered morphine sulfate. In a similar study, XPS was used to characterize the surface composition of poly(lactic acid) polymers and poly(lactic/glycolic acid) copolymers, but the surface wet ability was evaluated by other methods (26). The wet ability improvement of these surfaces by pluronic surfactants was then studied and found to be a function of surfactant composition.

X-RAY FLUORESCENCE SPECTROSCOPY

As illustrated in Figure 1, the electromagnetic radiation measured in an XRF experiment is the result of one or more valence electrons filling the vacancy created by an initial photoionization where a core electron was ejected upon absorption of X-ray photons. The quantity of radiation from a certain level will be dependent on the relative efficiency of the radiationless and radiative deactivation processes, with this relative efficiency being denoted at the fluorescent yield. The fluorescent yield is defined as the number of X-ray photons emitted within a given series divided by the number of vacancies formed in the associated level within the same time period.

As discussed in earlier sections, regardless of whether one uses the theory of Bohr or the quantum theory, the quantized energy difference between the states of a multielectron atom is given by:

$$\Delta E = \frac{Z^2 me^4}{2n^2 \hbar^2} - \left[\frac{1}{n_1^2} - \frac{1}{n_2^2} \right] \tag{7}$$

where Z is the atomic number of the element in question. In equation (7), one identifies the levels having $n = 1$ as being from the K shell, levels having $n = 2$ are from the L shell, levels having $n = 3$ are from the M shell, and so on. When an electron of the L shell is transferred into a vacancy in the K shell, the energy released in the process equals the energy difference between the states:

$$\Delta E_{K\alpha} = \frac{Z^2 me^4}{2n^2 \hbar^2} - \left[\frac{1}{1^2} - \frac{1}{2^2} \right] \tag{8}$$

This particular transition results in the emission of X-ray radiation known as the Kα line. For transfer from the M shell into the K shell, the energy of the Kβ line is given by:

$$\Delta E_{K\beta} = \frac{Z^2 me^4}{2n^2 \hbar^2} - \left[\frac{1}{1^2} - \frac{1}{3^2} \right] \tag{9}$$

Because both transitions described in equations (8) and (9) terminate in the K shell, they are known to belong to the spectrum of the K series. Similarly, the spectral lines that result when electrons fall into the L shell define the spectrum of the L series, and the energies of these lines are given by:

$$\Delta E_L = \frac{me^4}{2n^2 \hbar^2} - \left[\frac{1}{2^2} - \frac{1}{n^2} \right] \tag{10}$$

Equation (7) indicates that the energy difference between the two states involved in a XRF transition is proportional to the atomic number of the element in question, a fact realized some time ago by Mosley. The simple equations given previously do not account for the different orbitals that exist for each principal quantum number, and the inclusion of these levels into the spectral considerations explain the fine structure observed within almost all XRF bands. For instance, the L shell (defined by $n = 2$) contains the 2S and 2P levels and the transfer of electrons from these levels down to a vacancy in the K shell (defined by $n = 1$) results in the emission of X-ray radiation denoted as the Kα_1 and the Kα_2 lines.

A detailed summary of the characteristic XRF lines for the elements is available (27), as are the more detailed discussions of the phenomena of X-ray absorption and X-ray fluorescence (28–31).

Experimental Details

There are numerous types of instrumentation available for the measurement of XRF, but most of these are based either on *wavelength dispersive* methodology (typically referred to as WDX) or on the *energy dispersive* technique (typically known as EDX). For a detailed comparison of the two approaches for XRF measurement, the reader is referred to an excellent discussion by Jenkins (32).

The first XRF spectrometers employed the WDX methodology, which is schematically illustrated in the upper half of Figure 6. The X-rays emanating from the source are passed through a suitable filter or filters to remove any undesired wavelengths and collimated into a beam that is used to irradiate the sample. For instance, one typically uses a thin layer of elemental nickel to isolate the $K\alpha$ lines of a copper X-ray source from contamination by the $K\beta$ lines, because the K-edge absorption of nickel will serve to pass the $K\alpha$ radiation but not the $K\beta$ radiation.

The various atoms making up the sample emit their characteristic XRF at an angle of θ relative to the excitation beam, and the resulting X-rays are

Wavelength Dispersive XRF

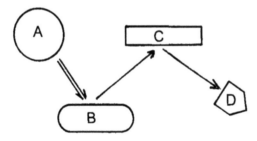

Energy Dispersive XRF

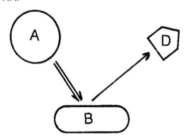

Figure 6 Basic components of the apparatus used for the measurement of XRF by the wavelength and energy dispersive methods. WDX: X-rays from the source (**A**) are allowed to impinge on the sample (**B**); the resulting XRF is discriminated by the crystal (**C**) and finally measured by the detector (**D**). *EDX*: X-rays from the source (**A**) are allowed to impinge on the sample (**B**), and the resulting XRF is measured by the detector (**D**). *Abbreviations*: WDX, wavelength dispersive XRF; EDX, energy dispersive XRF; XRF, X-ray fluorescence.

discriminated by a monochromator that uses a single crystal as a diffraction grating. The diffraction off the crystal must obey Bragg's law:

$$n\lambda = 2d\sin(\theta) \tag{11}$$

where n is the diffraction order, λ is the wavelength of X-rays undergoing diffraction, d is the distance between the planes of atoms acting as the diffraction grating, and θ is the angle of the excitation beam with the sample face. It is clear from equation (11) that no single crystal can serve as an efficient diffraction grating for all wavelengths of X-rays, and so WDX instruments typically contain a variety of different crystals that the user can select to optimize the XRF region of greatest interest.

The detector can be either a gas-filled tube detector or a scintillation detector, which is systematically swept over the sample and which measures the X-ray intensity as a function of the 2θ scattering angle. Through suitable calibration, each 2θ angle is converted into a wavelength value for display. The major drawback associated with WDX spectrometers is the reality derived from the Bragg scattering law that one cannot measure all wavelengths in a XRF spectrum in a single scan. Hence, one might be required to run multiple spectra if the range of elements to be studied is wide in terms of atomic number.

The EDX methodology was developed as a means to permit an analyst to acquire the entire XRF spectrum simultaneously, therefore eliminating the requirement to acquire data in parcels. A simple schematic of an EDX spectrometer is illustrated in the lower half of Figure 6. As with the WDX instrumentation, X-rays emanating from the source are filtered and collimated into a beam that is used to irradiate the sample. The XRF emitted by the sample is caused to fall onto a semiconductor diode detector, and one obtains the spectrum through the use of multichannel analysis of the detector output.

Although the instrumentation of the EDX method is simpler, it does not provide the same degree of wavelength resolution associated with WDX spectrometers. Jenkins (32) has provided an extremely useful set of criteria that permit a judicious choice as to which particular technique would be most appropriate for a given application:

1. A WDX spectrometer capable of simultaneously recording a span of wavelength information is most suitable for high-specimen throughput quantitative analysis, where speed of the essence and the high initial cost can be justified.
2. A sequential WDX spectrometer is most useful where time of analysis is not critical and where a moderately high initial cost can be justified.
3. An EDX system is usually chosen when initial cost is a major consideration or where the analyses to be run will be qualitative or semiqualitative in nature.

One of the problems faced in XRF spectroscopy is the fact that the absolute sensitivity of an element decreases with atomic number, and this decrease is most considerable for light elements. For quite some time, XRF spectrometers could not detect elements having atomic numbers less than 12, which did not permit analysis of any second row elements, such as carbon, nitrogen, oxygen, and fluorine. Fortunately, advances in detector design have been made, and now through the use of ultrathin beryllium window analysts, one can now obtain reliable results on the light elements. When EDX is combined with scanning electron microscopy, one can even use windowless detectors to maximize the response of the light elements (33).

Qualitative XRF Analysis

By virtue of their capacity to observe the entire range of X-ray fluorescence and speed of data acquisition, EDX spectrometers are admirably suited for qualitative analysis work. Equally well suited would be WDX spectrometers capable of observing XRF over a wide range of scattering angles. As discussed earlier and illustrated in Figure 3, the $K\alpha$ and $K\beta$ XRF of a given element is directly related to its atomic number, and extensive tables are available that provide accurate energies for the fluorescence (27). Such information can be stored in the analysis computer, and used in a peak-match mode to easily identify the elemental origins of the fluorescing atoms within the sample. The elemental identification is facilitated by the fact that XRF originates from core electron levels that are only weakly affected by the details of chemical bonding. It must be remembered, however, that the selectivity associated with XRF analysis ends with elemental speciation and that the technique is generally unable to distinguish between the same element contained in different chemical compounds.

X-ray fluorescence spectra of most elements consist largely of bands associated with the K, L, and M series. The XRF associated with the K-series is dominated by the α_1/α_2 doublet, although some very weak β XRF can often be observed at higher energies. The L-series XRF will consist of three main groups of lines, associated with the α, β, and γ structure. Ordinarily, the α_1 line will be the strongest, and the β_1 line will be the next most intense. The peaks within the M-series are observed only for the heavier elements and consist of an unresolved α_1/α_2 doublet as the strongest feature, followed by a band of peaks associated with the β XRF.

For instance, the $K\alpha_1$ and $K\alpha_2$ lines of elemental copper are observed at wavelengths of 1.540Å and 1.544Å, respectively, and $K\beta$ lines will be observed at 1.392Å and 1.381Å (33). The unresolved $L\alpha_1$ and $L\alpha_2$ lines are observed at a wavelength of 13.330Å, and $L\beta$ lines are observed at 13.053Å and 12.094Å. The presence of copper in a sample would be indicated by the presence of XRF peaks detected either at these wavelengths or at their corresponding energies.

XRF qualitative analysis entails the identification of each line in the measured spectrum. The analysis begins with the assumption that the most intense lines will be due to either Kα or Lα emission, and these are used to match the observed lines with a given element. Once the most intense α-lines are assigned, one then goes on to assign the β-lines, and these assignments serve to confirm those made from the analysis of the α-lines. Of course, this task has been made quite easy through the use of computer systems that store all the information and execute the peak matching analysis. For example, the EDX analysis of a fumarate salt of a drug substance was unexpectedly found to exhibit definite responses for chlorine, indicating that the sample actually consisted of mixed fumarate and hydrochloride salts (34).

In another qualitative study, EDX analysis was used to study the nature of the precipitate occasionally formed in Zn–insulin solutions (35). Identification of the EDX peaks obtained for the crystalline precipitates enabled the deduction that the solid consisted of a Zn–insulin complex, and a rough analysis of the peak intensities indicated that the composition of the precipitate was comparable with that existing in the starting materials. The combination of the EDX technique with scanning electron microscopy enabled the analyses to be conducted on relatively few numbers of extremely small particles.

Quantitative XRF Analysis

Owing to their superior fluorescent yield, heavy elements ordinarily yield considerably more intense XRF bands than the light elements. This feature can be exploited to determine the concentration of inorganic species in a sample or the concentration of a compound that contains a heavy element in some matrix. Many potential XRF applications have never been developed owing to the rise of atomic spectroscopic methods, particularly inductively coupled plasma atomic emission spectrometry (36). Nevertheless, under the right set of circumstances, XRF analysis can be profitably employed.

A number of experimental considerations must be addressed in order to use XRF as a quantitative tool, and these have been discussed at length in a number of texts (37–39). The effects on the usual analytical performance parameters (accuracy, precision, linearity, limits of detection and quantitation, and ruggedness) associated with the instrument are usually minimal. However, the effects associated with sample handling, preparation, and presentation cannot be ignored, as they have the potential to exert a major influence over the quality of the analysis.

The background emission detected in a XRF spectrum is usually due to scattering of the source radiation. Because the scattering intensity from a sample is inversely proportional to the atomic number of the scattering atom, it follows that background effects are more pronounced for samples consisting largely of second row elements (i.e., organic molecules of pharmaceutical interest). Hence, background correction routines play a major role in transforming raw XRF spectra into spectra suitable for quantitative analysis.

The performance of a quantitative XRF analysis requires that correction for the interelement interactions that are associated with the matrix of the sample. For homogeneous samples, matrix effects can serve to enhance or diminish the response of a given element. The intensity retardation effects usually originate from competition among scattering units in the sample for the incident X-ray radiation or from the attenuation of emitted XRF by its passage through the sample itself. Enhancement effects arise when emitted XRF is reabsorbed by other atoms in the sample, causing those atoms to exhibit a higher XRF intensity than would be associated with the incident X-ray radiation itself.

Because matrix and background effects are notoriously difficult to identify and control, quantitative XRF analysis usually entails the use of standards. One can write an expression relating the concentration of an analyte in a sample with the XRF intensity of one of its emission lines:

$$C_i = KI_iMB \tag{12}$$

where C_i is the concentration of the analyte exhibiting the intensity response I_i, K is a constant composed of instrumental factors, M is a constant composed of matrix factors, and B is a constant composed of background factors. Although it is conceivable that one could derive an equation describing how K would be calculated, it is clear that calculation of the M or B constants would be extraordinarily difficult.

However, when an internal standard containing an element of known concentration C_S is added to the sample, its concentration would be derived from its observed intensity I_S by:

$$C_S = KI_SMB \tag{13}$$

Because the analyte and standard are equally distributed in the analyzed sample, one can assume that the M and B constants would be the same for both species. In that case, dividing equation (12) by equation (13) and rearranging yields a relation enabling the calculation of the amount of analyte in the sample:

$$C_i = (I_i/I_S)C_S \tag{14}$$

This method of using an internal standard to deal with matrix and background effects has, of course, been used for ages in analytical chemistry. Equally useful would be another analytical method, namely that of standard additions. In this latter approach, the sample is spiked with successively increasing amounts of the pure analyte, and the intercept in the response–concentration curve is used to calculate the amount of analyte in the original sample.

There has always been great interest in developing algorithms that enable quantitative XRF spectroscopy to be performed without the use of standards, methodology that is often referred to as the fundamental parameters approach. The method was first proposed by Criss, who published a program suitable for its implementation (40). The advantage of the fundamental parameters approach is that it represents a more time-efficient and less tedious method of analysis that, under the right circumstances, does not sacrifice accuracy for the sake of expediency.

For example, fundamental parameter calculations have been used for the analysis of europium over concentration ranges of 0.1–30.0% w/w in alumina, calcia, magnesia, lanthania, and thoria oxidic catalyst supports (41). Results comparable with the multiple regression method were obtained when the matrix stoichiometry was defined as Eu_2O_3 and the catalyst oxide equally defined by its stoichiometry. It was found to be necessary to use two standards, which bracketed the Eu(III) concentrations in the actual samples. However, when these conditions were met, the results obtained were comparable with those obtained using a 10-point multiple regression calibration curve. It was also found that the percent relative difference between the fundamental parameters and multiple regression results was approximately 2%, but the fundamental parameter results were obtained with a considerable saving of standard preparation time.

Total reflection X-ray fluorescence (TXRF) has become very popular for the conduct of microanalysis and trace elemental analysis (42–44). TXRF relies on scatter properties near and below the Bragg angle to reduce background interference and to improve limits of detection that can amount to an order of magnitude (or more) over more traditional XRF measurements. As illustrated in Figure 7, if X-rays are directed at a smooth surface at a very small angle, virtually all of the radiation will be reflected at an equally small angle. However, few X-rays will excite atoms immediately at the surface, and those atoms will emit their characteristic radiation in all directions. One obtains very clean analytical signals when using the TXRF mode because there is essentially no backscatter radiation impinging on the detector.

TRXF was used to determine the trace elements in samples of lecithin, insulin, procaine, and tryptophan in an attempt to develop elemental fingerprints that could be used to determine the origin of the sample (45). It was reported that through the use of matrix-independent sample preparation and an internal standard, one could use TXRF to facilitate characterization of the samples without the need for extensive pretreatment. In another work, a study was made of the capability of TXRF for the determination of trace elements in pharmaceutical substances with and without preconcentration (46).

Trace amounts of bromine in sodium diclofenac, sodium {2-[(2,6-dichlorophenyl)amino]phenyl}acetate, have been determined using XRF (47), because the drug substance should not contain more than 100 ppm of organic bromine remaining after completion of the chemical synthesis. Pellets containing the analyte were compressed over a boric acid support, which yielded stable samples for analysis, and selected XRF spectra obtained in this study are shown in Figure 8. It was found that samples from the Far East contained

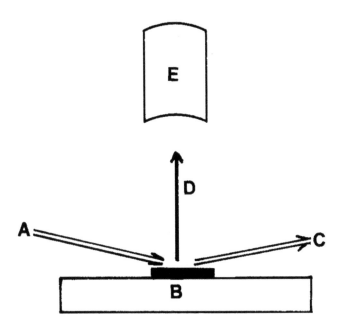

Figure 7 Instrumental arrangement for the measurement of total reflection XRF. The X-rays from the source (**A**) are allowed to impinge on the sample mounted on a reflector plate (**B**). Most of the incident radiation bounces off the sample (**C**) but some results in the production of XRF (**D**), which is measured by the detector (**E**). *Abbreviation*: XRF, X-ray fluorescence.

more than 4000 ppm of organic bromine, various samples from Europe contained about 500 ppm, and samples from an Italian source contained less than 10 ppm of organic bromine.

No sample pretreatment was required for analysis of the Cs^+, Ba^{2+}, Cu^{2+}, and La^{3+} content in poly(vinyl alcohol—vinyl sulfate) copolymers by TXRF (48). These assays facilitate the estimation of the efficiency associated with the ion-exchange preparation of polyelectrolytes and for determinations of the stoichiometry of the end product. The results of the analyses were cross-referenced against data obtained using ICP-AES and titrimetric technology, and good agreement was obtained between the various methods.

TXRF was used to characterize high-viscosity polymer dispersions (49), with special attention being paid to the different drying techniques and their effect on the uniformity of the deposited films. TXRF was also used as a means to classify different polymers on the basis of their incoherently scattered peaks (50). Dispersive XRF has been used to asses the level of aluminum in antacid tablets (51).

Probably the most effective use of XRF and TXRF continues to be in the analysis of samples of biological origin. For instance, TXRF has been used

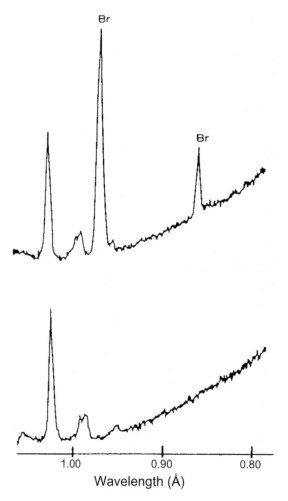

Figure 8   XRF spectra of pure diclofenac (*lower trace*), and diclofenac-containing organic bromine (*upper trace*). The bromine Kα peak is the marked peak at 1.04Å, whereas the bromine Kβ peak is the marked peak located at 9.93Å. *Source*: From Ref. 47.

without a significant amount of sample preparation to determine the metal cofactors in enzyme complexes (52). The protein content in a number of enzymes has been deduced through a TXRF of the sulfur content of the components methionine and crysteine (53). It was found that for enzymes with low molecular weights and minor amounts of buffer components that a reliable determination of sulfur was possible. In other works, TXRF was used to determine trace elements in serum and homogenized brain samples (54), selenium and other trace elements in serum and urine (55), lead in whole human blood (56), and the Zn/Cu ratio in serum as a means to aid in cancer diagnosis (57).

Thin sections of biological materials were prepared by means of a freezing microtome, placed on quartz carriers and analyzed by XXRF (58). Quantitation was effected by spiking the sections with an internal standard, and the time frame for the analysis was suitable for use as screening and monitoring methodology. The calcium and iodine content in gall bladder stones has been determined using EDX techniques, where quantitation was enabled by the method of standard additions (59). In other works, TXRF methods have been used to study the changes in trace elements present in different human tissues when these have be altered due to the presence of disease states (60–61).

REFERENCES

1. Siegbahn K, Nordling C, Fahlman A, Nordberg R, Hamrin K, Hedman J, Johansson G, Bergmark T, Karlsson SE, Lindgren I, Lindberg B. ESCA– –Atomic, Molecular, and Solid State Structure Studied by Means of Electron Spectroscopy. Uppsala, 1967.
2. Sleigh C, Pijpers AP, Jaspers A, Coussens B, Meier RJ. J Electron Spect Rel Phen 1996; 77:41.
3. Carlson TA. Photoelectron and Auger Spectroscopy. New York: Plenum Press, 1975:15–63.
4. Ghosh PK. Introduction to Photoelectron Spectroscopy. New York: John Wiley & Sons, 1983:23–52.
5. See, for example, Turner NH, Schreifels JA. Anal Chem 2000; 72:99R; ibid., 1998; 70:229R.
6. Steinhardt RG,. Granados FAD, Post GL. Anal Chem 1955; 27:1046.
7. Jolly WL, Bomben KD, Eyermann CJ. At Nucl Data Tables 1984; 31:433.
8. In: CRC Handbook of Chemistry and Physics. Lide DR, ed. 74th ed. Boca Raton, FL: CRC Press, 1993:10–272 through 10–277.
9. See Ref. 49, pp. 76–77.
10. Ghosh PK. Introduction to Photoelectron Spectroscopy. New York: John Wiley & Sons, 1983:55.
11. Wozniak K, He H, Klinowski J, Barr TL, Hardcastle SE. J Phys Chem 1996; 100:11408.
12. Wozniak K, He H, Klinowski J, Jones W, Barr TL. J Phys Chem 1995; 99:14667.
13. Wozniak K, He H, Klinowski J, Barr TL, Milart P. J Phys Chem 1996; 100:11420.
14. Sharma J, Garrett WL, Owens FJ, Vogel VL. J Phys Chem 1982; 86:1657.
15. Wang C-H, Sengothi K, Wong HM, Lee T. J Pharm Sci 1999; 88:221.
16. Lacasse FX, Hildgen P, McMullen JN. Int J Pharm 1998; 174:101.
17. Cassidy OE, Rowley G, Fletcher IW, Davies SF, Briggs D. Int J Pharm 1999; 182:199.
18. Bettencourt A, Calado A, Amaral J, Alfaia A, Vale FM. Monteiro J, Montemor MF, Ferreira MGS, Castro M. Int J Pharm 2004; 278:181.
19. Liu Z, Wu XY, Ballinger JR, Bendayan R. J Pharm Sci 2000; 89:807.
20. Davies MC, Wilding LR, Short RD, Khan MA, Watts JF, Melia CD. Int J Pharm 1989; 57:183.
21. Felton LA, Perry WL, Pharm Dev Tech 2002; 7:43.
22. Nakatani J, Ozawa S, Ogino Y. J Phys Chem 1989; 93:3255.

23. Burke GM, Wurster DE, Berg MJ, Veng-Pedersen P, Schottelius DD. Pharm Res 1992; 9:126.
24. Buckton G, Bulpett R, Verma N. Int J Pharm 1991; 72:157
25. Prestidge CA, Tsatouhas G. Int J Pharm 2000; 198:201
26. Kiss E, Bertoti I, Vargha-Butler EL. J Colloid Interfac Sci 2002; 245:91.
27. Cauchois Y, Senemand C. Wavelengths of X-Ray Emission Lines and Absorption Edges, Volume 18 of International Tables of Selected Constants, Oxford: Pergamon Press, 1978.
28. Herglotz HK, Birks LS. X-Ray Spectrometry. New York: Marcel Dekker, 1978.
29. Fabian DJ. Soft X-Ray Band Spectra. London: Academic Press, 1968.
30. Jenkins R. An Introduction to X-Ray Spectrometry. London: Heyden, 1970.
31. Van Griken RE, Markowicz AA. Handbook of X-Ray Spectrometry. New York: Marcel Dekker, 1992.
32. Jenkins R. Comparison of Wavelength and Energy Dispersive Spectrometers. Chapter 7 in X-Ray Fluorescence Spectrometry, 2nd edn. New York: Wiley-Interscience, 1999:111–121.
33. Newman AW, Brittain HG. Optical and Electron Microscopies. Chapter 5 in Physical Characterization of Pharmaceutical Solids. New York: Marcel Dekker, 1995:127–156.
34. Berridge JC. J Pharm Biomed Anal 1995; 14:7
35. Salemink PJ, Elzerman HJW, Stenfert JTh, Gribnau TCJ. J Pharm Biomed Anal 1989; 1989; 7:1261.
36. Wang T, Jia X, Wu J. J Pharm Biomed Anal 2003; 33:639.
37. Müller RO. Spectrochemical Analysis by X-Ray Fluorescence. New York: Plenum, 1972.
38. Betrin EP. Principles and Practice of X-Ray Spectrometric Analysis. 2nd edn. New York: Plenum, 1975.
39. Tertian R, Claisse F. Principles of Quantitative X-Ray Fluorescence Analysis. London: Heyden, 1982.
40. Criss JW. Adv X-Ray Anal 1980; 23:93
41. Kenny EM, Brittain HG. Adv X-Ray Anal 1985; 28:209.
42. Prange A. Spectrochim Acta 1989; B44:437
43. Aiginger H. Spectrochim Acta 1991; B46:1313
44. Wobrauschek P. J Anal Atom Spectromet 1998; 13:333.
45. Wagner W, Rostam-Khani P, Wittershagen A, Rittmeyer C, Kolbesen BO, Hoffmann H. Spectrochim Acta 1997; B52:961.
46. Kelki-Levai A, Varga I, Zih-Perenyi K, Laszity A. Spectrochim Acta 1999; B54:827.
47. Da Re MAP, Lucchini F, Parisi F, Salvi A. J Pharm Biomed Anal 1990; **8:975**.
48. Varga I, Nagy M. Spectrochim Acta 2001; 2001; B56:2229.
49. Vasquez C. Spectrochim Acta 2004; B59:1215.
50. Boeykens S, Vasquez C. Spectrochim Acta 2004; B59:1189.
51. Georgiades CA. J Assoc Off Anal Chem 1990; 73:385.
52. Wittershagen A, Rostam-Khani P, Klimmek O, Gross R, Zickermann V, Zickermann I, Gemeinhardt S, Kroger A, Ludwig B, Kolbesen BO. Spectrochim Acta 1997; B52:1033.
53. Mertens M, Rittmeyer C, Kolbesen BO. Spectrochim Acta 2001; B56:2157.
54. Marco LM, Greaves ED, Alvardo J. Spectrochim Acta 1999; B54:1469.

55. Bellisola G, Pasti F, Valdes M, Torboli A. Spectrochim Acta 1999;B54:1481.
56. Ayalaa RE, Alvarez EM, Wobrauschek P. Spectrochim Acta 1991; B46:1429.
57. Marco LM, Jimenez E, Hernandez EA, Rojas A, Greaves ED. Spectrochim Acta 2001;. B56:2195.
58. Klockenkamper R, von Bohlen A, Wiecken B. Spectrochim Acta 1989; B44:511.
59. Ekinci N, Sahin Y. Spectrochim Acta 2002; B57:167.
60. Benninghoff L, von Czarnowski D, Denkhaus E, Lemke K. Spectrochim Acta 1997; B52:1039.
61. von Czarnowski D, Denkhaus E, Lemke K. Spectrochim Acta 1997; B52:1047.

4

Molecular Orbital Theory and the Electronic Structure of Molecules

Harry G. Brittain

Center for Pharmaceutical Physics, Milford, New Jersey, U.S.A.

MOLECULAR ORBITAL (MO) THEORY OF ORGANIC MOLECULES

In the previous chapters, the electronic structure of atoms has been developed by first considering the orbitals, energy levels, and states of the hydrogen atom. Subsequently, these concepts were generalized and extended to deduce similar descriptions of the properties of multielectron atoms. Because electrons occupying completely filled shells of a given principal quantum number are localized around their respective nuclei, it was possible to discuss a variety of core electron spectroscopies in terms of atomic orbitals and the states derived from these.

In considering the electronic structure of polyatomic molecules, it is tempting to try a rigorous quantum mechanical approach, beginning with the Schrödinger equation:

$$\hat{H}\psi_i = E_i\psi_i \tag{1}$$

where E_i is the energy associated with the wave function ψ_i. As discussed in the previous chapter, the solution to the Schrödinger equation requires a specification of the form of the Hamiltonian operator:

$$\hat{H} = \hat{T} + \hat{V} \tag{2}$$

where $\hat{T}$ and $\hat{V}$ are the respective operators for the kinetic and potential energy terms. Although one may write the equations that describe all of the kinetic energy terms in a polyatomic molecule, great difficulty is encountered once equations for the potential energy terms are written. In the case of the hydrogen atom, the potential term for the electron was given by the attractive force between the negatively charged electron and the positively charged nucleus, that is, Coulomb's Law:

$$V = -e^2/r \tag{3}$$

where r is the distance of the electron from the nucleus, and e is the electron fundamental charge. For the hydrogen atom, the Hamiltonian operator took the form:

$$\hat{H} = \frac{-\hbar^2}{2m_e}\nabla^2 - \frac{e^2}{r} \tag{4}$$

where m_e is the mass of the electron. Through the use of appropriate mathematical techniques, one was able to substitute Equation (4) into Equation (1) and solve for the wave functions that described the electronic structure of the atom.

Let us now consider the case of simplest molecule possible, namely the hydrogen molecular ion (H_2^+). This species would consist of two nuclei and a single electron, and the Hamiltonian operator becomes considerably more complicated owing to the nature of the potential energy term:

$$\hat{H} = \frac{-\hbar^2}{2m_e}\nabla^2 + \left[\frac{-e^2}{r_A} - \frac{e^2}{r_B} + \frac{e^2}{r_{AB}}\right] \tag{5}$$

where r_A is the distance of the electron from nucleus A, r_B is the distance of the electron from nucleus B, and r_{AB} is the internuclear separation.

The Schrödinger equation obtained by the substitution of Equation (5) into Equation (1) can be solved analytically by transforming the system into one described by confocal elliptical coordinates. When this approach is taken, the resulting equations have been solved with great accuracy, yielding an exact solution that calculates the correct internuclear separation and electron dissociation energy (1). The calculated electron distribution for the ground state of H_2^+ indicates that the most probable region for the electron is to be located between the two nuclei. The localization of electron density between the two nuclei is described by a wave function known as a *molecular orbital* and represents the source of stability that holds the nuclei together to form the molecule.

Unfortunately, the ability to derive an exact solution to the Schrödinger equation for molecules containing more than one electron is not possible. Just as the addition of more electrons to the hydrogen made an exact solution of the atomic system impossible, the addition of more electrons and nuclei to the hydrogen molecular ion tremendously complicates the situation. The basic problem is that the Hamiltonian operator must contain terms that account for the attraction of each electron for each proton and a potential energy term that shows the effect of repulsion by each electron for one another:

$$V_{ee} = -e_1 e_2/r_{12} \tag{6}$$

where r_{12} is the interelectron separation. The resulting Schrödinger equation is an example of the classic multibody problem in physics, where the large number of interparticle potential energy terms makes solution of the equation impossible.

Because there can be no exact solutions of the Schrödinger equation for real molecules, it follows that rigorous wave functions for molecules cannot be derived. In order to develop theories of molecular bonding and electronic structure, approximation methods have been used to derive the best possible descriptions. One such theory has been termed as the valence-bond model, where covalent molecules are assembled on the basis of shared pairs of electrons (2). Although the valence-bond provides a qualitative understanding of molecular properties, true quantitative results require the use of variational and perturbation methods. There are better approaches, however, that yield useful qualitative and quantitative descriptions of molecular structure and electronic properties; these approaches will be discussed in the following sections.

LCAO-MO THEORY

The hydrogen molecule is formed by the combination of two hydrogen atoms, where shared electron density between the nuclei represents the bonding force that holds the two atoms together. However, it is evident that if one steadily increases the separation between the two nuclei, then at some internuclear separation the molecule must dissociate into two nonbonded hydrogen atoms. If these hydrogen atoms are sufficiently separated so that there is no significant interaction between them, then the result is a pair of simple hydrogen atoms whose wave functions can be obtained using the Schrödinger equation in the manner described in the previous sections. The following discussion is only a brief exposition of the basic principles of linear combination of atomic orbitals (LCAO)-MO theory, and interested readers should consult the standard references in the field (3–8).

The opposite concept of separation is the situation where two hydrogen atoms are brought successively closer to each other until electron density becomes shared between them. This suggests that one could reasonably approximate the MO of the hydrogen molecule as some sort of combination of the atomic orbitals of the constituent hydrogen atoms. The linear combination of atomic orbitals to form MO (commonly known as the LCAO-MO theory) is one such approach (3–8). The LCAO MOs of the hydrogen molecule are expressed as:

$$\Phi = N(a\psi_1 + b\psi_2) \tag{7}$$

where Φ is the wave function of the MO, ψ_1 and ψ_2 are the atomic orbitals of the individual atoms, N is a normalization factor, and a and b are the linear mixing coefficients.

Without specifying the form of the Hamiltonian operator, we can substitute the wave function of Equation (7) into the Schrödinger equation and solve for the energy of that wave function by a calculation of its expectation value:

$$E = \int \Phi^* \hat{H} \Phi \; d\tau \tag{8}$$

where the integration is to be performed over all space. Because E is a constant, Equation (8) can be rewritten as:

$$\int \Phi^* (\hat{H} - E) \Phi \; d\tau = 0 \tag{9}$$

Equation (9) can only be exactly satisfied for the Hamiltonian operator in question if the MO wave function is the exact wave function. But because we are working within an approximation framework, then the solution of Equation (9) will exceed zero by the residual, ρ:

$$\int \Phi^* (\hat{H} - E) \Phi \; d\tau = \rho \tag{10}$$

The smaller the value of ρ, the closer is the expectation value of the Hamiltonian expressing the correct energy, and the closer is the LCAO wave function expressing the true wave function.

If one substitutes the LCAO wave function of Equation (7) into Equation (10) and assumes the existence of real functions for Φ, then:

$$N^2 \int (a\psi_1 + b\psi_2)(\hat{H} - E)(a\psi_1 + b\psi_2) \; d\tau = \rho \tag{11}$$

The calculus of variations states that a function will be minimal whenever the partial derivatives with respect to each of its variables are simultaneously equal to zero. Considering the linear mixing parameters, a and b, to be the variables, then ρ will be at a minimum and the expectation value for the Hamiltonian over the LCAO wave function will be as close as possible to the true energy when:

$$(\partial\rho/\partial a) = 0 \tag{12}$$
$$(\partial\rho/\partial b) = 0 \tag{13}$$

Multiplying Equation (11) and collecting the terms in the mixing parameters, one obtains:

$$N^2 \left[a^2 \left(\int \psi_1 \hat{H} \psi_1 \; d\tau - E \int \psi_1 \psi_1 \; d\tau \right) + b^2 \left(\int \psi_2 \hat{H} \psi_2 \; d\tau - E \int \psi_2 \psi_2 \; d\tau \right) \right.$$
$$\left. + 2ab \left(\int \psi_1 \hat{H} \psi_2 \; d\tau - E \int \psi_1 \psi_2 \; d\tau \right) \right] = \rho \tag{14}$$

If the individual atomic orbitals are normalized in the usual manner, then:

$$\int \psi_1 \psi_1 \, d\tau = 1 \tag{15}$$

$$\int \psi_2 \psi_2 \, d\tau = 1 \tag{16}$$

It is usual practice to define the *Coulomb integral* as:

$$\alpha_i = \int \psi_i \hat{H} \psi_i \, d\tau \tag{17}$$

the *reasonance integral* as:

$$\beta_{ij} = \int \psi_i \hat{H} \psi_j \, d\tau \tag{18}$$

and the *overlap integral* as:

$$S_{ij} = \int \psi_i \psi_j \, d\tau \tag{19}$$

Substituting Equations (17), (18), and (19) into Equation (14), and dividing throughout by N^2, yields:

$$a^2(\alpha_1 - E) + b^2(\alpha_2 - E) + 2ab(\beta_{12} - ES_{12}) = \rho/N^2 \tag{20}$$

To apply the variational principle to Equation (20), one must minimize ρ/N^2 with respect to both a and b, yielding the two equations:

$$\partial(\rho/N^2)/\partial a = 2a(\alpha_1 - E) + 2b(\beta_{12} - ES_{12}) = 0 \tag{21}$$

$$\partial(\rho/N^2)/\partial b = 2a(\beta_{12} - ES_{12}) + 2b(\alpha_2 - E) = 0 \tag{22}$$

The result of the minimization process is two equations in the linear mixing coefficients. The only situation whereby one can have this set of linear equations simultaneously equalling zero is for the determinant, which multiplies the coefficients to equal zero:

$$\begin{vmatrix} \alpha_1 - E & \beta_{12} - ES_{12} \\ \beta_{12} - ES_{12} & \alpha_2 - E \end{vmatrix} = 0 \tag{23}$$

Because both atoms are equivalent in the case of the hydrogen molecule, the subscripts to the individual atoms are unnecessary. In this case, the value of the determinant becomes:

$$(\alpha - E)(\alpha - E) - (\beta - ES)(\beta - ES) = 0 \tag{24}$$

or

$$E^2(1 - S^2) - 2E(\alpha - S\beta) + \alpha^2 - S\beta^2 = 0 \tag{25}$$

Equation (25) can be directly solved using the quadratic equation, and the two solutions are found to be:

$$E_1 = (\alpha + \beta)/(1 + S) \qquad (26)$$
$$E_2 = (\alpha - \beta)/(1 - S) \qquad (27)$$

Evaluation of the integrals reveals that β has a negative value, whereas S has a relatively small positive value. One, therefore, finds that the two energy levels will be nearly symmetrically arrayed about the value of α, which is the energy of the state derived from the occupancy of one electron in the lowest energy atomic orbital of the isolated hydrogen atom. The implication of Equations (26) and (27) is that the LCAO method represents the bringing together of two hydrogen atoms and the consequent formation of two MOs from their respective atomic orbitals. As illustrated in Figure 1, one of these MOs will have a lower energy than the isolated atoms, and the other will have a higher energy than the isolated atoms.

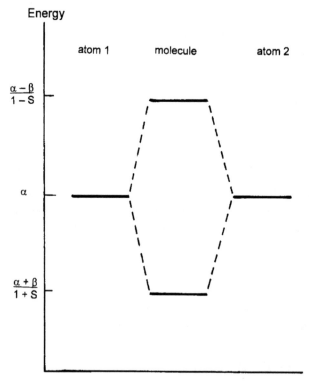

Figure 1   Energy level diagram for the two LCAO-MO energy levels resulting from the combination of the atomic orbitals of two atoms. *Abbreviation*: LCAO-MO, linear combination of atomic orbitals–molecular orbital.

It follows that if each hydrogen atom contributes one electron to the hydrogen molecule, then a more stable system is achieved if both of these electrons occupy the lower energy MO than if they remained in their atomic orbitals. For this reason, the lower energy MO is termed as a *bonding orbital*, and for reasons to be discussed next, the higher energy MO is termed as an *antibonding orbital*.

LCAO-MO THEORY AND THE SPECTROSCOPY OF DIATOMIC MOLECULES

In the previous discussion of atomic orbitals, it was recognized that the rules associated with quantum number possibilities placed restrictions on possible atomic wave functions. As a result, the electrons in the polyelectronic atoms of the first and second rows of the periodic table fill atomic orbitals in the order $1s$, $2s$, and then $2p$, a trend commonly referred to as the *Aufbau* principle. One, therefore, deduces the following electronic structures for these atoms:

H	1 electron	$\Rightarrow$	$1s^1$
He	2 electrons	$\Rightarrow$	$1s^2$
Li	3 electrons	$\Rightarrow$	$1s^2\,2s^1$
Be	4 electrons	$\Rightarrow$	$1s^2\,2s^2$
B	5 electrons	$\Rightarrow$	$1s^2\,2s^2\,2p^1$
C	6 electrons	$\Rightarrow$	$1s^2\,2s^2\,2p^2$
N	7 electrons	$\Rightarrow$	$1s^2\,2s^2\,2p^3$
O	8 electrons	$\Rightarrow$	$1s^2\,2s^2\,2p^4$
F	9 electrons	$\Rightarrow$	$1s^2\,2s^2\,2p^5$
Ne	10 electrons	$\Rightarrow$	$1s^2\,2s^2\,2p^6$

It follows that the LCAO MOs that can be formed from combination of these elements will necessarily involve electrons derived from s-orbitals and p-orbitals. The boundary surface of an s-orbital consists of a sphere, and so the mathematical sign of the orbital is the same everywhere. On the other hand, the boundary surfaces of p-orbitals are defined by a dumbbell shape, with one lobe having one type of sign and the other having the opposite sign. A much more detailed discussion of the electronic structure and states of diatomic molecules is available in Ref. (9).

As has been established during the discussion of the hydrogen atom, electron density that is shared between two atoms constitutes the attractive force that is commonly known as a chemical bond. Electron density can only be shared by the overlap of orbitals having the same mathematical sign, and Figure 2 show the types of bonds that can be formed by the combination of s- and p-orbitals. The bond types are further differentiated by the presence or absence of a node of electron density lying along the internuclear axis. Bonds formed by the overlap of two s-orbitals or one s-orbital and one p_X-orbital lack a node along this axis and are termed σ-bonds, with the two-electron wave functions describing these bond types are known as σ-orbitals. Bonds formed by the overlap of two p_Y-orbitals or two p_Z-orbitals have a single node along this axis and are termed π-bonds, and therefore the two-electron wave functions describing these bond types are known as π-orbitals. Antibonding orbitals are defined by the same symmetry types and are marked with an asterisk to distinguish them from bonding orbitals.

For homonuclear diatomic molecules, the MOs are further differentiated as being *gerade* (symmetric) or *ungerade* (antisymmetric) with respect to inversion through the center of gravity of the molecule. Hence, the bonding σ-orbitals and antibonding π-orbitals are symmetric in character (gerade), whereas the bonding π-orbitals and antibonding σ-orbitals are antisymmetric in character (ungerade).

The difference in energy between the component atomic orbitals also governs the types of MOs that can be formed between two atoms. For instance, the energies of the $2s$ electrons are considerably greater than the energies of the $1s$ electrons, so MOs cannot be formed by the overlap of the $1s$ orbital of one

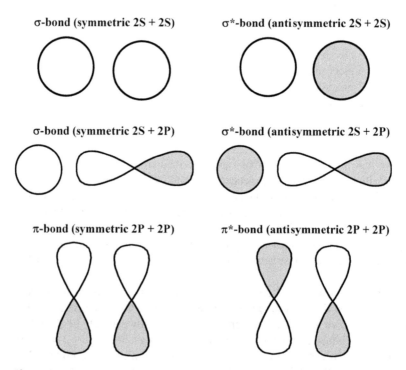

Figure 2   Types of bonds that can be formed by the combination of s- and p-orbitals. The *open* orbital types signify a positive wave function, and the *shaded* orbital types signify a negative wave function.

atom with the $2s$ orbital of the other. Likewise, the energy of the $2s$ orbital is sufficiently different from that of the p_X-orbital so that no MO is formed by these.

Considering both the symmetry of the MOs and the energetics of the component atomic orbitals, MOs are listed in the following sequence of increasing energy:

$$\sigma_g(1s) \ \sigma_u^*(1s) \ \sigma_g(2s) \ \sigma_u^*(2s) \ \sigma_g(2p) \ \pi_u(2p) \ \pi_g^*(2p) \ \sigma_u^*(2p)$$

In this sequence, it has been recognized that the $\pi_u(2p_Y)$ and $\pi_u(2p_Z)$ two-electron orbitals have exactly the same energy, and are therefore identified as the $\pi_u(2p)$ four-electron orbital. Similarly, the $\pi_g^*(2p)$ four-electron orbital is seen to consist of the isoenergetic pair of $\pi_g^*(2p_Y)$ and $\pi_g^*(2p_Z)$ two-electron orbitals. In addition, the $\sigma_g(2p)$ and $\sigma_u^*(2p)$ orbitals are understood to actually be the $\sigma_g(p_X)$ and $\sigma_u^*(p_X)$ orbitals.

The *Aufbau* principle can be applied to write the MO sequences of the homonuclear diatomic molecules formed by the elements of the first row of the periodic table:

$$
\begin{array}{lll}
H_2^+ & 1 \text{ electron} \implies & \sigma_g(1s)^1 \\
H_2 & 2 \text{ electrons} \implies & \sigma_g(1s)^2 \\
He_2^+ & 3 \text{ electrons} \implies & \sigma_g(1s)^2 \, \sigma_u^*(1s)^1 \\
He_2 & 4 \text{ electrons} \implies & \sigma_g(1s)^2 \, \sigma_u^*(1s)^2
\end{array}
$$

For the diatomic molecules of the second row, electrons in the $[\sigma_g(1s)^2 \, \sigma_u^*(1s)^2]$ kernel are the core electrons, and the electrons in the other MO are the valence electrons. In the absence of configurational interactions, the *Aufbau* principle can be applied to write the MO sequences of the homonuclear diatomic molecules formed by the elements of the first and second rows of the periodic table:

$$
\begin{array}{lll}
Li_2 & 6 \text{ electrons} \implies & [\sigma_g(1s)^2 \, \sigma_u^*(1s)^2] \ \sigma_g(2s)^2 \ \sigma_u^*(2s)^2 \\
Be_2 & 8 \text{ electrons} \implies & [\sigma_g(1s)^2 \, \sigma_u^*(1s)^2] \ \sigma_g(2s)^2 \ \sigma_u^*(2s)^2 \\
B_2 & 10 \text{ electrons} \implies & [\sigma_g(1s)^2 \, \sigma_u^*(1s)^2] \ \sigma_g(2s)^2 \ \sigma_u^*(2s)^2 \ \sigma_g(2p)^2 \\
C_2 & 12 \text{ electrons} \implies & [\sigma_g(1s)^2 \, \sigma_u^*(1s)^2] \ \sigma_g(2s)^2 \ \sigma_u^*(2s)^2 \ \sigma_g(2p)^2 \\
& & \pi_u(2p)^2 \\
N_2 & 14 \text{ electrons} \implies & [\sigma_g(1s)^2 \sigma_u^*(1s)^2] \ \ \sigma_g(2s)^2 \ \sigma_u^*(2s)^2 \ \sigma_g(2p)^2 \\
& & \pi_u(2p)^4 \\
O_2 & 16 \text{ electrons} \implies & [\sigma_g(1s)^2 \sigma_u^*(1s)^2] \ \ \sigma_g(2s)^2 \ \sigma_u^*(2s)^2 \ \sigma_g(2p)^2 \\
& & \pi_u(2p)^4 \, \pi_g^*(2p)^2 \\
F_2 & 18 \text{ electrons} \implies & [\sigma_g(1s)^2 \sigma_u^*(1s)^2] \ \ \sigma_g(2s)^2 \ \sigma_u^*(2s)^2 \ \sigma_g(2p)^2 \\
& & \pi_u(2p)^4 \, \pi_g^*(2p)^4 \\
Ne_2 & 20 \text{ electrons} \implies & [\sigma_g(1s)^2 \sigma_u^*(1s)^2] \ \ \sigma_g(2s)^2 \ \sigma_u^*(2s)^2 \ \sigma_g(2p)^2 \\
& & \pi_u(2p)^4 \, \pi_g^*(2p)^4 \sigma_u^*(2p)^2
\end{array}
$$

The *bond order* (BO) of a molecule is calculated as:

$$BO = (\# \text{ bonding electrons} - \# \text{ antibonding electrons})/2 \qquad (28)$$

The existence of a stable diatomic molecule is indicated by a bond order that exceeds zero. Any molecule that contains an equal number of bonding and antibonding electrons is characterized by a bond order of zero and should not exist, because there would be no stabilization mechanism to hold the atoms together. It is a fact that no evidence has been found implicating the existence of He_2, Be_2, or Ne_2 in their ground states.

The MO sequences of heteronuclear diatomic molecules follow a similar manner. Hence, for instance, the sequences for the following molecules are written as:

BeO   12 electrons $\Longrightarrow$ $[\sigma_g(1s)^2\sigma_u^*(1s)^2]\sigma_g(2s)^2\sigma_u^*(2s)^2\sigma_g(2p)^2\pi_u(2p)^2$

CN   13 electrons $\Longrightarrow$ $[\sigma_g(1s)^2\sigma_u^*(1s)^2]\sigma_g(2s)^2\sigma_u^*(2s)^2\sigma_g(2p)^2\pi_u(2p)^3$

CO   14 electrons $\Longrightarrow$ $[\sigma_g(1s)^2\sigma_u^*(1s)^2]\sigma_g(2s)^2\sigma_u^*(2s)^2\sigma_g(2p)^2\pi_u(2p)^4$

NO   15 electrons $\Longrightarrow$ $[\sigma_g(1s)^2\sigma_u^*(1s)^2]\sigma_g(2s)^2\sigma_u^*(2s)^2\sigma_g(2p)^2$
$\pi_u(2p)^4\ \pi_g^*(2p)^1$

NF   16 electrons $\Longrightarrow$ $[\sigma_g(1s)^2\sigma_u^*(1s)^2]\sigma_g(2s)^2\sigma_u^*(2s)^2\sigma_g(2p)^2$
$\pi_u(2p)^4\ \pi_g^*(2p)^2$

OF   17 electrons $\Longrightarrow$ $[\sigma_g(1s)^2\sigma_u^*(1s)^2]\sigma_g(2s)^2\sigma_u^*(2s)^2\sigma_g(2p)^2$
$\pi_u(2p)^4\ \pi_g^*(2p)^3$

It is tempting at this point to describe the electronic spectroscopy of one of these diatomic molecules as the promotion from a lower MO to one of the higher, unoccupied MOs, but this would be inappropriate. As was the case for atomic spectroscopy, electronic transitions are governed by the selection rules derived from their angular momenta properties, and the same situation holds for the electronic transitions of diatomic molecules. Only after the nature of the electronic states of the component atoms are established in the molecule, the nature of the electronic states of the molecule itself are established, and then only an understanding of the nature of the electronic transition can be developed.

It was established in an earlier section that the wave functions of a hydrogen-like atom are specified by the values of three quantum numbers. For a given value of the principal quantum number, n, the angular momentum quantum number, l, allows all possible values between 0 and $n-1$: orbitals for which $l=0$ are the s-orbitals, orbitals for which $l=1$ are the p-orbitals, orbitals for which $l = 2$ are the d-orbitals, and so on. Finally, for a given orbital type defined by its angular momentum quantum number, the magnetic quantum number, m, can take all values from $-l$ to 0 to $+l$. For atoms characterized by the existence of spherical symmetry, the energies must be the same for all orbitals having the same value 8 of n and l. It is important to understand that the value of the angular momentum quantum number implies an orbital angular momentum of $(h/2\pi)[l(l+1)]^{1/2}$, and the value of the magnetic quantum number implies a component of this angular momentum equal to $m(h/2\pi)$ along a specified direction. In a multielectronic atom, the energy of a given electron is found to depend on its value of all three quantum numbers.

For multielectronic atoms, there is a simple relation between the values of the angular momentum quantum numbers and the total angular momentum, L, of the atom. For an atom having two electrons characterized by l_1 and l_2, the permissible L values will be $(l_1 + l_2)$, $(l_1 + l_2 - 1)$, $(l_1 + l_2 - 1)$, ..., $(l_1 - l_2)$. In the same way, the atomic orbitals that are labeled as s, p, or d, indicate l values of 0, 1, or 2, and the electronic states for which L = 0, 1, or 2 are labeled as S, P, or D. The electron spin of an atomic orbital is described by the spin quantum number, m_s, for which the permissible values are $\pm 1/2$. The total spin angular momentum, S, for a two electron system will therefore be S = 0 (spins paired) or S = 1 (spins unpaired). The electronic state of an atom is defined by its L and S values, and the atomic spectroscopy discussed in the previous chapters is concerned with the selection rules associated with changes in L and S.

For diatomic molecules created by bonding of two separated atoms, the angular and spin momenta of the individual electrons combine to form the resultant orbital (Λ) and spin (S) angular momenta that now define the electronic states of the molecule. The following term symbols are used to identify electronic states by their resultant total orbital angular momentum values:

$\Lambda = 0$ Symbol $= \Sigma$

$\Lambda = 1$ Symbol $= \Pi$

$\Lambda = 2$ Symbol $= \Phi$

$\Lambda = 3$ Symbol $= \Delta$

Electronic states are further identified by their total spin angular momentum values, by placing the value of the spin multiplicity, $(2S + 1)$, as a left-hand superscript to the term symbol of the state. Any MO that contains its maximum number of allowed electrons (i.e., completely full) would be characterized by $\Lambda = 0$ and S = 0.

The number and types of possible electronic states will depend on the number and types of electrons occupying the various MOs. Equivalent electrons are defined as those occupying the same MO and therefore have the same n and l quantum numbers. The electronic terms that arise from the coupling of equivalent

σ- or π-electrons are:

$\sigma^1 \quad {}^2\Sigma$
$\sigma^2 \quad {}^1\Sigma$
$\pi^1 \quad {}^2\Pi$
$\pi^2 \quad {}^1\Sigma, {}^3\Sigma, {}^1\Delta$
$\pi^3 \quad {}^2\Pi$
$\pi^4 \quad {}^1\Sigma$

Hund's rule states that the ground states of an atom or molecule is that having the highest value for the spin multiplicity, and for states of equal spin multiplicity the state of the lower orbital angular momentum will be more stable. Diatomic molecules are described by the equivalent electron cases, and their descriptions summarized in Table 1. It should be noted that the effects of configuration interaction were neglected in deriving the ground state terms of Table 1, and therefore some of the states indicated for the diatomic molecules are not rigorously correct. However, they will suffice for the purposes of this discussion.

Nonequivalent electrons are those that occupy different MO and therefore have no restriction on their n and l quantum numbers. This case is of interest, because the excited states of diatomic molecules are described by the non-equivalent electron cases. The electronic terms arising from the coupling of nonequivalent σ- or π-electrons are:

$\sigma_i\sigma_j \quad {}^1\Sigma, {}^3\Sigma$

$\sigma_i\pi_j \quad {}^1\Pi, {}^3\Pi$

$\pi_i\pi_j \quad (2){}^1\Sigma, (2){}^3\Sigma, {}^1\Delta, {}^3\Delta$

Consider the ground electronic configuration of C_2:

$$GS = [\sigma_g(1s)^2\ \sigma_u^*(1s)^2]\ \sigma_g(2s)^2\ \sigma_u^*(2s)^2\ \sigma_g(2p)^2\ \pi_u(2p)^2$$

for which the term symbol of the ground state is $^3\Sigma$. This configuration would also cause the excited $^1\Sigma$ and $^1\Delta$ states. One excited state configuration for this diatomic molecule is:

$$ES_1 = [\sigma_g(1s)^2\ \sigma_u^*(1s)^2]\ \sigma_g(2s)^2\ \sigma_u^*(2s)^2\ \sigma_g(2p)^2\ \pi_u(2p)^1\ \pi_g^*(2p)^1$$

which would be characterized by two $^1\Sigma$, two $^3\Sigma$, one $^1\Delta$, and one $^3\Delta$ excited electronic states. Another excited state configuration for this diatomic molecule is:

$$ES_2 = [\sigma_g(1s)^2\ \sigma_u^*(1s)^2]\ \sigma_g(2s)^2\sigma_u^*(2s)^2\ \sigma_g(2p)^2\ \pi_u(2p)^1\ \sigma_u^*(2p)^1$$

which would be characterized by one $^1\Pi$ and one $^3\Pi$ excited electronic states. In principle, electronic transitions take place from the $^3\Sigma$ ground electronic state to any of these excited electronic states.

Table 1 Summaries of the Properties of Some Diatomic Molecules in Their Ground States, in the Absence of Configuration Interaction

	Electronic configuration	Bond order	Ground state
H_2^+	$\sigma_g(1s)^1$	$\frac{1}{2}$	$^2\Sigma$
H_2	$\sigma_g(1s)^2$	1	$^1\Sigma$
He_2^+	$\sigma_g(1s)^2\,\sigma_u^*(1s)^1$	$\frac{1}{2}$	$^2\Sigma$
He_2	$\sigma_g(1s)^2\,\sigma_u^*(1s)^2$	0	—
Li_2	$[\sigma_g(1s)^2\,\sigma_u^*(1s)^2]\,\sigma_g(2s)^2\,\sigma_u^*(2s)^2$	1	$^1\Sigma$
Be_2	$[\sigma_g(1s)^2\,\sigma_u^*(1s)^2]\,\sigma_g(2s)^2\,\sigma_u^*(2s)^2$	0	—
B_2	$[\sigma_g(1s)^2\,\sigma_u^*(1s)^2]\,\sigma_g(2s)^2\,\sigma_u^*(2s)^2\,\sigma_g(2p)^2$	1	$^1\Sigma$
C_2	$[\sigma_g(1s)^2\,\sigma_u^*(1s)^2]\,\sigma_g(2s)^2\,\sigma_u^*(2s)^2\,\sigma_g(2p)^2$ $\pi_u(2p)^2$	2	$^3\Sigma$
N_2	$[\sigma_g(1s)^2\,\sigma_u^*(1s)^2]\,\sigma_g(2s)^2\,\sigma_u^*(2s)^2\,\sigma_g(2p)^2$ $\pi_u(2p)^4$	3	$^1\Sigma$
O_2	$[\sigma_g(1s)^2\,\sigma_u^*(1s)^2]\,\sigma_g(2s)^2\,\sigma_u^*(2s)^2\,\sigma_g(2p)^2$ $\pi_u(2p)^4\,\pi_g^*(2p)^2$	2	$^3\Sigma$
F_2	$[\sigma_g(1s)^2\,\sigma_u^*(1s)^2]\,\sigma_g(2s)^2\,\sigma_u^*(2s)^2\,\sigma_g(2p)^2$ $\pi_u(2p)^4\,\pi_g^*(2p)^4$	1	$^1\Sigma$
Ne_2	$[\sigma_g(1s)^2\,\sigma_u^*(1s)^2]\,\sigma_g(2s)^2\,\sigma_u^*(2s)^2\,\sigma_g(2p)^2$ $\pi_u(2p)^4\,\pi_g^*(2p)^4\,\sigma_u^*(2p)^2$	0	—
BeO	$[\sigma(1s)^2\,\sigma^*(1s)^2]\,\sigma(2s)^2\,\sigma^*(2s)^2\,\sigma(2p)^2$ $\pi(2p)^2$	2	$^3\Sigma$
CN	$[\sigma(1s)^2\,\sigma^*(1s)^2]\,\sigma(2s)^2\,\sigma^*(2s)^2\,\sigma(2p)^2$ $\pi(2p)^3$	$1\frac{1}{2}$	$^2\Pi$
CO	$[\sigma(1s)^2\,\sigma^*(1s)^2]\,\sigma(2s)^2\,\sigma^*(2s)^2\,\sigma(2p)^2$ $\pi(2p)^4$	3	$^1\Sigma$
NO	$[\sigma(1s)^2\,\sigma^*(1s)^2]\,\sigma(2s)^2\,\sigma^*(2s)^2\,\sigma(2p)^2$ $\pi(2p)^4\,\pi^*(2p)^1$	$1\frac{1}{2}$	$^2\Pi$
NF	$[\sigma(1s)^2\,\sigma_u^*(1s)^2]\,\sigma(2s)^2\,\sigma^*(2s)^2\,\sigma(2p)^2$ $\pi(2p)^4\,\pi^*(2p)^2$	2	$^3\Sigma$
OF	$[\sigma(1s)^2\,\sigma^*(1s)^2]\,\sigma(2s)^2\,\sigma^*(2s)^2\,\sigma(2p)^2$ $\pi(2p)^4\,\pi^*(2p)^3$	$\frac{1}{2}$	$^2\Pi$

However, the rules developed by Hund state that electronic transitions characterized by $\Delta\Lambda = 0$ or ± 1, and $\Delta S = 0$ only are allowed. In that case, the plausible absorption spectroscopy of the C_2 molecule simplifies to the following:

$$^3\Sigma \longrightarrow {}^3\Sigma_A(ES_1),\ \ ^3\Sigma_B(ES_1),\ \ ^3\Pi(ES_2)$$

There are other Laporte and parity selection rules that would apply to the C_2 illustration (and which would further simplify the range possible for electronic transitions) but have not been considered in this discussion for the sake of simplicity.

Spectroscopy among electronic states is determined by the expectation value of the transition moment, R_e:

$$R_e = \int \psi_1^* \mu \psi_2 \, d\tau \tag{29}$$

where R_e is the electronic dipole moment, summed over the electronic coordinates. For those instances where the value of the transition moment equals zero, the transition is said to be forbidden. Hence, transitions among the electronic states of a molecule are referred to as electric-dipole transitions.

MO THEORY AND THE SPECTROSCOPY OF POLYATOMIC MOLECULES

The electronic spectra of diatomic molecules and their associated vibrational and rotation spectroscopies have been studied in far greater detail than most polyelectronic molecules. Because often, the energy of an electronically polyatomic molecule can be distributed so as to break weak chemical bonds, the spectroscopy of high-energy states is usually only studied through the use of photoelectron spectroscopy. As a result, the only electronic transitions routinely studied for polyatomic organic molecules are those that involve transitions between the ground state and its lowest-lying electronic states.

Nevertheless, the principles just developed to describe the electronic spectroscopy of diatomic molecules can be extended as a means to gain an understanding of the states of polyatomic molecules and their associated spectroscopy. Each MO, Φ_i, can be constructed as a linear combination of atomic orbitals, ψ_p, but now with the orbitals of each nuclei appearing in expansion:

$$\Phi_i = \Sigma c_i p \psi_p \tag{30}$$

and weighted by their respective linear mixing coefficients, c_{ip}. Hence, the MOs that can be constructed for polyatomic molecules are nonlocalized and extend over all atoms in the molecule.

Symmetry of Polyatomic Molecules and its Effect on their Electronic States

The problem with nonlinear polyatomic molecules is that there is no preferential direction (like the internuclear axis of a diatomic molecule) onto which the orbital angular momenta can be quantized. Although one can use the concept of σ- and π-orbitals to describe the character of individual electrons, it is not possible to define meaningful angular momenta for the whole system. However, by making use of the symmetry of the molecule and the associated science of group theory, it becomes possible to understand the electronic spectra of polyatomic molecules. The use of group theory in chemistry has been well documented elsewhere (10–14), and therefore only the essential aspects of the methodology will be presented here.

On considering the three-dimensional structures of molecules, it quickly becomes apparent that such molecules (especially small molecules) possess certain elements of recognizable symmetry. A *symmetry operation* is a movement of the atoms in a molecular species that, when performed, leaves the molecule in a configuration that is not distinguishable from the original configuration. A *symmetry element* is a property of a molecule about which symmetry operations can be carried out, such as a point, a line, or a plane. The action of a symmetry operation takes an atom having coordinates of (x, y, z) and transforms it to one having coordinates of (x', y', z'). For instance, if a molecule is rotated by $180°$ about the z-axis, there is the effect of changing the (x_i, y_i, z_i) coordinates of every atom in the molecule to $(-x_i, -y_i, -z_i)$. The actions of symmetry operations can be visualized through the use of 3×3 matrices, although a considerable simplification will be achieved by using the *characters* of these matrices.

The properties of the five fundamental symmetry operations can be illustrated through consideration of how their performance affects the coordinates of a hypothetical cyclic molecule consisting of four identical atoms. The first element is the *identity* operation (E), whose role sounds trivial but plays an essential part in the definition of a group:

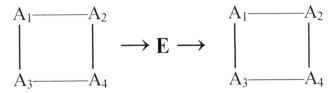

There are many *reflection planes* (σ) associated with this particular molecule, but the one that executes a reflection along the direction of this page can be illustrated as:

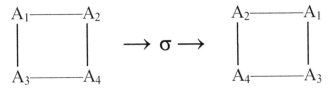

The performance of a second reflection of the molecule brings it back to its original position, an operation that is equivalent to the performance of the identity operation. The molecule also possesses an inversion center (i), whose action is illustrated as:

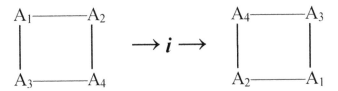

The performance of a second inversion operation on the molecule brings it back to its original position, an operation that is equivalent to the performance of the identity operation.

A *proper axis of rotation* (C_N) is defined as a rotation of the molecule by an angle of $360/N$ degrees about a defined axis. Proper rotation axes are often denoted by the number of times their performance brings the system back to its original position. For example, consider the fourfold axis of rotation (i.e., rotation by 90°) that is perpendicular to the molecular plane and passes through the center of gravity of the molecule:

$$A_1\text{------}A_2 \qquad \rightarrow C_4 \rightarrow \qquad A_3\text{------}A_1$$
$$A_3\text{------}A_4 \qquad\qquad\qquad A_4\text{------}A_2$$

After the C_4 operation is performed four times, the molecule is back to its original position, which is equivalent to the performance of the identity operation. Alternatively, consider the twofold axis of rotation (i.e., rotation by 180°) that passes through the line connecting atoms A_1 and A_4:

$$A_1\text{------}A_2 \qquad \rightarrow C_2 \rightarrow \qquad A_1\text{------}A_3$$
$$A_3\text{------}A_4 \qquad\qquad\qquad A_2\text{------}A_4$$

Because the axis passes through atoms A_1 and A_4, their positions are not altered upon performance of this symmetry operation. After the C_2 operation is performed twice, the molecule reverts to its original position, which is equivalent to the performance of the identity operation.

Finally, an *improper axis of rotation* (S_N) is defined as a rotation of the molecule by an angle of $360/N$ degrees about a defined axis, followed by a reflection operation through the plane. Improper rotation axes are also denoted by the number of times their performance brings the system back to its original position. We consider now the improper twofold axis (i.e., rotation by 180°, followed by a reflection) that passes through the line connecting atoms A_1 and A_4. The operation can be envisioned as consisting first of rotation about the twofold axis:

$$A_1\text{------}A_2 \qquad \rightarrow S_{2(a)} \rightarrow \qquad A_1\text{------}A_3$$
$$A_3\text{------}A_4 \qquad\qquad\qquad A_2\text{------}A_4$$

To complete the S_2 operation, the molecule is then subjected to a reflection oper-ation along the direction of the twofold axis of rotation:

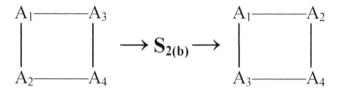

Interestingly, one finds that performance of the S_2 operation is equivalent to the performance of the identity operation.

In a more formal development of group theory, it can be shown that there are only 47 possible combinations of symmetry operations that can mutually coexist. These various symmetry combinations are known as *point groups*, because the symmetry elements of a molecule leave a point of the molecule fixed in space. This stands in contrast with the *space groups* found in crystals, where additional symmetry operations result in a translation of the molecule (or a unit cell) to a new location in the crystal. A more formal definition of a point group is that it is a complete set of nonredundant symmetry operations pos-sessing all the properties of a mathematical group.

Molecules can be systematically classified according to the number and types of symmetry elements and operations they possess. Because all molecules must be characterized by a finite set of symmetry operations, it follows that the observable properties of molecules will be described and governed by the point group that describes their symmetry properties. For instance, the application of group theory to a particular molecule can be used to understand its patterns of molecular vibration, its electronic properties, and the allowedness of the electronic transitions associated with its MO.

The characteristics of every point group are encapsulated in its *character table*. Readers who want to know how character tables are derived and assembled should consult the leading Refs. (10–14), but in this development we will focus exclusively on the use of character tables for our purposes. Character tables are derived from the transformation matrices that describe the three-dimensional effects of symmetry elements, with the character (the sum of the diagonal elements of each matrix) of each symmetry element making up the entries. It is a conse-quence of group theory that there is a finite set of *irreducible representations* that describe the symmetry properties of various combinations of symmetry elements. Each property of the molecule will have symmetry properties that cause it to be described by one of the irreducible representations of a point group.

All symmetry-related information is summed up in the character table of a point group, which is ordinarily divided into the four main areas. To illustrate the structure of a character table, the table for the C_{2v} point group is shown in Figure 3. Area I contains the names for the various irreducible representations, while Area II contains sets of characters relating the symmetry properties of

I		II			III	IV

C_{2v}	E	C_2	$\sigma_V(xz)$	$\sigma_V'(xz)$		
A_1	1	1	1	1	z	x^2, y^2, z^2
A_2	1	1	−1	−1	R_z	xy
B_1	1	−1	1	−1	x, R_Y	xz
B_2	1	−1	−1	1	y, R_X	yz

Figure 3 The character table of the C_{2v} point group, illustrating its subdivision into four areas.

each representation. Every wave function of a molecule will transform according to one of the representations in the point group, and therefore its properties will be defined by the characters of that group.

Area III indicates which representations appropriately describe motions along the x, y, and z coordinates and rotations about these same axes. The significance of Area III becomes apparent once one understands that the elements of the dipole moment operator transform according to the representations associated with x, y, and z. Then, because the symmetry properties of the group representations are the symmetry properties of MO, it will be possible to derive selection rules for allowing electric dipole transitions.

Finally, Area IV indicates the representations that describe squares and binary products of coordinates according to their properties. Area IV assumes significance because the elements of the polarizability tensor transform according to the representations of the cross products as indicated in Area IV of the character table. Because transitions in Raman spectroscopy entail changes in molecular polarizability, selection rules for this method of observing molecular motion can be derived as well.

Worked Example: The Electronic Spectroscopy of Formaldehyde

To illustrate the symmetry approach to MO theory and the electronic spectra of polyatomic molecules, we will first consider formaldehyde, H_2CO:

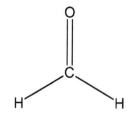

In chapter 2, the following atomic configurations were established:

H $1s^1$

C $1s^2 2s^2 2p^2$

O $1s^2 2s^2 2p^4$

We have also established in previous chapters that the oxygen $1s$ and carbon $1s$ core electrons will not be involved in the chemical bonding or in the electronic spectroscopy of formaldehyde. Thus, of the 16 electrons in formaldehyde, only 14 can be considered as being valence electrons.

In the conventional view, the central carbon is considered as being sp^2-hybridized to achieve the trigonal geometry for the molecule. In other words, its two $2s$ atomic orbitals are combined with one of its $2p$ atomic orbitals, yielding a hybridized atomic orbital holding three electrons. The remaining, non-hybridized, $2p$ orbital is the one perpendicular to the molecular plane. As a result, the new atomic configuration of the hybridized carbon can be written as:

$C(sp^2)$ $1s^2 2sp^3 2p^1$

In this view, one of the three electrons of the sp^2-hybrid will have the proper symmetry to form a σ-bond with the carbon atom.

In order to achieve the right type of orbital symmetry that enables formation of a double bond between the oxygen and carbon atoms, it is assumed that the valence atomic orbitals of the oxygen atom also become sp^2-hybridized. In such a case, the new atomic configuration of the hybridized oxygen can be written as:

$O(sp^2)$ $1s^2 2sp^3 2p^3$

In this view, the three electrons of the sp^2-hybrid will have the proper symmetry to form σ-bonds with the two hydrogen atoms and with the oxygen atom. It should be noted that one of the nonhybridized $2p$ orbitals (the one perpendicular to the molecular plane) will have the right symmetry to form a π-bond with the carbon atom, thus, explaining the double bond as consisting of one σ-bond and one π-bond.

Assuming that the high-energy $2p$ electrons of the oxygen atom drop down to fill the lower-energy orbitals of the sp^2 oxygen hybrids, the 16 electrons of formaldehyde can be placed into the following MO sequence describing the ground state of the molecule:

$H_2CO(G)$ [core] $(\sigma_{CH1})^2 (\sigma_{CH2})^2 (\sigma_{CO})^2 (\pi_{CO})^2 (sp_1^2)^2 (sp_2^2)^2$

The four 2-electron bonds of formaldehyde are represented by the σ_{CH1}, σ_{CH2}, σ_{CO}, and π_{CO} orbitals, and the four nonbonding electrons are represented by the sp_1^2 and sp_2^2 orbitals. For the sake of simplicity, the $(1s_C)^2$ and $(1s_O)^2$ electrons have been written simply as the core electrons.

Of course, the formation of every bonding orbital must be accompanied by the formation of its associated antibonding orbital. A large number of excited-state configurations are written, each of which corresponds to an excited state of the formaldehyde molecule. Although some of these are energetically unreasonable, the following three excited configurations and their states are of particular significance:

$$H_2CO(E_1) \quad [\text{core}] \ (\sigma_{CH1})^2 \ (\sigma_{CH2})^2 \ (\sigma_{CO})^2 \ (\pi_{CO})^2 \ (sp_1^2)^2 \ (sp_2^2)^1 \ (\pi_{CO}^*)^1$$

$$H_2CO(E_2) \quad [\text{core}] \ (\sigma_{CH1})^2 \ (\sigma_{CH2})^2 \ (\sigma_{CO})^2 \ (\pi_{CO})^1 \ (sp_1^2)^2 \ (sp_2^2)^2 \ (\pi_{CO}^*)^1$$

$$H_2CO(E_3) \quad [\text{core}] \ (\sigma_{CH1})^2 \ (\sigma_{CH2})^2 \ (\sigma_{CO})^2 \ (\pi_{CO})^2 \ (sp_1^2)^2 \ (sp_2^2)^1 \ (\sigma_{CO}^*)^1$$

Owing to the nature of the MO involved, the E_1 transition is commonly identified as the $n \rightarrow \pi^*$ transition, the E_2 transition is denoted as the $\pi \rightarrow \pi^*$ transition, and the E_3 transition is classified as the $n \rightarrow \sigma^*$ transition.

In order to determine whether any of these spectroscopic transitions can actually take place, one must determine the expectation value of the transition moment, $\int \psi_1^* \hat{\mu} \psi_2 d\tau$. Given the complexity of the involved MO wave functions, an exact calculation of the transition is not practical. Fortunately, group theory permits a ready determination of whether the expectation value will have a finite value if the integration is successfully performed. Because every electronic state of a molecule must possess symmetry properties that are described by one of the representations of its molecular point group, the representations in its character table can be used to determine whether the transition moment equals zero or not. Because the integration step necessary to evaluate the transition moment is performed over all space, any antisymmetric result must integrate to zero. Therefore, if one takes the direct product of the representations for ψ_1^*, μ, and ψ_2, unless the resulting representation contains the totally symmetric representation (i.e., the one where all characters equal $+1$), the transition moment of which must equal zero.

The task is greatly simplified when one recognizes that if the atomic orbitals of the ground state molecular wave function are all fully occupied, then that electronic state will transform as the totally symmetric representation of the point group. In the case of formaldehyde, whose point group is C_{2V} and whose character table is given in Figure 3, the A_1 representation summarizes the symmetry properties of the ground state. This recognition greatly reduces the complexity of the evaluation, because now one only needs to determine whether the direct product of the representation of the excited state and the representation of the components of the dipole moment operator contain the totally symmetric representation. But, it is a principle of group theory that only the direct product of a representation with itself can yield a totally symmetric result.

Therefore, as long as the ground state transforms as the totally symmetric representation of the point group (and it usually does for organic molecules), the transition moment will be nonzero if the representation of the excited state is the same as the representation of one of the dipole moment components. Consideration of the symmetry properties of the excited state orbitals enables one to deduce that the E_1 excited state transforms according to the A_2 representation, the E_2 excited state transforms according to the A_1 representation, and the E_3 excited state transforms according to the B_2 representation.

Examination of the C_{2V} character table of Figure 3 reveals that the x component of the dipole moment operator transforms as the B_1 representation, the y component of the dipole moment operator transforms as the B_2 representation, and the z component of the dipole moment operator transforms as the A_1 representation. One may, therefore, now deduce the following set of selection rules for the spectroscopy of formaldehyde:

1. Because the E_1 excited state transforms according to the A_2 representation and because none of the components of the dipole moment operator transform according to the A_2 representation, the $n \rightarrow \pi^*$ transition is forbidden.
2. Because the E_2 excited state transforms according to the A_1 representation and because the z component of the dipole moment operator transforms according to the A_1 representation, the $\pi \rightarrow \pi^*$ transition is allowed.
3. Because the E_3 excited state transforms according to the B_2 representation and because the y component of the dipole moment operator transforms according to the B_2 representation, the $n \rightarrow \sigma^*$ transition is allowed.

The power of group theory is that it permits one to determine whether a transition will or will not be allowed, but it does not yield information regarding the intensity of the transition.

We have not yet considered the effect of electron spin, and it is clear that the MO configurations of the E_1, E_2, and E_3 excited states could exist with either the spins of the unpaired electrons being paired or unpaired. For paired electrons, the multiplicity $(2S + 1)$ would equal one, and such states are known as singlet states. For unpaired electrons the multiplicity equals three, and such states are known as triplet states. Remembering that the multiplicity of a ground state whose MO is composed of completely filled atomic orbitals must equal unity and that the spin selection $\Delta S = 0$, the spin selection rule states that only transitions to singlet excited states will be allowed.

The final stage in a spectroscopic analysis would be to use the results of the theoretical study as a means to assign the nature of empirically observed electronic transitions. If assignments of the absorption bands to transitions between given states can be made, the relative energies of the various electronic states of the molecule can be deduced. It is a general rule that transitions allowed by symmetry considerations will be extremely strong, whereas transitions forbidden by symmetry will be extremely weak or not observed at all. For many carbonyl compounds, a weak band is observed around 280 nm, and this is attributed to the forbidden $n \rightarrow \pi^*$ transition. Clearly, some relaxation of the selection rule must have taken place to permit this transition to take place at all. Very strong absorption bands are observed in the 170–180 nm region, which are attributed to the overlapping $\pi \rightarrow \pi^*$ and $n \rightarrow \sigma^*$ transitions.

GENERAL FEATURES OF THE ABSORPTION SPECTROSCOPY OF ORGANIC MOLECULES

For the large molecules of pharmaceutical interest, the number of MOs becomes so unwieldy that the type of detailed analysis just conducted for formaldehyde is not feasible without the use of computers and semiempirical computer calculations. However, it has become established that although each MO extends over the entire molecule, many of these orbitals have the majority of their electron density localized at one particular part of the molecule. For instance, the π-orbital forming half of the double bond between two atoms will be mostly localized in the vicinity of those atoms. By their nature, nonbonding electrons will be localized on the atom, which furnished their atomic orbitals.

As a result of this electron localization, there are only a limited number of electronic transition types that a molecule might experience (15–20). For organic molecules, two of these involve transitions from bonding MOs ($\sigma \rightarrow \sigma^*$ or $\pi \rightarrow \pi^*$) and are generally allowed by the symmetry-based selection rules. The $\sigma \rightarrow \sigma^*$ transitions are generally observed in the far-ultraviolet (UV) region, whereas the $\pi \rightarrow \pi^*$ transitions occur in the visible (VIS) or UV regions of the spectrum. Two other transition types originate from nonbonding MOs ($n \rightarrow \sigma^*$ or $n \rightarrow \pi^*$). The fairly intense $n \rightarrow \sigma^*$ transition usually is observed in the middle to far UV region, whereas the weak $n \rightarrow \pi^*$ transition is usually observed in the VIS or near-UV regions of the spectrum.

For organic molecules of pharmaceutical interest, the $n \rightarrow \pi^*$ and $\pi \rightarrow \pi^*$ transitions are of the most interest and will therefore be discussed in more detail.

$n \rightarrow \pi^*$ Transitions

Unsaturated molecules containing atoms that have localized nonbonding electrons (oxygen, nitrogen, sulfur, etc.) often exhibit a weak band in their absorption spectra that can be assigned to a $n \rightarrow \pi^*$ transition. As described earlier for formaldehyde, and extendable to carbonyl groups in general, this transition is associated with an excitation from the lone-pair orbital localized on the oxygen to the π^*-orbital of the carbonyl group. The observation that the bands corresponding to the $n \rightarrow \pi^*$ transition are weak suggests that the transition is symmetry forbidden. The transition is completely forbidden in strict C_{2V} symmetry, but the transition gains intensity owing to the symmetry breaking of the carbonyl group caused by the presence of other atoms in the molecule. In the language of Beer's Law:

$$A = abc \tag{31}$$

where A is the total absorbance, a is the molar absorptivity, b is the path length of absorption, and c is the molar concentration, the molar absorptivity of a $n \rightarrow \pi^*$ transition will be in the range of $1-10$ L/mol cm.

In saturated aldehydes and ketones, the band due to the $n \rightarrow \pi^*$ transition is generally observed around $270-300$ nm, with a molar absorptivity of $1-2$ L/mol cm. In unsaturated carbonyl compounds, the absorption spectrum is usually a composition of ethylenic and carbonyl absorptions. If the carbonyl group forms part of a conjugated system, the $n \rightarrow \pi^*$ transition will be observed in the range of $300-350$ nm, with a molar absorptivity in the order of 10 L/mol cm.

The nature of the substituents in the immediate vicinity of the carbonyl group affects the wavelength of the $n \rightarrow \pi^*$ transition. Inductive (i.e., electro-static) effects tend to alter the energies of the ground n-state and the excited π^*-state by roughly an equivalent amount, and therefore the energy of the $n \rightarrow \pi^*$ transition tends not to be perturbed by inductive effects. Substituents that themselves possess nonbonding electrons act as electron donors to the π-system through a resonance effect, causing increase in the energies of all π-orbitals (bonding and antibonding) without affecting the energy of the n-orbital. The effect of resonance effects is to increase the energy of the $n \rightarrow \pi^*$ transition, which is manifested in a shift toward shorter wavelengths (commonly referred to as a "blue shift"). Replacement of the hydrogen atom in an aldehyde group by an alkyl group will have the same effect, and hence the $n \rightarrow \pi^*$ transition of a ketone is always blue-shifted relative to its aldehyde equivalent.

One of the simplest methods to assign a given band as being due to a $n \rightarrow \pi^*$ transition is to measure the spectrum of the corresponding molecule lacking the functional group. For instance, in the absorption spectrum of pyrazine, one observes a weak absorption band around 300 nm, which is not present in the spectrum of benzene. These findings enable one to deduce that the band must arise from a transition from the nonbonding orbitals of the nitrogen atom into the π^*-network of the aromatic system.

When the absorption spectra of compounds capable of exhibiting a $n \rightarrow \pi^*$ transition are measured in the solution phase, one finds that the solvent can exert a strong influence on the energy of the transition. For instance, the energy of the $n \rightarrow \pi^*$ transition will shift to shorter wavelengths when the solvent is changed from hydrocarbon to alcohol. This behavior has been explained that in the alcoholic solvent, the carbonyl group is hydrogen-bonded with the solvent as part of the overall solvation process, which stabilizes the ground state of the compound. The hydrogen bonding has no effect on the excited π^*-state, and hence the lowering of the energy of the ground n-state relative to the excited π^*-state is responsible for the increase in energy of the $n \rightarrow \pi^*$

transition in the interacting alcoholic solvent relative to the energy in the non-interacting hydrocarbon solvent.

Another characteristic of $n \to \pi^*$ transitions is that they are often strongly affected by the acidity of their solution. For instance, the $n \to \pi^*$ transition of pyridine disappears in acidic solutions owing to the reaction:

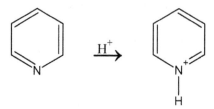

Once the pyridine nitrogen becomes protonated, the nonbonding electrons become involved in bonding, which in turn changes the overall pattern of molecular orbitals and their associated electronic transitions.

$\pi \to \pi^*$ Transitions

A $\pi \to \pi^*$ transition entails excitation from a π-electron system and can therefore occur in any molecule containing such a system if symmetry allows. Generally, such organic molecules can be characterized as containing two general types of π-systems, namely one where a localized π-bond exists and another where a delocalized π-bonding system exists.

Ethylene contains a total of 12 valence electrons, with 10 of these being contained within the σ-bonding system and the remaining two in the localized π-bonding orbital. The electronic configurations of the ground and excited states of ethylene can therefore be written as:

$$C_2H_4(G) \quad [\text{core}] \quad (\pi_{CC})^2$$

$$C_2H_4(E_1) \quad [\text{core}] \quad (\pi_{CC})^1(\pi_{CC}^*)^1$$

Ethylene belongs to the D_{2h} point group, whose character table is provided in Figure 4. A consideration of the symmetry of the molecular orbitals indicates that the π-orbital transforms as the B_{3u} representation and that the π^*-orbital transforms as the B_{2g} representation. Because all of the molecular orbitals in the ground state of the molecule are completely full, it must transform as the totally symmetric A_g representation. To derive the symmetry of the E_1 excited state, the direct product of the B_{3u} and B_{2g} representations is taken, which shows that the E_1 excited state transforms as the B_{1u} representation.

Examination of the D_{2h} character table of Figure 4 indicates that the $A_g \to B_{1u}$, $\pi \to \pi^*$, transition will be electric dipole allowed and polarized along the internuclear z-axis. Because this band is allowed by symmetry, one would anticipate that its degree of absorptivity would be large. In fact, ethylene

D_{2h}	E	$C_{2(Z)}$	$C_{2(Y)}$	$C_{2(X)}$	i	$\sigma(xy)$	$\sigma(xz)$	$\sigma(yz)$	
A_g	1	1	1	1	1	1	1	1	
B_{1g}	1	1	-1	-1	1	1	-1	-1	
B_{2g}	1	-1	1	-1	1	-1	1	-1	
B_{3g}	1	-1	-1	1	1	-1	-1	1	
A_u	1	1	1	1	-1	-1	-1	-1	
B_{1u}	1	1	-1	-1	-1	-1	1	1	Z
B_{2u}	1	-1	1	-1	-1	1	-1	1	Y
B_{3u}	1	-1	-1	1	-1	1	1	-1	X

Figure 4   The character table of the D_{2h} point group.

exhibits a strong absorbance band at 174 nm, characterized by a molar absorptivity of 16,000 L/mol cm.

Benzene contains a total of 30 valence electrons, with 24 of these being contained within the σ-bonding system and the remaining six the delocalized π-bonding system. Owing to its hexagonal and planar structure, it is not really possible to discuss the MO sequence of benzene without first considering its point group symmetry. The compound belongs to the D_{6h} point group, whose character table is shown in Figure 5. One of the interesting aspects of this point group is the existence of the *E*-type representations, because MOs having *E*-type symmetry and transforming according to one of the *E* representations will be capable of degeneracy (i.e., two MOs characterized by the same energy).

In benzene, the six $2p_z$ orbitals (perpendicular to the molecular plane) of the carbon atoms become six π-type molecular orbitals. The bonding π-orbital of lowest energy has A_{2u} symmetry, whereas the other two bonding π-orbitals of higher energy are found to be degenerate and transform according to the E_{1g} representation. The lowest energy antibonding π-orbital is of the degenerate E_{2u} symmetry, and the highest energy antibonding π-orbital transforms according to the B_{1g} representation. The electronic configurations of the ground and lowest excited states of benzene can therefore be written as:

$$C_6H_6(G) \quad [core] \quad (\pi_{A2u})^2(\pi_{E1g})^4$$
$$C_6H_6(E_1) \quad [core] \quad (\pi_{A2u})^2(\pi_{E1g})^3(\pi_{E2u}*)^1$$

It can be shown that the E_1 configuration will lead to the existence of three exhibited electronic states having E_{1u}, B_{2u}, and B_{1u} symmetry. Because the electrons in the $(\pi_{E1g})^3(\pi_{E2u}^*)^1$ configuration can occupy those orbitals either with

D_{6h}	E	$2C_6$	$2C_3$	C_2	$3C'_2$	$3C''_2$	i	$2S_3$	$2S_6$	σ_h	$3\sigma_d$	$3\sigma_v$	
A_{1g}	1	1	1	1	1	1	1	1	1	1	1	1	
A_{2g}	1	1	1	1	-1	-1	1	1	1	1	-1	-1	
B_{1g}	1	-1	1	-1	1	-1	1	-1	1	-1	1	-1	
B_{2g}	1	-1	1	-1	-1	1	1	-1	1	-1	-1	1	
E_{1g}	2	1	-1	-2	0	0	2	1	-1	-2	0	0	
E_{2g}	2	-1	-1	2	0	0	2	-1	-1	2	0	0	
A_{1u}	1	1	1	1	1	1	-1	-1	-1	-1	-1	-1	
A_{2u}	1	1	1	1	-1	-1	-1	-1	-1	-1	1	1	Z
B_{1u}	1	-1	1	-1	1	-1	-1	1	-1	1	-1	1	
B_{2u}	1	-1	1	-1	-1	1	-1	1	-1	1	1	-1	
E_{1u}	2	1	-1	-2	0	0	-2	-1	1	2	0	0	X,Y
E_{2u}	2	-1	-1	2	0	0	-2	1	1	-2	0	0	

Figure 5   The character table of the D_{6h} point group.

spins paired or unpaired, each of the electronic states may exist as either as a singlet (spins paired, so $2S + 1 = 1$) or as a triplet state (spins unpaired, so $2S + 1 = 3$). Because all of the MOs in the ground state of the molecule are completely full, it must transform as the totally symmetric A_{1g} representation.

The theory therefore predicts the existence of three possible spin-allowed $\pi \to \pi^*$ transitions for benzene, and the symmetry rules can be used to conclude that only one of these transitions is actually allowed. Nevertheless, various symmetry breaking mechanisms enable all three electronic transitions of benzene to be observed in its absorption spectrum. A weak band at approximately 255 nm (molar absorptivity of 200 L/mol cm) is assigned to the symmetry-forbidden $^1A_{1g} \to {}^1B_{2u}$ transition. The somewhat stronger band of the symmetry-forbidden $^1A_{1g} \to {}^1B_{1u}$, transition is located at approximately 200 nm (molar absorptivity of 6300 L/mol cm), and is in fact observed as a long-wavelength shoulder on the intense $^1A_{1g} \to {}^1E_{1u}$ transition, which has its maximum at 180 nm (molar absorptivity of 100,000 L/mol cm).

The effect of solvents on the energies of $\pi \to \pi^*$ transitions is opposite to that previously described for $n \to \pi^*$ transitions. In the case of benzene, the distribution of π-electrons in the ground state is symmetric and thus is not perturbed when the molecule is in a polar or nonpolar environment. On the other hand, the distribution of π-electrons in the excited state is antisymmetric, and hence the molecular orbital will be stabilized in a polar environment. Therefore, it is generally found that the energy of a $\pi \rightarrow \pi^*$ transition undergoes a shift to longer wavelength (known as a "red shift") when the polarity of the solvent dissolving an aromatic molecule is increased. However, the generality of this rule is sometimes broken when the solvent is capable of undergoing substantial hydrogen bonding with the solute. For example, the absorption spectrum of β-naphthol dissolved in dioxane exhibits additional absorption bands associated with hydrogen-bonded species relative to those of the compound dissolved in hexane.

Chromophores

As discussed earlier, the molecular orbitals of many functional groups are indeed localized on the grouping of atoms and are only secondarily perturbed by the remainder of the molecule. In other words, the absorption bands of a related series of molecules containing a particular chemical group are similar and occur in approximately the same region of the spectrum. Such groups are termed *chromophores* (literally "color carriers"). The transferability of a group absorption band and intensity over a given wavelength range from one molecule to another requires the existence of a lack of interaction between the chromophore and the rest of the molecule. The absorption bands of a number of chromophores were summarized by Guillory (19), and some of these have been abstracted in Table 2.

Recognizing that most transitions observed in the UV or VIS region of the spectrum are either $n \rightarrow \pi^*$ or $\pi \rightarrow \pi^*$ in character, Scott has provided an extensive summary of the properties of various chromophores (21). In addition, Scott has also developed a number of rules for the prediction of the absorption bands in substituted benzene derivatives. For instance, the absorption maximum of phenol is observed at 230 nm. Knowing that the effect of additional substituents on the phenol system would cause a red shift of the absorption, Scott developed tables for the estimation of the band maximum. For instance, the contribution for a amine group at the ortho, meta, and para positions would be $+15$, $+15$, and $+58$, so that the absorption maximum would shift to 245 nm for the ortho- and meta-substituted aminophenols and 288 nm for para-aminophenol.

Charge Transfer Transitions

For molecules containing two chromophores, or a molecule formed by the complexation of a donor and an acceptor, one often can measure broad featureless absorption bands that are not observed with either of the individual components. Because these bands often entail the redistribution of electron density from one chromophore to the other, they are frequently denoted as *charge transfer* bands. For example, the absorption spectra of 1,2,4,5-tetramethyl-benzene, tetra-cyano-ethylene, and their 1:1 mixture are shown in Figure 6, with the strong band at approximately 490 nm, which is due to the charge transfer absorption band.

Owing to the loose nature of the bonding between the donor and acceptor in the ground state of the complex, charge transfer bands are generally very broad

Table 2 Absorption Spectra of Selected Molecules Containing Chromophores

Group	Example	Wavelength (nm)	Molar absorptivity (L/mol cm)
C=C	$H_2C=CH_2$	182.5	250
		174.4	16,000
		170.4	16,500
		162.0	10,000
C=O	$H_2C=O$	295.0	10
		185.0	20,000
—OH	$CH_3—OH$	183.0	200
		150.0	1,900
—NO_2	$CH_3—NO_2$	277.5	10
		210.0	10,000
—NH_2	$CH_3—NH_2$	215.0	580
		190.5	3,200
Penyl	C_6H_6	255.0	200
		200.0	6,300
		180.0	100,000
Naphthyl	$C_{10}H_8$	211.0	250
		270.0	5,000
		221.0	100,000
Anthryl	$C_{14}H_{10}$	360.0	6,000
		250.0	150,000

Source: Adapted from Ref. 19.

and show no fine structure due to an underlying vibrational structure. Such bands will frequently change energy when the donor–acceptor complexes are crystallized, and the existence of polymorphic structures will often yield differently colored solids. For example, the charge-transfer complex formed by picric acid and 2-iodoaniline can be obtained in two polymorphs, depending on the crystallization process used (22). Monoclinic Form-I contains four cations (2-iodoanilinium) and four anions (picrate) per unit cell, and its crystals are yellow in color. The triclinic Form-II contains two iodoanilinium cations and two picrate anions per unit cell, and exists as a green solid. In Form-I, the anion–cation pairs are stacked to form segregated columns, in which the same ion types are stacked with each other. In Form-II, the anions and cations are alternately stacked to form continuous columns. The structural arrangement of Form-I does not allow for the generation of a charge-transfer absorption band, and hence, the solid maintains the yellow color of picric acid. On the other hand, the alternate stacking present in Form-II facilitates the overlap of suitable molecular orbitals to yield a new charge-transfer absorption band, which is equally as absorptive as the locally excited band of the picrate anion.

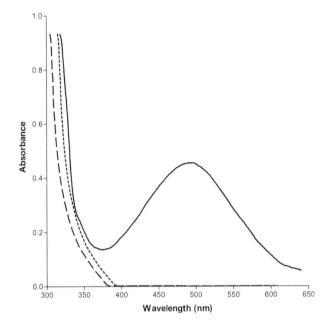

Figure 6   Absorption spectra of 1,2,4,5-tetramethylbenzene (*long dashed trace*), tetra-cyanoethylene (*short dashed trace*), and their 1:1 donor–acceptor complex (*solid trace*).

The compounds of transition elements often exhibit absorption bands that are due to intramolecular charge transfer transitions. The transitions can involve transfer of electron density from the coordinated ligands to the empty *d*-orbitals of the metal ions or transfer of electron density from metal ion *d*-orbitals to empty π^*-orbitals of the ligands.

REFERENCES

1. Jaffe GZ Physik 1934; 87: 535.
2. Pauling L. The Nature of the Chemical Bond, 3rd edn. Ithaca: Cornell University Press, 1960.
3. Kauzmann W. Quantum Chemistry. New York: Academic Press, 1957.
4. Slater JC. Quantum Theory of Molecules and Solids, volume 1. New York: Mc-Graw Hill Co., 1963.
5. Ballhausen CJ, Gray HB. Molecular Orbital Theory. Reading, MA: Benjamin/ Cummings Pub., 1964.
6. Flurry RL. Molecular Orbital Theories of Bonding in Organic Molecules. New York: Marcel Dekker, 1968.
7. McGlynn SP, Vanquickenborne LG, Kinoshita M, Carroll DG. Introduction to Applied Quantum Chemistry. New York: Holt, Rinehart and Winston, 1972.

8. Hanna MW. Molecular Quantum Mechanics in Chemistry. 3rd edn. Menlo Park, CA: Benjamin/Cummings Pub., 1981.
9. Herzberg G. Spectra of Diatomic Molecules, 2nd edn. Cincinnati, OH: Van Nostrand Reinhold Co., 1950.
10. Tinkham M. Group Theory and Quantum Mechanics. New York: McGraw-Hill Book Co., 1964.
11. Hochstrasser RM. Molecular Aspects of Symmetry. New York: W.A. Benjamin Inc., 1966.
12. Orchin M, Jaffe HH. Symmetry, Orbitals, and Spectra. New York: Wiley-Interscience, 1971.
13. Cotton FA. Chemical Applications of Group Theory, 2nd edn. New York: Wiley-Interscience, 1971.
14. Harris DC, Bertolucci MD. Symmetry and Spectroscopy. New York: Oxford University Press, 1978.
15. Herzberg G. Electronic Spectra and Electronic Structure of Polyatomic Molecules. Princeton, NJ: Van Nostrand Reinhold Co., 1967.
16. Barrow GM. Introduction to Molecular Spectroscopy. New York: McGraw-Hill Book Co., 1962.
17. King GW. Spectroscopy and Molecular Structure. New York: Holt, Rinehart and Winston, 1964.
18. Brittain EFH, George WO, Wells CHJ. Introduction to Molecular Spectroscopy Theory and Experiment. London: Academic Press, 1970.
19. Guillory WA. Introduction to Molecular Structure and Spectroscopy. Boston, MA: Allyn and Bacon, Inc., 1977.
20. McHale JL. Molecular Spectroscopy. Upper Saddle River, NJ: Prentice Hall, 1999.
21. Scott AI. Interpretation of the Ultraviolet Spectra of Natural Products. New York: Pergamon Press, 1964.
22. Tanaka M, Matsui H, Mizoguchi J-I, Kashino S. Bull Chem Soc Jap 1994; 67: 1572.

5

UV/VIS Reflectance Spectroscopy

Harry G. Brittain

Center for Pharmaceutical Physics, Milford, New Jersey, U.S.A.

INTRODUCTION

Everyone capable of sight is a practitioner of reflectance spectroscopy in the visible region of the spectrum. We are constantly bathed by light, some which is characterized by wavelengths between 400 nm (i.e., blue light) and 650 nm (i.e., red light). When our eyes perceive, for instance, a red piece of clothing, we are actually observing the reflected portion of the irradiation light. The clothing absorbs the blue light, reflects the unabsorbed red light, our eyes perceive the reflected red light, and our mind then interprets this measurement to conclude that the piece of clothing was actually red. However, our perception can be altered by changing the conditions of the experiment. If the same piece of clothing is now irradiated only with blue light, that light will be completely absorbed, and no light will be reflected. Because in that case our eye would not be able to detect any reflected color from the clothing, we would then conclude that the same piece of clothing was black.

Everyday experience therefore defines the process of reflectance spectroscopy. An object is irradiated with electromagnetic radiation, then the object absorbs whatever wavelengths it is capable of absorbing, the unabsorbed light is reflected, and ultimately quantitated by some sort of optical detector. From the discussions of the previous chapters, it is now completely clear that the criteria for wavelength absorption will be that dictated by the molecular orbitals of the molecules involved in the absorption process.

Transmission spectroscopy has been used for a long time to study the absorption spectroscopy of molecules dissolved in a solution phase, but owing to the opaque nature of most solids, such methodology cannot usually be used to study the electronic spectroscopy of molecules in the solid state. Reflectance spectroscopy takes advantage of the scattering of incident energy off a solid surface, so instead of measuring the light intensity before and after passing through a specimen, one measures the amount of light reflected off the sample surface with the aid of a suitably modified instrumental system.

The most important applications of reflectance spectroscopy in the ultraviolet and visible regions of the spectrum (UV/VIS) have been in the fields of color measurement and color matching, areas of great importance to the dye, paint, paper, textile, and printing industries. The pharmaceutical industry has similar interests in that the use of coloring agents in formulations requires specification. It is logical to conclude that because most pharmaceutical agents are administered as solid dosage forms, the most appropriate forms of UV/VIS spectroscopy would make use of reflectance techniques. As will be seen, reflectance spectroscopy has been used by a number of workers to study the kinetics and mechanisms associated with a variety of reactions, which were found to take place in the solid state.

The scope of this chapter will be mostly concerned with investigations of diffuse reflectance work performed in the UV/VIS region of the spectrum and with colors that can be perceived by the human eye. An enormous amount of very important work has been conducted using irradiation wavelengths in the near-infrared region of the spectrum, but that aspect will be covered elsewhere in this book. Several standard references on UV/VIS reflectance spectroscopy are available (1–5).

CATEGORIES OF LIGHT REFLECTANCE

When light impinges on a nonsmooth surface, its spatial distribution is altered so that it travels in numerous directions, a phenomenon known as *scatter*. One defines the *specular reflectance*, R_S, as the ratio of light reflected according to the laws of regular reflection to that of the incident light energy. The *diffuse reflectance*, R_D, is then defined as the ratio of light diffusely reflected in all directions (apart from that of direct reflection) to that of the incident light energy. The *total reflectance* is all of the light that is reflected by a solid and is defined as the sum of the specular and diffuse reflectances.

Specular reflectance (also called direct or mirror-type reflectance) represents the ordinary reflection of the incident light by the surface(s) of the interacting object(s) and is characterized by a complete absence of light transmission through the interiors of the object(s). The characteristics of reflected energy depend on wavelength, degree of polarization, and angle with respect to the surface. For instance, clear glass will reflect approximately 4% of the incident light if it impinges on the glass surface at a high angle. As the angle of incidence decreases, an increasingly higher percentage of the light becomes reflected until, at the grazing angle, the entire incident light is reflected. The theory associated with specular reflectance is fully outlined in chapter 2 of Ref. (2).

In contrast, diffuse reflectance is associated with the radiation, which penetrates into the particles to some extent and then emerges from the bulk solid. This light will exhibit spectral characteristics, which are modified from those of the incident beam by the wavelength-specific attenuation that took place as a result of induced electronic transitions caused by light absorption at the boundaries of the component particles. The attenuation of the diffuse part of the reflectance caused by absorption can be expressed by the Bouguer-Lambert law:

$$I = I_0 e^{-\varepsilon d} \tag{1}$$

where I is the intensity of diffusively reflected light, I_0 is the incident light intensity, ε is the molar extinction coefficient, and d is the mean thickness of penetration.

Specifically referring to powdered solids, diffusion of light is caused by the effects induced by multiple scattering of the individual particles within the interior of the sample. Equation (1) indicates that the distribution of wavelengths leaving the solid will be modified from that of the incident light by any wavelength-selective absorption within the sample. The electromagnetic radiation, which is diffusely reflected, must penetrate the sample before returning to the surface through multiple scattering. If one defines the depth of penetration of light into a layer as the distance required for the intensity to decrease by the factor, e, then the mean depth is defined by:

$$\bar{d} = 1/k \tag{2}$$

In Equation (2), k is the absorption coefficient of the solid in question and represents the contribution to the index of refraction due to absorption.

The strength of absorption of a sample can be related to the wavelength of the incident light, λ, as long as the same units are used throughout:

$$\text{Strong absorber} \Rightarrow \bar{d} < \lambda \tag{3}$$

$$\text{Weak absorber} \Rightarrow \bar{d} > \lambda \tag{4}$$

From these relationships, one can see that for strong absorbers (i.e., $\bar{d}$ is small), the degree of diffuse reflectance will be small so that the total reflectance will be dominated by the specular reflectance contribution. On the other hand, for weak absorbers (i.e., $\bar{d}$ is large), the degree of diffuse reflectance will be large because such materials can absorb a considerable amount of the incident light owing to the existence of numerous multiple reflections.

The size of the particles making up the interacting layer will also exert a strong effect on the diffuse reflectance. Because samples having a fine particle size distribution will scatter the incident radiation more efficiently than the larger particles, one finds that the absorption of a solid will be more pronounced for an ensemble of large particles than a similar ensemble of small particles. To illustrate this point, the reflectance spectra obtained for the xanthene dye eosin-Y absorbed onto fine particles of anhydrous lactose (i.e., material that passed through a 100-mesh sieve) and onto coarse anhydrous lactose particles (i.e., material that was retained on a 60-mesh sieve) are shown in Figure 1.

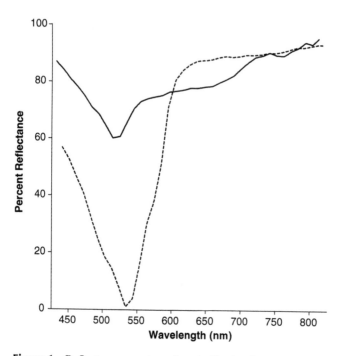

Figure 1 Reflectance spectra of eosin-Y absorbed onto anhydrous lactose passed through a 100-mesh sieve (*solid trace*) and onto anhydrous lactose retained on a 60-mesh sieve (*dashed trace*). The data were obtained using a Varian DMS 200 UV/VIS spectrophotometer fitted with a diffuse reflectance accessory. *Source*: Adapted from HG Brittain, (unpublished results).

Owing to the complementary nature of specular and diffuse reflectances, it is most essential to design experimental conditions for which the diffuse reflectance component of the total reflectance is maximized. High levels of specular reflectance are undesirable in this work, and both the collection optics and sample preparation must be optimized to minimize the effects of specular reflectance.

THEORY ASSOCIATED WITH DIFFUSE REFLECTANCE MEASUREMENTS

The theory associated with diffuse reflectance has been fully developed, and interested readers can obtain complete details in the texts by Wendlandt and Hecht (2), and Kortüm (3). The most useful theory concerning diffuse reflectance was developed by Kubelka and Munk (6,7), the derivation of which was given by Frei and MacNeil (4) in appendix 1 of their text. In this narrative, the essential aspects of the Kubelka-Munk theory will be summarized.

Consider a layer of substance, having a thickness equal to d, that is irradiated by a monochromatic beam of incident light having intensity equal to I_0.

The intensity of the light along the incident direction (taken to be the x-axis) is represented by I, and the light intensity in the negative x-direction (caused by scatter) is represented by J. If the maximum depth of penetration in the layer is dx, then the average path of the radiation $(d\phi)$ for these two components is given by:

$$d\phi_I = dx \int \left[\frac{1}{I \cos \phi} \right] \frac{\delta I}{\delta \phi} \, d\phi \tag{5}$$

and

$$d\phi_I = dx \int \left[\frac{1}{J \cos \phi} \right] \frac{\delta J}{\delta \phi} \, d\phi \tag{6}$$

In Equations (5) and (6), the integration limits are 0 to $\pi/2$, and the quantities $(\delta I/\delta \phi)$ and $(\delta J/\delta \phi)$ represent the angular distribution of the light. Assuming conditions for ideal diffuse radiation:

$$\frac{\delta I}{\delta \phi} = I \sin 2\phi \tag{7}$$

$$\frac{\delta J}{\delta \phi} = J \sin 2\phi \tag{8}$$

If one defines the absorption coefficient (k) as the efficiency of absorbance, and the scattering coefficient (s) as the efficiency of scattering, then the component absorbed in the layer will equal $(kI\,dx)$ and the back-scatter component will equal $(sI\,dx)$. The amount of scattering in the forward direction will equal $(sJ\,dx)$. Therefore, the change in light intensity along the direction of light propagation within the layer will be given by the sum of these factors, or:

$$dI = sJ\,dx - (k + s)I\,dx \tag{9}$$

and the change in light intensity in the negative direction is given by:

$$dJ = (k + s)J\,dx - sI\,dx \tag{10}$$

Integration of Equations (9) and (10) yields:

$$I = A(1 - \beta)e^{\sigma x} + B(1 + \beta)e^{-\sigma x} \tag{11}$$

$$J = A(1 + \beta)e^{\sigma x} + B(1 - \beta)e^{-\sigma x} \tag{12}$$

In Equations (11) and (12), the following definitions have been used:

$$\sigma = [k(k + 2s)]^{1/2} \tag{13}$$

$$\beta = [k/(k + 2s)]^{1/2} \tag{14}$$

The constants A and B of Equations (11) and (12) are determined by the limiting conditions of the experiment. If the entire thickness of the layer is d, then it can be shown that:

$$A = \frac{-[(1 - \beta)e^{-\sigma d}]I_0}{(1 + \beta)^2 e^{\sigma d} - (1 - \beta)^2 e^{-\sigma d}} \tag{15}$$

$$B = \frac{-[(1 + \beta)e^{\sigma d}]I_0}{(1 + \beta)^2 e^{\sigma d} - (1 - \beta)^2 e^{-\sigma d}} \tag{16}$$

The transmission (T) through the layer would be given by I (when $x = d$) divided by I_0, or:

$$T = \frac{4\beta}{(1 + \beta)^2 e^{\sigma d} - (1 - \beta)^2 e^{-\sigma d}} \tag{17}$$

$$= \frac{2\beta}{(1 + \beta)^2 \sin(h\sigma d) + 2\beta \cos(h\sigma d)} \tag{18}$$

The diffuse reflectance (R) of the layer would be given by I (when $x = 0$) divided by I_0, or:

$$R = \frac{(1 - \beta)^2 [e^{\sigma d} - e^{-\sigma d}]}{(1 + \beta)^2 e^{\sigma d} - (1 - \beta)^2 e^{-\sigma d}} \tag{19}$$

$$= \frac{(1 - \beta)^2 \sin(h\sigma d)}{(1 + \beta)^2 \sin(h\sigma d) + 2\beta \cos(h\sigma d)} \tag{20}$$

For the case of pure transmission, $s = 0$ and $\beta = 1$, Equation (17) reduces to the Bouguer-Lambert law quoted previously as Equation (1), and $R = 0$.

For a layer having infinite thickness, d approaches zero, and one obtains R_∞ (the absolute reflectance of an effectively infinitely thick layer) as:

$$R_\infty = \frac{s + k - [k(k + 2s)]^{1/2}}{s} \tag{21}$$

These conditions can be achieved experimentally if four conditions are met. The first of these is that the extent of the horizontal layer is so large relative to its thickness, and so the light diffused horizontally out of the edges of the sample can be neglected relative to the light moving perpendicular to the layer front. The second condition is that the material in the layer is homogeneous in its composition for the entire distance through which the light passes. The third condition is that the light incident on the top of the layer is so perfectly diffused that all points on the surface receive equal irradiation. The final condition is that the top of the sample has the same index of refraction as the medium (typically air) in contact with the sample. Equation (21) is usually rearranged to the more familiar Kubelka-Munk equation:

$$\frac{k}{s} = \frac{(1 - R_\infty)^2}{2R_\infty} \tag{22}$$

To study the UV or VIS absorption spectroscopy of a solid material, the radiation reflected from the surface of the sample is detected and recorded as a function of the incident wavelength. The choice of standard material is subject to the discretion of the investigator but is usually either MgO or $BaSO_4$. The most appropriate standard is one whose matrix permits the most useful data to be obtained on the system of interest. For such work, it is common practice to measure reflectance values obtained at each wavelength relative to the measured reflectance of a standard at that wavelength:

$$R_\infty' = R_\infty(\text{sample})/R_\infty(\text{standard}) \tag{23}$$

and to use R_∞' in the Kubelka-Munk equation. The term on the left side of Equation (22) is often termed the remission function (or the Kubelka-Munk function) and is frequently identified by $f(R_\infty)$. Equation (22) indicates that a linear relationship should exist between $f(R_\infty)$ and the sample absorption spectrum.

Taking the logarithm of the Kubelka-Munk function yields:

$$\log[f(R_\infty)] = \log k - \log s \tag{24}$$

It is concluded that when $\log[f(R_\infty)]$ is plotted against the analyzing wavelength, the resulting diffuse reflectance spectrum will be identical to the transmission spectrum of the compound. The only difference will be a displacement in the ordinate scale by a quantity equal to $-(\log s)$.

It needs to be reemphasized that the form of the Kubelka-Munk equation is valid only for weak absorbing systems, does not allow for any significant contribution from specular reflectance, and requires that the particle size of the powdered sample be relatively small. However, the theory for more complicated systems is available, and the equations pertaining to a wide range of possibilities are available (7). Other approaches to the theory of reflectance spectroscopy have been made using the statistical single particle theory (7–9).

INSTRUMENTATION FOR THE MEASUREMENT OF DIFFUSE REFLECTANCE

Probably, the first instrument designed for reflectance studies was that described by Nutting (10), which was improved by Taylor (11). Since that time, a large number of instruments have been proposed, and numerous modifications were suggested to continually improve signal-to-noise ratios (2–4).

The basic requirements for a diffuse reflectance spectrometer are a light source and a monochromator (to obtain monochromatic radiation), an integrating sphere, and a detection and recording system. The incident energy from the source need not be diffuse in itself, and good results are obtained using direct irradiation of the sample. The light is rendered monochromatic by either a monochromator or filter arrangement, split, and allowed to fall on both the standard and sample faces. The light diffusely reflected by the reference and the sample are collected by the internal reflections of the sphere, and the intensity is ultimately measured by the detector. The data are processed (usually according to the Kubelka-Munk equation) and displayed on a suitable device.

The essential and unique aspect of a diffuse reflectance spectrometer is the integrating sphere, the design of which is shown in Figure 2. The inner surfaces of the sphere are coated with highly reflective materials (typically either MgO or $BaSO_4$) to minimize absorption of the diffuse reflectance. The efficiency of the sphere is defined as the energy lost at the detector port divided by the total energy lost within the sphere. Detailed discussions are available regarding the design of and theory associated with integrating spheres (2–4). For an ideal sphere, the intensity due to reflected light at any point is independent of the spatial distribution within the sphere. Detection of the light intensity at any point is therefore an accurate determination of the diffuse reflectance.

A variety of sample-handling devices can be incorporated into the integrating sphere so as to permit the performance of necessary experiments. For instance, the addition of a hot stage to the sphere allows the reflectance of

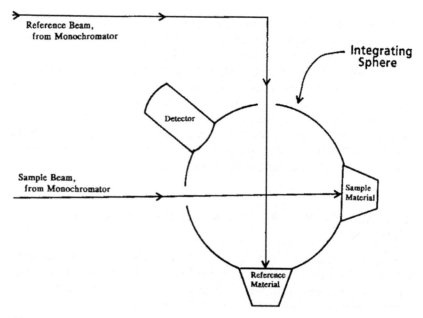

Figure 2 Schematic diagram of the integrating sphere portion of a diffuse reflectance spectrometer, illustrating the key elements of the optical train. Although the detector has been placed in the plane of the sample and reference materials, it would be mounted orthogonal to the plane created by the intersection of the optical beams in common practice.

materials to be obtained at elevated temperatures (12,13). This particular aspect has been found to be especially important in the detection and monitoring of the progress of solid-state chemical reactions. The technique of high-temperature diffuse reflectance spectroscopy becomes particularly useful with the combination of a gas evolution detection technique (14).

To meet the requirements of the Kubelka-Munk theory, the average particle size of the sample should be small and ideally between 0.1 and 1 μm. At particle sizes less than 0.1 μm, the scattering coefficient exhibits an unacceptably strong dependence on the frequency of the incident radiation. For particles larger than 1 μm, the specular reflectance becomes excessively large and depresses the magnitude of the diffuse reflectance spectrum. The effect of particle size on the quality of the diffuse spectrum was illustrated previously in Figure 1. The thickness of the sample layer should be between 1 and 5 mm, ensuring that the path length meets the requirements of the Kubelka-Munk theory so as to be infinitely deep.

QUANTITATIVE MEASUREMENT OF COLOR

Color is basically a perception that is developed in the mind of a given individual, and consequently different people can perceive a particular color in alternate fashions (15–17). Such variability in interpretation causes great difficulty in the evaluation of color-related phenomena and leads to subjective rather than objective judgments. For obvious reasons, the development of a quantitative method for color determination is highly desirable in that a good system would lead to elimination of the subjectivity associated with a visual interpretative measurement (18).

Although other systems have been developed for the quantitative expression of color (19,20), only the CIE (Commission Internationale de l'Éclairage) system and its multiple incarnations have gained general acceptance (22). The evolution of the CIE system and its alternates has been detailed in Refs. (20,23). The CIE system assumes that color may be expressed as the summation of selected spectral components in a three-dimensional manner, with the three primary colors normally added for such purposes being blue, green, and red. The CIE system is based on the fact that human sight is trichromatic in its color perception, and that two stimuli will produce the same color if each of the three tristimulus values (X, Y, and Z) are equal for the two:

$$X = k \int S(\lambda)R(\lambda)x(\lambda)\,d\lambda \tag{25}$$

$$Y = k \int S(\lambda)R(\lambda)y(\lambda)\,d\lambda \tag{26}$$

$$Z = k \int S(\lambda)R(\lambda)z(\lambda)\,d\lambda \tag{27}$$

where:

$$k = \frac{100}{\int S(\lambda)\underline{y}(\lambda)\,d\lambda} \tag{28}$$

In Equations (25–28), $S(\lambda)$ is the spectral power distribution of the illuminant, and $R(\lambda)$ is the spectral reflectance factor of the object. $x(\lambda)$, $y(\lambda)$, and $z(\lambda)$ are the color-matching functions of the observer. In the usual practice, k is defined so that the tristimulus value, Y, for a perfect reflecting diffusor [serving as the reference for $R(\lambda)$] equals 100. Using the functions proposed by the CIE in 1931, $y(\lambda)$ was made identical to the spectral photopic luminous efficiency function, and consequently its tristimulus value, Y, is a measure of the brightness of objects. The X and Z values describe aspects of color, which permit identification with various spectral regions.

The $x(\lambda)$, $y(\lambda)$, and $z(\lambda)$ terms were derived by the CIE from data obtained in visual experiments where observers matched colors obtained by the mixing of the blue, green, and red primary colors. The average result for human observers were defined as the CIE 1931 2° standard observer, and the wavelength dependencies of these color matching functions are illustrated in Figure 3 (23,24).

The proper implementation of the CIE system requires the use of a standard illumination source for calculation of the tristimulus values. Three standard sources were recommended in the 1931 CIE system and may be presented in terms of color temperatures (the temperature at which the color of a

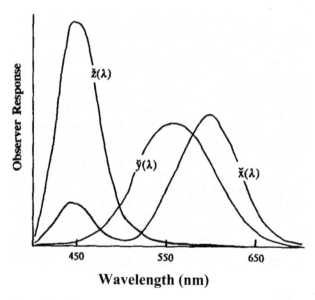

Figure 3   Color matching functions $\underline{x}(\lambda)$, $\underline{y}(\lambda)$, and $\underline{z}(\lambda)$ of the CIE 1931 2° standard observer. *Source*: Adapted from Ref. 23.

black-body radiator matches that of the illuminant). The simplest source is an incandescent lamp, operating at a color temperature of 2856 K. The other two sources are combinations of lamps and solution filters designed to provide the equivalent of sunlight at noon or the daylight associated with an overcast sky. The latter two sources are equivalent to color temperatures of 5000 and 6800 K, respectively.

Because color is a perceived quantity, a strict mathematical relation relating tristimulus values to a concept of color is not possible. However, an equation has been proposed that relates the perception of color to these values. The trichromatic equation for tristimulus values is normally put into the general form:

$$\text{color } (C) = x(X) + y(Y) + z(Z) \tag{29}$$

where x, y, and z represent the chromaticity coordinates of C, obtained via the following relations:

$$x = X/(X + Y + Z) \tag{30}$$
$$y = Y/(X + Y + Z) \tag{31}$$
$$z = Z/(X + Y + Z) \tag{32}$$

Only two of the three chromaticity coordinates need to be actually specified, since:

$$1 = x + y + z \tag{33}$$

A chromaticity diagram can therefore be drawn presenting colors in terms of their x and y coordinates, which are often termed as hue and saturation, respectively. A diagram of this type is presented in Figure 4 (23,24).

In terms of the chromaticity diagram, it can be said that two colors having the same chromaticities (x, y) and the same luminous reflectance (Y) are matched. Technically, no information is provided about the appearance of two matched colors, although subjective judgments about specific colors identified by a pair of (x, y) coordinates are often made.

The central point (marked **W** in Fig. 4) of a chromaticity diagram is termed the achromatic point and represents the combination of x and y values that yield white light. A straight line passing through **W** would connect complementary colors. Combination of the colors specified at the endpoints of this line in the relative amounts indicated by the distances from their respective points to **W** will result in the generation of white light.

The actual determination of color is made with a photoelectric tristimulus colorimeter fitted with a CIE illuminating source. Measurements are made relative to a standard (usually MgO or $BaSO_4$) taken through each of the provided filters. The experimental readings are converted to tristimulus values by means of instrumental factors provided by the instrument manufacturer, and taking the filter specifications into account. The tristimulus values are converted to chromaticity coordinates using Equations (30–32). The accuracy of this

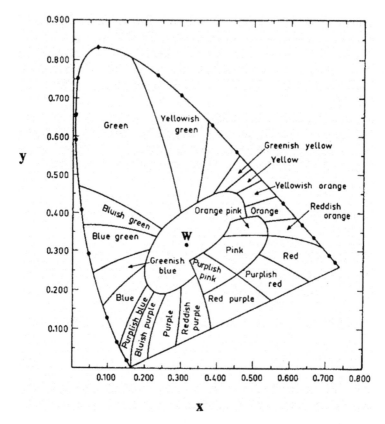

Figure 4 Chromaticity diagram presented in terms of the x and y coordinates. *Source*: Adapted from Ref. 22.

approach is determined by how well the combination of detector and tristimulus filters approximates the characteristics of the CIE standard observer.

With the introduction of computers and microprocessor-controlled instrumentation, it has become possible to use spectrophotometry to obtain far more accurate determinations of color and to obtain tristimulus values after integration of the data according to Equations (25–27). This degree of sophistication permits the use of more advanced methods of color quantitation, such as the 1976 CIE $L^*u^*v^*$ system (25) or other systems.

Raff has provided a discussion summarizing how the technology of the CIE system could be applied to the characterization of materials having pharmaceutical interest (26). At the time when this article was written, the U.S. Food and Drug Administration (FDA) was just becoming concerned about the stability of coloring agents and dyes, and the quantitative CIE system held a definite appeal for those seeking to eliminate subjectivity from such determinations. In this work, Raff provided the reflectance spectra of several pigments and attempted to study the changes in color that took place upon exposure to strong illumination.

In a subsequent work, Raff (27) used the CIE system to quantify the colors that could be obtained when using FD&C aluminum lakes as colorants in tablet formulations. He reported on the concentration dependence of the tristimulus values obtained when calcium sulfate dihydrate was compressed with various amounts of FD&C Blue No. 2 aluminum lake, and one example of the reported data is found in Table 1. The Y tristimulus value may be taken as the relative lightness of the tablet surface, with a value of 1.0 being the maximum. It is evident that increasing colorant concentrations decrease the apparent brightness of the tablets. The chromaticity coordinates, x and y, indicate that the apparent color of the tablets shifts as the colorant levels are raised, and that no limiting color can be reached solely through the use of FD&C Blue No. 2 aluminum lake. Because the perceived color cannot be saturated, careful control of the colorant level in this particular formulation is required to ensure that all batches yield equivalent appearances.

Goodhart et al. (28) used the CIE parameters to evaluate the chromaticity coordinates and brightness of nearly 50 pharmaceutical colorants compressed with lactose. The colorants were categorized as belonging to blue, black, brown, green, orange, red, violet, and yellow classes, and several iron oxide colorants were also studied. The aim of this work was to produce a data set that could be used to match any given color through the combination of pharmaceutically acceptable agents.

In an extension of their work, Goodhart et al. (29) developed a system whereby the final desired color of a compressed table formulation was first chosen from a standard color chart [such as the Munsell compilation (23)]. This color was then analyzed as to its CIE parameters, and these parameters were in turn used to develop a colorant combination that would produce a match of the desired color. The ultimate end of this work was to produce a database of sufficient depth that the empirical nature of color-matching could be eliminated.

Table 1 Tristimulus Values Obtained on a Series of Tablets Colored with FD&C Blue No. 2 Aluminum Lake

Percent dye (w/w)	X	Y	Z	x	y
0.065	0.744	0.762	1.012	0.296	0.303
0.29	0.568	0.582	0.860	0.282	0.289
0.90	0.439	0.447	0.718	0.274	0.279
2.60	0.279	0.280	0.518	0.259	0.260
6.20	0.175	0.172	0.362	0.247	0.243
10.00	0.142	0.138	0.304	0.243	0.236

Source: Adapted from Ref. 27.

Bogdansky (30) had used a combination of the CIE and alternate color systems to deduce color parameters associated with tablet colorants. In this work, thin color dispersions on chromatography paper were made, and the tristimulus values determined through appropriate colorimetry. The materials studied were ordered through their perceived color, and these judgments correlated with the quantitative color parameters. Most importantly, an evaluation procedure was established for the acceptability criteria to be used in judging the range of chromaticity values, which signified equivalence in the color of solid dosage forms produced as different lots. The compilation of tristimulus and chromaticity information was shown to provide a permanent description of a colored sample and to define any difference between an analyte and its standard.

The effect of particle size, and hence dispersion, on the coloring properties of aluminum lake dyes has been studied through quantitative measurement of color in compressed formulations (31). It was found that reduction in the particle size for the input lake material resulted in an increase in color strength, and those particles of submicron size contributed greatly to the observed effects. Analysis of the formulations using the parameters of the 1931 CIE system could only lead to a qualitative estimation of the effects, but the use of the 1976 CIE $L^*u^*v^*$ system provided a superior evaluation of the trends. With the latter system, the effects of dispersion on hue, chroma, lightness, and total color differences were quantitatively related to human visual perception.

The appearance testing of tablets through measurement of color changes has been automated through the use of fiber optic probes and factor analysis of the data (32). Good correlation between measured chromaticity parameters and visual subjective judgment was demonstrated, with samples of differing degrees of whiteness being used to develop the correlation. It was pointed out, however, that surface defects on the analyzed materials could compromise the quality of the correlation and that more sophisticated methods for data evaluation would be useful.

APPLICATIONS OF UV/VIS DIFFUSE REFLECTANCE SPECTROSCOPY TO THE STUDY OF SOLIDS HAVING PHARMACEUTICAL INTEREST

Although diffuse reflectance spectroscopy in the UV/VIS region of the spectrum has not played a major role in the characterization of pharmaceutical solids, it can be used to obtain basic information on formulated products. For instance, the color of formulations containing yellow nifumic acid with a variety of white excipients has been used as a means to evaluate the organization and compactibility of binary powder mixtures (33).

UV/VIS diffuse reflectance spectroscopy has been used, however, with great success to follow the course of reactions for which color changes were directly related to stability-indicating parameters. The most important of these have concerned the color stability of active pharmaceutical ingredients or excipients used in solid dose formulations, where the reflectance spectra were used to evaluate the relative rates of decomposition among the ingredients. Another area of investigation concerns the study of either drug–excipient or excipient–excipient reactions, and here the technique has been particularly useful in those instances where color either develops or disappears as the reaction proceeds.

Studies of Color Stability in Solid Dose Formulations

One of the most important uses for reflectance spectroscopy in the characterization of pharmaceutical solids concerned the stability of coloring agents in tablet formulations. With the description of a device, which enabled the surface of intact tablets to be studied (34), the photostability of various dyes and lakes in tablets was followed (35,36). Exposure of formulations to both normal and exaggerated light conditions was investigated, and the kinetics of the photodegradation evaluated. In most cases, the photoreactions appeared to follow first-order kinetics.

As an example of this type of work, the reflectance spectra of tablets formulated with FD&C Blue No. 1 and exposed to various light intensities are shown in Figure 5. Under normal illumination (45 foot candles), the decomposition kinetics was found to be modest. When the tablets were exposed to higher levels of illumination (550 foot candles), degradation proceeding by at least three different reaction pathways was observed, accompanied by very complicated kinetics (35).

In general, the kinetics associated with the fading of the various dyes studied was found to vary over wide ranges. Lakes prepared from several dyes were found to exhibit less photostability than did the corresponding dyes (36). In most instances, the initial rates of decomposition were significantly faster than the rates observed at later time points. The technique has also been applied to studies of the photostability of naturally sourced colorants in powder beds rather than in compressed products (37).

The ability of various protection schemes to retard the photodecomposition was investigated by means of diffuse reflectance spectroscopy. Glasses of different colors were evaluated as to their ability to retard the fading of tablets colored with FD&C Blue No. 1 and D&C Yellow No. 10 (38). The main conclusion of this work was that amber glass afforded the best protection against the effects of photoillumination, although any glass containing an effective UV-absorber could be equally effective. The incorporation of 2,4-dihydroxy-benzophenone into tablet formulations as an inherent UV-absorber was studied, and its effect on dye fading rates was evaluated (39). It was found that this compound could effectively protect against the photodecomposition reaction, as long as its absorption spectrum matched reasonably well with the dye under question.

In a detailed study of the color stability of tablets containing FD&C Red No. 3 dye, the ability of a variety of sunscreen agents to retard the color loss

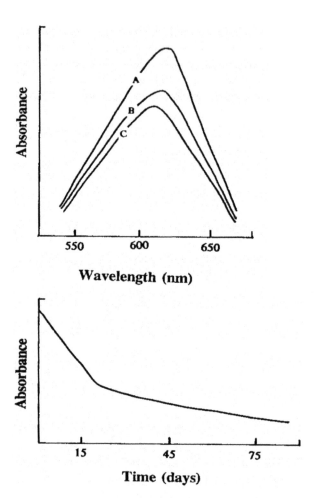

Figure 5   Reflectance spectra of tablets formulated with FD&C Blue No. 1 and exposed to ordinary illumination for various time periods. Full spectra are shown in the *upper box* for tablets (A) as initially prepared, and after (B) 42 days, and (C) 84 days of illumination. The dependence of dye absorbance on elapsed time is shown in the *lower trace*. *Source*: Adapted from Ref. 34.

was investigated (40). The agents were incorporated in films coated upon the model tablet formulation, and the tablets subsequently exposed to light of 1000 foot candles. The film coating method was not found to yield acceptable stability when using either glyceryl *p*-aminobenzoate or salicylates as sunscreens. The greatest protection against fading was observed when 2-ethoxyethyl *p*-methoxy-cinnamate was used as the protection agent.

The stability of a wide variety of certified dyes in tablets has been evaluated upon exposure to various illumination conditions (41–43). In these works, more detailed studies of dye concentration, exposure time, and irradiation light intensity were performed, and the data analyzed more rigorously. The effect of varying tablet excipients was also investigated with respect to the decomposition kinetics. This work finalized with the development of a systematic approach for the testing of color stability, and the investigators developed a ranking of colorant stabilities in tablets (43).

Studies of Solid-State Reactions in Formulated Materials

It was recognized very early that diffuse reflectance spectroscopy could be used to study the interactions of various compounds in a formulation, and the technique has been particularly useful in the characterization of solid-state reactions (44). Lach concluded that diffuse reflectance spectroscopy could also be used to verify the potency of a drug in its formulation. In addition, studies conducted under stress conditions would be useful in the study of drug–excipient interactions, drug degradation pathways, and alterations in bioavailability owing to chemisorption of the drug onto other components in the formulation (44).

In a series of works, Lach et al. (45–47) studied the solid-state interactions of a variety of compounds with various adjuvants. Working predominantly with compounds containing conjugated aromatic systems (oxytetracycline, anthracene, phenothiazine, salicylic acid, prednisone, and hydrochlorothiazide), it was deduced that the complexes formed with the adjuvants were of the donor–acceptor type. Large bathochromic and hyperchromic spectral shifts in absorption maxima upon complexation were inferred to be the result of strong charge transfer interactions. In the specific instance where the drug possessed a chelation center and the adjuvant contained a metallic center, the site-specific complexation could be identified as the source of the change in the reflectance spectrum (47,48).

The stability of ascorbic acid formulations has been successfully studied using diffuse reflectance spectroscopy because the degradation is accompanied by a color reaction, and this color change is directly related to the potency of the active component (49,50). Tablets stored at ambient conditions will gradually age from white to a yellowish-brown color, with the degree of stability being greatly affected by the excipients in the formulation. It was determined that materials containing metal ions (i.e., magnesium and calcium stearate, or talc) accelerated the color reaction, whereas metal-free excipients (such as stearic acid or hydrogenated vegetable oil) imparted maximal color stability (50). The effect of temperature and relative humidity on the stability of the formulations was also followed using diffuse reflectance spectroscopy, and it was found that the effect of elevated humidity was more important than the effect of elevated temperatures.

Diffuse reflectance spectroscopy was used to screen the possible interactions between a large number of adjuvants and several dyes (51). It was concluded that supposedly inert excipients (such as starch or lactose) were capable of undergoing significant reactions with the dyes investigated (Red No. 3, Blue No. 1, and Yellow No. 5). For adjuvants containing metal ions (zinc oxide, or calcium, magnesium, and aluminum hydroxides), the degree of interaction could be considerable. It was concluded from these studies that dye–excipient interactions could also be responsible for the lack of color stability in certain tablet formulations.

The reaction of a primary amine with lactose is accompanied by a browning of the solids, and the path of such reactions is easily followed by means of diffuse reflectance spectroscopy. For instance, the reaction of isonicotinic acid hydrazide (isoniazid) with lactose could be followed through changes in the reflectance spectrum (52). As may be seen in Figure 6, as the sample was heated for increasing amounts of time a steady decrease in reflectance was noted. The spectral data were used to deduce the rate constants for the browning reaction at various heating temperatures, and these rates could be correlated with those obtained through chemical analysis (formation of isonicotinoyl hydrazone of lactose) of the mixtures. In addition, the active was found to both chemisorb as well as physisorb onto magnesium oxide, with accompanying changes in the reflectance spectrum (31).

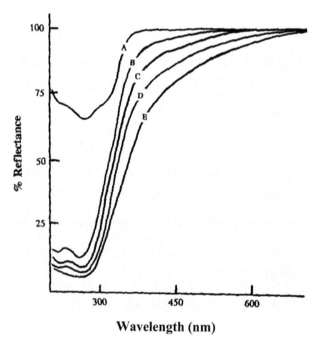

Wavelength (nm)

Figure 6 Reflectance spectra of the reaction products formed when isonicotinic acid hydrazide reacts with lactose. Reflectance spectra are shown for (A) the initially prepared material, and for samples illuminated for (B) 4 hours, (C) 10 hours, (D) 21 hours, and (E) 44 hours. *Source*: Adapted from Ref. 52.

A similar study has been conducted in which the interaction of *d*-amphetamine sulfate with spray-dried lactose was investigated (53). Upon storage at elevated temperatures, discoloration of the powder blends was noted, and the new absorption bands were characterized. A maximum was noted at 340 nm and was attributed to the chemisorption of the amine onto the lactose particles. The other band appeared at 295 nm and was attributed to the new compound (*d*-amphetamine-hydroxymethylfurfural) formed as a result of the reaction between the two. Through the use of Arrhenius plots, the browning rate anticipated for a temperature of 25°C was predicted.

The interaction of tetracycline and its derivatives with excipients has been studied by diffuse reflectance spectroscopy (54,55). Adjuvants containing metal ions were found to yield the largest degree of change in the reflectance spectra, with the effects of calcium being the most prominent. Complexation of the metal ion with specific functional groups on the drug molecules was postulated as the origin of these effects. The complexation would be anticipated to influence the bioavailability of the compound, and therefore the technique is demonstrated to provide useful information regarding the degree of drug–excipient compatibility.

Other workers have used the tristimulus parameters to study the kinetics of decomposition reactions. The fading of tablet colorants was shown to follow first-order reaction kinetics, with the source of the illumination energy apparently not affecting the kinetics (49). The effect of excipients on the discoloration of ascorbic acid in tablet formulations has also been followed through determination of color changes (57). In this latter work, it was established that lactose and Emdex® influenced color changes less than sorbitol.

Compounds known to undergo changes in their absorption spectra upon sorption onto a solid surface are termed adsorptiochromic, and such effects would be ideally studied by means of diffuse reflectance spectroscopy. In one such study, the absorption of various spiropyrans onto many different solids was investigated (58,59). For the compounds studied, the reflectance spectra were dominated by bands at 550 nm and in the range of 400–500 nm (most often at 472-nm). As an example, the reflectance spectra obtained for 6-nitrobenxospiropyran are shown in Figure 7. When the difference spectrum was taken between the spectrum of the pure compound and that obtained after sorption onto silicic acid, the bands characteristic of the adsorbed species were clearly evident.

The interaction between drug compounds and excipients, as these influence drug dissolution, can be successfully studied by means of reflectance spectroscopy. In one study concerning probucol and indomethacin, it was deduced that hydrogen bonding and van der Waals' forces determined the physisorption between the active and the excipients in several model formulations (60). Chemisorption forces were found to play only minor roles in these interactions. These studies indicated that surface catalytic effects could be important during the selection of formulation excipients.

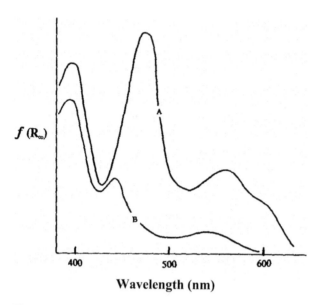

$f(R_\infty)$

Wavelength (nm)

Figure 7 Diffuse reflectance spectra of (A) pure 6-nitrobenxospiropyran, and (B) after its sorption onto silicic acid. *Source*: Adapted from Ref. 58.

The appearance of tablets and powders during accelerated stability testing can be quantified using tristimulus colorimetry (61). In this work, various formulations were stored under stress conditions, and the tristimulus parameters of the tablets were measured at various time points during the storage time. Because the particular systems studied were characterized by changes in tablet color as the decomposition reaction proceeded, the X, Y, and Z parameters were judged useful in deducing the kinetics of the reactions (Fig. 8). Both first-order or zero-order reactions were noted for the systems studied, and a formal Arrhenius treatment of the data could be used to predict shelf lives for the formulations.

Instrumental evaluation of the color of solid dosage forms has been incorporated into the stability testing protocol of a number of pharmaceutical products, including formulations of captopril, flucloxacillin sodium, cefoxitin sodium, and theophylline (62). The rate of color formulation in the dosage forms obeyed the Arrhenius equation is each case, and the CIE parameters were used as part of potency predictions when a relationship between the color change and degree of drug decomposition has been established.

Taking the application of quantitative color one step further, color parameters were shown to be a quantifiable variable for the degree of whiteness of uncoated tablets (63). Color intensity parameters related to yellowish or brownish colors were defined that were used to describe the discoloration kinetics with acceptable regression coefficients. The discoloration rates determined under several storage temperatures were found to follow Arrhenius kinetics, and

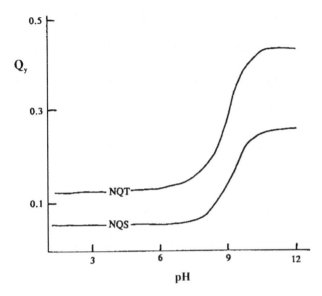

Figure 8 pH dependence of the Q_y chromaticity parameters obtained during the poten-tiometric titration of *Abbreviations*: NQS, 1,2-naphthoquinone-2-semicarbazone; NQT, 1,2-naphthoquinone-2-thiosemicarbazone. *Source*: Adapted from Ref. 61.

therefore the activation energy could be estimated for the products. It was deter-mined that through the use of the color intensity parameters, it was statistically possible to determine the period of time for which the uncoated tablets remained white.

USE OF ADSORBED INDICATORS FOR THE EVALUATION OF SURFACE ACIDITY AND BASICITY

Although the relative degree of acidity or basicity of a substance dissolved in an aqueous medium can be adequately defined in terms of the solution pH, a com-parable expression for the acidity or basicity of the surface of a solid is not as easily expressed. However, such an evaluation would be of great importance to a preformulation study, as it is well known that acidic or basic surfaces of solids are eminently suitable as catalytic agents (64). In the absence of such knowledge, formulators usually assume certain characteristics of the materials they are working with and then base their compounding on such assumptions.

The concept of the acid strength of a surface was first discussed in a sys-tematic manner by Walling (65), who sought to differentiate the acid strength of a surface from the stoichiometry of acidic sites. Walling extended the indicator theory developed by Hammett to explain the color changes associated with certain substances as the acidity of their medium was changed (66–68) to solids. In particular, Walling defined the acidity of a solid surface as the ability of the surface to convert an adsorbed neutral base into its conjugate acid. In order to see how the Walling model can be applied to solids, it is first necessary to work through some details of ionic equilibria.

Consider the hydrolysis reaction of a weak acid with water:

$$HA + H_2O \longleftrightarrow H_3O^+ + A^- \quad (34)$$

The position of equilibrium at a given temperature is defined by the magnitude of the equilibrium constant expression:

$$K = \frac{a_{H_3O^+} \cdot a_{a^-}}{a_{HA} a_{H_2O}} \quad (35)$$

where the amount of the ith interacting species is specified in terms of its activity, a_i. Owing to the fact that the mole fraction of the solvent in most aqueous solutions is effectively unity, the activity of water (a_{H_2O}) in such solutions equals one. In that case, the ionization constant expression becomes:

$$K_a = \frac{a_{H_3O^+} \cdot a_{A^-}}{a_{HA}} \quad (36)$$

The activity of the ith interacting species in a solution is related to its molar concentration, $[i]$, by the relation:

$$a_i = [i] f_i \quad (37)$$

where f_i is a dimensionless quantity known as the activity coefficient. The activity coefficient (and consequently the activity) of a species varies with the ionic strength of the medium such that the use of activities instead of molar concentrations in expressions of constant equilibrium frees the numerical value of the constant from its dependence on ionic strength. Therefore, the ionization constant expression for a hypothetical weak acid can also be written as:

$$K_a = \left[\frac{[H_3O^+][A^-]}{[HA]}\right]\left[\frac{f_{H_3O^+} \cdot f_{A^-}}{f_{HA}}\right] \quad (38)$$

The activity coefficients in Equation (38) vary with the overall ionic strength to keep the value of K_a numerically constant over a range of ionic strengths.

When the ionic strength of the medium is low (i.e., high dilution), all of the activity coefficients in Equation (38) approach unity, and the ionization constant expression becomes:

$$K_a' = \frac{[H_3O^+][A^-]}{[HA]} \quad (39)$$

It is important to recognize that when an ionization constant is expressed in terms of molar concentrations the value for K_a' will be a function of ionic strength. For instance, K_a' of acetic acid equals 1.75×10^{-5} at an ionic strength of 0.0,

2.85×10^{-5} at an ionic strength of 0.1, and 3.31×10^{-5} at an ionic strength of 0.5.

Using the "p" scale of Sørensen, where the factors in constant equilibrium expressions are expressed as the negative of their base 10 logarithms, Equation (39) can be arranged as:

$$-\log[H_3O^+] = pK'_a + \log\{[A^-]/[HA]\} \tag{40}$$

When the weak acid HA is an indicator, the concentration ratio $[A^-]/[HA]$ can be determined using colorimetry. In aqueous solutions, the left-hand side of Equation (40) is the pH, and Walling (65) defined that quantity as the acid strength, H_O, for solids. The H_O function can be equally defined as:

$$H_O = -a_{H_3O^+}(f_{A^-}/f_{HA}) \tag{41}$$

It is important to remember that decreasing values of H_O signify increasing acid strength. If the definition of acid strength is applicable to surfaces, then the acid strength of a surface should be readily measurable by the observation of the colors of absorbed indicators. In particular, the appearance of the color characteristic of the acid form of the indicator would be taken as indicating a value of the H_O function for the surface lower than the pK_a of the indicator used.

Walling (65) pointed out a number of requirements essential for the successful use of basic indicators for the determination of the acid strength of surfaces:

1. The indicators used are restricted to those for which the basic form is neutral and that are converted to their conjugate acids by the addition of a single proton.
2. The indicators must be capable of adsorbing onto the surface under investigation.
3. The solvent used to interact with the indicator with the solid surface must not perturb the properties of the surface itself.
4. The method requires substrate surfaces whose inherent color does not interfere with the color determination of the adsorbed indicator.

Walling then continued to define the indicator types that would eventually be termed the Hammett indicators. Any substance that is neutral or slightly basic in water, that has a different color in concentrated sulfuric acid, and that is an 1:1 electrolyte can act as a basic indicator at acidity strengths between those of water and sulfuric acid. Furthermore, the substance must owe its color change to protonation by a single proton and not by the addition of water or some other molecule. In the original work, 15 such indicators were indicated as having appropriate properties (65).

Benesi extended the methodology over a much wider range and used the color changes of adsorbed indicators to evaluate the acid strength of a variety of catalyst surfaces (69). Surprisingly, it was found that carefully dried samples of many solids yielded the acid response with all tested indicators, demonstrating that the tested material surfaces (metal oxides, acids supported on silica gel, clays, and cracking catalysts) were considerably more acidic than had been anticipated. In fact, the surfaces of some of the clay and catalysts were found to be so acidic that they could not be completely neutralized with sodium hydroxide.

Around the same time, the use of Hammett indicators in the quantitation of acid sites on the surfaces of acidic materials by amine titration became established (70,71). In this method, one first sorbs the indicator onto the surface of the substance to be titrated, which places the indicator into its acidic form as indicated by its characteristic color. One then titrates the solids with a stronger base that serves to displace the indicator and monitors the course of the reaction on the basis of the color change associated with the released indicator (which is then in its basic form, having a different color).

It follows that if the material surface under investigation is basic in character, then titration with a strong acid would be appropriate to evaluate the acid strength and stoichiometry of binding sites. For example, a series of solids have been suspended in benzene, and then titrated with trichloroacetic acid in the presence of Hammett indicators (72). This approach enabled the determination of the basic strength of a surface in terms of the Hammett acidity function, facilitating the evaluation of H_O across the entire range. It was found that the basicity of the studied solids increased across the series $ZnO > TiO_2 > \gamma\text{-}Al_2O_3 > BaO > B_2O3 > ZrO_2$.

Although the Hammett indicators could be used to provide visual determinations of the acidity or basicity of surfaces, the use of reflectance spectroscopy facilitated an extension of these studies. For instance, Pichat used UV reflectance spectroscopy to study the character of the binding sites in the adsorption of three diazine molecules on silica, alumina, and silica-alumina (73). In the case of silica or alumina, the spectral bands of adsorbed diazines were comparable with those obtained for the molecules dissolved in a hydrocarbon solvent, and it was concluded that the adsorption sites involved surface hydroxyl groups similar to those existing in aqueous or alcoholic solutions. However, significant shifts in the wavelength of the $n \rightarrow \pi^*$ transitions of the diazines were observed when the molecules were sorbed onto silica-alumina, confirming the strong protonic acidity of this material.

In a series of studies, Leermakers et al. (74–77) used UV spectroscopy to study the electronic spectra and structure of a number of molecules absorbed onto silica gel. Because the n,π^* excited state is less polar than the ground state molecule, then after sorption onto the support, the $n \rightarrow \pi^*$ transition would be observed at higher energy and therefore shifted to shorter wavelengths. Conversely, because the π,π^* excited is more polar than the ground state, then the $\pi \rightarrow \pi^*$ transition would be shifted to lower energies and hence longer wavelengths. This trend was rationalized on the basis that activated silica gel contains highly polar protonated sites capable of interacting with the oxygen atoms of the carbonyl groups, and these groups would cause a greater perturbation on the π^* excited state. This effect is illustrated in Figure 9 for the particular instance of biacetyl.

Limitations in the indicator methods exist, however, and have been found to be particularly important in the study of the surface acidity of zeolites (78,79). In these materials, the acid strength of the surface can be readily established through the use of Hammett indicators, but the quantitative titration of acid sites turned out to be strongly affected by the ability of the indicator dye to enter the zeolite cavities. A completely potassium-exchanged yttrium-zeolite was not found to exhibit observable acidity, but lanthanum- and gadolinium-exchanged zeolites were found to be strongly acidic even at low degrees of exchange (80).

Jozwiakowski and Connors (81) used a slightly different approach to study the surface polarity of silica samples. The samples to be characterized were suspended in glycerin and then allowed to interact with the 6-nitrobenzoindolino-pyran as the adsorption indicator. Because the powders remained in suspension, conventional UV spectroscopy was used to evaluate the color changes. It was found that the absorption maximum of the indicator could be related to the surface polarity of the silica samples by a relatively simple relationship.

Schmidt et al. have carried out a series of determinations of the surface acidity of pharmaceutical excipients, using a variety of absorption indicators to deduce the acid strengths. In the first study, the surface acidity of microcrystalline cellulose and dicalcium phosphate anhydrate was evaluated using thymol blue,

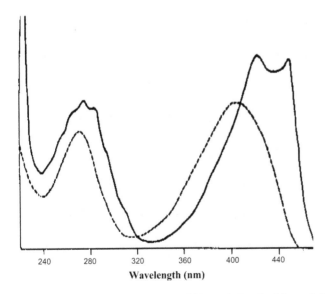

Figure 9 UV absorption spectra of biacetyl dissolved in cyclohexane (*solid trace*) and absorbed onto silica gel (*dashed trace*). *Source*: Adapted from Ref. 74.

bromocresol green, bromocresol purple, and phenol red as the absorption indicators (82). As shown in Figure 10, bromocresol green exists primarily in its acid form when adsorbed onto anhydrous dicalcium phosphate but is distributed between its acidic and basic forms when adsorbed onto microcrystalline cellulose. Using the pH dependence of dye absorbance published in the paper, one would conclude that H_O would be approximately 1.0 for dicalcium phosphate anhydrate and approximately 4.0 for microcrystalline cellulose. Granules were also prepared by the buffering of anhydrous dicalcium phosphate with a phosphate/citrate/borate system, and the H_O of the buffered granules could be estimated as approximately 7.0.

In a subsequent work, the effect of excipient surface acidity on the decomposition rate of acetylsalicylic acid was studied (83). The surface of anhydrous dicalcium phosphate was buffered to varying degrees, and then the surface acidity determined using the appropriate absorption indicators. A methanolic solution of acetylsalicylic acid was used to coat the granules, and then the dried coated granules were stored under various conditions to induce decomposition of the drug substance. It was found that the surface acidity decomposition-rate profiles were similar to the pH-rate profiles of acetylsalicylic acid in solution, indicating that the decomposition took place in water layers within the coated granules.

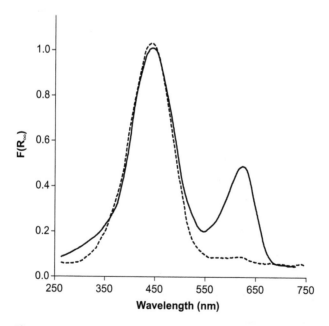

Figure 10   Diffuse reflectance spectra of bromocresol green adsorbed onto microcrystalline cellulose (*solid trace*) and onto dicalcium phosphate anhydrate (*dashed trace*). *Source*: Adapted from Ref. 82.

The surface acidities of a wide variety of excipients for solid dosage forms have been determined using absorption indicators (84). A surprisingly wide number of fillers, binders, disintegrants, and lubricants were found to exhibit H_O values in the range of 3–4, and only a few excipients exhibited H_O values larger than this. In fact, potassium bicarbonate ($H_O = 8.05$), sodium bicarbonate ($H_O = 8.27$), and sodium carbonate ($H_O = 8.58$) were found to be the only excipients studied whose surfaces exhibited basic properties.

Referring back to Figure 1, from the wavelength of the absorption maxima (approximately 525 nm) it can be concluded that the sorbed eosin-Y is present on the lactose samples in its acid form. Because the pKa of this dye is approximately 3.0, this finding implies that the H_O value of the lactose is less than 3.0.

SUMMARY

Although UV/VIS diffuse reflectance spectroscopy has not been used as extensively in the study of pharmaceutical solids as have other spectroscopic techniques, its applications have been sufficiently numerous that the power of the methodology is evident. The full reflectance spectra, or the derived colorimetry parameters, can be very useful in the study of solids, which are characterized by color detectable by the human eye. It is evident that questions pertaining to the colorants used for identification purposes in tablet formulations can be fully answered through the use of appropriately designed diffuse reflectance spectral experiments.

REFERENCES

1. Hardy AC. Handbook of Colorimetry. Cambridge, MA: M.I.T. Press, 1936.
2. Wendlandt WW, Hecht HG. Reflectance Spectroscopy. New York: Interscience Pub., 1966.
3. Kortüm G. Reflectance Spectroscopy: Principles, Methods, Applications. New York: Springer-Verlag, 1969.
4. Frei RW, MacNeil JD. Diffuse Reflectance Spectroscopy in Environmental Problem Solving. Cleveland OH: CRC Press, 1973.
5. Frei RW, Frodyma MM, Lieu VT. Diffuse reflectance spectroscopy. In: Svehla G, ed. Comprehensive Analytical Chemistry. Vol. IV, Chapter 3. Amsterdam: Elsevier, 1975: 263–354.
6. Kubelka P, Munk F. Z Tech Physik 1931; 12:593.
7. Kubelka P. J Opt Sci Am 1948; 38:448.
8. Melamed NT. J Appl Phys 1963; 34:560.
9. Simmons EL. Opt Acta 1971; 18:59.
10. Nutting PG. Trans Illum Eng Soc 1912; 7:412.
11. Taylor AH. J Opt Sci Am 1919; 4:9.
12. Wendlandt WW, Franke PH, Smith JP. Anal Chem 1963; 35:105.
13. Wendlandt WW. Pure Appl Chem 1971; 25:825.
14. Wendlandt WW, Bradley WS. Thermochim Acta 1970; 1:143.

15. Evans RM. An Introduction to Color. New York: John Wiley & Sons, 1948.
16. Committee on Colorimetry, Optical Society of America, The Science of Color. New York: Thomas Y. Crowell Co., 1953.
17. Wright WD. The Measurement of Color. New York: Macmillan Co., 1958.
18. Judd DB, Wyszecki G. Color in Business, Science, and Industry. 2nd edn. New York: John Wiley & Sons, 1963.
19. Kelly KL, Judd DB. The ISCC-NBS Method of Designating Colors and a Dictionary of Color Names. National Bureau of Standards Circular 553, Washington, D.C.: U.S. Government Printing Office, 1955.
20. Wyszecki G, Stiles WS. Color Science: Concepts and Methods, Quantitative Data and Formulae. 2nd edn. New York: John Wiley & Sons, 1982.
21. Hunter RS, Harold RW. The Measurement of Appearance. New York: Wiley-Interscience, 1987.
22. CIE, International Conference on Illumination, Proceedings of the Eighth Session, Cambridge, UK, Bureau Centrale de la CIE, Paris, 1931.
23. Billmeyer FW Jr, Saltzman M. Principles of Color Technology. New York: Wiley-Interscience, 1966.
24. Berns RS. Billmeyer and Saltzman's Principles of Color Technology. 3rd edn. New York: Wiley-Interscience, 2000.
25. CIE Publication 15.2, Colorimetry. 2nd edn. Central Bureau of CIE, Austria: Vienna, 1986.
26. Raff AM. J Pharm Sci 1963; 52:291.
27. Raff AM. J. Pharm Sci 1964; 53:380.
28. Everhard ME, Dickcius DA, Goodhart FW. J Pharm Sci 1964; 53:173.
29. Goodhart FW, Kelly MA, Lieferman HA. J Pharm Sci 1965; 54:1799.
30. Bogdansky FM. J Pharm Sci 1975; 64:323.
31. Wou LS, Mulley BA. J Pharm Sci 1988; 77:866.
32. Wirth M. J Pharm Sci 1991; 80:1177.
33. Barra J, Ullrich A, Falson-Rieg F, Doelker E. Pharm Dev Tech 2000; 5:87.
34. Urbanyi T, Swartz CJ, Lachman L. J Am Pharm Assoc Sci Ed 1960; 49:163.
35. Lachman L, Swartz CJ, Urbanyi T, Cooper J. J Am Pharm Assoc Sci Ed 1960; 49:165.
36. Lachman L, Weinstein S, Swartz CJ, Urbanyi T, Cooper J. J Pharm Sci 1961; 50:141.
37. Dehner EJ, Shiromani PK. Drug Dev Indust Pharm 1993; 19:1659.
38. Swartz CJ, Lachman L, Urbanyi T, Cooper J. J Pharm Sci 1961; 50:145.
39. Lachman L, Urbanyi T, Weinstein S, Cooper J, Swartz CJ. J Pharm Sci 1962; 51:321.
40. Hajratwala BR. J Pharm Sci 1974; 63:129.
41. Everhard ME, Goodhart FW. J Pharm Sci 1963; 52:281.
42. Goodhart FW, Everhard ME, Dickcius DA. J Pharm Sci 1964; 53:338.
43. Goodhart FW, Leiberman HA, Mody DS, Ninger FC. J Pharm Sci 1967; 56:63.
44. Pope DG, Lach JL. Pharm Acta Helv 1975; 50:165.
45. Lach JL, Bornstein M. J Pharm Sci 1965; 54:1730.
46. Bornstein M, Lach, JL. J Pharm Sci 1966; 55:1033.
47. Lach JL, Bornstein M. J Pharm Sci 1966; 55:1040.
48. Bornstein M, Lach JL, Munden BJ. J Pharm Sci 1968; 57:1653.
49. Carstensen JT, Johnson J, Valentine W, Vance J. J Pharm Sci 1964; 53:1050.
50. Wortz RB. J Pharm Sci 1967; 56:1169.
51. Bornstein M, Walsh JP, Munden BJ, Lach JL. J Pharm Sci 1967; 56:1410.

52. Wu W-H, Chin T-F, Lach JL. J Pharm Sci 1970; 59:1286.
53. Blaug SM, Huang W-T. J Pharm Sci 1972; 61:1770.
54. Lach JL, Bighley LD. J Pharm Sci 1970; 59:1261.
55. McCallister JD, Chin T-F, Lach JL. J Pharm Sci 1970; 59:1286.
56. Turi P, Brusco D, Maulding HV, Tausendfreund RA, Michaelis AF. J Pharm Sci 1972; 61:1811.
57. Vemuri S, Taracatac C, Skluzacek R. Drug Dev Indust Pharm 1985; 11:207.
58. Connors KM, Jozwiakowski MJ. J Pharm Sci 1987; 76:892.
59. Jozwiakowski MJ, Connors KM. J Pharm Sci 1988; 77:241.
60. Monkhouse DC, Lach JL. J Pharm Sci 1972; 61:1435.
61. Bosch E, Casassas E, Izquierdo A, Rosés M. Anal Chem 1984; 56:1422.
62. Stark G, Fawcett JP, Tucker IG, Weacherall IL. Int J Pharm 1996; 143:93.
63. Berberich J, Dee K-H, Hayauchi Y, Portner C. Int J Pharm 2002; 234:55.
64. Tanabe K. Solids Acids and Bases. New York: Academic Press, 1970.
65. Walling C. J Am Chem Soc 1950; 72:1164.
66. Hammett LP, Deyrup AJ. J Am Chem Soc 1932; 54:2721.
67. Hammett LP, Deyrup AJ. J Am Chem Soc 1932; 54:4239.
68. Hammett LP, Paul MA. J Am Chem Soc 1934; 56:827.
69. Benesi HA. J Am Chem Soc 1956; 78:5490.
70. Johnson O. J Phys Chem 1955; 59:827.
71. Benesi HA. J Phys Chem 1957; 61:970.
72. Yamanaka T, Tanabe K. J Phys Chem 1975; 79:2409.
73. Pichat P. J Phys Chem 1974; 78:2376.
74. Leermakers PA, Thomas HT. J Am Chem Soc 1965; 87:1620.
75. Leermakers PA, Thomas HT, Weis LD, James C. J Am Chem Soc 1966; 88:5075.
76. Evans TR, Toth AF, Leermakers PA. J Am Chem Soc 1967; 89:5060.
77. Weis LD, Evans TR, Leermakers PA. J Am Chem Soc 1968; 90:6109.
78. Kladnig WF. J Phys Chem 1979; 83:765.
79. Barthomeuf D. J Phys Chem 1979; 83:767.
80. Kladnig WF. J Phys Chem 1976; 80:262.
81. Jozwiakowski MJ, Connors KA. Pharm Res 1987; 4:398.
82. Glombitza BW, Oelkrug D, Schmidt PC. Eur J Pharm Biopharm 1994; 40:289.
83. Glombitza BW, Schmidt PC. Eur J Pharm Biopharm 1995; 41:114.
84. Scheef C-A, Oelkrug D, Schmidt PC. Eur J Pharm Biopharm 1998; 46:209.

6

Luminescence Spectroscopy

Harry G. Brittain

Center for Pharmaceutical Physics, Milford, New Jersey, U.S.A.

INTRODUCTION

Luminescence is a general term for the spontaneous emission of electromagnetic radiation by a material. This results from the transition of molecules from their electronically excited states down to their ground states. Different types of luminescence were historically differentiated on the basis of their respective timescales, but now their primary classification is derived from changes in spin angular momenta. Luminescence is most often thought of as being initiated by electromagnetic radiation having wavelengths in the X-ray, ultraviolet (UV), or visible (VIS) regions of the spectrum, with such processes being grouped under the category of photoluminescence. However, luminescence is initiated by a variety of other sources, such as ionizing radiation (scintillation), electron beams (cathodoluminescence), heat (thermoluminescence), chemical reaction (chemiluminescence), electrical current (electroluminescence), or mechanical impact (triboluminescence). Generally, as long as the resulting emission spectrum originates from transitions among electronic states of the molecule, it will exhibit the same wavelength dependence regardless of the mode of excitation.

The subject of molecular luminescence has been discussed at great length, with special emphasis being placed on the study of photophysical processes in fluid solution (1–25). Many of these general concepts are of great importance to any investigation of the luminescence phenomena in solids, such as the excitation spectrum, the emission spectrum, the quantum yield of luminescence, and the lifetime of emission. When dealing with spectroscopic investigations of fluorophores in single crystals, the polarization of the luminescence is also found to yield important information. A comprehensive summary of the older literature is available for those workers who are involved in the research of the early history of this field (26).

PHOTOPHYSICAL PROCESSES IN ISOLATED ORGANIC MOLECULES

In chapter 4, the LCAO-MO theory was used to develop an understanding of the ground and excited states of a molecule, and the conditions were established under which a given molecule might undergo an absorption transition from its ground electronic state to one of its excited electronic states. In that discussion, the consequences of the phrase "what goes up must come down" were not considered, but it is clear that if molecules only possessed the capability to absorb energy, then sooner or later the absorption process would cease when all of the molecules resided in their excited states. Although such population inversions can be created under certain conditions (i.e., for laser-associated processes), this situation is not ordinarily realized. Several mechanisms exist that facilitate deactivation of the molecule from its excited state and back to its ground state.

In the absence of excitation, the ground state of a molecule can be described as an electronic wave function onto which is superimposed a sequence

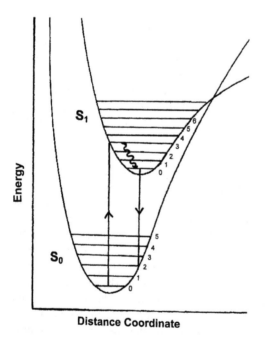

Figure 1 Jablonski-type energy level diagram of the ground and first excited state of a fluorescent organic molecule. Excitation is drawn as the $S_0(v = 0) \to S_1(v = 3)$ transition, radiationless deactivation of the initial state is the $S_1(v = 3) \to S_1(v = 0)$ transition, and fluorescence is the $S_1(v = 0) \to S_0(v = 2)$ transition. The *vertical lines* correspond to transitions characterized by the most favorable Franck-Condon overlap.

of vibrational wave functions. This has been illustrated in Figure 1 by horizontal levels proper to each curve (rotational states have not been included for the sake of clarity in the figure). Due to the magnitude of the energetics involved, a molecule in thermal equilibrium at ambient temperature will exist largely in the lowest vibrational state of its ground state (i.e., the $v = 0$ level). For the vast majority of organic molecules, this state will have all electron spins paired and hence its spin multiplicity will be $2S + 1 = 0$, which is identified as a singlet state. In this discussion, the singlet ground state will be identified as the state S_0.

When a molecule is irradiated with light that has an energy matching a separation between states, it is absorbed in a time frame of approximately 10^{-15} seconds. Because the strict spin selection rule is $\Delta S = 0$, excitation into the excited singlet states (S_1, S_2, S_3, etc.) is only possible, and only then the orbital angular momentum selection rules are also satisfied. Of course, these $S_0 \rightarrow S_i$ transitions are the same as those transitions among electronic states discussed in a previous chapter. Because the absorption process is the action that places molecules in excited electronic states, the wavelength dependence of absorption is termed the *excitation spectrum*.

Because the time frame for electronic absorption is very short, the atoms in the molecule do not have any opportunity to undergo any change in their positions or momenta within this time period. Consequently, the nuclear configuration and relative motions in the excited state immediately after absorption are the same as those of the ground state molecule prior to the light absorption. This understanding is termed as the *Franck-Condon Principle*, which states that the most probable transition will be the one in which there is no change in nuclear position or momentum between the two electronic states. Referring to Figure 1 again, it is seen that the most intense vibrational component of the transition will be the one that takes place at the same configuration of the ground state (drawn as the $v = 3$ level of state S_1). In general, the energy of the most probabe Franck-Condon peak will be the energy of the band maximum in an electronic absorption spectrum. In other words, for the $S_0 \rightarrow S_1$ transition of Figure 1, the position of maximal absorbance would correspond to the energy of the $S_0(v = 0) \rightarrow S_1(v = 3)$ transition.

A molecule raised into an upper vibrational level of the excited state rapidly loses its excess vibrational energy and decays into the lowest vibrational level of the excited state (i.e., the $v = 0$ level of state S_1) through a process known as *internal conversion*. For molecules dissolved in a solution phase, the vibrational deactivation is most commonly achieved through collisions with solvent molecules. In the solid state, the internal conversion process will involve transfer of energy between molecules until the requisite energy is lost to the system as heat. Many molecules existing in the $v = 0$ level of state S_1 return to the ground state through *radiationless deactivation*, and the excess electronic energy is liberated in the form of heat.

However, other molecules instead spontaneously emit a quantum of light from the excited state, and the wavelength dependence of the emitted light is known as the *emission spectrum*. When the emission takes place at the S_1 state and terminates in the S_0 state, the process is known as *fluorescence*. Fluorescence takes place in a time frame of approximately 10^{-9} seconds, and so the Franck-Condon principle still holds. Consequently, the most intense fluorescence transition of Figure 1 would be the one that takes place at the same geometrical configuration of the excited state, or the $S_1(v = 0) \rightarrow S_0(v = 2)$ transition.

Because room temperature absorbance takes place practically entirely from the $v = 0$ level of the ground S_0 state and because room temperature fluorescence takes place almost completely from the $v = 0$ level of the excited S_1 state, it follows that only one vibrational component of the two processes can take place at the same energy. That is, only the $S_0(v = 0) \leftrightarrow S_1(v = 0)$ transition can be observed in both the excitation and emission spectra. It is, therefore, evident that all excitation-transitions other than the $S_0(v = 0) \rightarrow S_1(v = 0)$ absorption must take place at higher energies, whereas all emission transitions other than the $S_1(v = 0) \rightarrow S_0(v = 0)$ fluorescence must take place at lower energies. This fact is formalized as *Stokes Law*, which states that the wavelength of fluorescence will always be longer (i.e., take place at lower energies) than that of the excited light.

Frequently, an emission spectrum will be a mirror image of the lowest energy absorption band because the pattern of vibrational energy levels in the two states is approximately the same. As an example, the excitation and emission spectra of sodium salicylate (27) dissolved in water are shown in Figure 2. The emission spectrum exhibits its maximum at 411.5 nm (24,301 cm^{-1}), the

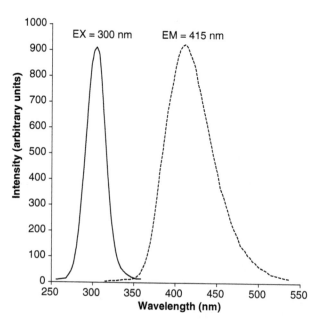

Figure 2 Excitation (*solid trace*) and emission (*dashed trace*) spectra of sodium salicylate dissolved in water. *Abbreviations*: EX, excitation; EM, emission. *Source*: Adapted from Ref. 27.

excitation spectrum exhibits its maximum at 304.3 nm (32,862 cm^{-1}), and so the Stokes shift equals 8561 cm^{-1}. The excitation and emission spectra interest at 349.9 nm (28,582 cm^{-1}), which becomes the best estimate of the $S_0(v = 0) \leftrightarrow S_1(v = 0)$ transition. The mirror image relationship between the spectra is indicated by the fact that during the excitation process the molecule is excited into a vibrational energy level that is 4280 cm^{-1} higher than the $S_1(v = 0)$ state and that during the emission process the molecule is in an excited vibrational state that is 4281 cm^{-1} higher than the $S_0(v = 0)$ state.

In the preceding discussion, only those excited states where the resultant spin angular momentum equalled zero were considered. However, through a process known as *intersystem crossing*, the excited molecule might enter into another excited state for which the resultant spin angular momentum equals one (i.e., spins unpaired). Because the multiplicity of this system would be $2(1) + 1$ or three, the state is a triplet state, for which direct excitation from the ground singlet state would be forbidden. When it is observed, luminescence from the excited triplet state down to the ground state (also a forbidden process) is known as phosphorescence.

Because the lowest-lying triplet state has less energy than the lowest-lying singlet state, the phosphorescence of a molecule will invariably be observed at

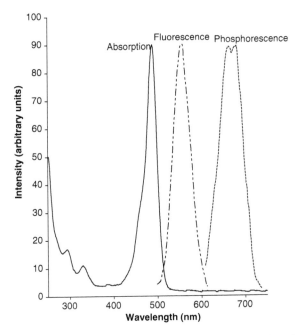

Figure 3 Room temperature absorption (*solid trace*) and fluorescence (*broken trace*) spectra of sodium fluorescein dissolved in water and the phosphorescence (*dashed trace*) spectrum obtained in an aqueous alcohol glass at 77K. *Source*: Adapted from Ref. 27.

lower energies than its fluorescence. This behavior has been illustrated in Figure 3, where the room temperature absorption and fluorescence spectra of sodium fluorescein dissolved in water are shown. Also shown is the phosphorescence spectrum of the same compound obtained in an aqueous alcohol glass at 77K (27).

The rate associated with the fluorescence process (Q) must equal the rate of absorption of excited excitation light (I_A, measured in quanta/sec) multiplied by the quantum efficiency (φ_F) of the fluorescence:

$$Q = I_A \varphi_F \tag{1}$$

or

$$Q = (I_0 - I_T)\varphi_F \tag{2}$$

where I_0 is the intensity of the incident light and I_T is the intensity of light that is transmitted through the sample. The quantum efficiency, or quantum yield, of the fluorescence process is defined as:

$$\varphi_F = N_F/N_A \tag{3}$$

where N_F is the number of photons emitted as fluorescence, and N_A is the total number of photons absorbed.

The absorption of excitation energy by the sample is given by the familiar Beer-Lambert law:

$$I_T/I_0 = e^{-kcb} \tag{4}$$

where c is the concentration of solute (in units of molecules/mL), k is the effective cross-section of one molecule (in units of cm^2), and b is the path length of the sample (in units of cm).

Substitution of Equation (4) into Equation (2) yields:

$$Q = I_0(1 - c^{-kcb})\varphi_F \tag{5}$$

The exponential of Equation (5) can be expanded as:

$$Q = I_0\varphi_F\left\{ kcb - \frac{(kcb)^2}{2} + \frac{(kcb)^3}{6} - \cdots \right\} \tag{6}$$

For dilute solutions, the amount of absorbance is usually small, and hence the term (kcb) will also be small. Under such circumstances, Equation (6) is simplified as:

$$Q = I_0\varphi_F kcb \tag{7}$$

and one concludes that the intensity of fluorescence will be proportional to both the intensity of the excitation energy and the magnitude of the absorptivity of the sample.

If a system initially consists of an ensemble of excited molecules, then the probability that any one of the molecules will emit electromagnetic radiation and return to the ground state is independent of the presence of other excited molecules. In this case, the change in the number of emissive molecules per unit time (dn/dt) will be given by the simple rate equation:

$$dn/dt = -k_F n \tag{8}$$

where n is the number of emissive molecules present at time $= t$. Integration of Equation (8) yields:

$$n = n_0 \exp(-k_F t) \tag{9}$$

where n_0 is the initial number of excited molecules. By the initial definitions:

$$Q = -dn/dt \tag{10}$$

So Equation (9) becomes:

$$Q = Q_0 \exp(-k_F t) \tag{11}$$

It is therefore concluded that the *decay* of fluorescence will be exponential in nature. One defines the mean radiative lifetime (τ_R) of the fluorescence as the reciprocal of the fluorescence rate constant:

$$\tau_R = 1/k_F \tag{12}$$

Bowen (28) has provided a simple equation for estimation of the radiative lifetime:

$$1/\tau_R = 2900 \, n^2 v_0^2 \int a \, dv \tag{13}$$

where v_0 is the wavenumber of the maximum in the absorption band, n is the refractive index of the solvent, and $\int a \, dv$ is the area under the curve when the molar absorptivity is plotted against the wavenumber. More complicated equations for evaluation of radiative lifetimes have been developed (29).

Typical lifetimes of fluorescence are in the order of 10^{-9} seconds, or on a nanosecond scale. On the other hand, the typical lifetimes of phosphorescence are in the order of 10^{-3} seconds (or higher), or typically on a millisecond scale.

PHOTOPHYSICAL PROCESSES ASSOCIATED WITH INTERACTING MOLECULES

The photophysics of well-behaved systems can usually be described using the isolated molecule principles outlined in the previous section, but it should be recognized that even isolated molecules can exhibit deviations from their ideal nature. For example, if the molecule undergoes some type of change in its configuration during the photophysical process, this will cause sufficient alteration in the dipole moment or polarizability so as to alter the energies of the respective states. The most typical example of this behavior is encountered in molecules for which the ground and excited states are differently solvated. The situation is most important for polar solutes dissolved in polar solvents, and in fact measurements of the 0–0 band shift in differing solvents have been used to deduce the dipole moment of the molecule in its excited state.

However, the deviations from isolated molecule ideality become highly important for organic molecules in the solid state. One key difference is that once absorbed, excitation energy might undergo transfer between molecules if such interaction is facilitated by the details of the solid-state environment. Another possibility is that cooperative phenomena existing in the solid state can create excited-state dimeric (or greater) species that will take up luminescence characteristics of their own.

A basis for understanding solid-state phenomena can be built on the model of the isolated molecule to include different types of interactive processes that might take place among the molecules subsequently after their excitation.

Intermolecular Energy Transfer

One of the processes competing with fluorescence is a nonradiative deactivation of the excited state through transfer of the absorbed molecule to some other species. The primary excitation step of a hypothetical molecule is described in the usual kinetic scheme as:

$$D + h\nu \rightarrow D^* \tag{14}$$

where D is the molecule in its ground state, and D^* represents that molecule in its excited state. The fluorescence process is then expressed as:

$$D^* \rightarrow D + h\nu' \tag{15}$$

and the rate constant for this process is denoted at k_e. Alternatively, the excited molecule can return to its ground state by a radiationless process:

$$D^* \rightarrow D + \Delta \tag{16}$$

where the process is characterized by the rate constant k_n, and Δ is the amount of heat liberated by the nonradiative deactivation of the excited state. Should the molecule undergo internal conversion to its triplet state (T^*):

$$D^* \rightarrow T^* + \Delta' \tag{17}$$

The corresponding rate constant would be k_g, and Δ' would be the amount of heat liberated by the internal conversion process.

However, if during the lifetime of its excited state, D^* encounters another species capable of accepting energy from one of the D^* excited states, then the following reaction would occur:

$$D^* + Q \rightarrow D + Q^* \tag{18}$$

This process is known as *quenching*, and characterized by the rate constant k_q. The rate of the energy transfer obviously will also have a dependence on the concentration of the quenching species, Q. Once its excited state becomes populated, Q^* will return to the ground state either by the radiative mechanism characterized by the rate constant k'_e

$$Q^* \rightarrow Q + h\nu \tag{19}$$

or by the non-radiative mechanism characterized by the rate constant k'_n, and release of an amount of heat equal to Δ':

$$Q^* \rightarrow Q + \Delta \tag{20}$$

Consider now a luminescent material undergoing irradiation with excitation energy for an illumination period, which is longer in comparison with its emission lifetime. Under these conditions, a steady state is set up in which the rate of production of excited molecules is exactly balanced by the rate of disappearance of these species. In the absence of any photochemical reaction or intersystem crossing from any of the upper excited states, one may express the rate of production of excited molecules as being equal to the rate of light absorption, I_a:

$$I_a = (k_e + k_n + k_g)[D^*] \tag{21}$$

where $[D^*]$ represents the concentration of molecules in the excited luminescent state. When the energy transfer process is operative, an additional source of excited state deactivation becomes operative, and Equation (21) becomes:

$$I_a = (k_e + k_n + k_g + k_q[Q])[D^*] \tag{22}$$

where $[Q]$ is the concentration of the quencher.

Earlier in Equation (3), the quantum efficiency was defined as the number of photons emitted in fluorescence divided by the total number of photons absorbed. This quantity is clearly related to the steady-state rate constants, and in the absence of any quenching process the luminescence quantum efficiency (φ_F^0) is given as:

$$\varphi_F^0 = k_e/(k_e + k_n + k_g) \tag{23}$$

When luminescence quenching is also present, Equation (23) becomes:

$$\varphi_F = k_e/(k_e + k_n + k_g + k_q[Q]) \tag{24}$$

Taking the dividend of Equations (23) and (24) yields:

$$\varphi_F^0/\varphi_F = 1 + \frac{k_q[Q]}{(k_e + k_n + k_g)} \tag{25}$$

If we define a new constant as:

$$k_{SV}{}^\varphi = k_q/(k_e + k_n + k_g) \tag{26}$$

then Equation (25) becomes:

$$\varphi_F{}^0/\varphi_F = 1 + k_{SV}{}^{\varphi}[Q] \tag{27}$$

Equation (27) is of the form developed for luminescence quenching by Stern and Volmer (30), and k_{SV}^{φ} is termed the Stern-Volmer constant for luminescence quenching.

In a similar manner, one may develop a similar Stern-Volmer expression to describe the shortening of luminescence lifetime that would be caused by the quenching agent:

$$\tau^0/\tau = 1 + k_{SV}{}^{\tau}[Q] \tag{28}$$

In Equation (28), τ^0 and τ are the emission lifetimes in the absence and presence, respectively, when the quencher concentration equals [Q].

It is important to note that Equations (27) and (28) are valid only as long as the donor and quencher species encounter each other via collisional processes. This mechanism is termed *dynamic quenching* and would be anticipated when the donor and quencher are not physically bound. The diffusion of monomeric donors and acceptors, which encounter each other solely through collisional processes, is seen to fulfill these requirements. Under these conditions, it must be that:

$$k_{SV}{}^{\varphi} = k_{SV}{}^{\tau} \tag{29}$$

A comparison of energy transfer as studied simultaneously through intensity and lifetime quenching can, therefore, serve to identify the conditions under which no association of donor and quencher takes place. If the conditions of Equation (29) are found to be met, then the presence of pure dynamic quenching may be taken as indicating the lack of physical association between the donor and acceptor compounds. This situation has been demonstrated as being characteristic of luminescence processes that take place in the vapor phase (31).

A very different situation would result when the donor and acceptor species are bound in close proximity as part of an association-type structure. In this case, the energy transfer process would not require the encounter of Q and D during the excited state lifetime of the donor, as all geometrical considerations would be fulfilled as a result of formation of the dimer species. One would anticipate that if the energy transfer was sufficiently efficient, then the DQ pairs would be nonluminescent as far as the donor is concerned. The only luminescence that could be observed in the system would therefore have to originate either from Q in either the DQ complex or from the intrinsic emission from uncomplexed D or Q. When the quenching mechanism involves complex formation, the process is termed *static* quenching. Static quenching would also result in the observation of reduced emission intensities and in the observation of a linear Stern-Volmer dependence. However, the significance of the quenching constant would necessarily be very different when compared with the constant obtained from pure collisional quenching.

Pure static quenching cannot affect the observed luminescence lifetime, as this quantity can only be determined for the species responsible for the emission. As long as Q and DQ are not luminescent, any experimentally measured lifetime would only correspond to free D. Therefore, in the absence of collisional quenching, it would follow that $\tau^0 = \tau$, and therefore from Equation (28) it would be necessary that $k_{SV}^\tau[Q] = 0$. Thus, the presence of pure-static quenching in a system can be demonstrated by the collection of lifetime data during the course of intensity experiments. If the luminescence intensities decrease with increasing concentrations of the quencher and yet no change in the lifetime of luminescence is noted, then this indicates the existence of pure static quenching.

However, in most situations the energy transfer process would entail a combination of simultaneous dynamic and static quenching. Whatever quantity of donor/acceptor pairs that could be formed would do so prior to the absorption of excitation energy by the donor:

$$D + Q \leftrightarrow DQ \tag{30}$$

The association constant, K_C, corresponding to formation of the DQ pair would be given by the usual equilibrium constant expression:

$$K_C = \frac{[DQ]}{[D][Q]} \tag{31}$$

where [D], [Q], and [DQ] are the respective activities of the donor, the quencher, and the dimer complex.

The intensity of light that would be emitted from a solution in which both static and quenching processes were operative would be given by:

$$I = A F_D F_{LUM} \tag{32}$$

where A is a proportionality constant, F_D is the fraction of excitation energy that actually excites the free D, and F_{LUM} is the fraction of free D, which is not dynamically deactivated by the quencher Q. The assumptions behind Equation (32) are that D is dynamically quenched and any DQ pair that might be formed is nonluminescent but capable of absorbing excitation energy. In addition, free Q is assumed not to absorb any of the incident radiation, and it is also assumed that the solution is optically dense so that all excitation energy will be absorbed by either D or DQ.

F_D is defined from the Beer's Law contributions of the D and DQ species:

$$F_D = \frac{a_D[D]}{a_D[D] + a_{DQ}[DQ]} \tag{33}$$

where a_D and a_{DQ} are the molar absorptivities of the donor and dimer complex, respectively. Unless the association of donor and acceptor leads to the formation of a charge-transfer complex, it usually follows that:

$$a_D = a_{DQ} \tag{34}$$

and, therefore, Equation (33) is simplified to a relation based on concentration only:

$$F_D = \frac{[D]}{[D] + [DQ]} \tag{35}$$

The fraction of free D that is not dynamically quenched by Q is simply the reciprocal of Equation (28) and is therefore given by:

$$F_{LUM} = \frac{1}{1 + k_{SV}{}^\tau [Q]} \tag{34}$$

Substitution of Equations (34) and (35) into Equation (32) yields:

$$I = A \left[\frac{[D]}{[D] + [DQ]} \right] \left[\frac{1}{1 + k_{SV}{}^\tau [Q]} \right] \tag{35}$$

From Equation (31), it is known that:

$$[DQ] = KC[D][Q] \tag{36}$$

Therefore, Equation (35) becomes:

$$I = A \left[\frac{1}{1 + k_C[Q]} \right] \left[\frac{1}{1 + k_{SV}{}^\tau [Q]} \right] \tag{37}$$

The proportionality constant, A, must equal the incident light intensity (I_0), because when $[Q] = 0$, it follows that:

$$A = I_0 \tag{38}$$

A general Stern-Volmer equation covering simultaneous dynamic and static quenching may be obtained by dividing Equation (38) by Equation (37). One then finds:

$$I_0/I = 1 + (k_{SV}{}^\tau + k_C)[Q] + (k_{SV}{}^\tau\, k_C)[Q]^2 \tag{39}$$

Several publications have appeared that treat nonlinear Stern-Volmer quenching kinetics and the subject of complexation (32–35). In one interesting study, the critical micelle concentration of a fluorescent detergent was measured through determinations of Stern-Volmer quenching parameters as a function of detergent concentration (36). Besides enabling the determination of the quenching mechanism, one of the practical and useful consequences of simultaneous measurements of emission intensity and lifetime quenching is in the calculation of association constants.

When the luminescence quenching takes place entirely through the static mechanism, it follows that $k_{SV}^{\tau} = 0$, and, therefore, Equation (39) reduces to:

$$I_0/I = 1 + k_C[Q] \tag{40}$$

The form of the Stern-Volmer relation for pure static quenching is seen to be the same as that obtained for pure dynamic quenching, except that now the Stern-Volmer quenching constant is identified as the formation constant of the donor/quencher complex. It can, therefore, be noted that the observation of a linear Stern-Volmer relationship does not indicate the existence of either pure static or dynamic quenching, and that identification of mechanism can only be deduced using simultaneous studies of luminescence intensity and lifetime quenching. Of course, the observation of a non-linear Stern-Volmer relation would necessarily imply the existence of both types of quenching in the system.

It is worth noting that the bimolecular rate constant corresponding to the quenching process may be calculated by:

$$k_q = k_{SV}/\tau_0 \tag{41}$$

In many studies of energy transfer among organic or donor/quencher systems, when the quenching mechanism is fully dynamic (or nearly so), the reactions are observed to proceed in the diffusion-controlled limit.

Photophysics of Associated Molecules

It will be shown in the subsequent section that the excitation and emission spectra of organic molecules in the solid state is often dominated by the cooperative effects associated with energy transfer. It is, therefore, informative to briefly consider the transient excited state of dimeric species that can exist in the solution phase as a means to introduce the phenomena.

The primary photoexcitation step of a molecule was described earlier by the following equation:

$$D + h\nu \rightarrow D^* \tag{14}$$

and the excited D^* molecule was described as being capable of undergoing a number of subsequent reactions involving energy and its transfer. One possible reaction of the excited D^* molecule would be to form an excited-state complex species with an unexcited molecule:

$$D^* + D \rightarrow \{DD\}^* \tag{42}$$

The $\{DD\}^*$ species is known as an *excimer*, and an extensive review of excimer photophysics is available (37). In general, quenching by formation of excimers is favored in nonpolar solvents, and quenching via electron transfer is the mechanism observed in polar solvent systems.

The polynuclear aromatic hydrocarbon pyrene represents one of the best known examples of excimer fluorescence and was first studied by Forster and

Kasper (38,39). These workers found that at low concentrations of pyrene in benzene, only the structured fluorescence characteristic of monomeric molecules was observed. As the concentration of pyrene in the solution was increased, the monomer fluorescence decreased in intensity and became replaced by a featureless emission observed at much longer wavelengths. In their work, Forster and Kasper proved that no ground state association existed between pyrene molecules.

To illustrate the photophysical characteristics of excimers, Figure 4 shows the emission spectrum that was obtained for pyrene dissolved in cyclohexane at concentrations of 0.01 mg/mL (27). The structured emission is characteristic of the well-known fluorescence of the monomeric form of the solute. Also shown in Figure 4 is the emission spectrum of pyrene but dissolved at a concentration of 0.1 mg/mL. In this spectrum, the broad, red-shifted, fluorescence spectrum is characteristic of the excimer species.

From a detailed kinetic analysis of the photophysics associated with the monomeric and excimer species of pyrene, several deductions were made that were found to be experimentally verifiable (40). From a consideration of the "half-value concentration" (the concentration at which the fluorescence of the monomer has been quenched by 50%, and the concentration at which the fluorescence of the excimer has risen to 50% of its maximal value), it was deduced that the association reaction to form the {DD}* species would be diffusion-controlled. In

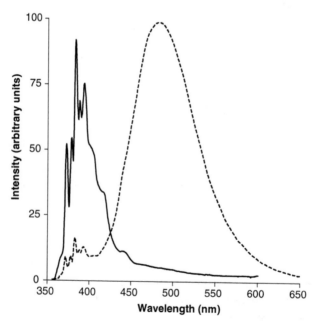

Figure 4   Fluorescence spectra of pyrene dissolved in cyclohexane. Spectra are shown in arbitrary units for solute dissolved at a concentration of 0.01 mg/mL (*solid trace*) and 0.1 mg/mL (*dashed trace*). *Source*: Adapted from Ref. 27.

such a case, the reciprocal of the associated rate constant would be proportional to the viscosity (41,42). This relationship was shown to exist in a number of different solvent systems, which confirms the mechanism.

Doller and Forster studied the temperature dependence of the emission efficiencies for pyrene in hydrocarbon solutions and were able to demonstrate that the excimer formation was a reversible reaction that could be described by thermodynamics (42). It was found that in dilute solution, the fluorescence intensity decreased with increasing temperature, demonstrating the effect of increasing amounts of radiationless deactivation. For mixtures of monomer and excimer, a complicated relationship between temperature and the fluorescence efficiencies of both species was observed, and the fitting of these results enabled a calculation of the enthalpy of formation for the excimer species.

Since the original work, a large number of molecules have been shown to be capable of forming excimer species at suitable concentrations in appropriate solvent systems. One of the most studied systems are molecules containing two identical fluorophores separated by an alkane chain. Excimer emission has been detected in systems, such as the diphenylalkanes (43), dinaphthylalkanes (44), and dipyridylalkanes (45), as long as the separating carbon chain consisted of three methylene units.

When the pyrene molecule contains a dissymmetric center, the excimer can exhibit enormous optical activity relative to the monomer molecule. The excited-state optical activity (as measured by the circularly polarized luminescence) of the two enantiomers of 1-(1-hydroxyhexyl)pyrene was found to be negligible at low concentrations but became very intense as the solute concentration was increased to the point where significant amounts of the excimer were formed (46). These findings were interpreted as demonstrating a strong stereoselective preference for the stacking of the pyrene aromatic rings in the excimer. The intensity of the circularly polarized luminescence of the enantiomeric excimers was found to be equal in magnitude, but opposite in sign, at equivalent concentrations, suggesting that the sense of chiral stacking was opposite for the excimer species.

Studies of excimer fluorescence have become important in the study of certain polymeric systems (47). For example, intramolecular excimers have been studied in aqueous solutions of poly(styrenesulfonic acid) and its salts (48,49). For the Li(I), Na(I), Cs(I), and Mg(II) salts, the excimer-to-monomer ratios were found to be independent of the salt concentrations, indicating that condensation with these counter ions did not result in a conformational change in the polyelectrolyte. On the other hand, for the K(I), Ca(II), Sr(II), and La(III) salts, the excimer-to-monomer ratios increased as the salt concentration was increased, indicating the existence of a conformational change that facilitated adjacent phenyl groups to become more interactive.

Another possible reaction of the excited D* molecule would be to form an excited-state complex species, but with an unexcited molecule that has a different chemical identity:

$$D^* + Q \rightarrow \{DQ\}^* \tag{43}$$

When D and Q are different molecules, the term *exciplex* is used to describe the luminescent {DQ}* species. Of course, if the {DQ}* species is not luminescent, it follows that the previous discussion pertaining to luminescence quenching applies. Studies of exciplexes have been thoroughly reviewed (50).

To illustrate the spectral characteristics of exciplex formation, the experiments of Knibbe et al. (51) have been reproduced. Figure 5 shows the fluorescence spectra of anthracene dissolved at a concentration of 1.8 μg/mL in toluene and the spectra of the same solution when increasing amounts of N, N-diethylaniline were added (27). It is obvious from the figure that the addition of diethylaniline quenches the fluorescence of anthracene and that the fluorescence of a new entity increases in intensity with the amount of diethylaniline added. It was shown by Knibbe et al. (51) that the broad fluorescence originated from the 1:1 exciplex and that the degree of formation of ground-state complex species was very small.

The ability of amines to form exciplexes with a multitude of fluorescent organic molecules has become an extremely well-studied phenomenon, and the family of stilbene-like molecules represents one such research interest (52). For example, *trans*-3-aminostilbene and its N-methyl derivatives are strongly fluorescent when dissolved in cyclohexane, and form exciplexes when alkylamines are added (53). Modeling of the concentration dependence of the quenching of

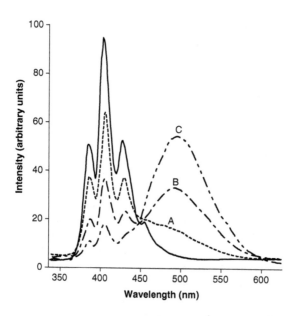

Figure 5 Fluorescence spectra of anthracene dissolved at a concentration of 1.8 μg/mL in toluene (*solid trace*). Spectra are also shown for this solution but also containing N,N-diethylaniline at concentrations of 0.75 mg/mL (A), 3.75 mg/mL (B), and 15.0 mg/mL (C). *Source*: Adapted from Ref. 27.

the monomer fluorescence, and the growth of exciplex fluorescence, with primary amine concentration led to the conclusion that excited-state complexes having 1:1 and 1:2 stilbene-amine stoichiometries could be formed. Secondary amines tended to yield only the 1:1 exciplexes. The formation of the exciplex was attributed to a Lewis acid/base interaction between the lone pair acceptor of the stilbene moiety with the lone pair donor of the alkylamine.

Stereoselectivity has been detected in the fluorescence quenching of $1,1'$-binaphthyl by chiral amines (54). It was reported that specific geometrical requirements were associated with the fluorescence quenching of $1,1'$-binaphthyl by the resolved enantiomers of *N*, *N*-dimethyl-α-phenethylamine, although the overall details of the quenching were similar to those of other quenching studies of aromatic molecules by dialkylamines. In this study, the degree of stereoselectivity was found to be strongly dependent on the solvent used, with solvents of low dielectric constant yielding the highest enantiomeric discrimination values. One part of the solvent dependence was attributed to alterations in the quenching mechanism and another part to stabilization of certain conformers of the $1,1'$-binaphthyl donor.

Photophysics in the Solid State

It is evident from the discussion of excimers and exciplexes that the photophysics of organic compounds can become complex when the molecules are brought into close proximity and allowed to interact with each other. In fact, many compounds that are highly fluorescent in solution become absolutely nonfluorescent in the solid state. For example, the fluorescence quantum yield of the sodium salt of fluorescein is close to unity, but no fluorescence can be detected for solid uranine. In other instances, a nonfluorescent material can become luminescent if the defect sites in the crystal can serve as energy traps. A number of texts have appeared that treat the solid-state photoluminescence of inorganic and organic systems, and the coverage in these is of considerable historical interest (55–60).

The repetitive three-dimensional structure of the crystalline state of a luminescent molecule locks the component molecules in well-defined orientations with respect to each other that are defined by the details of the unit cell and its translational symmetry. If the degree of interaction between molecules is weak, excitation into low-lying electronic states results in the population of excited states that are similar to those for an individual molecule in the vapor phase. However, this excited state cannot be a stationary state of the crystal, and, over time, the probability of finding the excitation on the initial molecule decreases, and the probability on the adjoining molecules increases.

As a result, energy transfer between molecules, via the common radiation field, is often facilitated, and such radiationless transfer of energy can take place over appreciable distances in the crystal. Although the ground electronic state is localized on individualized molecules even in the crystalline state, the excited electronic states of these can be so strongly interactive that excitation energy becomes delocalized among the coupled molecules. The excitation energy, therefore, spreads outward through the crystal like a wave, or alternatively, as the motion of a quasiparticle through the crystal. Frenkel first termed such a delocalized excited state as an exciton (61), and the theory of excitons has been thoroughly applied to organic crystals by Davydov (62). Additional discussions of exciton theory can be found in the texts by McClure (63) and Knox (64), and only a brief summary will be given here.

Craig and Walmsley (65) have summarized a number of characteristics pertaining to excitons. Because the transfer of electronic energy between the excited molecules is rapid relative to the time frame of fluorescence, the delocalization of excitation energy has the effect of extending the excited state molecular orbital over the ensemble of molecules involved in the energy transfer. For a pair of molecules, the interaction leads to a splitting of the single-molecule energy level into a pair of levels, where the magnitude of the splitting is determined by the strength of the coupling. This type of splitting is commonly referred to as Davydov splitting and becomes manifest in the appearance of new bands in the excitation spectrum. In addition, the theory predicts that the mean frequency of the Davydov components would be displaced to lower energies from that of the free molecule value as a result of cooperative interactions in the crystalline state.

For an ensemble of N identical, decoupled, molecules, the wave functions (ϕ_i) of the excited states where one molecule is excited and the others are in the ground state can be written as:

$$\phi_1 = \psi^*(1) \ \psi(2) \ \psi(3) \cdots \psi(n)$$
$$\phi_2 = \psi(1) \ \psi^*(2) \ \psi(3) \cdots \psi(n)$$
$$\phi_3 = \psi(1) \ \psi(2) \ \psi^*(3) \cdots \psi(n) \tag{44}$$
$$\cdots\cdots\cdots\cdots\cdots$$
$$\phi_N = \psi(1) \ \psi(2) \ \psi(3) \cdots \psi^*(n)$$

where $\psi(i)$ is the wave function of the molecule it its ground state and $\psi^*(j)$ is that of a molecule in the excited state. If there is no interaction between the molecules, then all of the excited state wave functions will have the same energy. But, if transfer of excitation energy between the excited states takes place, then those states are not stationary states of the system and must be mixed. The mixing process yields a new series of states (Ψn) that will be formed as linear combinations of the ϕ_i functions and that will differ from each other by the values of the coefficients, C_{nm}:

$$\Psi_1 = C_{11}\phi_1 \ C_{12}\phi_2 \ C_{13}\phi_3 \cdots C_{1m}\phi_N$$
$$\Psi_2 = C_{21}\phi_1 \ C_{22}\phi_2 \ C_{23}\phi_3 \cdots C_{2m}\phi_N$$
$$\Psi_3 = C_{31}\phi_1 \ C_{32}\phi_2 \ C_{33}\phi_3 \cdots C_{3m}\phi_N \tag{45}$$
$$\cdots\cdots\cdots\cdots\cdots\cdots$$
$$\Psi_N = C_{n1}\phi_1 \ C_{n2}\phi_2 \ C_{n3}\phi_3 \cdots C_{nm}\phi_N$$

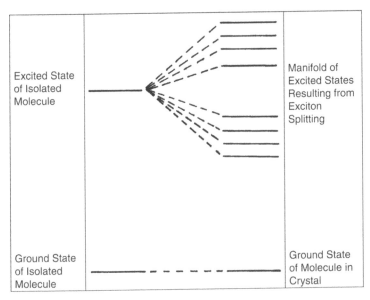

Figure 6 Schematic energy level diagram illustrating the effect of exciton coupling that results in a splitting of the excited state into a manifold of states.

As illustrated in Figure 6, the result of exciton coupling is to produce an excitation multiplet corresponding to a band of n levels, each differing in energy by small amounts. The overall spread of the excitation band is determined by the strength of the coupling, and one can actually observe a separation between the symmetric linear combinations and the antisymmetric linear combinations if the coupling is sufficiently strong.

In molecular crystals, the magnitude of the interaction energy and the degree of exciton coupling would necessarily be dependent on the relative orientation of the molecules in the crystal and on their distance separation. It follows that because polymorphic crystal forms are characterized by the existence of differing structural properties, the nature of the exciton coupling in the various forms would be dictated by the particular structural characteristics of each form. As a result, one would anticipate that the magnitude of the Davydov splitting, and the degree of shifting of the levels, would be dependent on the exact structural details existing in the different polymorphic forms.

The effect of the crystalline state on the fluorescence of an organic compound can be illustrated using the example of 4-methylsalicylic acid, and Figure 7 shows the normalized solution-phase and solid-state excitation and emission spectra obtained for this compound (66). A considerable amount of energy transfer quenching must exist in the solid state, as these spectra were weaker on the order of 100 times than the corresponding solution phase spectra. The emission spectrum obtained for the solution phase exhibited a

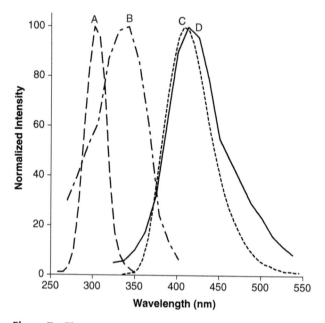

Figure 7   Fluorescence spectra of 4-methylsalicylic acid. Spectra are shown for the solution-phase excitation (A) and emission (C) obtained at a concentration of 15 μg/mL and for the solid-state excitation (B) and emission (D) spectra. All of the spectra have been normalized to a common intensity factor. *Source*: Adapted from Ref. 66.

maximum at a wavelength of 408 nm, whereas the emission maximum of the solid-state spectrum was only slightly red-shifted to 413 nm, indicating that the fluorescent state of 4-methylsalicylic acid was equivalent in both phases. On the other hand, the maximum of the excitation spectra strongly differed, being observed at 302 nm in the solution phase and at 342 nm in the solid state. This strong red-shift of the excitation spectra is a direct consequence of the intermolecular energy transfer that was also associated with the overall loss in fluorescence intensity in the solid state.

INSTRUMENTATION FOR THE MEASUREMENT OF LUMINESCENCE SPECTRA

Steady-State Photoluminescence Spectroscopy

At a minimum, a fluorescence system will contain the essential components of a source of excitation energy, a holder for the sample, a device to analyze the resulting emission, and a detector for its quantitative measurement. Figure 8 shows a block diagram that illustrates the most commonly employed right-angle method of measurement. Very detailed descriptions of steady-state luminescence

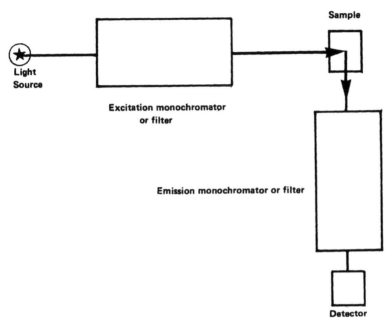

Figure 8   Block diagram of a steady-state fluorescence spectrometer.

spectrometers are available, and those needing more detailed discussions of individual topics should consult these literature sources (7–12,15–19,22,67).

In the most simple fluorescence system, the wavelengths of excited and emitted light are selected by broad-pass filters that allow measurements to be made at any pair of fixed groups of wavelengths. Such devices are termed filter fluorimeters and have the advantage that their use facilitates the greatest degree of analytical sensitivity. For maximum sensitivity, the sample can be excited by a broad range of wavelengths that would span its entire absorption band, and the intensity of the entire emission spectrum can be measured at once.

More sophisticated fluorescence spectrometers are based upon the monochromator discrimination of either the excitation or emission optical trains. In such systems, one may set the emission analyzer to a fixed wavelength and systematically vary the excitation wavelength to measure the *excitation spectrum.* Alternatively, one could set the excitation monochromator at a fixed wavelength and systematically vary the analyzing emission wavelength to measure the *emission* or *fluorescence spectrum.*

Commonly employed sources in fluorescence spectrometry are characterized either as having spectral outputs as a continuum of energy over a wide range of wavelengths or as a series of discrete lines. An example of the latter type of source that has received much use in the past as an excitation source is the mercury lamp, where the output from a low-pressure lamp would be concentrated in the UV region of the spectrum. Medium- and high-pressure lamps will have an output that effectively covers the entire range of wavelengths in the UV and VIS regions.

However, the use of a monochromator to select the proper excitation wavelength really requires the use of a continuum source, and the most useful source of this type is undoubtedly xenon arc source. When used as light sources for steady-state spectroscopy, these sources are operated on a continuous basis, but they can also be operated stroboscopically for time-resolved investigations. The output is essentially a continuum of electromagnetic energy, onto which is superimposed a number of sharp lines (8,10). The xenon source effectively allows the use of any wavelength throughout the UV and VIS regions of the spectrum.

Arc lamps are inherently more unstable than other sources, and a method to compensate for the drift is usually employed when long-term stability is required. Typically, this is accomplished by splitting the excitation beam so that a small portion is quantitated by a reference detector. This reference signal is used to ratio the signal from the detector observing the sample, and the arrangement serves to account for variations in source intensity. Arc lamps can produce ozone from the photolysis of atmospheric oxygen, which ordinarily requires their venting. However, ozone-free lamps are available that minimize this problem, although at the expense of the lowest wavelengths of the source.

A filter fluorimeter system uses fixed filters to isolate both the excited and emitted wavelengths, and often a pair of cut-off filters is used to isolate one particular wavelength from a source emitting a line spectrum. The filters may be either commercial glass slides or can be solutions contained in cuvettes. The emission filter must be chosen so that any Rayleigh or Tyndall scattered light is not passed, but that the light emitted by the sample is transmitted. Such a simple filter system can be used for routine quantitative work, but only after the sufficient characterization of the photophysics of the system to know the most appropriate combination of filters.

When selectivity in excitation or emission is required, monochromators containing diffraction gratings are used to select the desired wavelengths. These types of fluorescence spectrometers are then capable of recording both excitation and emission spectra and are, therefore, the most versatile systems. Although solution-phase fluorescence studies do not require high resolution (and in fact are usually operated under low-resolution conditions to achieve sensitivity), the study of luminescent materials in the solid state can require the use of high resolution and narrow monochromator slit widths.

In the past, commercial fluorescence systems made use of photomultiplier tubes as their detectors (68). A wide variety of tube types are available, which differ mainly in the substance from which the photocathode is made. Photomultiplier tubes having S5 response are responsive to light having wavelengths ranging from the UV to wavelengths as long as 650 nm. For studies conducted in the near-infrared region (NIR), a photomultiplier tube having S20 response is ordinarily used. The sensitivity of a photomultiplier tube is limited by the level of its dark current (the signal from the tube in the absence of light), but this can be reduced by cooling the tube.

The spectral response of all photomultiplier tubes varies with wavelength, but when it is necessary to determine the actual quantum intensity of the incident radiation, or obtain a response-free spectrum, the detector response is corrected. This is most conveniently achieved through the use of a quantum counter, which is a substance for which the quantum yield of fluorescence is unity. One such quantum counter is a concentrated solution of Rhodamine 101 in ethylene glycol, which will emit the same number of quanta of light as it absorbs but over a very wide wavelength range. Thus, by measuring the output of the quantum counter at one wavelength, the number of incident quanta over a wide wavelength range can be measured.

The versatility of luminescence spectroscopy has been greatly enhanced through the use of multichannel analyzers, because such devices permit the acquisition of complete spectroscopic data almost instantaneously over a range of wavelengths (69,70). Among the most commonly used detectors are diode arrays, charge-coupled and charge-injection detectors, and vidicons.

Sample configurations for solid-state work usually involve front-face excitation of the sample, with the luminescence being recorded at a grazing angle with respect to the irradiation beam. For single crystals, films, and supporting matrices, the sample can be mounted on a suitable holder and placed at the proper geometry in the optical train. Powdered materials can be placed in a thin-walled tube (preferably made of vitreous quartz, but glass can serve as well) prior to its placement in the spectrometer. Whatever the arrangement, care must be paid to the problem of spurious signals being detected that originated from inadvertent reflection of either the excitation or emission beams.

Time-Resolved Luminescence Spectroscopy

In many instances, the basic techniques of steady-state luminescence spectroscopy suffice to solve a problem of physical, analytical, or other scientific interest. When these methodologies prove to be insufficient, the next stage in sophistication would be to use synchronous scanning of excitation and emission spectroscopies as a means to achieve additional selectivity. Synchronous fluorimetry entails the simultaneous scanning of both the excitation and emission monochromator wavelengths, but where the wavelength difference between the two settings is maintained at a constant value. This combination of parameters results in a simplification of the observed spectra features and a reduction in the observed peak widths, both factors that lead to enhanced selectivity (71–74).

Synchronous luminescence spectroscopy has proven to be very useful in the characterization of mixtures of compounds for which overlapping of spectroscopic bands is a problem, and for which an evaluation of the entire excitation–emission matrix is not necessary. For example, the combination of synchronous fluorimetry and derivative data analysis has been used to determine the levels of salicylic acid and its metabolites in urine samples (75,76). Additional advances have included the use of variable angle (77) and constant-energy (78) synchronous scanning fluorescence methods. However, all of the methods suffer from the same limitation, namely the requirement for the optimal value of the scanning wavelength offset that yields the greatest photophysical advantage to be determined, which can sometimes be a lengthy process.

In the search for additional analytical selectivity and sensitivity in luminescence spectroscopy, a considerable amount of work has been conducted where the time scale of the photophysical processes is incorporated into the measurement. In preceding sections of this chapter, a number of spectroscopic transitions (both radiative and nonradiative) have been discussed, but until now the time evolution of these has not been discussed. It is evident that every transition must take place within a finite amount of time, and, therefore, the incorporation of the time element into luminescence techniques will necessarily impart a selectivity into the methodology that can be exploited (79,80).

Two methods have become widely used for the determination of time-resolved luminescence properties, with these operating in the time domain and in the frequency domain. In time-domain luminescence spectroscopy, the sample is excited using pulsed excitation energy, and the spectroscopy observable is measured at some finite time after the pulse is complete. In frequency-domain spectroscopy, the sample is excited with modulated excitation energy, and the spectroscopy observable is found in the phase shift of the measured signal with respect to the initial excitation. Each approach has its strengths and limitations (81), and these will be briefly reviewed in the following sections.

Photophysical Measurements in the Time Domain

In time domain fluorimetry, the initial excitation of the sample is achieved using a light pulse of sufficiently short duration so that the width of the pulse is short relative to the decay time of the sample. As discussed earlier, the change in the number of luminescent molecules per unit time (dn/dt) is given by:

$$dn/dt = -k_F n \tag{46}$$

where n is the number of emissive molecules present at time $= t$. Defining Q as $-dn/dt$, and integrating Equation (46) yields the familiar equation for the exponential decay of fluorescence:

$$Q = Q_0 \exp(-k_F t) \tag{47}$$

The mean radiative lifetime (τ_R) of a fluorescence decay equals the reciprocal of the fluorescence rate constant, so therefore:

$$Q = Q_0 \exp(-t/\tau_R) \tag{48}$$

In the more general instance, an experimentally observed luminescence lifetime, τ, will be determined by both the nonradiative (k_{NR}) and radiative (k_F) decay constants

$$\tau = 1/(k_F + k_{NR}) \tag{49}$$

Methods for determination of luminescence lifetimes are available in all of the general references, but the most complete discussion is found in the monograph by Demas (20).

The fluorescence decay curve of a hypothetical fluorescent substance exhibiting a 25 nsec emission decay is illustrated in Figure 9, along with the decay curves of a pulsed flashlamp characterized by a 2 nsec pulse. It can be concluded that if one sought to measure the spectra characteristics of the compound free from contamination by energy from the source, then one would delay the recording of data for a time period of at least 15 nsec after the initial pulse.

One approach for the detection of authentic luminescence is to use a gating method, where the detector is only activated after a definite time period following the excitation pulse. Alternatively, various investigators have made use of boxcar integrators, transient signal recorders, multichannel analyzers, and video fluorimeters to monitor the time dependence of fluorescence.

The power of using time resolution to enhance the selectivity of a photophysical measurement is illustrated in Figure 10, where the observed decay

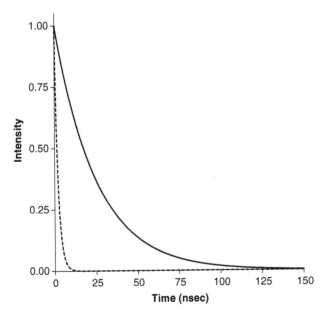

Figure 9 Decay curves of a 2 nsec pulsed flashlamp (*dashed trace*) and of a hypothetical fluorescent substance exhibiting a 25 nsec emission decay.

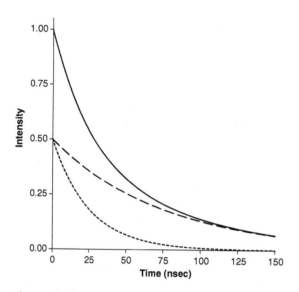

Figure 10   Decay curves of an equimolar mixture of hypothetical fluorescent substances, one exhibiting a 25 nsec emission decay (*short-dashed trace*) and the other exhibiting a 75 nsec emission decay (*long-dashed trace*). The total observed decay curve is the *solid trace.*

curve of an equimolar mixture of fluorescent substances is shown along with the individual decay curves of the components. For the two hypothetical substances exhibiting fluorescence lifetimes of 25 and 75 nsec, it is clear that one could observe the fluorescence of only the 75 nsec component in the mixture by employing a 100 nsec gate time. As long as this particular substance was the analyte of interest, one could obtain the full range of information accessible with luminescence techniques without the need to first conduct a separation step.

The technique of time-correlated single-photon counting is probably the most widespread method used to acquire photophysical information in the time domain (82,83). In this approach, the sample is excited using a pulse source, and only those photons that are time-correlated to the excitation pulse are counted to develop the fluorescence signal. By using a repetitive pulsed source, the detection system builds up a histogram of detector counts in bins of time channels. Taken together, this histogram has the form of a fluorescence decay. The observed data are then deconvoluted into contributions from the system and associated with optical elements to ultimately obtain the decay curve of the sample itself (84).

Photophysical Measurements in the Frequency Domain

Another approach to the measurement of time-resolved luminescence spectroscopy involves modulated light sources and phase shift measurements and, hence, is typically known as phase-resolved fluorescence spectroscopy. This technique uses a source whose intensity is sinusoidally modulated and yields the lifetime of the sample from the phase shift between the excitation waveform and that of the sample emission. It has been the subject of numerous reviews (85–87).

The time-dependent intensity of the excitation energy, I_T, is described by:

$$I_T = I_0[1 + M_{EX} \sin(\omega t)] \tag{50}$$

where I_0 is the DC intensity component of the excitation beam, and ω is the angular modulation frequency related to the linear modulation frequency, f, by the relation $\omega = 2\pi f$. The degree of modulation, M_{EX}, is the ratio of the amplitude of the AC intensity to the DC intensity component.

After excitation by the modulated exciting beam, the resulting fluorescence will be characterized by its own time dependence, Q_T and will also be demodulated and phase shifted by an amount related to the fluorescence lifetime of the emissive species:

$$Q_T = Q_0[1 + M_{EX}m \sin(\omega t - \varphi)] \tag{51}$$

In Equation (50), Q_0 is the DC component of the fluorescence, φ is the phase shift of the emission relative to the excitation, and m is the demodulation factor ($m = \cos \varphi$). The demodulation factor can also be evaluated as the ratio of the degree of AC modulation of the emissive species to the degree of AC modulation of a pure scattering reference.

Modulation of the source can be effected using either a Pockels cell or a photoelastic modulator, and the magnitude of the phase shift is measured using phase-sensitive detection (typically with a lock-in amplifier). Additional precision in the measurements can be achieved through the use of cross-correlation techniques.

From knowledge of the phase shift in the fluorescence, the luminescence lifetime of the fluorescent species can be calculated using:

$$\tau = (1/\omega) \tan \varphi \tag{52}$$

The lifetime can also be determined using the demodulation factor:

$$\tau = \frac{1}{\omega}\left[\frac{1}{m^2} - 1\right] \tag{53}$$

The considerable advances in instrumental sophistication, and methods for the suppression of interfering signals, have made the use of phase-resolution luminescence spectroscopy as the most versatile of the time-resolved methods. Advances in data analysis and handling have facilitated the study of complex mixtures, and the technique has become the investigative tool of choice for the study of complicated biomolecular systems.

PHOTOLUMINESCENCE AND THE STRUCTURE
OF ORGANIC MOLECULES

As stated earlier, not all organic molecules are capable of exhibiting fluorescence in either the solid state or in the solution phase, and the subject of molecular structure and luminescence efficiency has been discussed by a number of authors in the general texts cited earlier. Particularly useful compilations of fluorescence spectra are those of Berlman (6), Schmillen and Legler (88), and Maeda (89). Other general discussions include fluorescence and phosphorescence of pharmaceutical compounds (90) and the fluorescence of organic natural products (91).

When considering the molecules, or functional groupings within molecules that lead to the observation of measurable fluorescence, one must keep in mind the photophysical reactions described earlier in Equations (14) through (20). Molecules in their lowest-energy excited state (typically the lowest-singlet state of an organic molecule) can be deactivated through fluorescence, but that state can also become depopulated through radiationless deactivation (internal conversion or intersystem crossing) or by the conduct of a photochemical reaction. The relative rates of the reaction for the different processes determine the outcome of a photoexcitation process, and the existence of fluorescence requires that the rate constant for radiative transitions be large relative to those for nonradiative decay or photodecomposition.

Guibault has summarized the characteristics of molecules that should exhibit strong fluorescence (92), and these are summarized as follows.

- The spin-allowed electronic absorption transition of lowest energy is very intense. Because the intensity of absorption is directly proportional to the rate constant for the radiative transition and because fluorescence is the reverse of absorption, it follows that the more probable the absorption transition, the more probable will be the fluorescence transition.
- The energy of the lowest spin-allowed absorption transition should be reasonably low. High energy excitation of a molecule is more likely to result in its photodissociation rather than in the initiation of the fluorescence process.
- The terminal state of the absorption transition (i.e., the excitation process) should not be strongly involved in chemical bonding. If the state is strongly bonded, its excitation could result in bond dissociation rather than fluorescence.
- The molecule should not contain structural features or functional groups that enhance the rates of radiationless transitions. Certain structural features greatly increase the rates of radiationless processes and therefore adversely affect fluorescence efficiencies.

Guibault goes on to use these simple considerations to understand why aromatic hydrocarbons are usually very intensely fluorescent. In molecules of this type, it is well known that the states derived from orbitals in the π-bonding system are higher in energy than are the states derived from orbitals in the σ-bonding system. Thus, aromatic hydrocarbons can be excited at relatively low energies to undergo the strongly allowed $\pi \rightarrow \pi^*$ transition without significant alteration in the chemical bonding patterns of the molecule. Excitation into the high energy $\sigma \rightarrow \pi^*$ transitions does not usually result in observable fluorescence, owing to the large perturbations in chemical bonding that accompany the excitation process.

The situation becomes more complicated in molecules containing functional groups that contain nonbonding electrons, such as the carbonyl groups present in aldehydes, ketones, and carboxylic acids. In these instances, the energy of the $n \rightarrow \pi^*$ transition will usually be lower relative to the energy of the $\pi \rightarrow \pi^*$ transition. However, the intensity of the $n \rightarrow \pi^*$ transition is invariably less than the intensity of the $\pi \rightarrow \pi^*$ transition, and therefore one would anticipate that aromatic carbonyl compounds exhibit lower intensities of fluorescence than the parent aromatic hydrocarbons.

Interested readers seeking more information regarding the effects of molecular structure on the fluorescence of aromatic molecules should consult the monograph by Birks (14).

STUDIES OF LUMINESCENCE IN THE CRYSTALLINE STATE

The first investigations into the luminescence phenomena of pure organic molecules in their crystalline state dealt with aromatic hydrocarbons, such as naphthalene and anthracene. Many of the earliest studies were flawed by the presence of impurities, which acted as energy sinks, leading to photoluminescence from the impurity but not from the host crystal. For example, Shpak and Sheka showed that many previously recorded fluorescence spectra of naphthalene in its crystalline state were actually the fluorescence spectra of the β-methyl-naphthalene impurity (93).

The crystalline naphthalene system has been used in a number of chemical physics studies to examine the flow of excitons in both pure and mixed crystals. For example, the energy transfer phenomena have been formulated in terms of a static percolation theory, where the dynamic motion of excitons has been described as being a function of the excitation lifetime, jump time, coherence time, and trapping time (94). In another model, a random-walk formulism was used to investigate the anisotropic interactions associated with the energy transfer of excitons and to better understand the first singlet excited state of naphthalene (95). The flow of excitation energy, and the concept of exciton energy funnels, has been studied in different mixed naphthalene crystals, where the luminescence of the impurity species has been used as a means to probe the photophysics in the crystal (96,97).

Crystalline anthracene represents another system where the photophysical phenomena in the crystalline state are very well understood, and the absorption (98,99) and fluorescence (100,101) spectra in the crystalline state have been well detailed. A considerable amount of work has been expended in an effort to understand the 25,000 cm^{-1} (approximately 400 nm) band system and the nearly 500 cm^{-1} gap between the absorption and fluorescence maxima. From low temperature studies conducted on zone-refined material, the origin of the fluorescence spectrum (i.e., the $v = 0$ to $v = 0$ transition) has been identified as having an energy of 25,103 cm^{-1}, which correlates well with the *b*-polarized absorption transition at 25,150 cm^{-1}. In one study, the energy of the Davydov split state was found to be 24,705 cm^{-1}, indicating that the magnitude of the Davydov splitting was approximately 400 cm^{-1}.

Photophysical phenomena have been studied in a number of other molecular crystals, largely taking the chemical physics approach of understanding the structure of energy levels, monitoring the flow of excitation energy, and evaluating the resulting photoluminescence. For example, investigations of the photophysics in crystalline biphenyl (102,103), *trans*-stilbene (104), dimethylbenzophenone (105,106), and dichlorobenzophenone (107) have been reported. Fluorescence studies have been used to study the phase transition, which takes place at a temperature of 191 K ($-82°$C) in crystals of *p*-terphenyl (108).

It is probably true, however, that pharmaceutical scientists are more interested in spectroscopic studies that provide insights into the structures of substances, rather than in studies that are conducted to probe photophysical pathways. Such work would entail investigations of the relationship between fluorescence and molecular geometry, where X-ray structural analysis results would be correlated with solid-state luminescence properties. For example, the red-shifted, structureless emission associated with crystalline spiroanthronyl-substituted dihydrobenzanthracenes has been attributed to the presence of excimer luminescence in the solid state (109). In other works, this same group studied the structure-mediated intramolecular energy transfer in a series of anthronylanthracenes (110), and the effect of molecular stereochemistry on the excimer phenomena of crystalline dihydroanthracenes (111).

The utility of solid-state photoluminescence as a means to study compounds of pharmaceutical interest can be illustrated through a recent study conducted on erythromycin-B derivatives (27). Figure 11 shows the excitation and emission spectra of erythromycin-B dihydrate, where it may be observed that the excitation spectrum consists of two bands having maxima at approximately 350 and 390 nm. The maxima of the fluorescence resulting in the excitation of these bands are nearly equivalent, exhibiting maxima at 482 and 488 nm, respectively. Dehydration of the substance causes the fluorescence intensity to be reduced by a factor of two, and a substantial red-shift in the maxima of the fluorescence bands. Excitation of the dehydrated material at 350 nm now yields fluorescence having a maximum at 459 nm, while excitation at 390 nm yields fluorescence having a maximum at 472 nm.

Quite different behavior was observed for the hydrochloride salt of erythromycin-B, as shown in Figure 12. Once again, the excitation spectrum consists of two bands having maxima at approximately 350 and 390 nm, and the maxima of the fluorescence obtained on excitation of these bands are found to exhibit maxima at 455

Erythromycin-B dihydrate

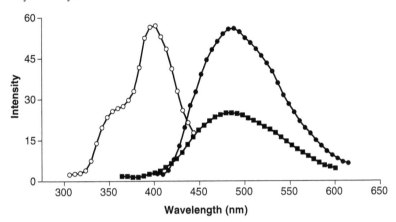

Desolvated Erythromycin-B dihydrate

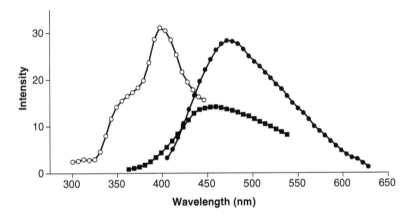

Figure 11 Solid-state fluorescence spectra of erythromycin-B dihydrate (*upper spectra*) and on its dehydrated product (*lower spectra*). The excitation spectra (*open circle*) were obtained using an analyzing emission at wavelength of 465 nm, whereas the emission spectra were obtained upon excitation at 350 nm (*solid square*) and at 390 nm (*solid circle*). Although the intensity scales are arbitrary, the spectra are scaled to permit comparisons to be made. *Source*: Adapted from Ref. 27.

and 470 nm. The intensity of fluorescence obtained from the hydrated hydrochloride salt was found to be more than twice that of the hydrated free base. Dehydration of the hydrated hydrochloride salt was found to have little effect on the emission maxima (455 nm for excitation at 350 nm, and 472 nm for excitation at 390 nm), but the intensity of emission decreased by more than a factor of 15.

Hydrated Erythromycin-B hydrochloride

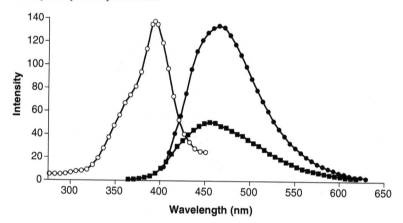

Dehydrated Erythromycin-B hydrochloride

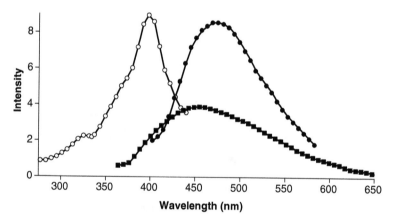

Figure 12 Solid-state fluorescence spectra of the hydrochloride salt of erythromycin-B, obtained on the precipitated hydrate (*upper spectra*) and on its dehydrated product (*lower spectra*). The excitation spectra (*open circle*) were obtained using an analyzing emission at wavelength of 465 nm, while the emission spectra were obtained upon excitation at 350 nm (*solid square*) and at 390 nm (*solid circle*). Although the intensity scales are arbitrary, the spectra are scaled to permit comparisons to be made. *Source*: Adapted from Ref. 27.

The erythromycin-B example clearly illustrated the sensitivity of photophysics to details of crystal structure. In another example, the different forms of the disodium salt of a HMG-CoA reductase inhibitor (known to be capable of existing in a number of hydrate species) were found to exhibit varying fluorescence properties in their respective solid states (112). Type I was known to be a monohydrate species and was characterized by an excitation maximum at 345 nm and a main fluorescence maximum at 371 nm. Type II was a dihydrate whose photophysical parameters were equivalent to those of Type I but for which the overall levels of fluorescence were greatly reduced. Incorporation of additional waters of hydration into the substance produced Type III material, which exhibited an excitation maximum at 400 nm and a fluorescence maximum at 485 nm. Finally, upon binding of nine water molecules, the structure relaxed into a liquid crystalline form, which exhibited photophysical properties equivalent to those of type III, but which were considerably reduced in intensity.

The trihydrate phases of ampicillin and amoxicillin are almost isostructural, and yet their photoluminescence spectra have been found to be quite sensitive to small differences in solid-state structure (113). The solid-state excitation and emission spectra of both solids were dominated by energy transfer and exciton effects, which were manifested as decreases in the energy of the excitation and emission bands of the solid-state systems relative to those of the free molecule in solution. The photoluminescence data revealed that in spite of the known structural similarity of ampicillin trihydrate and amoxicillin trihydrate, the magnitude of the Davydov splitting and the degree of band energy shifting differed between the two systems. This finding indicated that the small differences in crystal structure existing between the two compounds leads to measurable differences in the patterns of energy transfer.

When the crystal structures of different polymorphic forms of the same compound are sufficiently different, these differences can be manifested in the photophysics of the system. It was found that exciton effects dominated the excitation spectra of four polymorphic forms of diflunisal, leading to a decrease in the energy of the excitation bands relative to that observed for the free molecule in fluid solution, and in a splitting of the excitation peak into two Davydov components (114). The trends in the excitation and emission spectra led to the grouping of diflunisal Forms I, II, and III into one category, and diflunisal Form-IV into a separate category. This categorization is illustrated in Figure 13, where the excitation spectra of Forms III and IV are plotted. Because it had been established that Form-IV was characterized by the highest crystal density and degree of intermolecular interaction, the magnitude of the exciton coupling could be deduced as being derived from the face-to-face overlap of the fluorescent salicylate-type fluorophores.

The anhydrous Form-III and dihydrate phases of carbamazepine were found to exhibit identical maxima in their excitation spectra (347 nm), but the intensities of the fluorescence associated with these solvatomorphs differed widely (115). The peak maxima associated with the emission spectra were also found to differ between the two forms, with 407 nm being observed for Form-III and 416 nm for the dihydrate phase. Because the emissive intensity of the dihydrate phase was approximately 3.15 times that of the anhydrate phase, this difference was exploited to develop a method for the study of the kinetics of the aqueous solution-mediated phase transformation between these forms. Studies were conducted at temperatures in the range of $18-40°C$, and it was found that the phase transformation was adequately characterized by

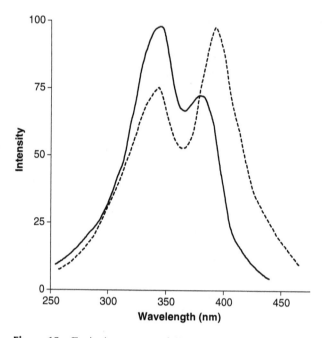

Figure 13   Excitation spectra of diflunisal, Form-III (*solid trace*) and Form-IV (*dashed trace*), normalized to a common intensity factor. *Source*: Adapted from Ref. 93.

first-order reaction kinetics. The temperature dependence in the calculated rate constants was used to calculate an activation energy of 11.2 kCal/mol (47.4 cal/g) for the anhydrate-to-dihydrate phase conversion.

PHOTOLUMINESCENCE FROM MOLECULES ADSORBED ON SURFACES

As has been discussed earlier, the photophysics of molecules in their bulk crystalline state is often significantly changed by the effects of intermolecular energy transfer, and the resulting excitation and emission spectra are often quite different from those of the corresponding isolated molecule. It has been found that the photophysics of luminescent molecules can also be altered in an entirely different way by adsorbing these molecules onto a solid surface (116–118). In this latter instance, the photophysics of the adsorbates will generally reflect details of their interaction with the host surface and can even provide information regarding the properties of the surface itself.

The types of research conducted on the photoluminescence from molecules adsorbed onto a solid surface can be divided into two broad classes. One of these is primarily chemical physics-oriented, where one typically uses the excitation and emission spectroscopy of an adsorbed molecule to probe its intrinsic photophysics as perturbed by the surface interaction. In this manner, one can obtain information on either the probe molecule itself or on the surface contacting the probe molecule. The other broad class of investigation is more analytical in nature, where an analyst seeks to enhance some photophysical property of the absorbed molecules in order to obtain qualitative or quantitative information on their composition or concentration.

Photophysical Studies of Fluorophore–Surface Interactions

Xanthene Dyes as Surface Probes

Owing to their favorable photophysical properties, and the sensitivity of these to their immediate environment, xanthene dyes have been used to study a number of solid surfaces. Lendvay published a comprehensive paper that illustrated how the adsorption of fluorescein onto various basic surfaces perturbed the photophysics of the system to differing degrees depending on the identity of the adsorbing substrate (119). In this study, fluorescein was adsorbed onto the surfaces of powdered basic metal hydroxides, and the resulting changes in fluorescence spectra were documented. Some of these experiments have been repeated, and the resulting emission spectra are shown in Figure 14 (27). Even though fluorescein contains only basic functional groups, its interactions with the surface hydroxyl groups of the solids causes a red-shifting of the luminescence that is highly dependent on the substrate. As evident in the figure, the shift of the emission band toward longer wavelengths followed the order of $Zn(OH)_2 > Mg(OH)_2 > Ca(OH)_2$, undoubtedly reflecting differing degrees and types of hydrogen bonding between the surface and the adsorbed dye.

When dye molecules are adsorbed onto crystal surfaces, it has been found that the presence of distorted adsorption sites can lead to the observation of reduced fluorescence lifetimes and altered photophysics. In one study, this phenomenon was interpreted in terms of a quenching mechanism where rhodamine B probe molecules at the defect sites of glass and naphthalene crystals acted as energy traps and as subsequent photoluminescent sites (120). In a more comprehensive study, the interactions of rhodamine B, rhodamine 101, and pyronine B with single crystals of naphthalene and phenanthrene were studied (121). In this latter work, the additional mechanism of enhanced internal conversion was also found to shorten the fluorescence lifetimes. In a detailed analysis of the decay kinetics, it was learned that the shortest lifetime fluorescence was associated with the most heavily distorted adsorption site.

From studies conducted on rhodamine B in the solution phase, it was learned that the compound could be induced to dimerize under the effect of hydrostatic pressure (122). The monomer and dimer peaks of rhodamine B were found to shift to lower energy levels with increasing pressure, as was anticipated for a $\pi \rightarrow \pi^*$ excitation process. When the dye was adsorbed onto the (0001) face of single crystals of zinc oxide, the presence of dimeric dye molecules was evident in the emission spectra as a second red-shifted peak, whose

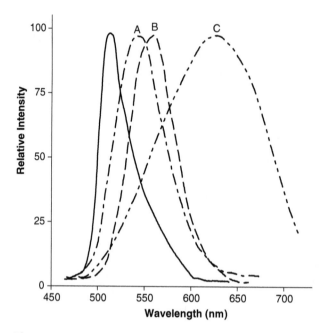

Figure 14 Emission spectra of fluorescein, normalized to a common intensity factor. Spectra are shown for the solution phase emission (*solid trace*), and for the fluorophore adsorbed onto $Zn(OH)_2$ (C), $Mg(OH)_2$ (A), and $Ca(OH)_2$ (B). *Source*: Adapted from Ref. 27.

intensity could exceed that of the monomer emission that was observed at low applied pressures. Through a study of the changes in molecular volume upon compression, it was deduced that the mechanism whereby rhodamine B interacted with the crystalline ZnO differed from the mode of interaction for cresyl violet adsorbed onto the same substrate (123).

The effect on dye photophysics upon adsorption onto amorphous phases has also been investigated, with picosecond photon counting fluorescence profiles being used to evaluate the properties of cresyl violet adsorbed onto amorphous quartz (124). The fluorescence decays were found to be consistent with the presence of monomeric cresyl violet molecules, and no conclusive evidence for dimeric species could be obtained. In another study, time-resolved fluorescence depolarization was measured by time-correlated photon counting for submonolayers of rhodamine 3B absorbed onto amorphous quartz (125). In this work, it was found that when the surface coverage exceeded 0.4 times that of a monolayer, aggregates of rhodamine 3B that formed were perturbed the photophysics in a less readily interpretable manner.

The phosphorescence of erythrosin B (tetraiodofluorescein) dispersed in amorphous thin films of maltose and maltitol has been used to monitor the molecular mobility of these matrices over the temperature range of $-25°C$ to over $110°C$ (126). Analysis of the energies of the emission peaks and their associated bandwidths, together with time-resolved intensity decay parameters, was used to deduce information about the modes of thermally induced mobility, because this mobility affected the rate of dipolar relaxation of the triplet state and the rate of intersystem crossing to the ground state. The mobility measured in mixed matrices was found to be less than that of either of the pure substance, a phenomenon that was considered to be comparable with the antiplasticization seen in mixtures of small molecule plasticizers with synthetic polymers and starch.

Pyrene as a Surface Probe

Another probe molecule that has been used in many studies is the polynuclear aromatic hydrocarbon, pyrene. This compound is known to form excimers in both the solution phase and in its crystalline state (14), and the details of its vibrational structure within its monomer emission are known to be dependent on the polarity of its environment (127). In a particular work, it was found from time-resolved spectral studies that both pyrene and naphthalene exhibited excimer-like fluorescence after being adsorbed onto silica gel (128). Because no ground-state complex was known to exist, this finding was viewed as being indicative of substantial interaction between the adsorbed molecules in their excited states.

A detailed study of the excitation and emission spectra, and the associated fluorescence decay, of pyrene adsorbed on silica gel, porous Vycor glass, alumina, and calcium fluoride have been reported (129). When silica was used as the solid substrate, monomer emission and a longer-wavelength, broadband, emission that decayed in a nonexponential manner was observed. The distribution of pyrene on the silica surfaces was deduced not to be random with its bimolecular association and its associated excimer-type fluorescence was stabilized at certain preferential sites. The data also indicated that the phenomenon was probably related to peculiarities in the hydrogen-bonding interaction between the surface and the π-electron system of the polynuclear aromatic hydrocarbon.

The interaction of surfactants with solid surfaces can be studied by first forming the substrate, then adsorbing the luminescent probe, and finally using the photophysical characteristics of the system to evaluate the surface characteristics. For example, the interaction of the surfactant Tritron X-100 [i.e., octylphenol poly(oxyethylene)] with solid silica surfaces has been investigated using pyrene as the fluorescent probe molecule (130). The approach in this work was to solubilize the probe molecule within the hydrophobic regions of the adsorption layer and to use the decay kinetics to understand the spatial extent to which the probe was free to move. It was concluded that the surfactant aggregates formed on the surface greatly resembled those of regular solution-phase micelles, although the regimes of structure type were found to be concentration dependent.

Because 1,3-di-*l*-pyrenylpropane is capable of forming intramolecular excimers, its use as a bifunctional pyrene probe molecule has been advanced as a means to avoid experimental complications associated with ground-state aggregation (131). The compound was adsorbed onto silica in the presence and absence of coadsorbed octanol, and also onto octadecylsilica, and then the photophysics studied using steady-state and single-photon counting methods. Evidence was obtained for dynamic intramolecular excimers on silica surfaces in the presence of octanol but not when the probe was adsorbed onto untreated silica. This latter effect on untreated silica was attributed to complete immobilization of the fluorescent probe, which prevented the occurrence of the excited-state association. The presence of a monolayer of adsorbed octanol on the silica surface served to provide the necessary molecular mobility to form excimer species, which paralleled those formed by pyrene in the same system.

The polarity of both polymeric and monomeric C18 chromatographic stationary phases has been evaluated from the fluorescence vibronic fine structure of pyrene adsorbed onto the surfaces of the packing materials (132). It was learned that the intercalation of the organic modifier into the stationary phase produced an inverse relationship between the polarity of the particle surface and that of the mobile phase, and that when the mobile phase contained the highest water content to adsorbed probe became partially exposed to the solution phase. This latter effect became manifested in perturbations of the photophysics of the probe molecule.

The illustrations given here for studies involving the characterization of xanthene dye and pyrene fluorescence probes adsorbed onto various solid surfaces represent only a few of the countless works published in this area. It is safe to state that significant advances in colloid science have been derived from the study of fluorescent probe molecules and their interactions with various interfacial structures, and interested readers are referred to the literature for discussions of these applications (133–135).

RTP of Adsorbed Luminescent Molecules

It is well established that techniques associated with luminescence spectroscopy have great analytical potential owing to the high degrees of sensitivity that can be achieved, and the selectivity that can be derived from a judicious choice of excitation and emission wavelengths. When luminescence studies are performed on solid surfaces, a variety of mechanisms serve to enhance the photophysical processes, and it is not uncommon to achieve limits of detection in the nanogram or subnanogram range. Because the instrumentation required for such work can be relatively simple and inexpensive, and because the analysis time can be exceedingly rapid, solid-surface luminescence spectroscopy has been extensively investigated for its potential as analytical methodology.

One of the more interesting consequences of adsorbing a sample onto a solid surface is that intersystem crossing from the manifold of singlet states can be facilitated into the lowest energy triplet state. This process is illustrated in the Jablonski-type energy level diagram of Figure 15, where the primary excitation into the lowest energy singlet state does not result in fluorescence but is instead followed by intersystem crossing into the lowest energy triplet state. The process of *phosphorescence* can then take place from the triplet state, in the form of the spin-forbidden transition down to the ground state. Phosphorimetry can, of course, be practiced in the solution phase but usually requires that the sample be cooled to cryogenic temperatures before the effect can be detected. When the luminescent material is absorbed onto certain solid surfaces, typically coabsorbed with an agent containing an atom of high atomic number that can promote the intersystem crossing process, room temperature phosphorescence can often be observed.

Vo-Dinh has summarized a number of general trends regarding the room temperature phosphorescence (RTP) of organic compounds and has provided extensive tables of organic compounds and their RTP properties (117). Polynuclear aromatic compounds are often intrinsically weakly phosphorescent due to the large singlet-triplet energy gap and the weak spin-orbit coupling between the $\pi\pi^*$ singlet and triplet states. Upon adsorption onto a solid surface in the

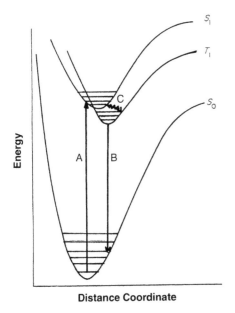

Figure 15   Jablonski-type energy level diagram of the ground and excited states of a fluorescent organic molecule, illustrating (A) the primary excitation into the lowest energy singlet state, (B) intersystem crossing into the lowest energy triplet state, and finally (C) phosphorescence taking place from the triplet state in the form of the spin-forbidden transition down to the ground state.

presence of a heavy atom, RTP is often observed. Nitrogen heterocyclic compounds normally do not exhibit phosphorescence but can exhibit a highly appreciable RTP upon adsorption on cellulose surfaces in the presence of silver or thallium salts. Several other substantial reviews of the phenomenon are available (116–118,136–138).

Photophysical and Mechanistic Studies

During a study of the decomposition of benzoyl peroxide in tetrachloroethylene by thin-layer chromatography, strong blue-green phosphorescence was observed from dried spots corresponding to the ammonium salts of biphenylcarboxylic acids (139). In studying this effect further, it was found that RTP could be obtained from the salts of a wide variety of the salts of polynuclear carboxylic or sulfonic acids, phenols, and amines adsorbed onto paper, silica, alumina, and other supports. A general trend reported in this paper was that the emission and excitation spectra of the adsorbed species were very similar to the spectra obtained from frozen solutions at cryogenic temperatures and that the emission lifetimes of the adsorbed molecules on the millisecond time scale were known to be typical for phosphorescence.

The effects of moisture, oxygen, substrate on the photophysical properties of the sodium salts of 4-biphenylcarboxylate and 1-naphthoate adsorbed onto cellulose paper, sucrose, starch, glass fiber paper, silanized paper, and stainless steel have been studied (140). It was learned that RTP could only be generated when the surface contained large quantities of available hydroxyl groups, presumably to appropriately bind and orient the adsorbed fluorophores through hydrogen bonding. At the same time, it was found that moisture and oxygen quenched the RTP, and hence an appropriate sample matrix would be one that was set up to resist the penetration and subsequent quenching effects of these agents.

The effect of temperature on the quantum yield for solid-surface luminescence of p-aminobenzoic acid adsorbed onto sodium acetate has been reported over the temperature range of $-180°C$ to $+23°C$, with these data being supplemented by measurements of the phosphorescence lifetime over a slightly wider temperature range (141). It was found that although the fluorescence quantum yields were almost temperature independent, the quantum yield for phosphorescence exhibited a strong temperature dependence. The results supported the model of a rigidly held fluorophore adsorbed onto the substrate, but where the adsorbed acid had been converted into its anion.

The effect of oxygen on the solid-matrix phosphorescence of perdeuterated phenanthrene adsorbed on different grades and types of paper has been reported (142). Effects associated with molecular diffusion on the surfaces were found to be minimal, and diffusional quenching did not contribute greatly to the photophysics. Static quenching by oxygen was found to be more significant on partially hydrophobic paper than on the usual Whatman filter papers, and this effect is illustrated in Figure 16. The luminescence decay results were evaluated using both linear and nonlinear fitting routines but yielded comparable results.

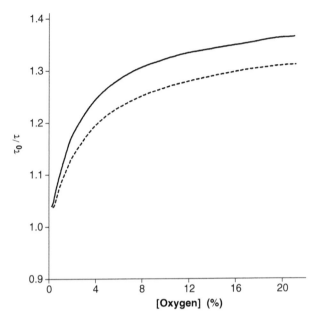

Figure 16 Ratios of the solid-matrix phosphorescence lifetimes obtained in the absence (τ_0) and presence (τ) of oxygen for perdeuterated phenanthrene adsorbed onto partially hydrophobic filter paper (*solid trace*) and on Whatman No. 1 filter paper (*short dashed trace*). *Source*: Adapted from Ref. 142.

The effect of coadsorbed alkali halide salts on the RTP of 2-naphthalenesulfonate adsorbed onto filter paper has been examined to determine the contribution made by the heavy atom effect (143). It was found that the RTP of this fluorophore was enhanced in the anticipated order of NaI > NaBr > NaCl > NaF, indicating that the external heavy atoms increase the radiative triplet decay constant more than the competing nonradiative rate constant. The results were consistent with the existence of an exchange interaction model between the heavy atom perturber and the fluorophore, where singlet ↔ triplet transition intensity was borrowed from electronic transitions of the perturbing halide ion.

The RTP characteristics of several aromatic ketones adsorbed onto filter paper were established, and compared with the phosphorescence spectra obtained at low temperatures in frozen matrices (144). It was found that ketones for which the cryogenic luminescence originated from the $^3(n\pi^*)$ state did not exhibit appreciable RPT, but that when the glassy phosphorescence originated from the $^3(\pi\pi^*)$ or triplet charge-transfer state, the room temperature phosphorescence was fairly intense. In addition, strong hydrogen bonding of the fluorophore to the hydroxyl groups of the filter paper was a prerequisite for the observation of RTP.

Both room temperature phosphorescence and delayed charge-transfer fluorescence were observed from coronene absorbed onto thermally activated γ-alumina (145). The fluorescence was only observed at low temperatures and was deduced to originate from the charge-transfer singlet state of an electron donor–acceptor complex between the adsorbed coronene and Lewis acceptor centers on the surface. On the other hand, the phosphorescence efficiency was found to be almost constant over the entire range of temperatures studied (100 to 500 K) and was attributed to a locally excited triplet state of the adsorbed coronene.

A method for the determination of quantum yield efficiencies of room temperature fluorescence or phosphorescence of organic molecules adsorbed onto a variety of solid surfaces has been described (146). The method makes use of a conventional scanning luminescence spectrometer and entails careful measurement of instrumental and sample blanks for proper referencing. The quantum yield of sodium salicylate adsorbed onto solid sodium acetate was established as a reference standard, with a value of 0.52 ± 0.03 being reported for a 1% dispersion in the absence of a nitrogen flow. The quantum yields of unknowns could then be easily determined as relative values through comparison with the standard, and such values were reported for a variety of adsorbed fluorophores.

Hurtubise has contributed an extensive review on the trace analysis of compounds using solid-matrix luminescence analysis (147). In this review, he compared the quantum yields for fluorescence and phosphorescence of several emissive compounds adsorbed onto a variety of solid matrices and compared many of these with the analogous solution phase values. Also discussed were the importance of rigidity in the substrate, the effects of water and gases on the photophysics, the effects of heavy atoms, and the heat capacity of the various systems.

Studies of Substrate Surfaces Appropriate for RTP Work

When adsorbed fluorophores are incapable of interacting in a site-specific manner with functional groups on the substrate surface, RTP will not be observed. As a result, a number of studies have been conducted where specific fluorophore types were adsorbed onto a wide variety of solid surfaces, and the resulting photophysical properties of the adsorbates were used to characterize the nature of the fluorophore–substrate interaction. For example, three molecules having known spectroscopic properties (benzo[a]phenazine, 1,4-diazatriphenylene, and Michlers ketone) were adsorbed onto Whatman 4 filter paper, and it was learned that the paper provided a very polar environment with a high degree of hydrogen bonding activity for the adsorbed molecules (148).

The RTP phenomenon has been investigated for eight indolecarboxylic acids adsorbed onto several ion-exchange filter papers (149). It was found that the largest RTP signals were obtained when using DE-81 anion-exchange paper, and substantial heavy atom effects were noted when using sodium iodide as a coadsorbate. In a related study, various Whatman filter and anion-exchange papers were evaluated for their utility as RTP substrates (150). The intensity of the RTP signals on these papers was found to be largely independent of the method used for the sample preparation, and also generally were more intense than the RTP intensities obtained for the same fluorophores when those were adsorbed onto silica gel thin-layer chromatography plates.

In another study, the RTP characteristics of a large number of aromatic ketones adsorbed onto a variety of solid substrates have been reported (151). In this work, the ability of silica gel chromaplates, filter paper, filter paper containing poly(acrylic) acid, and sodium chloride doped with 0.5% poly(acrylic) acid to promote RTP from adsorbed fluorophores was studied. Although it was found that no one surface yielded optimal RTP for all of the model compounds used, good analytical data were obtained from the majority of the surfaces investigated. However, the choice of the substrate still remained somewhat empirical, as no strong predictive model emerged from the work. For example, 2-acetonaphthone did not yield RTP after adsorption onto filter paper but did so from the other substrates studied. Were one to restrict the choice of surfaces to the family of filter papers (as has been recommended elsewhere), one would reach the erroneous conclusion that this particular compound could not exhibit the RTP phenomenon.

The RTF and RTP quantum yields, intersystem crossing efficiencies, and phosphorescence lifetime values have been obtained for p-amino-benzoic acid adsorbed onto sodium acetate or onto mixtures of sodium acetate and sodium chloride (152). It was deduced that the efficiency of RTF and RTP were determined primarily by how efficiently the support matrix became packed with the fluorophore molecules, although the effect was more important for the fluorescence process than it was for the phosphorescence process. Key to the luminescence induction process appeared to be the formation of a salt upon interaction of the p-aminobenzoic acid with the sodium acetate substrate, with the resulting anion being the species adsorbed into the matrix.

The luminescence properties of the p-aminobenzoate anion adsorbed onto sodium acetate and filter paper were compared to evaluate the interactions that determined the relative yields for fluorescence and phosphorescence (153). A relationship was found between the reciprocal of the phosphorescence lifetimes and the thermal processes that led to deactivation of the triplet state. Analysis of the results indicated that the p-aminobenzoate anion was incorporated into the sodium acetate crystal structure, and that its triplet energy was lost primarily into the skeletal vibrations of the support. For the anion adsorbed onto filter paper, it was concluded that although some of the triplet energy was lost into the vibrational modes of the filter paper, other factors were involved in the deactivation of the triplet state.

Interesting adsorption effects were observed when generating analytical calibration curves for benzo[f]quinoline adsorbed onto a silica thin-layer chromatography place under different pH conditions (154). As the concentration of analyte was increased, the calibration curves for the RTF were found to eventually level off at a constant value, whereas those for RTP increased to a maximum value and then decreased later. These findings were interpreted to indicate that the fluorescence process could take place from molecules adsorbed onto the surface and from multilayers of these molecules, but that the phosphorescence process could only originate from molecules adsorbed onto the surface itself and not in the multilayers.

Mixtures of cyclodextrins with sodium chloride have been found to constitute good substrates for the generation of RTP. Fifty-five organic compounds containing various functional groups were adsorbed onto an 80% α-cyclodextrin–NaCl matrix, and a large number of these yielded photophysical properties that were deemed to be appropriately selective and sensitive for development of analytical methods (155). In another study, it was found that use of a matrix consisting of 30% β-cyclodextrin–NaCl yielded appreciable RTP in most of the molecules studied without the use of an external heavy atom (156). In this latter work, it was shown that the luminescence spectra obtained using the cyclodextrin matrix possessed more fine structure than did the corresponding spectra obtained using filter paper as the substrate and that the degree of fine structure approach that was observed for the fluorophores was frozen in cryogenic media.

The enhancement of RTP from molecules adsorbed onto filter paper by the co-adsorption of α-, β-, or γ-cyclodextrins in the presence of heavy atoms has been studied (157). For most compounds, the use of α-cyclodextrin did not yield appreciable RTP, and it was also found that the RTP intensities could be amplified if the filter paper was pretreated with the cyclodextrin. However, it was clear that once the substrate surface sites were saturated with the cyclodextrin, the degree of RTP enhancement reached a limiting value.

The energy gap law showing an inverse correlation between the rates of non-radiative transitions from the lowest excited triplet states to the ground states and the differences in energies between the lowest vibrational energy levels of the excited and ground states has been shown to be applicable for the RTP of polynuclear aromatic molecules adsorbed onto cyclodextrin-salt matrices (158). Changes in the phosphorescence lifetime of the adsorbed fluorophore were correlated with the magnitude of the nonradiative rate constants, which were in turn correlated with the energy gap between the excited triplet and the ground states. Thus, upon measurement of the phosphorescence lifetime for a polynuclear aromatic hydrocarbon and the energy of its triplet state, one could predict whether the magnitude of the observed RTP intensity would be strong or weak. For example, given the phosphorescence lifetimes characteristic for polynuclear aromatic hydrocarbons, those molecules whose triplet energy was less than 12,000 cm^{-1} should not exhibit RTP.

Glasses prepared by the controlled evaporation of glucose or trehalose solutions have been found to function as matrices for the induction of RTP and suitable for use in trace analysis (159). The advantage of immobilizing the analyte in a glassy matrix is that the photophysics is essentially unaffected by moisture or oxygen, thus yielding a more robust analytical procedure. The disadvantage of the procedure is that the RTP signals tended to decrease over time, suggesting that the glassy matrices were not sufficiently rigid to withstand UV-induced damage. However, under the right set of circumstances, the easily prepared glucose glasses would be a superior choice for solid-matrix luminescence studies.

Glucose glasses have also been obtained from the molten state, and the solid-matrix luminescence of heterocyclic aromatic amines in these solids has been reported (160). The melt technique that permitted the exclusion of water from the system enhanced the luminescence quantum yields in most cases. In another work, the effects of glass preparative method and role of heavy atoms in the matrices were studied (161). It was deduced that the phosphorescence intensity was a function of how efficiently the triplet state could be populated and how quickly the triplet state became depopulated through phosphorescence. When sodium iodide was included in the glass matrix as the heavy atom, the triplet state formation became temperature independent, making the phosphorescence intensity dependent only on its lifetime.

The quenching of the solid-matrix phosphorescence associated with heterocyclic aromatic amines by water was studied and additional characterization performed by NIR spectroscopy and differential scanning calorimetry (DSC) (162). The NIR method was used to determine the water content in the glasses, and a modified Stern-Volmer equation was used to describe the changes in phosphorescence intensities and lifetimes. The DSC method was used to evaluate the glass transition temperatures of the matrices, and these values correlated with the photophysical results. The microenvironment of the glucose glasses was evaluated using a polarity probe, facilitating development of a model that described the effects of the different parameters on the photophysical properties. For example, water was found to be able to diffuse though channels in the glasses to cause dynamic quenching and also served to disrupt the hydrogen bonding network that structured the glasses.

Analysis of Mixtures

As discussed earlier, luminescence spectroscopy contains an extra element of selectivity in that even for steady-state work, the analyst has control over both the excitation and emission wavelengths. The selectivity possible with the technique becomes even more advantageous when one adds in additional refinements, such as time resolution or mathematical processing of the data. For example, derivative techniques were used for the identification of components in binary and ternary mixtures of nitrogen heterocyclic compounds (160). Direct, first, and second derivative RTP spectra were used for such identifications by matching spectral wavelengths from mixtures with the spectral wavelengths of standards.

Different nitrogen heterocyclic compounds and a polycyclic aromatic hydrocarbon were combined to form various binary and ternary mixtures, and then the components were determined at the nanogram level by RTF or RTP in conjunction with selective excitation or emission monitoring (161). Multicomponent mixtures of polynuclear aromatic hydrocarbons were absorbed onto α-cyclodextrin/sodium chloride mixtures, and then characterized by their solid-state photoluminescence (162). Using a protocol involving extraction, selective excitation, and photoluminescent detection, mixtures containing as many as nine components could be analyzed at the nanogram level.

The RTP obtained from eight benzoquinone isomers can be resolved into the linear or angular subgroups on the basis of the conventional luminescence spectra, but the selectivity can be improved using second derivative and synchronous scanning techniques (163). The approach was applied to determining the levels of acridine, benzo(*h*)quinoline, and phenanthridine in a coal tar fraction. In this work, substantial differences were noted between luminescence and HPLC measurements, suggesting that although photoluminescence can be used as a screening tool, it should be combined with a separation step to obtain optimal results.

In another report, the use of combined solid-surface fluorescence and phosphorescence for the determination of several luminescence compounds in different mixtures has been studied (164). Parameters—such as the pH of the sample environment, the nature and preparation of the solid substrates used, and the choice of heavy atom perturbers—were used to selectively determine the components. It was found that even more selectivity could be introduced into the RTF and RTP approaches by working with the optimal substrate and heavy atom, effectively removing contributions made by substances other than the analytes.

An analysis method, based on solid-matrix luminescence, has been reported for the determination of 1-hydroxypyrene, tetrol, and 9-hydroxyphenanthrene in urine fractions (165). The greatest analytical sensitivity was obtained in the phosphorescence mode, with the compounds being absorbed onto Whatman 1PS paper with thallium nitrate as the heavy-atom agent. The 9-hydroxyphenanthrene metabolite was found to undergo photobleaching under the analysis conditions, so its limit of detection could only be estimated. However, by taking suitable advantage of the differing photophysical parameters, it was possible to identify all compounds in spiked urine fractions.

Compounds Having Pharmaceutical Interest

The utility of RTP spectroscopy to the determination of 28 pharmaceutical compounds has been reported, with deliberate changes in experimental conditions being used to maximize the sensitivity (166). It was reported that the best heavy atom perturber and promoter of RTP was iodide ion and that Ag(I) appeared to have little effect. Generally, when RTP was observed, the emission bands were shifted to lower energies relative to the phosphorescence bands obtained in frozen cryogenic matrices. Interestingly, a connection between molecular structure and phosphorescence characteristics could not be drawn, with closely related compounds often exhibiting vast different RTP intensities. However, the RTP methodology was shown to be simple and selective and, therefore, suitable for use in clinical analysis.

The effects of anionic, nonionic, and cationic surfactants on the RTP of 11 phenothiazine derivatives have been described (167). For these compounds, strong RTP could be observed once the compounds were adsorbed onto filter paper, but in the presence of coadsorbed thallium(I) nitrate as a heavy atom perturbing agent, a tripling of the RTP signal could be obtained. Interestingly, use of the Tl(I) heavy atom in the presence of the surfactants resulted in a quenching of the RTP. As part of a much larger study of 30 compounds having pharmaceutical interest, a study of eight phenothiazine compounds was conducted (168). A wide variety of photophysical properties were noted for the various compounds, and the intensity of RTP for a given substance was not readily predictable. In fact, more compounds were found to phosphoresce at room temperature when adsorbed onto filter paper that could phosphoresce in frozen ethanol solutions.

Room temperature phosphorimetry of a mixture of two methylxanthines has been reported, but because the spectral characteristics of caffeine and theophylline are so similar, the prior use of a separation procedure was required (169). In another work, RTP was used as an alternative detection technique for the chromatographic determination of caffeine, theophylline, and theobromine in urine samples (170). Although the analytical performance parameters associated with the RTP and conventional UV absorbance methods were comparable, superior selectivity was achieved with the RTP. One of the more interesting aspects of the latter method was the idea that chromatograms could be stored to document the assay results, but this proposal has become dated in this day of automated data documentation.

The RTP properties of diazepam were investigated as a function of the type of filter paper used as the solid substrate, the choice of heavy atom used to enhance the triplet state yield, and the pH of the initial spotting solution (171). It was concluded that the largest RTP signal for diazepam was obtained when the substance was adsorbed onto Whatman No. 1 filter paper from an acidic solution, and using Hg(II) as the heavy atom. Tablets containing 10 mg of active ingredient could be assayed by the RTP method with good analytical performance parameters, and the method could also be used to obtain good results for the determination of diazepam in serum. Spiking of diazepam into serum yielded low recoveries because the substance is known to be highly protein-bound, but through the use of an extraction procedure the analyte could be quantitatively recovered.

The mutagenic and carcinogenic properties of polycyclic aromatic hydrocarbons are well documented, and therefore there is a continuing interest in methodology to evaluate their toxic effects. Some of these compounds can be metabolized into compounds that are able to form adducts with DNA, and these adducts have been the focus in cancer research. One compound that has received extensive interest is benzo[*a*]pyrene, and not surprisingly it has been found that solid-matrix RTP a useful technique for the characterization of its adducts with DNA.

Solid-matrix fluorescence and phosphorescence were obtained for DNA adducts of benzo[*a*]pyrene when these complexes were adsorbed on Whatman No. 1 and 1PS filter papers together with thallium nitrate as the heavy-atom inducer (172). Using selective excitation, it was possible to excite different forms of the adducts, and moisture quenching of the phosphorescence was used to detect quasi-intercalated adducts. In other studies, solid-matrix phosphorescence was used study the adducts formed by DNA with benzo[*a*]pyrene-7,8-dihydrodiol-9,10 epoxide (173) and racemic *anti*-dibenzo[*a,1*]pyrene-11,12-diol-13,14-epoxide (174).

It has been established that isomeric tetrols of benzo[*a*]pyrene are formed by the acid hydrolysis of the benzo[*a*]pyrene-DNA adducts, and consequently study of the spectroscopic properties of these tetrols has proven important. In one study, four tetrols were separated by HPLC, and the chromatographic fractions characterized as to their solid-matrix phosphorescence properties (175). It was reported that the compounds could be detected at the pictogram levels, and that the luminescence method could be used to identify levels of compounds that were undetectable by HPLC. Additional luminescence studies have been reported for tetrols formed from benzo[*a*]pyrene-DNA adducts and that have been supported on α-cyclodextrin/NaCl (176) and β-cyclodextrin/salt mixtures (177).

The effects of a heavy-atom salt on the solid matrix luminescence properties of the racemic *trans*-7,8-dihydroxy-*anti*-9,10-epoxy-7,8,9,10-tetrahydrobenzo[*a*]pyrene-DNA adducts and tetrol I-1 were studied on several substrates (178). Although thallium nitrate was effective in promoting RTP, the RTF was quenched. It was futher reported that the most superior substrates for inducing phosphorescence were 1% α-cyclodextrin/TlNO$_3$, 10% trehalose/TlNO$_3$, and 30% TlNO$_3$/sodium acetate. Subsequently, the quenching of the solid matrix fluorescence by sodium iodide and thallium nitrate was used to investigate the nature of the adducts (179). When using thallium nitrate as the quenching agent, a two-site model with two independent quenching sites was applicable to the data. However, for sodium iodide quenching, the data for tetrol I-1 were fit to the sphere of action model, but the data for the DNA adducts was qualitatively related to a BET isotherm. Additional insight into the quenching mechanisms was derived in a subsequent study, where it was also learned that the DNA in the adducts were held very rigidly to the matrix, thus facilitating their interaction with the heavy atom (180).

The RTP approach has been applied to the study of other polynuclear aromatic hydrocarbons. The solid matrix spectroscopy of benzo[*f*]quinoline and benzo[*h*]-quinoline adsorbed on silica gel chromatography subsequently submerged in mixed chloroform/hexane solvents (181). The phosphorescence data were interpreted to indicate that the emissive compounds were protected from collisional deactivation by the matrix and that adsorbed chloroform minimally disrupted the interactions between the adsorbate and the substrate. Fluorescence and phosphorescence quantum yields, polarization values, and emission lifetimes were obtained for benzo[*f*]quinoline adsorbed onto β-cyclodextrin/NaCl matrices over a wide range of temperatures (182). As is often noted in systems of this type, the fluorescence data exhibited only a minor temperature dependence, but temperature was found to exert a strong influence on the phosphorescence parameters.

REFERENCES

1. Pringsheim P. Luminescence of Liquids and Solids. New York: Interscience Pub., 1943.
2. De Ment J. Fluorochemistry. New York: Chemical Pub. Co., 1945.
3. Pringsheim P. Fluorescence and Phosphorescence. New York: Interscience Pub., 1949.

4. Radley JA. Fluorescence Analysis in Ultraviolet Light. London: Chapman & Hall, 1954.
5. Forster T. Fluoreszenz Organischer Verbindungen. Gottingen: Vandenhoeck und Ruprecht, 1951.
6. Berlman IB. Handbook of Fluorescence Spectra of Aromatic Molecules. New York: Academic Press, 1965.
7. Hercules DM. Fluorescence and Phosphorescence Analysis. New York: Interscience Pub., 1966.
8. Calvert JG, Pitts JN. Photochemistry. New York: John Wiley & Sons, 1966.
9. Guilbault GG. Fluorescence: Theory, Instrumentation, and Practice. New York: Marcel Dekker, 1967.
10. Parker CA. Photoluminescence of Solutions. New York: Elsevier Pub., 1968.
11. Zander M. Phosphorimetry. New York: Academic Press, 1968.
12. Becker RS. Theory and Interpretation of Fluorescence and Phosphorescence. New York: Wiley-Interscience, 1969.
13. Glynn SP, Azumi T, Kinoshita M. Molecular Spectroscopy of the Triplet State. Englewood Cliffs, NJ: Prentice-Hall, 1969.
14. Birks JB. Photophysics of Aromatic Molecules. New York: J. Wiley & Sons, Inc., 1970.
15. White CE, Argauer RJ. Fluorescence Analysis: A Practical Approach. New York: Marcel Dekker, 1970.
16. Winefordner JD, Schulman SG, O'Haver, TC. Luminescence Spectrometry in Analytical Chemistry. New York: John Wiley & Sons, 1972.
17. Guilbault GG. Practical Fluorescence: Theory, Methods, and Techniques. New York: Marcel Dekker, 1973.
18. Schulman SG. Fluorescence and Phosphorescence Spectroscopy: Physicochemical Principles and Practice. Oxford: Pergamon Press, 1977.
19. Lakowicz JR. Principles of Fluorescence Spectroscopy. New York: Plenum Press, 1983.
20. Demas JN. Excited State Lifetime Measurements. New York: Academic Press, 1983.
21. Baeyens WRG, De Keukeleire D, Korkidis K. Luminescence Techniques in Chemical and Biochemical Analysis. New York: Marcel Dekker, 1991.
22. Lakowicz JR. Principles of Fluorescence Spectroscopy. 2nd ed. New York: Kluwer Academic/Plenum Pub., 1999.
23. Udenfriend S. Fluorescence Assay in Biology and Medicine. New York: Academic Press, volume 1 (1962); volume 2 (1969).
24. Wehry EL. Modern Fluorescence Spectroscopy. New York: Plenum Press, volume 1 (1976); volume 2 (1976); volume 3 (1981); volume 4 (1981).
25. Schulman SG. Molecular Luminescence Spectroscopy. New York: John Wiley & Sons, volume 1 (1985); volume 2 (1988); volume 3 (1993).
26. Passwater RA. Guide to the Fluorescence Literature. New York: Plenum Press, volume 1 (1967); volume 2 (1970); volume 3 (1974).
27. Brittain HG. unpublished fluorescence results obtained using a Perkin-Elmer model LS-5B luminescence spectrometer, whose sample compartment has been modified to facilitate front-surface excitation and emission spectroscopies of solids contained in thin-walled glass tubes.
28. Bowen EJ. Proc Royal Soc London 1936; A154:349.
29. Strickler SJ, Berg RA. J. Chem Phys 1962; 37:814.

30. Stern O, Volmer M. Phys Z 1919; 20:183.
31. Bowen EJ. Trans Far Soc 1954; 50:97.
32. Boaz H, Rollefson GK. J Am Chem Soc 1950; 72:3435.
33. Bowen EJ, Metcalf WS. Proc Royal Soc London 1951; A206:437.
34. Keizer J. J Am Chem Soc 1983; 105:1494.
35. Blatt E, Mau AW-H, Sasse WHF, Sawyer WH. Aust J Chem 1988; 41:127.
36. Gabor G, Turro NJ. Photochem Photobiol 1985; 42:447.
37. See reference [14], pp. 420–491.
38. Forster T, Kasper K. Z Physik Chem (Frankfurt) 1954; 1:275.
39. Forster T, Kasper K. Z Elektrochem 1955; 59:977.
40. For a detailed discussion of excimer kinetics, see reference [10], pp. 347–356.
41. Kasper K. Z Physik Chem (Frankfurt) 1959; 12:52.
42. Doller E, Forster T. Z Physik Chem (Frankfurt) 1962; 34:132.
43. Hirayama F. J Chem Phys 1965; 42:3363.
44. Chandross EA, Dempster CJ. J Am Chem Soc 1970; 92:3586.
45. Handa T, Utena Y, Yajima H. J Phys Chem 1984; 88:5150.
46. Brittain HG, Ambrozich DL; Saburi M, Fendler JH. J Am Chem Soc 1980; 102:6372.
47. Anufrieva EV, Ghiggino KP, Gotlig YY, Phillips D, Roberts AJ. Luminescence, Advances in Polymer Science. 40:1981.
48. Ander P, Mahmoudhagh MK. Macromolecules 1982; 15:213.
49. Turro NJ, Ukubo T. J Phys Chem 1982; 86:1485.
50. Froehlich P, Wehry EL. "The Study of Excited-State Complexes ('Exciplexes) by Fluorescence Spectroscopy", chapter 8 in Modern Fluorescence Spectroscopy. Wehry EL. ed., volume 2, New York: Plenum Press, 1976, pp. 319–438.
51. Knibbe H, Rehm D, Weller A. Ber Bunsenges Phys Chem 1968; 72:257.
52. Lewis FD. Acc Chem Res 1979; 12:152.
53. Lewis FD, Kalgutkar RS, Kurth TL. J Phys Chem 2004; 108:1425.
54. Irie M, Yorozu T, Hayashi K. J Am Chem Soc 1978; 100:2236.
55. Kroger FA. Some Aspects of the Luminescence of Solids. New York: Elsevier, 1948.
56. Garlick GFJ. Luminescent Materials. Oxford: Clarendon Press, 1949.
57. Leverenz HW. An Introduction to Luminescence of Solids. New York: John Wiley & Sons, 1950.
58. Kallmann HP, Spruch GM. Luminescence of Organic and Inorganic Materials. New York: John Wiley & Sons, 1962.
59. Curie D. Luminescence in Crystals. London: Methuen & Co., 1963. This book is a translation of Luminescence Cristalline. Paris: Dunod Pub., 1960.
60. Goldberg P. Luminescence of Inorganic Solids. New York: Academic Press, 1966.
61. Frenkel JI. Phys Rev 1931; 37(17): 1276.
62. Davydov AS. Theory of Molecular Excitons. New York: McGraw-Hill, 1962.
63. McClure DS. Electronic Spectra of Molecules and Ions in Crystals. New York: Academic Press, 1959.
64. Knox RS. Theory of Excitons. New York: Academic Press, 1963.
65. Craig DP, Walmsley SH. Excitons in Molecular Crystals. New York: Benjamin WA, 1968.
66. Brittain HG, unpublished fluorescence results obtained using high-resolution spectrometers constructed at the Center for Pharmaceutical Physics. The excitation spectrometer used a 0.5 meter monochromator (Spex model 1870) to discriminate

the output of a 300 watt xenon arc lamp at a resolution of 0.6 nm. The fluorescence was obtained using front-face excitation, allowed to pass through a long-pass filter (1% w/v solution of sodium nitrite in a 1 cm cell) to remove scattering light, and then detected by an end-on photomultiplier tube (Thorn EMI type 6256QB having S-11 response). The emission spectrometer used a combination of glass and solution filters to obtain the desired output of a 250 watt xenon arc lamp. The filtered light impinged on the sample in the front-face excitation mode, and the fluorescence analyzed by a 0.5 meter monochromator (Spex model 1870) at a resolution of 0.8 nm. The fluorescence was detected by an end-on photomultiplier tube (Thorn EMI type 9558QB having S-20 response).

67. Mielenz KD. Measurement of Photoluminescence. New York: Academic Press, 1982.
68. Budde W. Physical Detectors of Optical Radiation. New York: Academic Press, 1983.
69. Talmi Y. Multichannel Image Detectors. Washington, DC: ACS Symposium Series 102, American Chemical Society, 1979.
70. Christian GD, James B. Callis, Davidson ER. "Array Detectors and Excitation-Emission Matrices in Multicomponent Analysis", chapter 4 in Modern Fluorescence Spectroscopy. volume 4, Wehry EL. ed., New York: Plenum Press, 1981, pp. 111–165.
71. Vo-Dinh T. Anal Chem 1978; 50:396.
72. Lloyd JBF, Evett IW. Anal Chem 1978; 50:1710.
73. Green GL, O'Haver TC. Anal Chem 1974; 46:2191.
74. John P, Soutar I. Anal Chem 1976; 48:520.
75. Munoz de la Pena A, Salinas F, Duran-Meras I. Anal Chem 1988; 60:2493.
76. Munoz de la Pena A, Duran-Meras I, Duran MS. Analyst 1990; 115: 1007.
77. Miller JM. Analyst 1984; 109, 191.
78. Inman EL, Winefordner JD. Anal Chim Acta 1982; 138, 245.
79. Warner IM, Patonay G, Thomas MP. Anal Chem 1985; 57, 463A.
80. Ndou TT, Warner IM. Chem Rev 1991; 91:493.
81. Demas JN. "Time-Resolved and Phase-Resolved Emission Spectroscopy", chapter 2 in Molecular Luminescence Spectroscopy, volume 2, Schulman SG. ed., New York: John Wiley & Sons, 1988, pp. 79–127.
82. Cundall RB, Dale RE. Time-Resolved Fluorescence Spectroscopy in Biochemistry and Biology. New York: Plenum Press, 1983.
83. O'Connor DV, Phillips D. Time-Correlated Single Photon Counting. London: Academic Press, 1984.
84. Reference 22, pp. 101–110.
85. McGown LB, Bright FV. Anal Chem 1984; 56:1400A.
86. Gratton E, Jameson DM, Hall RD. Ann Rev Biophys Bioeng 1984; 13:105.
87. McGown LB, Bright FV. Crit Rev Anal Chem 1987; 18:245.
88. Schmillen A, Legler R. Luminescence of Organic Substances. Berlin: Springer-Verlag, 1967.
89. Maeda M. Laser Dyes. Tokyo: Academic Press, 1984.
90. Baeyens WRG. "Fluorescence and Phosphorescence of Pharmaceuticals," Molecular Luminescence Spectroscopy. Volume 1, Schulman SG. ed., New York: J. Wiley & Sons, Inc., 1985, pp. 29–166.
91. Wolfbeis OS. "Fluorescence of Organic Natural Products," Molecular Luminescence Spectroscopy. Volume 1, Schulman SG. ed., New York: J. Wiley & Sons, Inc., 1985, pp. 167–370.

92. Reference [17], pp. 80–81.
93. Shpak MT, Sheka EF. Optics and Spectroscopy 1960; 9:29.
94. Kopelman R. J Phys Chem 1976; 80:2191.
95. Argyrakis P, Blumen A, Kopelman R, Zumofen G. J Phys Chem 1984; 88:1973.
96. Argyrakis P, Hooper D, Kopelman R. J Phys Chem 1983; 87:1467.
97. Gentry ST, Kopelman R. J Phys Chem 1984; 88:3170.
98. Craig DP, Hobbins PC. J Chem Soc 539 (1955); Craig DP. *ibid.*, 2302 (1955); Craig DP, Hobbins PC. *ibid.*, 2309 (1955).
99. Bree A, Lyons LE. J Chem Soc 2658 (1956); Bree A, Lyons LE. *ibid.*, 2662 (1956); Lyons LE, Morris GC. *ibid.*, 1551 (1959); Bree A, Lyons LE. *ibid.*, 5206 (1960).
100. Lacey AR, Lyons LE. J Chem Soc 5393 (1964).
101. Akon CD, Craig DP. Trans Far Soc 1966; 62:1673.
102. Taliani C, Bree A. J Phys Chem 1984; 88:2351.
103. Taen S. J Phys Chem 1988; 92:107.
104. Ikeyama T, Azumi T. J Phys Chem 1985; 89:5332.
105. Bonno B, Laporte JL, Rousset Y. J Luminescence 1983; 28:443.
106. Grahm DJ. J Phys Chem 1985; 89:5330.
107. Singham SB, Pratt DW. J Phys Chem 1982; 86:507.
108. Wakayama NI. J Luminescence 1982; 27:299.
109. Becker H-D, Sandros K, Skelton BW, White AH. J Phys Chem 1981; 85:2923.
110. Becker H-D, Sandros K, Skelton BW, White AH. J Phys Chem 1981; 85:2927.
111. Becker H-D, Sandros K, Skelton BW, White AH. J Phys Chem 1981; 85:2930.
112. Brittain HG, Ranadive SA, Serajuddin ATM. Pharm Res 1995; 12:556.
113. Brittain HG. AAPS PharmSciTech 2005; 5. In press.
114. Brittain HG, Elder BJ, Isbester PK, Salerno AH. Pharm Res 2005; 22:999.
115. Brittain HG. J Pharm Sci 2004; 93:375.
116. Hurtubise RJ. Solid Surface Luminescence Analysis. New York: Marcel Dekker, 1981.
117. Vo-Dinh, T. Room Temperature Phosphorimetry for Chemical Analysis. New York: John Wiley & Sons, 1984.
118. Hurtubise RJ. "Luminescence from Solid Surfaces," chapter 1 in Molecular Luminescence Spectroscopy. volume 2, Schulman SG. ed., New York: John Wiley & Sons, 1988, pp. 1–77.
119. Lendvay E. J Phys Chem 1965; 69:738.
120. Klemnitz K, Murao T, Yamazaki I, Nakashima N, Yoshihara K. Chem Phys Lett 1983; 101:337.
121. Klemnitz K, Tamai N, Yamazaki I, Nakashima N, Yoshihara K. J Phys Chem 1987; 91:1423.
122. Roberts ER, Drickamer HG. J Phys Chem 1985; 89:3092.
123. Clark FT, Drickamer HG. J Chem Phys 1984; 81:1024.
124. Anfinrud P, Crackel RL, Struve WS. J Phys Chem 1984; 88:5873.
125. Anfinrud P, Hart DE, Hedstrom JF, Struve WS. J Phys Chem 1986; 90:3116.
126. Shirke S, Takhistov P, Ludescher RD. J Phys Chem B 2005; 109:16119.
127. Kalyanasundaram K, Thomas JK. J Am Chem Soc 1977; 99:2039.
128. Hara K, de Mayo P, Ware WR, Weedon AC, Wong GS, Wu KC. Chem Phys Lett 1980; 69:105.
129. Bauer RK, de Mayo P, Ware WR, Wu KC. J Phys Chem 1982; 86:3781.
130. Levitz P, van Damme H, Keravis D. J Phys Chem 1984; 88:2228.

131. Avnir D, Russe R, Ottolenghi M, Wellner E, Zachariasse KA. J Phys Chem 1985; 89:3521.
132. Carr JW, Harris JM. Anal Chem 1986; 58:626.
133. Thomas JK. Chemistry of Excitation at Interfaces. Washington, DC: ACS Monograph 191, American Chemical Society, 1983.
134. Fendler JH. Membrane Mimetic Chemistry. New York: John Wiley & Sons, 1982.
135. Thomas JK. J Phys Chem 1987; 91:267.
136. Vo-Dinh T, Winefordner JD, Appl Spect Rev 1977; 13:261.
137. Hurtubise RJ. Anal Chem 1989; 61:889A.
138. Gunshefski M, Santana JJ, Stephenson J, Winefordner JD. Appl Spect Rev 1992; 27:143.
139. Schulman EM, Walling C. J Phys Chem 1973; 77:902.
140. Schulman EM, Parker RT. J Phys Chem 1977; 81:1932.
141. Ramasamy SM, Hurtubise RJ. Anal Chem 1987; 59:432.
142. Hurtubise RJ, Ackerman AH, Smith BW. Appl Spectrosc. 2001; 55:490.
143. White W, Seybold PG. J Phys Chem 1977; 81:2035.
144. Scharf G, Winefordner JD. Talanta 1986; 33:17.
145. Honnen W, Krabicher G, Uhl S, Oelkrug D. J Phys Chem 1983; 87:4872.
146. Ramasamy SM, Senthilnathan VP, Hurtubise RJ. Anal Chem 1986; 58:612.
147. Hurtubise RJ. Anal Chim Acta 1997; 351, 1.
148. Suter GW, Kallir AJ, Wild UP, T Vo-Dinh. J Phys Chem 1986; 90:4941.
149. Andino M, Aaron JJ, Winefordner JD. Talanta 1986; 33:27.
150. Fidanza J, Aaron JJ. Talanta 1986; 33:215.
151. Citta LA, Hurtubise RJ. Microchem J 1986; 34:56.
152. Ramasamy SM, Hurtubise RJ. Anal Chem 1987; 59:2144.
153. Ramasamy SM, Hurtubise RJ. Anal Chem 1990: 62:1060.
154. Burrell GJ, Hurtubise RJ. Anal Chem 1987; 59:965.
155. Bello J, Hurtubise RJ. Appl Spect 1986; 40:790.
156. Richmond MD, Hurtubise RJ. Anal Chem 1989; 61:2643.
157. Alak AM, Contolini N, Vo-Dinh T. Anal Chim Acta 1989; 217:171.
158. Ramasamy SM, Hurtubise RJ. Appl Spect 1996; 50:115.
159. Wang J, Hurtubise RJ. Appl Spect 1996; 50:53.
160. Mendonsa SD, Hurtubise RJ. Appl Spect 2000; 54:456.
161. Mendonsa SD, Hurtubise RJ. J Lumin. 2002; 97:19.
162. Bello LM, Hurtubise RJ. Anal Chem 1988; 60:1285.
163. Vo-Dinh T, Miller GH, Abbott DW, Moody RL, Ma CY, Ho C-H. Anal Chim Acta 1985; 175:181.
164. Asafu-Ajaye EB, Su SY. Anal Chem 1986; 58:539.
165. Smith BW, Hurtubise RJ. Appl Spectrosc, 2000; 54:1357.
166. Bower ELY, Winefordner JD. Anal Chim Acta 1978; 101:319.
167. MCG Alvarez-Coque, Ramos GR, O'Reilly AM, Winefordner JD. Anal Chim Acta 1988; 204:247.
168. Khasawneh IM, Alvarez-Coque MCG, Ramos GR, Winefordner JD. J Pharm Biomed Anal 1989; 7:29.
169. Perry LM, Shao EY, Winefordner JD, Talanta 1989; 36:1037.
170. Campiglia AD, Iaserna JJ, Berthod A, Winefordner JD. Anal Chim Acta 1991; 244:215.

171. Andino MM, Winefordner JD. J Pharm Biomed Anal, 1986; 4:317.
172. Tjioe SW, Hurtubise RJ. Appl Spectrosc 1998; 52:414.
173. Li M, Hurtubise RJ, Weston A. Anal Chem 1999; 71:4679.
174. Thompson AJ, Hurtubise RJ. Appl Spectrosc 2005; 59:126.
175. Tjioe SW, Hurtubise RJ. Talanta 1995; 42, 59.
176. Corley J, Hurtubise RJ. Anal Chem 1993; 65:2601.
177. Richmond MD, Hurtubise RJ. Anal Chim Acta 1991; 255:335.
178. Chu Y, Hurtubise RJ. Appl Spectrosc 1996; 50:476.
179. Smith BW, Hurtubise RJ. Appl Spectrosc 2003; 57:943.
180. Smith BW, Hurtubise RJ. Anal Chim Acta 2004; 502:149.
181. Burrell GJ, Hurtubise RJ. Anal Chem 1988; 60:564.
182. Richmond MD, Hurtubise RJ. Anal Chem 1991; 63:169.

_____ *7* _____

Molecular Motion and Vibrational Spectroscopy

Harry G. Brittain
Center for Pharmaceutical Physics, Milford, New Jersey, U.S.A.

INTRODUCTION

The energies associated with the fundamental vibrational modes of a chemical compound lie within the range of $400-4000$ cm^{-1}, a spectral region corresponding to midinfrared (mid-IR) electromagnetic radiation. Transitions among vibrational energy levels can be observed directly through their absorbance in the infrared (IR) region of the spectrum or through inelastic scattering of incident energy via the Raman effect. Overtones and combination bands of vibrational modes are observed in the near-infrared (NIR) region of the spectrum (4000 to 13,350 cm^{-1}).

IR absorption spectroscopy, especially when measured by means of the Fourier transform method (FTIR), is a powerful technique for the physical characterization of pharmaceutical solids. When the structural characteristics of a solid perturb the pattern of vibrational motion for a given molecule, one can use these alterations as a means to study the solid-state chemistry of the system. FTIR spectra are often used to evaluate the type of polymorphism existing in a drug substance and can be very useful in studies of water contained within a hydrate species.

The vibrational energy levels of a molecule in its solid form can also be studied using Raman spectroscopy. Because most compounds of pharmaceutical interest are characterized by low molecular symmetry, the same bands observed in the IR absorption spectrum will also be observed in the Raman spectrum. However, the fundamentally different nature of the selection rules associated with the Raman effect leads to the observation of significant differences in peak intensity between the two methods. In general, symmetric vibrations and nonpolar groups yield the most intense Raman scattering bands, whereas antisymmetric vibrations and polar groups yield the most intense IR absorption bands.

Spectral transitions that are observed in the NIR region of the spectrum are associated with overtone and combinations of the fundamental bands detected in IR absorption or Raman spectroscopies. Some of the spectral features that have found the greatest utility are those functional groups that contain unique hydrogen atoms. For example, studies of water in solids can be easily performed through systematic characterization of the characteristic —OH band, usually observed around 5170 cm^{-1}. The determination of hydrate species in an anhydrous matrix can easily be performed using NIR analysis.

In this chapter, an overview of the theoretical foundations for all three types of vibrational spectroscopy will be presented. It goes without saying that interested readers can considerably find more complexity and details in the literature, with the two books by Gerhard Herzberg forming the best possible references to consult (1,2). Support for the discussed quantum mechanical presentations are available in references (3–8), and more specific details of molecular spectroscopy can be found in references (9–13). Finally, a listing of significant books covering vibrational spectroscopy is located in references (14–24).

MOTION OF NUCLEI IN MOLECULES

Certain types of molecular spectroscopy have their origins in the patterns of motion executed by the atoms making up the molecule, and it is intuitive that the energies associated with these motions are the key to a theoretical interpretation of the observed spectra. It is therefore important to consider the nature of these motions and the manner in which energy can be distributed among the various types of motion. In order to simplify this discussion, the perturbing effects of other molecules in a condensed state will not be considered, but instead we will develop the state of a molecule that is essentially free to move through space. The perturbing effects of the solid state on the properties of the individual molecule will be developed in the succeeding chapters.

At the outset, it is essential to define a useful term in this connection, namely the *degree of freedom*. Essentially, the number of degrees of freedom that any body night possess is related to the number of independent coordinates that are needed to specify its position in space. For instance, a sphere that is able to move randomly in space would require the specification of all three of the Cartesian coordinates (x, y, and z) to define its position. It can therefore be stated that this free object has three degrees of freedom. However, if the object was constrained to move only in the xy-plane, then the value of the z-coordinate would be fixed and only the x and y coordinates would have to be specified during the motion. In that case, the object has only two degrees of freedom. If it is further required that the object moves along a line in this plane, then only the coordinate along which the object moved (i.e., one degree of freedom) needs to be specified.

If one now considers a molecule consisting of N atoms, it is clear that one must specify the x, y, and z coordinates for each atom, and therefore the total number of degrees of freedom possible for this molecule would be equal to $3N$. At the same time, the atoms in the molecule are bound, and this bonding will affect how the $3N$ degrees of freedom would be distributed among the different types of motion that the molecule can perform.

For a single particle, only one type of motion is possible, which corresponds to its translation through space. For unrestrained motion, this situation is completely described when all three coordinates of the particle are specified and the object has three degrees of translational freedom. In a molecule, however, each atom is not free to move independently of the others, but instead the molecule translates as a whole. Hence, the translational motion of the entire molecule can be described in terms of the three coordinates of the center of mass, namely the point in the molecule where all the mass can be considered to be concentrated. The translational motions for a nonlinear triatomic molecule are shown in Figure 1. It can be concluded from this discussion that both monatomic and polyatomic molecules possess three translational degrees of freedom.

For polyatomic molecules consisting of N atoms, there are still $(3N - 3)$ degrees of freedom remaining to be specified. These are termed internal modes of motion, which consist of rotational and vibrational motions. The rotational motions may be most easily visualized by considering the atoms to be held fixed by rigid bonds so that the molecule as a whole rotates about axes through its center of mass. As shown in Figure 2 for a simple diatomic, a linear molecule can rotate about two mutually perpendicular axes through the center of mass. Motion about the molecular axis will not contribute to the rotational energy of the molecule, so only two angles of rotation are required to describe the rotational motion of a linear molecule. Therefore, linear molecules possess two rotational degrees of freedom. On the other hand, a nonlinear polyatomic molecule can rotate about three mutually perpendicular axes through its center of mass (Fig. 3). Consequently, three angles of rotation must be specified, and non-linear molecules, therefore, possess three degrees of rotational freedom.

Now $3N$ is the total number of degrees of freedom for a molecule with N atoms, of which three are translational and two and three for linear and nonlinear molecules, respectively, are rotational. For a linear molecule the remaining number of degrees of freedom must equal $(3N - 5)$, and for a nonlinear molecule the remaining number of degrees of freedom must equal $(3N - 6)$. These remaining degrees of freedom must be due to another form of internal motion and is known as vibrational motion.

For a simple diatomic molecule, the number of degrees of freedom is given by $[3(2) - 5]$, or 1. It is easy to see that the only nontranslational or non-rotational type of motion accessible to a diatomic molecule is the simple stretching of the bond connecting the two atoms. This type of molecular motion is often termed as *stretching mode of vibration* and consists of a periodic lengthening and shortening of the distance between the two atoms about some equilibrium position. Because of this motion, the diatomic molecule is seen to possess both vibrational kinetic and potential energies, and therefore the vibrational modes of motion will have two energy terms associated with it.

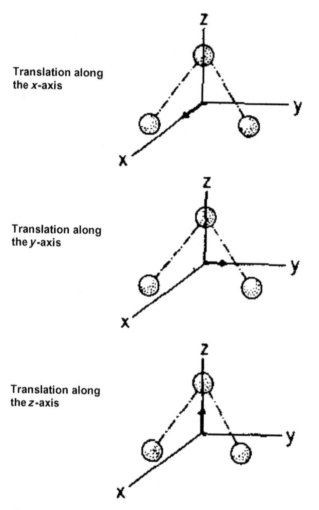

Figure 1 Translational motions of the center of mass for a nonlinear triatomic molecule.

As discussed earlier, a linear polyatomic molecule consisting of N atoms will be characterized by $(3N - 5)$ nontranslational and nonrotational vibrational motions. For a linear triatomic molecule, the number of authentic vibrations will therefore be equal $[3(3) - 5]$, for four vibrational normal modes of motion. These are depicted in Figure 4, where the atoms are shown in their equilibrium positions and the arrows indicate the directions of motion of the atoms away from these positions for one phase of the vibration. The four vibrational modes of a linear triatomic molecule are termed as the *symmetric stretching* motion, the *antisymmetric stretching* motion, and two *bending motions*. The symmetric stretching mode involves atomic movement where the bonds stretch

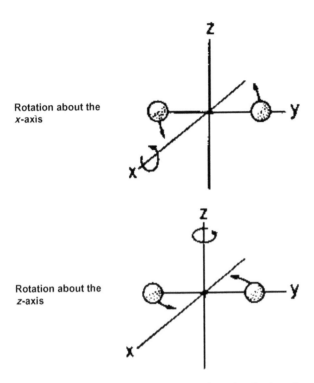

Figure 2 Rotational motion in a diatomic molecule about its center of mass.

symmetrically from the equilibrium position. One phase of the anti-symmetric stretching vibration involves the left hand and center atoms moving together and the right-hand atom moving away from the center atom, effecting a compression of the bond on the left and a stretching of the bond on the right. Finally, one bending vibration arises from a slight displacement of the end atoms above (or below) the bond axis, while the central atom moves simultaneously below (or above) this axis. The second bending vibration (the fourth vibration for this molecule) is obtained from the first by rotating the molecule by 90° about the molecular axis. The result is that the end atoms move into (or out of) the plane of the paper while the central atom moves out of (or into) this plane.

On the other hand, a nonlinear polyatomic molecule consisting of N atoms will be characterized by $(3N - 6)$ nontranslational and nonrotational vibrational motions. For a nonlinear triatomic molecule, the number of authentic vibrations will therefore be equal $[3(3) - 6]$ for three vibrational normal modes of motion, which are depicted in Figure 5. To preserve the pure vibrational nature of the symmetric stretching motion, the central atom must move an equal amount in the direction opposite to the motion of the end atoms, so that the net result of the simultaneous motions of the end and central atoms is a symmetric stretching of

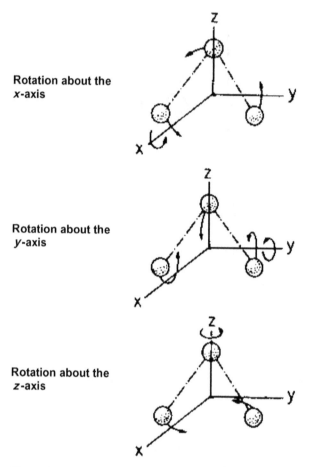

Figure 3   Rotational motion in a non-linear triatomic molecule about its center of mass.

the bonds. The atomic motions for the asymmetric stretching vibration gives rise to a compression of one bond and a stretching of the other bond, whereas the atomic motions for the bending vibration yield an opening and closing of the bond angle.

ROTATION AND VIBRATION IN DIATOMIC MOLECULES

In order to develop a more rigorous treatment of the wave functions associated with rotational and vibrational motion, the application of wave mechanics to diatomic molecules will be considered. In the approximation of Born and Oppenheimer, it is recognized that the time frame for nuclear motion is very slow when compared with that of electronic motion, and hence the two may be separated

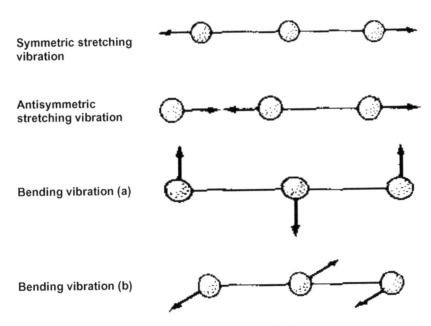

Symmetric stretching
vibration

Antisymmetric
stretching vibration

Bending vibration (a)

Bending vibration (b)

Figure 4 Vibrational motion in a linear triatomic molecule.

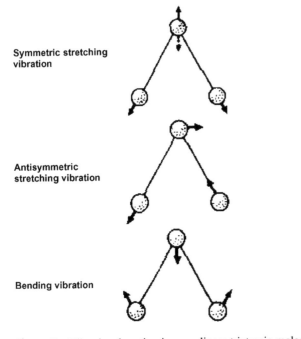

Symmetric stretching
vibration

Antisymmetric
stretching vibration

Bending vibration

Figure 5 Vibrational motion in a nonlinear triatomic molecule.

from each other. In other words, the molecule will remain in the same electronic state during the entire lifetime of nuclear motion. As a result, one can develop the wave mechanical expressions for nuclear motion independent of the expressions for electronic motion.

For a diatomic molecule, the Hamiltonian operator for the nuclear motion is given by the sum of the kinetic and the potential energy operators. One can then write the Schrödinger equation for nuclear motion as:

$$[T_N + V(r)]\psi = E\psi \tag{1}$$

where T_N is the kinetic energy operator, $V(r)$ is the potential energy function, and it is understood that ψ is the wave function for nuclear motion. For the diatomic molecule A-B, Equation (1) becomes:

$$\frac{1}{m_A}\nabla_A^2\psi + \frac{1}{m_B}\nabla_B^2\psi + 2[E - V(r)]\psi = 0 \tag{2}$$

where m_A and m_B are the masses of the two atoms comprising the molecule.

To eliminate terms associated with translation of the nuclei, the motion of the atoms will be described in terms of a set of axes that move with the molecule. In that case, the only forces to be considered are due to the effect of the potential field that is directed along the internuclear axis. Mathematically, this situation is equivalent to that of a single particle moving in the field defined by $V(r)$, where the mass of this equivalent particle is given by the reduced mass (μ) of the two atoms:

$$\mu = \frac{m_A m_B}{m_A + m_B} \tag{3}$$

The wave equation now becomes:

$$\nabla^2\psi + \frac{2\mu}{\hbar^2}[E - V(r)]\psi = 0 \tag{4}$$

which is analogous to Equation (5) of chapter 2, except that the form of the potential energy function has not yet been specified. As previously, after the wave function is transformed into spherical polar coordinates, it can be factored into functions of a single variable:

$$\psi(r, \theta, \phi) = \mathbf{R}(r)\Theta(\theta)\Phi(\phi) \tag{5}$$

Once the Schrödinger equation is solved in the manner described in the chapter, the resulting wave function will have the form:

$$\psi(r, \theta, \phi) = \mathbf{R}(r)\mathbf{Y}(\theta, \phi) \tag{6}$$

where the $\mathbf{Y}(\theta, \phi)$ aspect are the surface harmonics defined in Equation (23) of chapter 2. The radial functions are solutions of the equation:

$$\frac{1}{r^2}\frac{d}{dr}\left[r^2\frac{d\mathbf{R}}{dr}\right] + \left[\frac{2\mu}{\hbar^2}[E - V(r)] - \frac{C}{r^2}\right]\mathbf{R} = 0 \tag{7}$$

where C equals the constant of chapter 2, Equation (13). If we now substitute:

$$C = J(J+1) \tag{8}$$
$$\mathbf{S}(r) = r\mathbf{R}(r) \tag{9}$$

where $\mathbf{S}(r)$ is a radial distribution function. The meaning of J will be made clear shortly. Equation (7) now becomes:

$$\frac{d^2\mathbf{S}}{dr^2} + \frac{2\mu}{\hbar^2}\left[E - V(r) - \frac{J(J+1)\hbar^2}{2\mu r^2}\right]\mathbf{S} = 0 \tag{10}$$

Equation (10) describes two types of nuclear motion. One of these is termed vibrational motion, because the wave functions that are a function of r describe the motion of the nuclei directed along the axis of the diatomic molecule. In other words, Equation (10) is the wave equation for one-dimensional motion, where the potential field is given by:

$$V(r) + \frac{J(J+1)\hbar^2}{2\mu r^2} \tag{11}$$

The second term in Equation (10) accounts for the centrifugal potential due to rotation of the molecule with respect to the space fixed coordinate axes.

The solution to Equation (10) yields a set of vibrational wave functions and their energies, each of which is distinguished by a quantum number, v. The vibrational wave functions are further defined by the permissible values of J, the quantum number for the angular momentum of the system.

VIBRATIONAL ENERGY LEVELS AND SPECTROSCOPY OF DIATOMIC MOLECULES

In order to understand the quantization of vibrational energy, it is useful to consider the model where the molecule is not permitted to undergo rotational motion. More simply stated, one merely sets $J = 0$ in Equation (10) to obtain:

$$\frac{d^2\mathbf{S}}{dr^2} + \frac{2\mu}{\hbar^2}[E - V(r)]\mathbf{S} = 0 \tag{12}$$

Equation (12) cannot be considered further until one chooses a form for the potential function, $V(r)$.

The Harmonic Oscillator Model

In the simplest model, $V(r)$ is assumed to have the form associated with simple classical harmonic motion:

$$V(r) = \tfrac{1}{2}kr^2 \tag{13}$$

where k is the force constant associated with the magnitude of the restoring force on the particle.

In the classical theory, the vibrational frequency (υ_{OSC}) of the harmonic oscillator is:

$$\upsilon_{OSC} = (1/2\pi)(k/\mu)^{1/2} \tag{14}$$

The solution to Equation (12) is simplified if three new variables are defined:

$$\alpha = 2\mu E/\hbar^2 \tag{15}$$
$$\beta = (\mu k)^{1/2}/\hbar \tag{16}$$
$$\gamma = r/(\beta)^{1/2} \tag{17}$$

After these substitutions are made, Equation (12) becomes:

$$\frac{d^2S}{d\gamma^2} + \left[\frac{\alpha}{\beta} - \gamma^2\right]S = 0 \tag{18}$$

One approach to a solution for Equation (18) is to see how the equation behaves as very large values of γ, namely $\gamma \gg 1$. In that case, $\gamma \gg \alpha/\beta$, and the solution becomes:

$$S(\gamma) = C\exp(\pm\gamma^2/2) \tag{19}$$

where C is a constant. In fact, only the equation having the negative exponent is acceptable, because the positive exponent solution has the unacceptable result of $S \to \infty$ as $\gamma \to \infty$. Therefore, at large values of γ, the wave equation has the form:

$$S(\gamma) = C\exp(-\gamma^2/2) \tag{20}$$

To obtain wave functions that are acceptable at all values of γ and that yield Equation (20) at large values of γ, it is useful to introduce another function $U(\gamma)$, which is a correction function. In that case:

$$S(\gamma) = CU(\gamma)\exp(-\gamma^2/2) \tag{21}$$

The most convenient form of $U(\gamma)$ is that of a power series in γ that approaches unity as $\gamma \to \infty$. If one substitutes Equation (21) into Equation (18), then:

$$\frac{d^2U}{d\gamma^2} - 2\gamma\frac{dU}{d\gamma} + \left[\frac{\alpha}{\beta} - 1\right]U = 0 \tag{22}$$

It happens that Equation (22) is in the form of a well-known standard equation, namely that of Hermite. The solutions to the equation have the form of a complex power series, but recognizing that $x = \gamma$ and $2V = [(\alpha/\beta) - 1]$, then the equation takes the general form:

$$\frac{d^2Y}{dx^2} - 2x\frac{dY}{dx} + 2VY = 0 \tag{23}$$

The solutions to Equation (23) are known as the Hermite polynomials, $H_V(x)$, and are given by the following equation as long as $V = 0$, or if V is a positive number:

$$H_V(x) = (2x)^V - \frac{V(V-1)(2x)^{V-2}}{1!}$$
$$+ \frac{V(V-1)(V-2)(V-3)(2x)^{V-4}}{2!} + \cdots \tag{24}$$

The relationship between Equations (22) and (23) is apparent if:

$$Y = U(\gamma) \tag{25}$$
$$X = \gamma \tag{26}$$
$$2V = (\alpha/\beta) - 1 \tag{27}$$

The Hermite polynomials provide solutions to Equation (22):

$$U(\gamma) = C'H_V(\gamma) \tag{28}$$

where C' is another constant, and the $H_V(\gamma)$ functions are defined by a vibrational quantum number, namely:

$$v = 0, 1, 2, 3, \ldots \tag{29}$$

The vibrational eigenfunctions of the Schrodinger equation are in turn defined by their characteristic v quantum number and have the general form:

$$S_V = N_V H_V(\gamma) \exp(-\gamma^2/2) \tag{30}$$

where N_V is the normalization factor.

The lowest energy vibrational wave functions have the form:

$$S_0 = (\alpha/\pi)^{1/4} \exp(-\alpha\gamma^2/2) \tag{31}$$
$$S_1 = (4\alpha^3/\pi)^{1/4} \gamma \exp(-\alpha\gamma^2/2) \tag{32}$$
$$S_2 = (\alpha/4\pi)^{1/4}(1 - 2\alpha\gamma^2) \exp(-\alpha\gamma^2/2) \tag{33}$$
$$S_3 = (9\alpha^3/\pi)^{1/4}(\gamma^{-2/3}\alpha\gamma^3) \exp(-\alpha\gamma^2/2) \tag{34}$$

Spectroscopy of a Harmonically Oscillating Diatomic Molecule

If the definitions of α from Equation (15) and β from Equation (16) are substituted into Equation (27), then the energies (E_V) of the vibrational wave functions are functions of the reduced mass and the bond force constant:

$$E_V = \hbar (k/\mu)^{1/2}(v + \tfrac{1}{2})$$ (35)

Using the definition of Equation (14), the energies (in units of ergs) are also equal to:

$$E_V = \hbar v_{OSC}(v + \tfrac{1}{2})$$ (36)

Alternatively, one can express the energy of a given vibrational state in units of reciprocal centimeters (cm^{-1}) by dividing Equation (36) by the speed of light:

$$G_V = \frac{\hbar v_{OSC}}{c}(v + \tfrac{1}{2})$$ (37)

It is conventional to define the oscillator energy as ω (in units of cm^{-1}), and in that case:

$$G_V = \omega(v + \tfrac{1}{2})$$ (38)

For most molecules, ω will have values in the range of 200–4000 cm^{-1}.

Because a homonuclear diatomic molecule has no permanent dipole moment, it cannot have an IR absorption spectrum. The reason for this is that the electric dipole spectral transition probability, P_{ij}, from a lower vibrational state (S_i) to an upper vibrational state (S_j) must equal zero:

$$P_{ij} = \int S_i \mu S_j \, d\tau$$ (39)

For heteronuclear diatomic molecules that have a permanent dipole moment, it can be shown that Equation (39) will equal zero unless the following selection rule is met:

$$\Delta v = \pm 1$$ (40)

The energetics of vibrational energy levels is such that the ground vibrational state of a given molecule will be characterized by $v = 0$. The implication of the harmonic oscillator selection rule is that the only allowed vibrational transition would be $S_0 \to S_1$, which is known as the *fundamental* vibrational transition. In the harmonic oscillator model, transitions from S_0 to higher excited states S_V are forbidden. It will be shown in a later section that the harmonic oscillator selection rule of $\Delta v = \pm 1$ can be relaxed by anharmonic effects.

In the harmonic oscillator model, therefore, the energy of the $E(v = 0)$ state is given by:

$$E_0 = \frac{\hbar}{2}(k/\mu)^{1/2} \tag{41}$$

and the energy of the $E(v = 1)$ state is given by:

$$E_1 = \frac{3\hbar}{2}(k/\mu)^{1/2} \tag{42}$$

Therefore, energy of the $(v = 0) \rightarrow (v = 1)$ transition is given by:

$$E_1 - E_0 = \hbar(k/\mu)^{1/2} \tag{43}$$

or

$$G_1 - G_0 = \omega \tag{44}$$

A considerable insight into vibrational spectroscopy can be gleaned through a consideration of the equations derived previously. According to Equation (43), the energy of a vibrational transition may be seen to be partially defined by the magnitude of the force constant (which is interpreted as a measure of the bond strength). In particular, it can be shown that one can calculate a value for the force constant of a harmonic oscillator using the relation:

$$k = \mu(4\pi^2\omega^2c^2) \tag{45}$$

where the reduced mass μ is give in units of grams/molecule, ω is in units of cm^{-1}, and the speed of light c equals 2.9979246×10^{10} cm/sec. Wilson, Decius, and Cross have compiled a listing of force constants for various bond types (24).

The series of diatomic hydrogen halides constitutes a good illustration of the effect of force constant on the fundamental stretching frequency of a diatomic molecule. The reduced masses of these molecules are dominated by the hydrogen atom, being and equal to 0.957159, 0.979726, 0.995384, and 0.999998 atomic units for HF, HCl, HBr, and HI, respectively. From the extensive compilation of measured vibrational frequencies of diatomic molecules published by Herzberg (2), it is known that the energies of the fundamental vibrational transitions decrease in the order of 4138.52 cm^{-1} (HF), 2989.74 cm^{-1} (HCl), 2649.47 cm^{-1} (HBr), and 2309.5 cm^{-1} (HI). Using Equation (45), the calculated harmonic oscillator force constants are found to be 9.659×10^5 dynes/cm for HF, 5.160×10^5 dynes/cm for HCl, 4.117×10^5 dynes/cm for HBr, and 3.143×10^5 dynes/cm for HI. These calculated bond force constants correlate well with the known bond energies (25) for these molecules, which is illustrated in Figure 6.

As discussed previously, the energy of a vibrational transition is also critically dependent on the reduced mass of the atoms involved in the vibrational motion. This can be illustrated by various interhalogen diatomic molecules for which the force constants would be fairly similar but which would differ significantly in the respective reduced masses. The reduced masses of ClBr, ClI, and BrI are 24.3389, 27.4338, and 49.0317 atomic units, respectively, and the energies of the respective vibrational absorption bands are 430.0, 384.2, and 268.4 cm^{-1}. The harmonic oscillator force constants are roughly equivalent, being calculated as 2.65 × 10^5 dynes/cm for ClBr, 2.39 × 10^5 dynes/cm for ClI, and 2.08 × 10^5 dynes/cm for BrI.

The diatomic molecule harmonic oscillator has one additional lesson to teach regarding the effect of reduced mass and force constant on the energies of vibrational transitions. Using the average bond stretching force constants of Wilson, Decius, and Cross (24), and the relation:

$$\omega = \frac{(k/\mu)^{1/2}}{2\pi c} \tag{46}$$

it becomes possible to calculate the transition energies for the fundamental vibrational transitions of some chemically relevant bond types. For instance, the

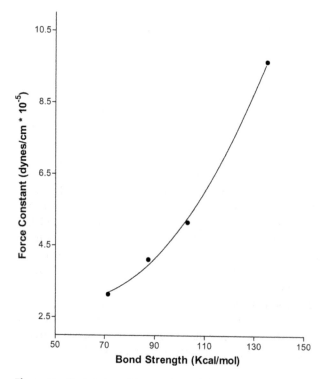

Figure 6 Correlation of harmonic oscillator force constants with bond energies for the series of hydrogen halide molecules.

energy of the C—H stretching mode is calculated to be 3444 cm^{-1}, and the energy of the C—C stretching mode is calculated to be 1189 cm^{-1}. In addition, the energy of the N—H stretching mode is calculated to be 3425 cm^{-1}, and the energy of the N—N stretching mode is calculated to be 1044 cm^{-1}. Finally, the energy of the O—H stretching mode is calculated to be 3737 cm^{-1}, and the energy of the C—C stretching mode is calculated to be 950 cm^{-1}. From this analysis, the general trend emerges that the stretching frequencies of molecules containing a hydrogen atom will be much higher than the stretching frequencies of molecules that do not contain a hydrogen atom. The spectroscopic consequence of this trend (which will be developed further in forthcoming sections) is that vibrational transitions associated with atom–hydrogen stretching modes will be observed at much higher energies than the transitions of other functional groups.

VIBRATIONAL ENERGY LEVELS AND SPECTROSCOPY OF THE FUNDAMENTAL TRANSITIONS OF POLYATOMIC MOLECULES

Even for polyatomic molecules, the harmonic oscillator model can be used with a fair degree of adequacy to describe their fundamental vibrational transitions. In fact, one may gain considerable insight into the vibrational spectra of complicated models by considering them as summations of a number of harmonic oscillators. In addition, many vibrational wave functions turn out to be essentially localized on the atoms of a functional group and its nearest neighbors, a fact that will ultimately lead to the development of the concept of group frequencies.

Vibrational Modes and Wave Functions of Polyatomic Molecules

As discussed earlier, the number of remaining degrees of freedom is not limited for a nonlinear polyatomic molecule, and therefore the number of possible vibrational modes will equal $3N - 6$. For instance, a nonlinear triatomic molecule will exhibit $[3(3) - 6]$ genuine modes of vibrational motion, of three vibrational modes. These motions are illustrated in Figure 5, but now, one can identify the symmetric stretching mode as v_1, the "scissors" bending mode as v_2, and the antisymmetric stretching mode as v_3.

Because the individual vibrational modes are independent of each other, the vibrational wave function describing the states of this nonlinear triatomic molecule will be:

$$\psi(v_1, v_2, v_3) = \psi(v_1)\psi(v_2)\psi(v_3) \tag{47}$$

The ground state of this non-linear triatomic molecule will be characterized by $\psi(v_1)$, $\psi(v_2)$, and $\psi(v_3)$ each being in their respective $v = 0$ states, and its wave function will be abbreviated as ψ_{000}. The lowest energy excited states will be characterized by two of the $\psi(v_1)$, $\psi(v_2)$, or $\psi(v_3)$ functions each being in the $v = 0$ state and the third function being in the $v = 1$ state. These three possible excited states would be identified by the functions ψ_{100}, ψ_{010}, and ψ_{001}.

The effects of force constant and reduced mass on the vibrational frequencies of the constituent groups of atoms in a molecule leads to certain generalizations about the vibrational spectroscopy of polyatomic molecules. Because the force constants of most bonds are fairly constant from molecule to molecule and because the reduced masses are often comparable as well, it follows that the vibrational frequency of a given functional group is primarily determined by the identity of the atoms involved. This concept has led to the concept of *group frequencies*, and this understanding has greatly facilitated the interpretation of vibrational spectra.

The vibrations associated with atoms in a molecule linked by covalent bonds are classified as being due to either *stretching* or *deformation* (i.e., bending) modes of molecular motion. The force constants associated with the deformation modes are usually weaker than those of stretching modes by approximately one order of magnitude and therefore consistently observed at lower energies. The weaker nature of the deformation modes also makes these more sensitive to details of chemical bonding, a fact that has proven to be very useful to vibrational spectroscopists over the years. For complex molecules, the same atom may be involved with both stretching and deformational modes of motion.

The constancy of force constants (and therefore of group frequencies) is valid only for similar chemical environments, and therefore one observes that changes in molecular bonding can yield shifts in the vibrational frequency of a given functional group. As a result, the published correlation tables usually present a range of frequencies as being characteristic of a particular functional group. For instance, the bending mode of C—H is observed around 1400 cm^{-1}, but nevertheless exhibits variability depending on the remainder of the molecule (18). In the —O—CO—CH$_3$ fragment, the frequency of the C—H bending mode will be in the range of $1380-1365 \text{ cm}^{-1}$. In the —CO—CH$_3$ fragment, the frequency of the C—H bending mode decreases from $1360-1355 \text{ cm}^{-1}$, and in the —COO—CH$_3$ fragment, the frequency of the C—H bending mode increases from 1440 to 1435 cm^{-1}.

Practically, every book covering vibrational spectroscopy features some type of summary regarding the frequencies associated with various functional groups, but the most complete compilation of characteristic group frequencies may be found in monographs developed especially for this purpose (26–28).

Electric Dipole Absorption Spectroscopy of the Fundamental Modes of Polyatomic Molecules

Although one can readily determine that the nonlinear triatomic molecule mentioned previously must exhibit three genuine modes of vibrational motions, it remains to be determined whether transitions can take place within the states associated with these vibrational modes. Because the harmonic oscillator selection rule is still $\Delta v = \pm 1$, it follows that the only possible fundamental vibrational spectroscopic transitions would correspond to:

$$\psi_{000} \longrightarrow \psi_{100} \tag{48}$$

$$\psi_{000} \longrightarrow \psi_{010} \tag{49}$$

$$\psi_{000} \longrightarrow \psi_{001} \tag{50}$$

As previously discussed in the case of electric dipole transitions, the value of the transition moment:

$$P_{ij} = \int \psi_i \mu \psi_j \, d\tau \tag{51}$$

for an allowed transition cannot equal zero. Fortunately, one can determine whether or not a given transition equals zero or (or not) solely on the basis of molecular symmetry-based group theory. This is because the integration step necessary to evaluate P_{ij} is to be performed over all space, and it follows that any antisymmetric result must integrate to zero.

In practice, one uses the symmetry properties (i.e., its representation) of ψ_i, μ, and ψ_j to deduce whether P_{ij} does or does not equal zero. As discussed in chapter 4, the ground state of a molecule (denoted as ψ_{000} for the nonlinear triatomic molecule of the illustration) must transform as the total symmetric representation in its point group. To achieve a symmetric end result, the representation of the excited state wave function must be the same as that of the representation for one of the elements of the dipole moment operator (29–32).

For the nonlinear triatomic molecule, the ν_1 symmetric stretching mode transforms according to the A_1 representation, the ν_2 bending mode also as the A_1 representation, and the ν_3 antisymmetric stretching mode as the B_1 representation. Examination of the character table for the C_{2v} point group (Fig. 3 of chap. 4) reveals that the z-component of the dipole moment operator transforms as the A_1 representation and that the x-component of the dipole moment operator transforms as the B_1 representation. Consequently, it can be concluded that all three vibrational modes of a nonlinear triatomic molecule will exhibit absorbance in the IR region of the spectrum.

For water (a typical nonlinear triatomic molecule), the energies of the vibrational spectroscopic transitions in the gas phase are:

$$\psi_{000} \longrightarrow \psi_{100} \quad (\nu_1) = 3652 \, \text{cm}^{-1} \tag{52}$$

$$\psi_{000} \longrightarrow \psi_{010} \quad (\nu_2) = 3756 \, \text{cm}^{-1} \tag{53}$$

$$\psi_{000} \longrightarrow \psi_{001} \quad (\nu_3) = 1545 \, \text{cm}^{-1} \tag{54}$$

As per the spectrum of bulk water shown in Figure 7, these energies are substantially shifted owing to the extensive degree of hydrogen bonding.

Analysis of the vibrational spectrum of a polyatomic molecule becomes progressively more complicated as the number of atoms increases. As another example, we will consider ethylene, C_2H_4, a molecule having D_{2h} symmetry and

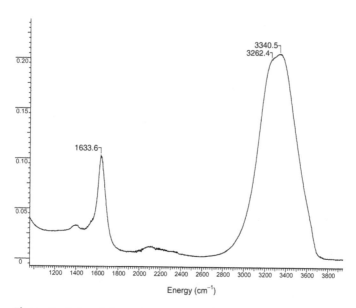

Energy (cm^{-1})

Figure 7 Infrared absorption spectrum of bulk water (the intensity scale is in Kubelka-Munk units).

whose character table is found in Figure 4 of chapter 4. This molecule contains a total of six atoms, and so the number of genuine molecular vibrational modes will be given by [3(6) − 6], or 12 modes of vibration. Each vibrational mode must transform according to one of the irreducible representations of the D_{2h} point group as summarized in Table 1. Because the three elements of the dipole moment operator transform according to the B_{1u}, B_{2u}, and B_{3u} irreducible representations, it follows that only those vibrational modes that transform according to one of these representations will exhibit allowed transitions in the IR spectrum. As a result, only five out of 12 vibrational modes will exhibit an IR absorption band.

For very complicated organic molecules, there is usually a sufficient lack of molecular symmetry that all possible vibrational modes will yield a fundamental absorption band in the IR spectrum. However, the intensity of the resulting bands will still be determined by the magnitude of the transition moment, and not all will exhibit strong degrees of absorption. For this reason, one ordinarily includes an intensity estimation (i.e., very strong, strong, moderate, weak, very weak) in any listing of peak energies.

Raman Spectroscopy of the Fundamental Modes of Polyatomic Molecules

Alternatively, the vibrational modes of a compound may be measured using Raman spectroscopy, where one measures the inelastic scattering of radiation by a nonabsorbing medium (19–23). In the Raman effect, when an intense

Table 1 Molecular Motion Descriptions and Symmetry of the Vibrational Modes of Ethylene, a Molecule Having D_{2h} Symmetry

Description of molecular motion		Corresponding irreducible representation in the D_{2h} point group
ν_1	Carbon–carbon stretching mode	A_g
ν_2	Carbon–hydrogen stretching mode (1)	A_g
ν_3	Carbon–hydrogen stretching mode (2)	B_{1u}
ν_4	Carbon–hydrogen stretching mode (3)	B_{2g}
ν_5	Carbon–hydrogen stretching mode (4)	B_{3u}
ν_6	Hydrogen–carbon–hydrogen bending mode (1)	A_g
ν_7	Hydrogen–carbon–hydrogen bending mode (2)	B_{1u}
ν_8	Wagging mode (1)	B_{2g}
ν_9	Wagging mode (2)	B_{3u}
ν_{10}	Out-of-plane bending mode (1)	A_u
ν_{11}	Out-of-plane bending mode (2)	B_{2u}
ν_{12}	Out-of-plane bending mode (3)	B_{3g}

Source: Adapted from Ref. 30.

beam of light is passed through a medium, approximately one in every million incident photons is scattered with a loss or gain of energy. The inelastic-scattered radiation can occur at lower (Stokes lines) and higher (anti-Stokes lines) frequencies relative to that of the incident (or elastically scattered) light, corresponding to the vibrational transition frequencies of the sample or of the medium. The actual intensities of the Stokes and anti-Stokes lines are determined by the Boltzmann factor characterizing the vibrational population. For high-frequency vibrations, the Stokes lines are relatively intense relative to the anti-Stokes lines, so conventional Raman spectroscopy makes exclusive use of the Stokes component.

The Raman effect originates from the interaction of the oscillating-induced polarization or dipole moment of the medium with the electric field vector of the incident radiation, and a significant degree of understanding can be obtained by considering the classical theory of scattering. The theory of scattering begins with the fact that because light consists of electromagnetic radiation, the oscillating electric field can induce a dipole moment in an irradiated molecule. This molecule will in turn radiate light having the same frequency as the incident radiation, but which is instead scattered in directions that are different from that of the incident light.

The magnitude of the induced dipole moment, M, will depend both on the intensity of the incident radiation and on the polarizability of the molecule. Thus, one may consider the polarizability as a measure of how readily the electrons in the molecule are displaced by the electric field of the light wave. In other words,

$$M = \alpha E \tag{55}$$

where α is the tensor describing the polarizability molecule, and E is the intensity of the electric field of the incident radiation.

When electromagnetic radiation interacts with a molecule, the induced electric field is given by:

$$E = E_0 \cos(2\pi\nu_0 t) \qquad (56)$$

where E_0 is the amplitude of the electric vector of the incident light, and ν_0 is the frequency of incident light interacting with the molecule. The electric moment induced in the molecule will have the value:

$$M = \alpha E_0 \cos(2\pi\nu_0 t) \qquad (57)$$

Equation (57) describes *Rayleigh scattering*, or the situation where the frequency of the scattered light is the same as that of the incident light.

The polarizability, α, is a scalar quantity only for isotropic systems, and the full tensor description must be used for anisotropic molecules. For such molecules, the following equations are required to relate the nonequal polarizability along the principal axes of the molecule:

$$M_X = \alpha_{XX}E_X + \alpha_{XY}E_Y + \alpha_{XZ}E_Z \qquad (58)$$
$$M_Y = \alpha_{YX}E_X + \alpha_{YY}E_Y + \alpha_{YZ}E_Z \qquad (59)$$
$$M_X = \alpha_{ZX}E_X + \alpha_{ZY}E_Y + \alpha_{ZZ}E_Z \qquad (60)$$

where M_i is the component of the induced dipole moment induced along the ith axis.

For small amplitudes of molecular vibration, the polarizability of the molecule can be expressed in terms of vibrational coordinates by the equation:

$$\alpha = \alpha_0 + \left[\frac{\partial\alpha}{\partial Q_V}\right]_0 Q_V \qquad (61)$$

where α_0 is the molecular polarizability at the equilibrium configuration of the molecule, and $(\partial\alpha/\partial Q_V)_0$ is the rate of change of the polarization with change in bond length (evaluated at the equilibrium configuration). The vibrational displacement function, Q_V, is a function of time because the molecule vibrates with a fundamental frequency equal to ν_V:

$$Q_V = Q_V^0 \cos(2\pi\nu_V t) \qquad (62)$$

where Q_V^0 is the vibrational amplitude.

Substitution of Equations (61) and (62) into Equation (57) yields:

$$M = E_0\alpha_0 \cos(2\pi\nu_0 t) + E_0 Q_V^0 + \left[\frac{\partial\alpha}{\partial Q_V}\right]_0 \cos(2\pi\nu_0 t)\cos(2\pi\nu_V t) \qquad (63)$$

Through the use of the trigonometric identity:

$$\cos(\gamma)\cos(\beta) = \frac{1}{2}[\cos(\gamma+\beta) + \cos(\gamma-\beta)] \tag{64}$$

Equation (63) can be arranged to:

$$M = E_0\alpha_0\cos(2\pi\nu_0 t) + \frac{1}{2}E_0 Q_V^0 + \left[\frac{\partial\alpha}{\partial Q_V}\right]_0 \cos[2\pi(\nu_0+\nu_V)t]$$

$$+ \frac{1}{2}E_0 Q_V^0 + \left[\frac{\partial\alpha}{\partial Q_V}\right]_0 \cos[2\pi(\nu_0-\nu_V)t] \tag{65}$$

The first term on the right-hand side of Equation (65) corresponds to Rayleigh scattering, where the frequency of the scattered radiation is the same as that of the incident light. The second term will give rise to weak scattering of light whose frequency equals $(\nu_0+\nu_V)$, whereas the third term gives rise to weak scattering of light whose frequency equals $(\nu_0-\nu_V)$. The weakness of the scattering associated with the second and third terms of Equation (65) originates in the fact that the polarizability changes are relatively less during the process.

The scattering of electromagnetic radiation at frequencies shifted from the incident radiation by amounts equal to vibrational frequencies is known as the *Raman effect*. The family of scattering bands observed at frequencies of $(\nu_0-\nu_V)$ are known as the *Stokes* lines, whereas the family of scattering bands observed at frequencies of $(\nu_0+\nu_V)$ are known as the *anti-Stokes* lines. The classical theory predicts that the Stokes and anti-Stokes lines should be observed with equal probability (and hence with equal intensity), but in reality the anti-Stokes lines are found to be much weaker in intensity than the Stokes lines, which are themselves much weaker than the Rayleigh scattering intensity. As will be shown in the succeeding chapter, the quantum mechanical treatment of the Raman effect correctly predicts the relative intensities of the Stokes and anti-Stokes lines.

A vibrational mode will only appear in the Raman spectrum if, and only if, the displacements in atomic position change the polarizability of the molecule. Because this criterion is completely different from that of the electric dipole absorption, it is possible that some vibrational modes will be seen in the Raman spectrum and not in the IR absorption spectrum (or vice versa). When a vibrational mode is allowed under both electric dipole and polarizability selection rules, the observed frequency will be identical. Owing to the different nature of the selection rules, however, the intensities of corresponding bands may exhibit considerable differences.

The intensity of a Raman transition is governed by the magnitude of its transition moment:

$$P_{ij} = E\int \psi_i \alpha \psi_j \, d\tau \tag{66}$$

The transition moment can be divided into its various components:

$$P_{ij}(x) = E \int \psi_i \alpha_{XX} \psi_j \, d\tau + E \int \psi_i \alpha_{XY} \psi_j \, d\tau + E \int \psi_i \alpha_{XZ} \psi_j \, d\tau \qquad (67)$$

$$P_{ij}(y) = E \int \psi_i \alpha_{YX} \psi_j \, d\tau + E \int \psi_i \alpha_{YY} \psi_j \, d\tau + E \int \psi_i \alpha_{YZ} \psi_j \, d\tau \qquad (68)$$

$$P_{ij}(z) = E \int \psi_i \alpha_{ZX} \psi_j \, d\tau + E \int \psi_i \alpha_{ZY} \psi_j \, d\tau + E \int \psi_i \alpha_{ZZ} \psi_j \, d\tau \qquad (69)$$

If any one of the transition moment components of Equations (67) through (69) are nonzero, then the transition will be allowed in the Raman spectrum.

As discussed previously, one can use molecular symmetry-based group theory to determine whether or not a given transition component will equal zero or (or not). This process uses the symmetry properties (i.e., its representation) of ψ_i, α, and ψ_j to deduce whether any of the P_{ij} components is nonzero. As before, the ground state of a molecule (which would be the state for which all vibrational wave functions are in their $v = 0$ level) must transform as the totally symmetric representation in its point group. To achieve a symmetric end result, the representation of the excited state wave function must be the same as that of the representation for one of the elements of the polarizability operator (29–32). As discussed in chapter 4, the elements of the polarizability tensor will transform as the squares and binary products of coordinates listed in area IV of the point group character table.

Returning to the example of the nonlinear triatomic molecule, it may be recalled that the ν_1 symmetric stretching mode also transforms according to the A_1 representation, the ν_2 bending mode also transforms as the A_1 representation, and the ν_3 antisymmetric stretching mode as the B_1 representation. Examination of the character table for the C_{2v} point group (Fig. 3 of chap. 4) reveals that the x^2, y^2, and z^2 coordinates of the polarizability tensor transforms as the A_1 representation, and that the xy cross-product of the polarizability tensor transforms as the B_1 representation. Consequently, it can be concluded that all three vibrational modes of a nonlinear triatomic molecule will exhibit Raman scattering.

Turning to the previously considered D_{2h} symmetry example of ethylene (C_2H_4), examination of the character table of Figure 4 of chapter 4 indicates that the x^2, y^2, and z^2 coordinates of the polarizability tensor transform as the A_g representation and that the xy, xz, and yx cross-products of the polarizability tensor transform as the B_{1g}, B_{2g}, and B_{3g} representation. As shown in Table 1, each of the genuine molecular vibrational modes transforms according to one of the irreducible representations of the D_{2h} point group. In order for a vibrational mode to exhibit Raman scattering, the symmetry of the molecular motion must transform according to one of the B_{1g}, B_{2g}, or B_{3g} irreducible representations. As a result, only six out of the 12 vibrational modes will exhibit Raman scattering.

The D_{2h} symmetry example reveals an interesting fact regarding vibrational spectroscopy. When a molecule contains an inversion center as one of its symmetry elements, vibrational modes will be active either in electric dipole absorption spectroscopy or in Raman scattering but not in both types of spectroscopy. In the case of ethylene, only the ν_1, ν_2, ν_4, ν_6, ν_8, and ν_{12} vibrational modes of Table 1 will exhibit Raman scattering, and only the ν_3, ν_5, ν_7, ν_9, and ν_{11} vibrational modes will exhibit absorption in the IR spectrum. Most interestingly, because the vibrational mode identified as ν_{10} (one of the out-of-plane bending modes) transforms as the A_u representation, it is forbidden by both the electric dipole and polarizability selection rules, and therefore its fundamental cannot be observed in any mode of vibrational spectroscopy.

It should be noted that the same summaries of characteristic group frequencies that are found in vibrational spectroscopy books and compilations for correlation of IR absorption bands apply equally well to the interpretation of Raman spectra. Generally, symmetric vibrations and nonpolar groups yield the most intense Raman scattering bands, whereas antisymmetric vibrations and polar groups yield the most intense IR absorption bands. Thus, even though low-symmetry organic molecules (typically those of pharmaceutical interest) will feature all vibrational modes in both the IR absorption and Raman spectra, the spectra often look quite different owing to the different origins of the transitions. Nevertheless, the measured frequency of a given vibrational mode will be the same whether it is detected through its IR absorption or through its Raman scattering.

SPECTROSCOPY OF TRANSITIONS ASSOCIATED WITH OVERTONES AND COMBINATIONS OF FUNDAMENTAL VIBRATIONAL MODES

As discussed previously, the harmonic oscillator selection rule states that only transitions for which $\Delta v = \pm 1$ will be allowed. Because at room temperature, the majority of molecules will be in the $v = 0$ state, this implies that only the fundamental transitions are observed in either the IR absorption or Raman spectra. However, it should be remembered that the harmonic oscillator model pertains only to one type of potential energy function substituted into Equation (1).

A more realistic description of the energy level sequence is achieved through the use of potential energy functions that permit molecular distortion in highly excited vibrational states. The use of more realistic potential energy functions serve to weaken the $\Delta v = \pm 1$ selection rule as a result of anharmonic effects. Three requirements are associated with more realistic potential energy functions. The first of these is that as the magnitude of the vibration approaches infinity, the magnitude of the potential function must approach a constant value. This constant value is often termed the dissociation energy, D_e, because when that much energy is absorbed by the system the bond must break. Second, as the magnitude of the vibration approaches zero, the magnitude of the potential function must approach infinity. In other words, the nuclei are not allowed to touch one another no matter what the degree of molecular motion. Finally, the potential energy function must exhibit a minimal value at the equilibrium position of the nuclei.

The most widely used potential energy function is the Morse potential:

$$V(r) = D_e[1 - e^{-a(r-re)}]^2 \tag{70}$$

The rigorous approach would then be to substitute this form for the potential energy function back into Equation (1) and to then solve the resulting Schrödinger equation in the usual manner. However, this solution leads to great mathematical difficulty, and hence more empirical approaches are usually taken.

The Anharmonic Oscillator Model

A more practical approach toward vibrational spectroscopy is to expand the harmonic oscillator Equation (38) in an infinite power series based on $(v + \frac{1}{2})$ to include anharmonic effects:

$$G'_V = \omega_e(v + \tfrac{1}{2}) - \omega_e X_e(v + \tfrac{1}{2})^2 + \omega_e Y_e(v + \tfrac{1}{2})^3 + \cdots. \tag{71}$$

where ω_e is the harmonic oscillator constant, and where X_e and Y_e are anharmonic constants. Above the energy of the dissociation limit, there exists a continuum of levels that are infinitely close together in energy, and these describe the breakup of the molecule. This dissociation energy, D_e, is given by the sum:

$$D_e = \Sigma \Delta G'(v + \tfrac{1}{2}) \tag{72}$$

where the summation is performed over the range of $v = 0$ to the highest vibrational level for which the bond in the molecule is still intact.

Usually, the $\omega_e X_e$ term will be considerably larger than the $\omega_e Y_e$ term, and thus most experimentally determined vibrational spectra can be fitted to the simplified equation:

$$G'_V = \omega_e(v + \tfrac{1}{2}) - \omega_e X_e(v + \tfrac{1}{2})^2 \tag{73}$$

For most molecules, the $\omega_e X_e$ term will be positive in character, so the separation between successively higher excited states will decrease as v increases.

In the simplified model, the separation between the G'_V state and the G'_{V+1} state will be given by:

$$\Delta G'_V = \omega_e - \omega_e X_e(2v + 2) \tag{74}$$

In this simplified model of the anharmonic oscillator, the dissociation energy is given by:

$$D_e = \omega_e / 4X_e \tag{75}$$

Equation (73) indicates that progressions of experimentally observed vibrational transitions that can be envisioned as a sum of a harmonic and an anharmonic part. However, at ambient conditions the majority of molecules are in the $v = 0$ state because the thermal energy available at $25°C$ is not sufficient to raise more than a few percent of molecules into the $v = 1$ state. In other words, for a transition from $v = 0$ to $v = 1$, the contribution from the $\omega_e X_e (2v + 2)$ term in Equation (73) is relatively minor, and consequently the fundamental transition can be adequately understood in terms of the harmonic oscillator model.

Overtone Vibrational Spectroscopy

Probably the most important consequence of including anharmonic effects in vibrational energy states is that the strict harmonic oscillator selection rule of $\Delta v = \pm 1$ becomes relaxed. This situation causes transitions from the ground vibrational state to higher vibrationally excited states to gain some allowedness in their character. However, the overtone transitions (i.e., where $\Delta v = 2, 3, 4$, etc.) will generally be less intense than the corresponding fundamental transition because the overtone transitions can take place only through relaxation of the primary selection rule. Nevertheless, overtone transitions are easily detected, and are observed in the NIR region of the electromagnetic spectrum (33–36).

For example, in the vapor phase, HCl exhibits its fundamental ($v = 0 \rightarrow v = 1$) absorption at an energy of 2885.9 cm^{-1} (3465.1 nm). The energy of the first overtone ($v = 0 \rightarrow v = 2$) transition is observed at 5668.0 cm^{-1} (1764.3 nm), and the energy of the second overtone ($v = 0 \rightarrow v = 3$) transition is observed at 8347.0 cm^{-1} (1198.0 nm).

Returning to the illustration of the nonlinear triatomic molecule, the number of genuine vibrational modes was found to be $[3(3) - 6]$, or three. As discussed earlier, the individual vibrational modes are independent of each other, and so the vibrational wave function describing the states of this nonlinear triatomic molecule will be:

$$\psi(v_1, v_2, v_3) = \psi(v_1)\psi(v_2)\psi(v_3) \tag{76}$$

The ground state of this non-linear triatomic molecule will be characterized by $\psi(v_1)$, $\psi(v_2)$, and $\psi(v_3)$ each being in their respective $v = 0$ states (ψ_{000}). The lowest energy excited states will be characterized by two of the $\psi(v_1)$, $\psi(v_2)$, or $\psi(v_3)$ functions each being in the $v = 0$ states and the third function in the $v = 1$ state (i.e., ψ_{100}, ψ_{010}, or ψ_{001}), and so the fundamental vibrational bands would be $\psi_{000} \rightarrow \psi_{100}$, $\psi_{000} \rightarrow \psi_{010}$, and $\psi_{000} \rightarrow \psi_{001}$.

Once the effects of anharmonicity are factored into the transition moment calculations, almost any type of vibrational excitation becomes possible. One could observe sequences of simple *overtone* transitions, such as:

$$\psi_{000} \longrightarrow \psi_{200}, \psi_{300}, \psi_{400}, \text{etc.} \tag{77}$$

$$\psi_{000} \longrightarrow \psi_{020}, \psi_{030}, \psi_{040}, \text{etc.} \tag{78}$$

$$\psi_{000} \longrightarrow \psi_{002}, \psi_{003}, \psi_{004}, \text{etc.} \tag{79}$$

or any one of a number of *combination* transitions, some of which can be written as:

$$\psi_{000} \longrightarrow \psi_{110}, \psi_{101}, \psi_{011}, \text{etc.} \tag{80}$$

$$\psi_{000} \longrightarrow \psi_{120}, \psi_{201}, \psi_{021}, \text{etc.} \tag{81}$$

It should be remembered that the overtone and combination transitions are still of the electric dipole type, and will therefore not be observed unless the value of the transition moment

$$P_{ij} = \int \psi_i \mu \psi_j \, d\tau \tag{82}$$

does not equal zero. As before, one can use molecular symmetry-based group theory to determine whether or not a given transition moment will equal zero. This is because the integration step necessary to evaluate P_{ij} is to be performed over all space, and it follows that any antisymmetric result must integrate to zero.

It was established earlier that for a nonlinear triatomic molecule, the v_1 symmetric stretching mode transforms according to the A_1 representation, the v_2 bending mode also transforms as the A_1 representation, and the v_3 antisymmetric stretching mode as the B_1 representation. The ground ψ_{000} state must transform as the totally symmetric representation (A_1). To determine the symmetries of the overtone and combination excited states, one takes the direct product of the representations of the two excited modes and determines the representation of the point group at which the resulting product transforms. The simplicity of the C_{2v} point group, and the symmetries of the fundamental vibrations, make this a relatively simple task. For instance, the ψ_{110} state will transform as the A_1 representation, the ψ_{101} state as the B_1 representation, and the ψ_{011} state also as the B_1 representation. Because it has already been established that the z-component of the dipole moment operator transforms as the A_1 representation and that the x-component of the dipole moment operator transforms as the B_1 representation, it can be concluded that the $\psi_{000} \rightarrow \psi_{110}$, $\psi_{000} \rightarrow \psi_{101}$, and $\psi_{000} \rightarrow \psi_{011}$ transitions will be allowed by group theory and would exhibit absorbance in the NIR region of the spectrum.

A summary of the overtone and combination bands of water in the vapor phase is given in Table 2, and a portion of its liquid phase absorption spectrum is shown in Figure 8. For bulk water, the moderate band observed at 1455 nm can be assigned to the $\psi_{000} \rightarrow \psi_{101}$ transition, the weak band at 1806 nm can be assigned to the $\psi_{000} \rightarrow \psi_{021}$ transition, and the strong bad at 1936 nm can be assigned to the $\psi_{000} \rightarrow \psi_{011}$ transition.

Table 2 Overtone and Combination Vibrational Transitions of Water in the Vapor Phase

Upper state					
ν_1	ν_2	ν_3	Energy (cm^{-1})	Wavelength (nm)	Intensity
0	1	1	5,332.0	1875.5	Medium
0	2	1	6,874.0	1454.8	Weak
1	0	1	7,251.6	1379.0	Medium
1	1	1	8,807.1	1135.4	Strong
2	0	1	10,613.1	942.2	Strong
0	0	3	11,032.4	906.4	Medium
2	1	1	12,151.2	823.0	Medium
3	0	1	13,830.9	723.0	Weak
1	0	3	14,318.8	698.4	Weak

SUMMARY

Not only are the vibrational modes of a compound affected by the identity and character of the rest of the atoms in the molecule, the molecular motion can also be further perturbed by details of the solid-state structure. This feature of vibrational spectroscopy makes the various techniques highly useful as adjunct

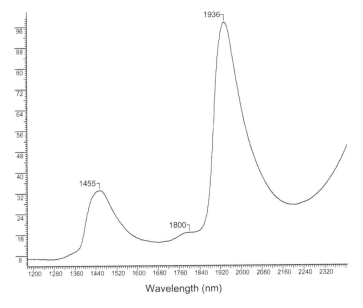

Figure 8 Near-infrared absorption spectrum of bulk water (the intensity scale is in arbitrary units).

methods to the standard procedures of crystallography. As will be seen in the following chapters, through the use of appropriate sample manipulation, one can obtain valuable information regarding molecules in their solid state.

REFERENCES

1. Herzberg G. Infrared and Raman Spectra of Polyatomic Molecules. New York, NY: Van Nostrand Reinhold Co., 1945.
2. Herzberg G. Spectra of Diatomic Molecules, 2nd edn. New York, NY: Van Nostrand Reinhold Co., 1950.
3. Pauling L, Wilson EB. Introduction to Quantum Mechanics. New York: McGraw-Hill, 1935.
4. Eyring H, Walter J, Kimball GE. Quantum Chemistry. New York: John Wiley & Sons, 1944.
5. Kauzmann W. Quantum Chemistry. New York: Academic Press, 1957.
6. Davis JC. Advanced Physical Chemistry. New York: Ronald Press, 1965.
7. Levine IN. Quantum Chemistry. Boston: Allyn and Bacon, 1970.
8. Hanna MW. Quantum Mechanics in Chemistry. Menlo Park: Benjamin/Cummings, 1981.
9. Barrow GM. Molecular Spectroscopy. New York: McGraw-Hill, 1962.
10. King GW. Spectroscopy and Molecular Structure. New York: Holt, Rinehart and Winston, 1964.
11. Brittain EFH, George WO, Wells CHJ. Introduction to Molecular Spectroscopy. London: Academic Press, 1970.
12. Guillory WA. Introduction to Molecular Structure and Spectroscopy. Boston: Allyn and Bacon, 1977.
13. McHale JL. Molecular Spectroscopy. Upper Saddle River. NJ: Prentice Hall, 1999.
14. Rao CNR. Chemical Applications of Infrared Spectroscopy. New York: Academic Press, 1963.
15. Kendall DN. Applied Infrared Spectroscopy. New York: Reinhold Pub., 1966.
16. Conley RT. Infrared Spectroscopy. Boston, MA: Allyn and Bacon, 1966.
17. Colthup NB, Daly LH, Wiberley SE. Introduction to Infrared and Raman Spectroscopy, 2nd ed. London: Academic Press, 1975.
18. Nakanishi K, Solomon PH. Infrared Absorption Spectroscopy. San Francisco, CA: Holden-Day, 1977.
19. Bhagavantam S. Scattering of Light and the Raman Effect. New York: Chemical Publishing Co., 1942.
20. Grasselli JG, Snavely MK, Bulkin BJ. Chemical Applications of Raman Spectroscopy. New York: Wiley-Interscience, 1981.
21. Grasselli JG, Bulkin BJ. Analytical Raman Spectroscopy. New York: John Wiley & Sons, 1991.
22. Ferraro JR, Nakamoto K. Introductory Raman Spectroscopy. New York: Academic Press, 1994.
23. Lewis IR, Edwards HGM. Handbook of Raman Spectroscopy. New York: Marcel Dekker, 2001.

24. Wilson EB, Decius JC, Cross PC. Molecular Vibrations: The Theory of Infrared and Raman Vibrational Spectra. New York: McGraw-Hill Book Co., 1955.
25. Weast RC. Handbook of Chemistry and Physics, 50th edn. Cleveland OH: Chemical Rubber Co., 1969:F-158–F-161.
26. Flett MC. Characteristic Frequencies of Groups in the Infra-Red. Amsterdam: Elsevier Pub., 1963.
27. Bellamy LJ. Advances in Infrared Group Frequencies. London: Methuen & Co., 1968.
28. Socrates G. Infrared and Raman Characteristic Group Frequencies—Tables and Charts, 3rd edn. Chichester: John Wiley & Sons, 2001.
29. Hochstrasser RM. Molecular Aspects of Symmetry. New York: WA Benjamin Inc., 1966.
30. Orchin M, Jaffe HH. Symmetry, Orbitals, and Spectra. New York: Wiley-Interscience, 1971.
31. Cotton FA. Chemical Applications of Group Theory, 2nd ed. New York: Wiley-Interscience, 1971.
32. Harris DC, Bertolucci MD. Symmetry and Spectroscopy. New York: Oxford University Press, 1978.
33. Fearn T, Osborne BG. Near Infrared Spectroscopy in Food Analysis. Dublin, Ireland: Longman Scientific & Technical Pub., 1986.
34. Williams P, Norris K. Near-Infrared Technology in the Agricultural and Food Industries. Saint Paul, MN: American Association of Cereal Chemists, 2001.
35. Siesler HW, Ozaki Y, Kawata S, Heise HM. Near-Infrared Spectroscopy: Principles, Instruments, Applications. Weinheim: Wiley-VCH Pub., 2002.
36. Burns DA, Ciurczak EW. Handbook of Near-Infrared Analysis, 2nd edn. New York: Marcel Dekker, 2001.
37. Ciurczak EW, Drennen JK. Pharmaceutical and Medical Applications of Near-Infrared Spectroscopy. New York: Marcel Dekker, 2002.

8

Infrared Absorption Spectroscopy

David E. Bugay
SSCI, Inc., West Lafayette, Indiana, U.S.A.

Harry G. Brittain
Center for Pharmaceutical Physics, Milford, New Jersey, U.S.A.

INTRODUCTION

The applications of infrared (IR) absorption spectroscopy are one of the most fundamental and useful spectroscopic techniques available in the arsenal of a pharmaceutical scientist. Typically, when most investigator scientists think of vibrational spectroscopy, studies in the IR region of the spectrum come to mind, owing to the commercial availability of spectrophotometers, the common understanding of the technique, the ease of spectral interpretation with the aid of correlation tables, and the widespread reference to the technique in pharmaceutically based regulatory guidance documents. Within this framework, the IR spectroscopy is routinely used as an identification assay method for various intermediate compounds, active pharmaceutical ingredients, excipients, and formulated drug products, and the methodology can also be developed as a quantitative technique for any of these.

As important as IR spectroscopy might be, however, it must be remembered that the full vibrational analysis of a molecule would also include Raman spectroscopy. Although IR and Raman spectroscopies are complementary techniques, the widespread use of the Raman technique in pharmaceutical investigations has been limited. Before the advent of Fourier transform (FT) techniques and lasers, experimental difficulties tended to limit the use of Raman spectroscopy. However, over the last 25 years a renaissance of the Raman technique has been seen, due mainly to instrumentation development. Applications of the Raman effect in solid-state pharmaceutics will be the subject of the following chapter.

The use of vibrational spectroscopy, specifically IR spectroscopy, in pharmaceutical applications has become very diverse. Bulk drug substance characterization uses traditional techniques, including structure elucidation, routine compound identification, and solid-state characterization (1–5). In addition, hyphenated techniques, such as liquid chromatography-IR spectroscopy (LC-IR) (6) and thermogravimetric (TG) analysis-IR spectroscopy (TG-IR) (7,8) have entered the arsenals of analytical and pharmaceutical scientists. Drug–excipient interactions, drug and/or excipient interactions with storage vessels, particulate identification, contaminant analysis, and mapping of pharmaceutical tablets are just a few of the microscopy-related applications of IR spectroscopy. Infrared absorption spectroscopy has also found extensive use in biopharmaceutics as a technique for the evaluation of peptide secondary structure (9), in reaction monitoring systems for chemical process engineers (10) and in on-line or at-line systems for monitoring chemical processes, manufacturing operations, or clinical packaging procedures (11).

It follows that in order for an investigator to extract the highest level of information from the conduct of a vibrational analysis, it is imperative that the basic concepts of the effect, the limitations of the instrumentation, and sampling techniques be fully understood. In some cases, Raman spectroscopy may be the more appropriate technique but might be cost prohibitive. Infrared spectroscopy might be applicable, but the scientist must first appreciate the limitations of the technique.

Since the advent of FT techniques and advances in laser technology and computer systems, a myriad of sampling techniques have become available for the vibrational spectroscopist. Analysis can now be performed on virtually any type of sample, such as single crystals, bulk particulate material, slurries, creams, films, solutions (aqueous or organic), oils, gas-phase samples, and on-process streams. Additional advantages of vibrational analysis include their typically nondestructive nature (with the ability to recover the material for further characterization), quantitative technique under proper sampling conditions, and complementary to other characterization techniques.

SOME FUNDAMENTAL PRINCIPLES OF IR ABSORPTION SPECTROSCOPY

A brief description of the IR spectroscopic theory will be presented so that the important basics associated with the spectroscopic effect are available to the reader. A more complete description of the underlying theory to IR spectroscopy is found in the previous chapter and in the literature (12–17).

Upon irradiation of a sample by a broad-band source of IR energy, the absorption of light having appropriate frequencies results in transitions between molecular vibrational and rotational energy levels. As discussed in the previous chapter, a vibrational transition might be approximated by the consideration of atoms bonded together within a molecule as a harmonic oscillator. Based on Hooke's law, the vibrational frequency between two atoms might be approximated as:

$$\nu = \frac{1}{2\pi}\sqrt{\frac{k}{\mu}} \tag{1}$$

where μ is the reduced mass of the two atoms, such that $\mu = (m_1 m_2)/(m_1 + m_2)$, and k is the force constant of the bond (dynes/cm). Quantum mechanical analysis of the harmonic oscillator model reveals a series of equally spaced vibrational energy levels (defined by the vibrational quantum number v, where $v = 0, 1, 2, 3 \ldots$) that are expressed as:

$$E_v = \left(v + \frac{1}{2}\right)h\nu_0 \tag{2}$$

where E_v is the energy of the vth level, h is Planck's constant, and ν_0 is the fundamental vibrational frequency.

It must be noted that the fundamental vibrational frequencies in a polyatomic molecule do not correspond to the vibrations of single pairs of atoms, but rather originate from the motions of groups of linked atoms. The absorption of IR energy by a molecule corresponds to approximately 2–10-kcal/mole, which in turn equals the stretching and bending vibrational frequencies of most bonds in covalently bonded molecules.

The number of fundamental vibrational modes of a molecule is equal to the number of degrees of vibrational freedom. As discussed in the previous chapter, for a nonlinear molecule consisting of N atoms, $3N - 6$ of vibrational freedom exist, and so there would be $3N - 6$ fundamental vibrational modes. For a linear molecule, the number of fundamental vibrational modes equals $3N - 5$, and so in a total vibrational analysis of a linear molecule by complementary IR and Raman techniques, a total of five vibrational transitions would be observed. Additional, but weak, absorption bands can also be observed due to transitions not allowed by the harmonic oscillator model but that gain intensity from anharmonic effects.

For a fundamental vibrational mode to be able to absorb IR energy, a change in the molecular dipole must take place during the molecular vibration. Typically, atoms in a molecule that possess different electronegativities and that are chemically bonded will undergo a change in the net dipole moment during normal molecular motion. As a result, antisymmetric vibrational modes and vibrations due to polar groups are more likely to exhibit prominent IR absorption bands.

The percent transmission (%*T*) of an IR absorption spectrum might be calculated using:

$$\text{Percent Transmission} = \frac{I}{I_0} \times 100 \tag{3}$$

where *I* equals the intensity of transmitted IR radiation, and I_0 equals the intensity of the incident IR beam. In this representation, the abscissa of a percent transmission IR spectrum would be in units of wavenumbers (cm^{-1}) and %*T* on the ordinate (Fig. 1).

The IR absorption spectra is presented in absorbance units using:

$$\text{Absorbance} = \log\left(\frac{I_0}{I}\right) = abc \tag{4}$$

where *a* is the absorptivity, *b* is the sample cell thickness, and *c* is the concentration. When there is an intention to use IR spectroscopy for quantitative purposes, one uses the absorbance presentation.

As previously mentioned, one of the most important pharmaceutical applications of IR spectroscopy is for the identification of molecular entities. This arises from the fact that the fundamental vibrational frequencies of a small group of atoms might be approximated by the harmonic oscillator model, indicating that the energy of the vibration is determined by the force constant of the bond and the reduced mass of the atoms in motion. For example, if the force constant (k) for a double bond is 10×10^5 dynes/cm, then by the harmonic oscillator equation

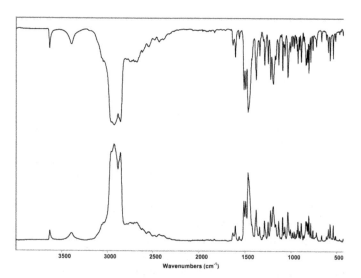

Figure 1 The infrared vibrational spectrum of a compound represented displayed as percent transmittance (*upper trace*) and absorbance (*lower trace*).

one can approximate the vibrational frequency for C=C as:

$$\nu = 4.12\sqrt{\frac{k}{\mu}} = 4.12\sqrt{\frac{10 \times 10^5}{[12 \times 12/12 + 12]}} = 1682\,\text{cm}^{-1} \tag{5}$$

A fairly good agreement exists between the calculated value of 1682 cm^{-1} and the empirically observed value of 1650 cm^{-1}. Based on the harmonic oscillator approximation, numerous correlation tables are generated that allow one to estimate the characteristic absorption frequency of a specific functionality (16). On this basis, one might recognize contributions by the various functional groups in a molecule in the IR spectrum, ultimately enabling the use of IR spectroscopy to identify a molecular entity. This understanding can subsequently enable the performance of quantitative analysis on the molecule.

The far-IR region might be considered an extension of the mid-IR region, because this area of the spectrum is typically used to investigate the fundamental vibrational modes of heavy atoms. The region extends from 400 to 10 cm^{-1}, and the low frequency molecular vibrations arise from the vibrational motion of heavy atoms (such as metal–ligand bonds). In addition, this region of the spectrum is used to study the intermolecular vibrations of crystalline materials. The motion of molecules relative to one another gives rise to lattice modes, which are at times referred to as external modes. These low-frequency modes are very sensitive to changes in the conformation or structure of a molecule and are observed in the range 200 to 10 cm^{-1}. Although this spectral region is important for the study of polymorphism, very few literature reports on such phenomena have been published. From a practical standpoint, investigations in the far-IR region require nontraditional optical windows, such as polyethylene or polypropylene because alkali halide windows do not transmit in this region. The spectral quality also is compromised due to the existence of relatively poor sources and insensitive detectors.

INSTRUMENTATION FOR THE MEASUREMENT OF IR SPECTRA IN THE SOLID STATE

The instrumentation for the measurement of IR spectra varies almost as widely as do its applications. Historically, mid-IR instrumentation made use of the conventional dispersive optical layout, where a source was rendered monochromatic, allowed to impinge on a sample, and the amount of beam attenuation measured as a function of incident wavelength (18). The dispersive methodology suffers from a lack of sensitivity and worked as well as it did owing to the strong degree of molar absorptivity associated with the electric dipole allowed vibrational transitions. Although dispersive mid-IR instruments remain in use, only a few manufacturers still produce such instrumentation. However, dispersive mid-IR instruments have found a niche in the process-monitoring field.

The acquisition of high-quality IR spectra on solid materials has been made possible by the introduction of the FTIR method, because this approach minimizes transmission and beam attenuation problems. Essentially all FTIR spectrometers use a Michelson interferometer, where IR energy entering the interferometer is split into two beams by means of a beam splitter. One beam follows a path of fixed distance before it gets reflected back into the beam splitter, while the other beam travels a variable distance before being recombined with the first beam. The recombination of these two beams yields an interference pattern, where the time-dependent constructive and destructive interferences have the effect of forming a cosine signal. Each component wavelength of the source will yield a unique cosine wave, having a maximum at the zero pathlength difference (ZPD) and which decays with increasing distance from the ZPD. The position of the detector is such that the radiation in the central image of the interference pattern will be incident upon it, and therefore the intensity variations in the recombined beam are manifest as phase differences. The observed signal at the detector is a summation of all the cosine waves, having a maximum at the ZPD and decaying rapidly with increasing distance from the ZPD. If the component cosine waves can be resolved, then the contribution from individual wavelengths are observed. The frequency domain spectrum is obtained from the interferogram by performing the Fourier transformation mathematical operation. The review in the classic book by Griffiths and DeHaseth (19) gives a full description of the components of an IR spectrophotometer.

Nevertheless, the greatest challenge for the IR spectroscopist is that of sample preparation. Since the development of the FTIR spectrophotometers, an abundance of sampling techniques for analytical purposes has been developed, and it is most appropriate to examine the most important of these that are widely used for pharmaceutical problem solving and method development.

Alkali Halide Pellet

The classic and most time-honored IR spectroscopic sampling technique is the preparation of alkali halide pellet (20). This technique involves mixing the sample of interest with powdered and dry alkali halide (typically KBr or KCl) at a 1–2% w/w ratio of sample to alkali halide. The mixture is pulverized until it is finely ground and homogeneous, placed into a die (typically made of stainless steel), and subjected to approximately 10,000-psi pressure for a sufficient period to produce a glassy pellet. The pellet (with the sample finely dispersed throughout the glass) might then be placed into the IR spectrophotometer for spectral data acquisition.

This sampling technique is widely used for the preparation of samples for chemical identity testing and is commonly incorporated into a regulatory submission as an identification test for bulk drug substances or excipients (21). The advantage of the pellet technique is that only a small amount of sample is required (usually 1 mg) and a high-quality spectrum is obtained in a matter of minutes. However, the investigator should know that disadvantages, such as solid-state transformation of the sample due to the pressure requirements to form the glass pellet and possible halide exchange between KBr or KCl and the sample of interest (22,23) may take place. The disadvantages become critical on considering the IR spectroscopy as one of the investigative tools in the study of solid-state properties, because the act of sample preparation can serve to alter the characteristics of interest.

Quantitation of mixtures was attempted with fair success utilizing the alkali halide pellet sampling technique (24). However, owing to the disadvantages just discussed, it is appropriate that this sampling technique be used only for simple compound identification assays.

Mineral-Oil Mull

Another classical sampling technique for solids is the suspension of the analyte in mineral oil (20). In this technique, a small amount of the sample (approximately 1 mg) is placed into an agate mortar, a small amount of mineral oil is added, and the sample and oil are mixed to achieve an even consistency. The mixture that is placed onto an IR-transmissive optical window is sampled by the IR spectrophotometer. One advantage to this technique is that there is no likelihood of solid-state transformations due to mixing and/or grinding, and hence the use of mineral-oil mulls represents a good technique for the qualitative determination of the polymorphic identity of a compound. Unfortunately, the mineral oil itself exhibits a number of intense absorption bands (observed at 2952, 2923, 2853, 1458, and 1376 cm^{-1}) that may overlap important absorption bands associated with the sample of iterest. Typically, this technique is used for qualitative identification assays whenever the alkali halide pellet technique is inappropriate.

Diffuse Reflectance Technique

The diffuse reflectance (DR) technique is one of the most important solid-state sampling techniques for pharmaceutical problem solving and method development (25). Sometimes referred to as DRIFTS (diffuse reflectance infrared Fourier transform spectroscopy), this technique has found extensive use in the mid- and near-IR spectral regions. The technique involves irradiation of the powdered sample by an IR beam, where the incident radiation undergoes absorption, reflection, and diffraction by the particles of the sample and only the incident radiation that undergoes DR contains absorptivity information about the sample. Figure 2 shows a schematic diagram for a DR sampling accessory.

A number of significant advantages are associated with the DR method of sampling. Some samples may be investigated in their as-received condition, or at worst would be studied after dilution with a nonabsorbing matrix, such as KBr or KCl (typically at a 1–5%-w/w active to nonabsorbing matrix material ratio). Macro- and micro-sampling cups are usually provided with the DR accessory, with approximately 400 and 10 mg of sample being required for each cup,

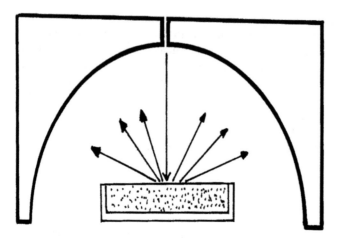

Figure 2 Schematic diagram of a diffuse reflectance sampling accessory. The incident infrared radiation is shown as entering from an opening in the collection mirrors and impacting with the surface of the sample. The diffuse reflectance from the sample is collected by the mirrors and ultimately passed on to the detector.

respectively. The sample also is recovered completely, enabling other solid-state investigations to be conducted on the same material. By the use of an environmental chamber attachment, variable-temperature and -humidity DR experiments can be performed (26). By varying the temperature of the sample, information about temperature-mediated crystal form transformations and the nature of the interaction of a solvate with the parent molecule can be determined.

The DR technique lends itself to studies of polymorphic composition because it is noninvasive, its use causes no changes in polymorph character due to its inherent limited sample handling, and the technique can be used for quantitative purposes (27). One disadvantage of DR IR spectroscopy is that the spectral results are sometimes affected by the particle size of the analyte (25,28). The development of DRIFTS quantitative assays for polymorphic composition must take the particle size of each component into account. Variations in absorption band intensity may be falsely interpreted as concentration changes in the analyte when, in reality, the variation is actually due to particle size effects.

Attenuated Total Reflectance

Attenuated total reflectance (ATR) is a sampling technique suitable for the study of an analyte in any physical form other than the vapor phase (29,30). In the ATR technique, IR radiation is passed through a crystal at an angle less than the critical angle, which causes the light to undergo total internal reflection. At each such reflection, the radiation penetrates a small distance beyond the crystal surface, and if a substance capable of absorbing IR energy is in physical contact with the crystal, then the internally reflected energy will be attenuated at those frequencies corresponding to changes in molecular vibrational states. As illustrated in Figure 3, the IR beam passes through the optically denser crystal and reflects at the surface of the sample.

An advantage of this technique is that it requires very little sample preparation, because one simply needs to place the sample in contact with the crystal. Typically, the sample is clamped onto the surface of the crystal with moderate pressure to ensure a sufficient degree of optical contact with the irradiating IR beam. The simplicity of sample handling makes the IR-ATR method most attractive for the study of polymorphic and solvatomorphic solids, because there is no opportunity for the sample to encounter any outside substances or sources of energy that might perturb its nature and composition.

Obviously, the analyst must be scrupulous in the cleaning of the crystal surface between analyses in order to avoid cross contamination. Because the internal reflectance process does not permit the IR beam to pass very deeply into the sample, it is typical to determine the composition or an analyte up to a sampling depth in the range of $5-10$ μm.

Microspectroscopy

The first linkage between a microscope and an IR spectrophotometer was reported in 1949 (31), but now every manufacturer of IR spectrophotometers offers an optical/IR microscope sampling accessory. Because optical microscopy can be used to obtain significant information about a sample, such as its crystalline or amorphous nature, particle morphology, and size, the inclusion of IR microspectroscopy in a program of study is appropriate for any solid-state investigation. Interfacing the microscope to an IR spectrophotometer enables

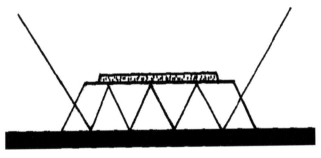

Figure 3 Schematic diagram of a sampling accessory suitable for multiple reflection attenuated total reflectance. The incident infrared beam enters the crystal and undergoes multiple reflections at the crystal surfaces before ultimately leaving the crystal on its way to the detector. At each reflection from the sample mounted at the crystal upper surface, the beam is sufficiently attenuated by sample absorbance so that the absorption spectrum can be derived from the beam.

investigators to obtain molecular identification of individual particles or crystals and to provide a valuable means for the identification of particulate contamination in bulk or formulated drug products. Other advantages for the technique are nondestructive sampling, the ability to make reflectance and/or transmittance measurements, and minimal sample requirements (32).

The IR microspectroscopy sampling technique can be considered to be the ultimate sampling technique, because the entire IR spectrum can be obtained on only a single particle. However, due to the limitation of diffraction, the particles of interest must typically be larger than 10×10 μm. The sample of interest is positioned upon an IR optical window, and the slide placed onto the microscope stage for its visual inspection. Once the sample is in focus, the field of view is apertured down to the sample, where, depending upon sample morphology, thickness, and transmittance properties, a reflectance and/or transmittance IR spectrum may be acquired by the IR microscope accessory.

By using the hot-stage microscopy, one may monitor the IR spectra of products being generated during the course of solid-state transformations. Alternatively, one can use an environmentally controlled chamber to change the atmosphere about a small sample and simultaneously observe the sample using both optical and spectroscopic means. This approach would be particularly useful when observing temperature-mediated crystal form changes that are accompanied by changes in crystal morphology.

TG/IR Analysis

In this sampling technique, a TG analyzer is interfaced to an IR spectrophotometer so that the evolved gas from the heated sample in the TG furnace is directed to an IR gas-phase cell. This IR sampling technique lends itself to the identification and quantitation of residual solvent content for a pharmaceutical solid (7) and for the investigation of pharmaceutical solvates/hydrates (8).

Analogous to standard TG analysis procedures, the sample of interest for TG/IR analysis is placed into a TG sample cup and introduced into the TG furnace. The TG balance then monitors the weight loss of the sample as a function of temperature (typical heating rate of $10°C/min$). In ordinary practice, one deduces the residual solvent content of a sample from the TG weight loss data, that is, the amount of weight loss over a specific temperature range. However, by acquiring the IR absorption data in addition to the weight loss measurement, one can make an unequivocal identification of the evolved gases. This enables the analyst to identify the species responsible for the weight loss and facilitates the determination of the residual solvent level.

Although the residual solvent content of a sample can be obtained by the TG/IR analysis, gas chromatography techniques usually outperform TG/IR experiments because of their lower detection limits and ease of quantitation. One area where the TG/IR technique excels is in the investigation of pharmaceutical solvates and hydrates. Conventional thermal analysis methods, such as thermogravimetry and differential scanning calorimetry are used to determine the presence of solvatomorphs after correlating a TG weight loss with a DSC endotherm over the same range of temperatures. However, the use of these methods cannot yield the identity of the solvent of crystallization, but such information is obtained by measuring the IR spectrum analysis of the gas evolved within the studied temperature range (8).

Photoacoustic Detection

The photoacoustic effect was first discovered by Alexander Graham Bell in the early 1880s, and the instrumentation for its detection was once known as the spectrophone (33). It took nearly a century for the detection scheme to be applied to FTIR spectroscopy (34,35), when it was recognized that photoacoustic spectroscopy (PAS) could be used to acquire spectra on opaque materials (which are commonly found in pharmaceutical formulations). It was also recognized that minimal sample preparation is necessary for measurement of the effect and that depth profiling of the sample is possible.

The PAS phenomenon involves the selective absorption of modulated IR radiation by the sample and subsequent measurement of the heat released when the excited molecules return to their ground states. Once the IR radiation is absorbed (obviously at frequencies that correspond to the fundamental vibrational frequencies of the sample) and reradiated as IR energy having shifted frequencies, the evolved heat escapes from the solid sample and heats a boundary layer of gas. Typically, this conversion from modulated IR radiation to heat involves only a small degree of temperature increase at the surface of the sample (approximately $10^{-6 \circ}$C). Because the sample is placed into a closed cavity cell, which is filled with a coupling gas (usually helium), the increase in temperature produces pressure changes in the surrounding gas that can be detected by a very sensitive microphone as sound waves. Because the incident IR radiation is modulated, the pressure changes in the coupling gas take place at the same modulation frequency, causing the acoustic wave to vibrate at the same frequency. After detection by the microphone, the resulting signal is amplified by phase-sensitive detection and Fourier-processed to yield the PAS spectrum.

The depth profiling of a solid sample may be performed by varying the interferometer moving mirror velocity. By increasing the mirror velocity, the sampling depth varies and surface studies might be performed. Limitations do exist, but the technique has proven to be quite effective for solid samples (36). In addition, particle size has a minimal effect upon the photoacoustic measurement, thus representing an advantage over DR sampling techniques.

Fiber Optics

The development of fiber optics in IR spectroscopy grew out of need to analyze hazardous materials from a remote location. Solids are analyzed with a DR arrangement of optical fibers, whereas liquids can be analyzed with an ATR arrangement of optical fibers on the end of the probe. Much of the early development of IR spectroscopy using fiber optic probes was carried out in the near-IR spectral region due to the limitations of the materials available for use as optical fibers. With the development of chalcogenide fibers, the use of fiber optics has extended into the mid-IR spectral region. Transmission measurements with chalcogenide fibers are suitable over the range of 4000 to 1000 cm^{-1}, but absorptions due to hydrogen-containing impurities in the fiber limit the spectral region of 2250–2100 cm^{-1}. The normal use of the chalcogenide fibers is limited to fiber lengths of approximately two meters in length (37).

APPLICATIONS OF IR SPECTROSCOPY TO AREAS OF PHARMACEUTICAL INTEREST

Infrared absorption spectroscopy has a significant history in the chemical and physical characterization of pharmaceutical solids. Tremendous hardware developments in IR spectrophotometers took place after World War II. IR analysis is the most widely reported spectroscopic technique utilized in the study of solid-state crystal form issues because of the hardware developments and because the IR spectroscopy is widely understood and the spectrophotometers are relatively inexpensive (as compared to, e.g., solid-state nuclear magnetic resonance spectrometers). Additionally, because solid-state IR spectroscopy can be used to probe the nature of polymorphism on the molecular level, the method is particularly useful in instances where full crystallographic characterization of polymorphism is not possible. Additionally, recent generations of hardware design allows for relatively high spatial resolution of IR data acquisition through microspectroscopy. This technique allows for drug product mapping applications and for contaminant analysis.

Qualitative Identification of Pharmaceutical Compounds

The identity testing of compounds of pharmaceutical interest represents one very important role of a spectroscopy laboratory. This mode of testing can be accomplished using methods associated with the mid-IR or near-IR spectral regions, and FT-Raman spectroscopy has begun to play a similar role. As stated in the official general test of the U.S. Pharmacopeia (21), the IR absorption spectrum of a substance provides the most conclusive evidence of identity that can be obtained from a single test. For bulk drug substances, the identity test most often used is that of the general test <197k>, where the sample is processed as a KBr pellet.

The identity testing of solid dosage forms of drugs usually involves a number of wet chemical extraction steps, followed the processing of the extract into a KBr pellet by the general method <197k>. For example, in the identity testing of leucovorin calcium tablets, one begins by grinding tablets, dissolving the powder in water, sonicating the dispersion, and filtering. The filtrate is mixed with ammonium oxalate, shaken, and centrifuged. Methanol is added to the preparation until it is clear, and then the preparation is placed in a 0°C freezer until a precipitate is formed. The supernatant is decanted, methanol is added to dissolve the precipitate, and the solution is allowed to evaporate. After the remaining precipitate is dried, the KBr pellet is prepared, and the IR spectrum obtained is compared with the IR absorption spectrum obtained from a reference standard prepared by the same procedure.

Studies of the physical properties of pharmaceutical solids, and especially the active pharmaceutical ingredient, are a very important aspect of the drug development process. To that end, IR spectroscopy has become a widely utilized physical characterization technique for the qualitative identification of compounds. For example, the IR absorption spectra of benzoic acid and sodium benzoate are shown in Figure 4, where the differentiation between the two is self-evident (38). The identity of the salt is readily confirmed by the shift in its carbonyl band frequency to 1595 cm^{-1} from the value of 1678 cm^{-1} that is characteristic of the free acid.

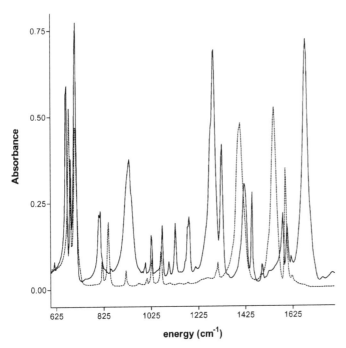

Figure 4 The FTIR–ATR spectra in the fingerprint region of benzoic acid (*solid trace*) and sodium benzoate (*dashed trace*). *Abbreviation*: FTIR–ATR, Fourier transform infrared–attenuated total reflectance. *Source*: Adapted from Ref. 38.

IR Spectral Studies of Polymorphic Systems

When coupled with a nonperturbing sampling mode, such as DR or ATR, the IR spectroscopy can also be used to establish the phase identity of a potentially polymorphic substance. For example, as shown in Figure 5, the polymorphic identity of famotidine can be readily established using FTIR-ATR spectroscopy, because the spectra are sufficiently different in the fingerprint region to enable such a distinction to be made (38). It should be noted, however, that the magnitude of peak shifting between two forms can often be as little as $10\,\mathrm{cm}^{-1}$, hence careful attention needs to be paid to sample handling and peak measurement.

   In the case of the monotropic polymorphic system formed by DuP-747, it was found that distinctly different DR, Raman, and solid-state ^{13}C-NMR spectra existed for each physical form (39). The complementary nature of the IR and Raman spectroscopies provided evidence that the polymorphic pair was roughly equivalent in conformation, and it was deduced that the crystal structure differences between the polymorphs of DuP-747 resulted from different modes of molecular packing even though more accurate crystallographic information could not be obtained.

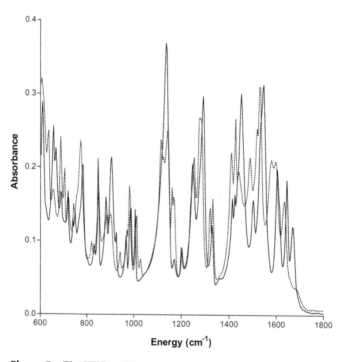

Figure 5   The FTIR–ATR spectra in the fingerprint region of famotidine Form-A (*solid trace*) and Form-B (*dashed trace*). *Abbreviation*: FTIR–ATR, Fourier transform infrared–attenuated total reflectance. *Source*: Adapted from Ref. 38.

The structure of only one of the two polymorphs of fosinopril sodium was solved, because it has not been found possible to grow useful crystals of the other form (40). Consequently, the two nonsolvated crystal forms were studied using a multidisciplinary approach consisting of XRPD, IR, NMR, and thermal analysis. Through detailed studies of the IR absorption and solid-state ^{13}C—NMR spectra, it was possible to conclude that the environment of the acetal side chain of the fosinopril moiety differed in the two forms. In addition, possible *cis-trans* isomerization about the C_6—N peptide bond was also identified. In the absence of single-crystal structural results, these conformational differences were postulated as the origin of the observed polymorphism in fosinopril sodium.

A common theme throughout many of the reports on polymorphic characterization is the use of a number of different physical analytical techniques. In the majority of cases, XRPD is the primary method of choice for determining the number of solid-state forms, because the effect originates from crystallographic differences. Thermal analysis and solid-state spectroscopy were used to support the XPRD results, and further characterize the structural aspects of the solid-state system. A number of examples where a multidisciplinary approach to solid-state form characterization was used are provided in Table 1. Outside of pure crystallographic investigations, the majority of studies on polymorphic systems make use of multiple techniques, very often including solid-state IR absorption spectroscopy.

In one such multidisciplinary study, five differently colored, solid-state forms of 5-methyl-2-[(2-nitrophenyl)amino]-3-thiophenecarbonitrile were discussed (41). It has been deduced that the color of the polymorphic crystal forms has a molecular origin and is related to the angle between the planes of the phenyl and thiophene rings, which changes the characters of the molecular orbitals of the systems and affects the degree of electron delocalization. The IR spectra for three of the solid-state forms illustrate shifts in frequencies that originate from the structural differences. For example, the nitrile stretching frequency varies by approximately 10 cm^{-1} each for the yellow, orange, and red solid-state forms of the compound. In the solid-state ^{13}C-NMR spectra, the C-3 carbon of the thiophene ring shows a chemical shift difference of about 5 ppm for the same three forms.

In a different type of multidisciplinary study, a theoretical and experimental study have been carried out on the three polymorphs of octahydro-1,3,5,7-tetranitro-1,3,5,7-tetrazocine (82). The density functional theory and scaled force-field methods were used to calculate the gas phase vibrational spectroscopy for the molecule in different conformations and molecular symmetries, and the authors compared these calculated spectra with experimentally observed data to interpret the various crystal spectra. It was concluded that the molecule existed in different conformations in each of the three polymorphs, with the molecule in the β-phase exhibiting C_i symmetry and those in the α- and δ-phases being characterized by C_{2V} symmetry.

Table 1 Polymorphic Systems Studied Using Infrared Spectroscopy Techniques

Compound	Character of the study	References
AG-337	Qualitative characterization	(42)
Carbamazepine anhydrate/dihydrate	Qualitative and quantitative characterization	(43–45)
Carbovir	Qualitative characterization	(46)
Cefamandole nafate	Qualitative characterization	(47)
Cefazolin	Qualitative and quantitative characterization	(48,49)
Cefepime·2HCl	Quantitative characterization	(27)
Compound H	Quantitative characterization	(50)
Cortisone acetate	Qualitative characterization	(51–53)
Delavirdine mesylate	Quantitative characterization	(54–56)
Diflunisal	Qualitative characterization	(57)
1,2-Dihydro-6-neopentyl-2-oxonicotinic	Qualitative characterization	(58)
11-α-Dimethylamino-2-β-ethoxy-3-α-hydroxy-5-α-pregnan-20-one	Qualitative characterization	(59)
Dirithromycin	Qualitative characterization	(60)
DuP-747	Qualitative characterization	(39)
β-Estradiol	Qualitative characterization	(61)
Fluocinolone acetonide	Qualitative characterization	(62)
Fosinopril sodium	Qualitative characterization	(40)
Frusemide	Qualitative characterization	(63)
Indomethacin	Qualitative characterization	(64)
Lactose	Qualitative characterization	(65)
Losartan	Qualitative characterization	(66)
Mefloquine HCl	Qualitative characterization	(67)
4'-Methyl-2'-nitroacetanilide	Qualitative characterization	(68)
5-methyl-2-[(2-nitrophenyl)amino]-3-thiophenecarbonitrile	Qualitative characterization	(41)
Nedocromil magnesium	Qualitative characterization	(69)
Nicardipine HCl	Qualitative characterization	(70)
Nimodipine	Qualitative characterization	(71)
Paroxetine HCl	Qualitative characterization	(72)
Ranitidine HCl	Qualitative and quantitative characterization	(73,74)
RG-12525	Qualitative characterization	(75)
SC-41930	Quantitative characterization	(76)
Spironolactone	Qualitative characterization	(77)
SQ-33600	Qualitative characterization	(78)

(Continued)

Table 1 Polymorphic Systems Studied Using Infrared Spectroscopy Techniques
(*Continued*)

Compound	Character of the study	References
Sulfamethoxazole	Qualitative and quantitative characterization	(79)
Sulfaproxiline	Qualitative characterization	(80)
Testosterone	Qualitative characterization	(81)

In a study of the amorphous salt formed by the coprecipitation of cimetidine and diflunisal, the solid-state IR absorption spectroscopy was used to prove the existence of the salt species (83). The prominent carbonyl absorption band observed at $1650\,cm^{-1}$ in crystalline diflunisal could not be observed in the spectrum of the amorphous salt, but a new peak was noted at $1580\,cm^{-1}$ that was assigned to an asymmetric stretching mode of a carboxylate group. These findings were taken to indicate that the amorphous character of precipitates formed by cimetidine and other nonsteroidal antiinflammatory agents were due to salt formation and not nonbonding intermolecular interactions.

IR Spectral Studies of Solvatomorphic Systems

When the solid-state properties of solvates and hydrates are catalogued by means of the multidisciplinary approach, one often finds that IR absorption spectroscopy yields particularly useful information. For example, as shown in Figure 6, the characteristic absorption bands in the fingerprint and high-frequency regions permit the ready differentiation of lactose anhydrate from lactose monohydrate (38). For instance, the presence of a strong absorption band at an energy of $1068\,cm^{-1}$ readily identifies a lactose sample as consisting of the monohydrate phase. In the high-frequency region, the well-defined absorbance band located at $3522\,cm^{-1}$ demonstrates the presence of crystalline water in the monohydrate phase.

Fluconazole has been isolated in a number of solvated and nonsolvated forms, and the IR spectroscopy has proven to be an important tool in the characterization of these. In one study involving two nonsolvated polymorphs (Forms I and II) and several solvatomorphs (the $\frac{1}{4}$-acetone solvate, a $\frac{1}{4}$-benzene solvate, and a monohydrate), the IR spectra of the different forms showed differentiation in bands associated with the triazole and 2,4-difluorobenzyl groups, and in the propane backbone (84). In another study, the diagnostic IR spectral characteristics of nonsolvated Form-III and two solvatomorphs (the $\frac{1}{4}$-ethyl acetate solvate, and a monohydrate) were used to demonstrate the novelty of the new forms relative to those in the literature (85).

During a study of the physicochemical properties of niclosamide anhydrate and two of its monohydrates, it was found that the carbonyl frequency was particularly sensitive to the crystal form of the drug substance (86). Previously, the drug

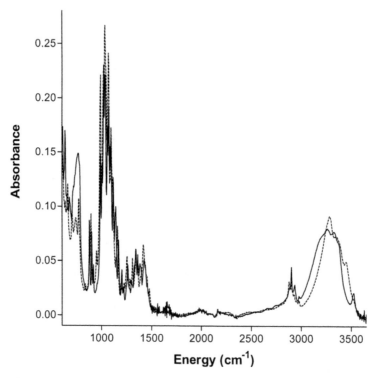

Figure 6   The FTIR–ATR spectra of lactose monohydrate (*solid trace*) and lactose anhydrate (*dashed trace*). *Abbreviation*: FTIR–ATR, Fourier transform infrared–attenuated total reflectance. *Source*: Adapted from Ref. 38.

substance was known to be able to exist in two conformations, and the IR spectra were used to demonstrate the conformational state of the compound in the solvatomorphic forms. Similarly, the IR spectra obtained in the fingerprint and high-frequency regions for four nonsolvated polymorphs of tenoxicam, and of its acetonitrile, dioxane, dimethyl formamide, ethyl acetate, acetone, and isopropanol solvates facilitated a differentiation between the various crystal forms (87). On the other hand, the mid-IR spectra of three solvatomorphs of tetroxoprim were found to be virtually identical, indicating that vibrational spectroscopy is not always the method of choice for such work (88).

The SQ-33600 was investigated by a variety of characterization techniques, including solid-state ^{13}C-NMR and IR absorption spectroscopies (78). Humidity-dependent changes were noted in the crystal properties of this HMG-CoA reductase inhibitor, where different types of hydrate formation could be achieved depending upon the relative humidity of the storage condition. In each hydrate type, a distinct IR vibrational frequency could be detected in the high frequency region, indicating that different hydrogen-binding effects perturbed the vibrational frequency of the crystalline water in the hydrates. It was postulated that the hydration continuously varied within lattice channels.

Norfloxacin is somewhat unusual in that the hydrate form is more soluble in water than is the anhydrate form. To investigate its hydration behavior, FTIR microspectroscopy was used to examine the structural changes associated with the hydration/dehydration processes that took place at different relative humidities (89). It was found that when norfloxacin anhydrate was converted to its hydrate, the intensities of the absorption bands at 1732 and 1253 cm^{-1} (assigned to the C=O and C=O moieties of the carboxylic acid group) decreased gradually with increasing water content. On the other hand, the intensities of the absorption bands at 1584 and 1339 cm^{-1} (corresponding to asymmetric and symmetric carboxylate modes) increased with the water content. In addition, the peak at 2553 cm^{-1} (assigned to the NH_2^+ group) shifted from 2558 cm^{-1} as the water content increased. These spectral changes correlated with transformations of COOH to COO$^-$ and from NH to NH_2^+, attributable to a proton transfer from carboxylic acid group and suggestive that hydration can induce proton transfer processes in the solid state.

IR Spectroscopy Studies of Phase Transformations

As long as the pattern of the vibrational modes is sufficiently perturbed by differences caused by crystallographic variations between polymorphs or solvatomorphs, the IR absorption spectroscopy can be used to study the phase transformation process. For example, it is well known that granulation of anhydrous lactose with water will result in the conversion of the anhydrate phase to the monohydrate phase, and a change from the β-anomer to the α-anomer. Figure 7 shows the FTIR–ATR spectra obtained in the high-frequency region of lactose monohydrate and at different stages in its solution-mediated conversion to the monohydrate phase (38). It is amply evident that, if desired, one could use the absorbance band at 3522 cm^{-1} to follow the kinetics of the formation of the monohydrate.

The solution-mediated phase transformation of the metastable β-phase of glycine to its stable α-phase was followed by IR absorption spectroscopy (90). A number of changes in the IR spectra were found to take place during the phase change, most notably the disappearance of the weak peak at 1660 cm^{-1} once the transformation was complete. Other bands that were observed to shift in energy during the process were the antisymmetric stretch of the carboxyl group (1580–1590 cm^{-1}), the deformation of the NH_3^+ group (1515–1500 cm^{-1}), and the O—C=O group bend (around 700 cm^{-1}).

The IR absorption spectroscopy played an important role in the characterization of the two polymorphs of racemic [2-[4-(3-ethoxy-2-hydroxypropyl)phenylcarbamoyl]ethyl]trimethyl-ammonium benzene sulfonate, and in the studies of the phase conversion of the least stable γ-phase into the most stable ϵ-phase (91). Very large band frequencies were observed for the O—H stretching modes (3340–3600 cm^{-1}) and the S—H stretching modes (1160–1260 cm^{-1}),

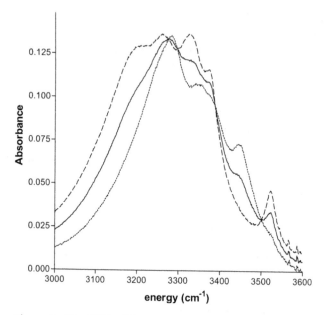

Figure 7 The FTIR-ATR spectra in the high-frequency region of lactose monohydrate (*dotted trace*), after approximately 50% conversion to the monohydrate phase (*solid trace*), and full conversion to the monohydrate phase (*dashed trace*). Abbreviation: FTIR–ATR, Fourier transform infrared–attenuated total reflectance. *Source*: Adapted from Ref. 38.

indicating that the bonding modes of the hydroxyl and benzene sulfonate groups were distinctly different in the two crystal forms. Because the kinetics of the phase transition was exceedingly rapid, the process could only be followed using in situ FTIR–ATR techniques.

By the use of FTIR microspectroscopy, the phase transformation of acetaminophen polymorphs could be studied (92). The as-received substance (Form-I) was found to melt around 165°C but to solidify into an amorphous state upon cooling. If this glassy substance was heated again, it was found to crystallize at 85°C into Form-III. Continued heating of the newly formed phase led to observation of another phase transformation into Form-II around 118°C. Although the changes could be followed visually through the microscopy, changes in the profiles and bandshapes of the IR absorption peaks proved to be a much more selective means to deduce the temperatures of the phase transitions.

The dehydration, rehydration, and solidification processes of trehalose dihydrate have been studied using vibrational spectroscopy (93). Trehalose dihydrate undergoes a dehydration transition around 100°C, and the IR spectra obtained during this process were found to change significantly over the 1500–1800 cm^{-1} region. For example, the band intensities at 1640 and 1687 cm^{-1} underwent sharp decreases around 65°C but remained relatively constant at higher temperatures.

These phenomena were taken to imply the existence of a phase transition of trehalose dihydrate over the temperature range of $64-67°C$, reflecting a transformation where the lattice water became liquid-like prior to the eventual dehydration. During the rehydration process where trehalose anhydrate became trehalose dihydrate, the liquid-like water became solid-like water over the same temperature range. It was postulated that this phase transition could be related to the protective effect of trehalose in preserving protein stability.

IR Spectroscopic Studies of Manufacture and Processing

When the vibrational spectra of polymorphic or solvatomorphic systems are suitable, IR absorption spectroscopy and ATR sampling can be used to follow and study the crystallization of the respective forms. Févotte (94) has published an excellent summary of perspectives for the use of on-line monitoring of pharmaceutical crystallization processes with the use of in situ IR absorption spectroscopy. As long as the spectroscopy is appropriate, one can insert an IR–ATR probe into a crystallizing vessel, and determine the solubility profiles of drug substances, as well as the limits of their metastable zones. In addition, one can evaluate appropriate seeding policies and the parameters required to obtain a real increase in the quality of the final product. Finally, the use of in situ IR absorption spectroscopy should have potential in following the impurity content in a vessel, and the results are used to minimize such species.

For example, the FTIR–ATR methodology was used to monitor the supersaturation level during a batch cooling crystallization of succinic acid (95). The system was set up to monitor the absorbance ratio of the peak area within the $1806-1675 \text{ cm}^{-1}$ region (carboxylate carbonyl absorption) to the peak area of the $1671-1496 \text{ cm}^{-1}$ region (water absorption) and could be used with prior knowledge of kinetic data. Using a feedback control loop, the system could maintain the supersaturation level close to the equilibrium solubility during the cooling process and, hence, could be optimized to yield the appropriate crystallite size.

In other studies of this type, the FTIR–ATR methodology and suitable in situ probes have been used to study the crystallization of mandelic acid (96), paracetamol (97), and glycine (98) from water. These and other examples amply demonstrate the utility of in situ IR absorption spectroscopy as a general method applicable for process analytical control.

The effect of secondary processing on the solid-state properties of substances having pharmaceutical interest is well known, and a variety of methodologies have been used to study such transformations (99). For example, when benzoic acid was compressed with methylated cyclodextrins, the carbonyl band normally observed at 1688 cm^{-1} was found to shift up to 1720 cm^{-1} (100). It was determined that the compression process induced proton transfer in the hydrogen-bonded eight-membered ring of dimeric benzoic acid pairs and that this proton transfer was responsible for the frequency shift.

The effect of particle pulverization on the properties of glisentide was studied using a multitude of techniques, including IR absorption spectroscopy (101). The grinding of the drug substance, or cogrinding it with polyvinylpyrrolidone, was not found to induce a change in the polymorphic composition, but it did result in the formation of an amorphous content and a consequent increase in the dissolution rate. The various peaks in the IR spectrum reflected this trend in that no shifting of peak maxima was observed, and the bands underwent a general reduction of intensity and a loss of spectral resolution.

The effect of secondary processes, such as size reduction, wet granulation, consolidation, and compression on celecoxib and its solvatomorphs, was studied (102). Although the polymorphic state of the nonsolvated form and the dimethylacetamide solvate did not change with milling, shifts in the IR spectrum that were caused by milling the dimethylformamide solvate indicated a loss of bound solvent and transformation to the nonsolvated form. In the DMF solvate form the N—H stretching doublet is absent, but upon milling (and the accompanying desolvation), the band characteristic of the nonsolvate developed at 3235 cm^{-1}. Wet granulation of celecoxib did not lead to any change in polymorphic condition, but the solvates were converted into the nonsolvated form by the action of the aqueous granulation fluid. The same type of desolvation process could be induced by the compression of the solvates into a compact.

The effect of grinding on the photostability of two polymorphs of nicardipine hydrochloride was studied by a number of techniques, including IR reflectance spectroscopy (103). The nicardipine content at the surface of a solid dosage form could be readily determined on the basis of its carbonyl absorbance at 1700 cm^{-1}, and the formation of photodegradation products could be detected by their absorption at 1734 cm^{-1}. It was found that the photodegradation rate constant of the β-form was greater than that of the α-form, although both phases were found to decompose much more quickly when being ground to a fine powder.

Quantitative Analytical Applications

In the studies of celecoxib and nicardipine hydrochloride just discussed, qualitative IR spectroscopy was used to obtain information about the phase transformations involved. However, in both of these works the investigators took the qualitative aspects discussed earlier and validated quantitative methods for the determination of the species involved. In the case of celecoxib, the ratio of the intensity of the doublet associated with the S$=$O symmetric stretching mode (1135 and 1165 cm^{-1}) to the intensity of the singlet associated with the aromatic C$=$C stretching mode (1498 cm^{-1}) was found to be a linear function of the free celecoxib concentration in its mixtures with the dimethylacetamide solvate (102). For nicardipine hydrochloride, the intensity of carbonyl absorbance at 1700 cm^{-1} was used as a quantitative indicator of the amount of residual drug substance after photodecomposition (103).

When working with the DR technique, two critical factors must be kept in mind when developing a quantitative assay, namely the production of homogeneous calibration and validation samples and a consistent particle size for all components, including subsequent samples for analysis. During the assay development and method validation, one must be cognizant of mixing techniques available to achieve homogeneous samples, because inhomogeneity in the calibration and validation samples can lead to inaccurate band absorption values and subsequent prediction errors. In addition, variation in the particle size of the nonabsorbing matrix or of the sample can influence the DR IR spectrum, which can also lead to prediction errors.

These considerations were fully taken into account during the development of a quantitative DR assay for cefepime dihydrochloride dihydrate in a matrix of cefepime dihydrochloride monohydrate (27). The validation process was hampered by difficulties encountered in mixing reference materials having very different particle morphologies, but this was overcome by slurrying the materials in hexane, mixing thoroughly, and then evaporating off the inert solvent. A working range of 1.0–8.0% was established, with a detection limit of 0.3% and a quantitation limit of 1.0%. The effect of particle size on the analytical parameters was also evaluated, and it was established that the IR method was valid for particles larger than 125 μm but smaller than 590 μm.

Sulfamethoxazole is known to exist in at least two polymorphic forms that have been fully characterized (79). Because the different forms exhibit distinctly different DR mid-IR spectra, it was possible to quantitate one form in the presence of the other. During the assay development, a number of mixing techniques were investigated in an effort to achieve homogeneous samples. After mixing and particle size factors were optimized, a quantitative DR IR spectroscopic assay method was developed in which independent validation samples were predicted within 4% of theoretical values.

In some pharmaceutical spectroscopy laboratories, TG/IR analysis is routinely used for solvate identification of pseudopolymorphic compounds. For example, a bulk drug substance exhibited distinctly different X-ray powder diffraction and differential scanning calorimetry data when compared to the reference standard data (8). Figure 8 displays the TG analysis weight loss curve, and the mid-IR spectra that were acquired at various time points and temperatures for the sample in question. Based upon the existence of absorption bands at 2972 and 1066 cm^{-1}, an organic species containing aliphatic C—H and C—C—O moieties, respectively, was deduced to be present in the evolved gas. Subsequent analysis of the IR spectra collected at the completion of the weight loss (spectral library matching) revealed that an ethanol solvate was present. Thus, TG/IR provided an unequivocal characterization as to the origin of the solvatomorphic character of this particular compound (8).

The IR spectroscopy and ATR have been used as a means to detect concentrations of principal species during the titration of malic acid (104). After subtracting the various types of water (pure, acidic, or basic water) from the spectra, factor analysis was used to deduce the three malic species as being nonionized malic acid, a monosodium malate, and a doubly ionized salt. Although these studies were conducted in aqueous media, there is no reason

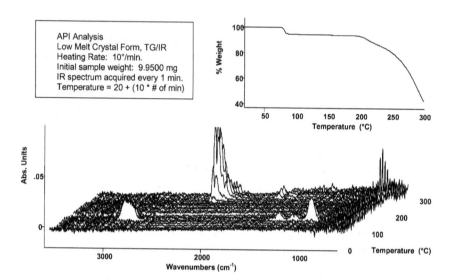

API Analysis
Low Melt Crystal Form, TG/IR
Heating Rate: 10°/min.
Initial sample weight: 9.9500 mg
IR spectrum acquired every 1 min.
Temperature = 20 + (10 * # of min)

Figure 8 Thermogravimetric weight loss curve and subsequent infrared spectra measured as a function of temperature. A slight lag time exists between the TG weight loss and the infrared spectral acquisition due to the evolved gas having to be swept into the gas sampling cell by the helium carrier gas *Abbreviation*: TG, thermogravimetric.

why an investigator could not use the results of a similar solid-state study to evaluate the salt species existing in a bulk drug substance. In fact, IR absorption spectroscopy is one of the methods whereby one can differentiate the patented hemisodium salt of valproic acid from the nonpatented monosodium salt (105).

IR Spectroscopic Studies During the Preformulation Stage of Drug Development

During preformulation, the possible interactions between a drug substance and its proposed excipients are typically studied using either thermal analysis (106) or chromatographic (107) techniques. Unfortunately, the use of chromatography requires extraction and/or dissolution techniques that may destroy critical physical and chemical information, and the use of thermal analysis is often accompanied by false results associated with the measurement technique itself. There is no doubt that a methodology with the ability to study drug–excipient interactions in a noninvasive manner is of great value, and IR absorption spectroscopy certainly can meet the need.

The utility of IR absorption spectroscopy in support of drug interaction studies can be illustrated through a study of intentionally selected incompatible materials. It is well known that omeprazole is unstable under acidic conditions, and that the substance undergoes rapid decomposition when allowed to interact with an acid (108). A 50%-w/w blend of omeprazole and salicylic acid was prepared, and as would be expected the IR spectrum of the blend consisted of the amalgamation of the spectra of the individual components (38). After this blend was exposed to an environment of 40°C and 75% relative humidity for 14 days (Fig. 9), the IR spectrum was now found to reflect almost entirely of the spectrum of salicylic acid. This finding demonstrates the practically complete decomposition of the omeprazole during this accelerated stability study and the incompatibility of the two substances.

The reaction of isoniazid (isonicotinic acid hydrazide) with magnesium oxide was studied by several techniques, including IR absorption spectroscopy (109). Chemisorption and physisorption led to new ultraviolet absorption bands, imparting color into stressed mixtures. Through a comparison of the IR spectrum of a genuine coordination complex of isoniazid with those of the starting material and the MgO product, it was concluded that no formation of a

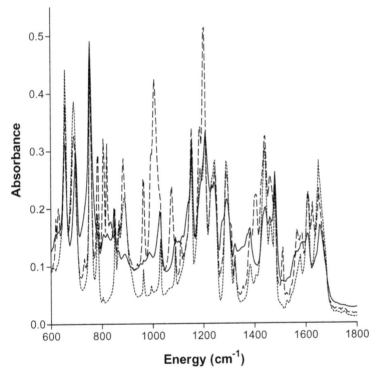

Figure 9 The FTIR–ATR spectra in the fingerprint region of a 50%-w/w blend of omeprazole with salicylic acid (*dashed trace*), and the same sample after being exposed to 40°C and 75% relative humidity storage condition for 14 days (*solid trace*). Also shown for comparison purposes is the spectrum of salicylic acid (*dotted trace*). Abbreviation: FTIR–ATR, Fourier transform infrared–attenuated total reflectance. *Source*: Adapted from Ref. 38.

Mg(II)-isoniazid coordination complex took place. Rather, the incompatibility between the drug and excipient was attributed to a donor–acceptor mechanism, with the pyridine ring of the drug substance acting as the donor.

The interaction of *dextro*-amphetamine sulfate with lactose has been studied in detail, with the degree of reaction being more pronounced with spray-dried lactose owing to the higher amounts of 5-hydroxymethylfurfual in the excipient (110). The presence of a C=N vibrational band, which was not present in the initial materials, in the IR absorption spectrum of the decomposition product was used to assist in an identification of the brown product as being the amphetamine-hydroxymethylfurfual Schiff base reaction product commonly associated with the Maillard-type reaction.

Differential scanning calorimetry and DR IR spectroscopy have been used to study the yellow or brown color that develops when aminophylline is mixed with lactose (111). The DSC thermogram of the aminophylline/lactose mixture is not a direct superposition of the individual components, and this observation was taken as an indication of an incompatibility. After the complete analysis of the IR spectra of the individual components, physical mixtures of these, and various blended samples subjected to stress conditions (60°C for three weeks), it was concluded that ethylenediamine is liberated from the aminophylline complex and reacts with lactose through the usual Schiff base intermediate. This reaction, in turn, results in brown discoloration of the sample.

Combinatorial chemistry is rapidly becoming a reliable way for pharmaceutical companies to discover and identify drug candidates and to study their interactions with potential excipient materials. For example, an IR spectroscopic method for the analysis of solid-phase organic reactions that occur on resin beads was described (112,113). This method represents an improvement over previous IR spectroscopic methods that required preparing a KBr pellet from at least 10 mg of resin beads. In the described method, a drop of resin solution was removed from the reaction vessel, washed in organic solvents, and dried under vacuum. The dried resin beads are placed on an NaCl window, flattened to 10–15 μm thickness, and spectra acquired in the transmission mode of the IR microscope. The authors demonstrated that the flattened beads provided superior results to nonflattened beads.

Bioanalytical Applications

Pharmaceutical companies are showing great interest in developing products based on proteins, enzymes, and peptides, and with the development of such products come the need for methods to evaluate the purity and structural nature of these biopharmaceuticals. However, in the solution phase these biopolymers are susceptible to a variety of chemical and physical degradation processes (114), which has caused many formulators to developed lyophilized dosage forms. Apart from a wide range of chemical instabilities (115), protein pharmaceuticals are prone to physical changes in their secondary structure that lead to the deleterious effects of denaturation, aggregation, surface adsorption, and precipitation (116). Consequently, methods to evaluate the secondary structure content of pharmaceutically active biopolymers are required, and IR absorption spectroscopy has proven to be one of the important methods (117–119).

The secondary structure of proteins is reflected in the amide-I (1600–1700 cm^{-1}), amide-II (1500–1600 cm^{-1}), and amide-III (1225–1300 cm^{-1}) absorption bands, as these are primarily associated with the stretching vibrations of peptide carbonyl groups. Of these, the amide-I and amide II bands were found to be particularly useful for IR-based work, and the measurement of the various band energies enable an identification of the peptide secondary structure type. For example, poly(L-lysine) exists completely in the α-helix conformation at pH 11.2, exhibiting an amide-I band at 1644 cm^{-1} and a split amide-II band located at 1521 and 1548 cm^{-1} (120). Heating this solution at 30–40°C causes a conformational change to the antiparallel β-sheet, whereupon the amide-I band splits to absorptions at 1615 and 1691 cm^{-1}, and the split amide-II bands shift to 1535 and 1563 cm^{-1}. At pH 6.5, poly(D,L-alanine) is known to exist in the random coil conformation, exhibiting its amide-I band at 1650 cm^{-1} and its amide-II band at 1551 cm^{-1}.

The effect of lyophilization on the secondary structure of α-chymotrypsin was reported, and a comparison of the solution phase spectrum within the amide-I band with that of the dried solid is shown in Figure 9 (121). Quantitative analysis of the solid-phase spectrum yielded values for β-sheet, α-helix, β-turn, and random coil percentages that agreed closely with the values derived from an X-ray crystallographic analysis. However, the lyophilization process was found to lead to a reduction in the percentages of random coil and β-turn percentages relative to the solution phase composition, and an increase in the content of β-sheet and α-helix. It was concluded that even though these differences were not large, the protein was not in its native form in the lyophilized solid.

As mentioned, maintaining a biopolymer as close as possible to its native state during lyophilization is crucial for that substance to maintain its efficacy, and for this reason formulators normally include a cryoprotectant in their formulations. It should be recognized, however, that the freezing and subsequent dehydration steps of lyophilization present different types of stress to the secondary structure of a protein, and these should be treated separately for optimal protein stabilization. The effect of dimethyl sulfoxide, glycerol, and ethylene glycol on the secondary structures of cytochrome C and lysozyme were studied using IR absorption spectroscopy (122). Although the secondary structure of cytochrome C seemed to remain unchanged by the cosolvents, shifts in the amide-I band of lysozyme indicated the formation of additional α-helical content at the expense of the β-turn and random coil contents.

The secondary structures of two labile enzymes, lactate dehydrogenase and phosphofructokinase, have been evaluated using IR absorption spectroscopy in the aqueous and lyophilized states (123). The addition of 10-mM mannitol, lactose, or trehalose or 1% polyethylene glycol (PEG) to the enzyme solutions attenuated the protein unfolding, but the IR spectra still indicated the existence of significant spectral differences for the enzymes between the dried state and the aqueous conformation. However, when a combination of 1% PEG and either 10-mM mannitol, lactose, or trehalose was added, the native structure was preserved during lyophilization, and essentially full enzymatic activity was recovered upon reconstitution. The authors concluded that for labile proteins, the preservation of the native structure during lyophilization is esssential for the recovery of activity following rehydration.

The FTIR spectroscopy was used to characterize the interaction of stabilizing carbohydrates with dried proteins (124). The lyophilization of lysozyme with trehalose, lactose, or myoinositol resulted in substantial alterations of the IR spectra of the dried carbohydrates. It was also found that dehydration-induced shifts in the positions of the lysozyme amide-I and amide-II bands could be partially and fully reversed, respectively, when the protein was lyophilized in the presence of either trehalose or lactose. The results indicated that not only did hydrogen bonding exist between the dried proteins and the carbohydrates but also that carbohydrate binding to the biopolymers was required for the carbohydrates to induce protein stabilization during the processes of lyophilization and rehydration.

The interactions between proteins and carbohydrates in the dried state have been further studied using DR FTIR methods (125). Recombinant human deoxyribonuclease I and recombinant human insulin were lyophilized with mannitol, sucrose, trehalose, and dextrans of different molecular weights, and the IR spectra demonstrated that changes in the protein secondary structure accompanied the lyophilization process. For both proteins, the presence of intermolecular β-sheets due to aggregation was detected, and the α-helical content decreased significantly. The use of carbohydrates as cryoprotectants was found to inhibit rearrangements in the protein secondary structure, with sucrose and trehalose being the most efficient for this purpose. However, whatever conformational changes were noted for these proteins appeared to be reversible after rehydration.

The IR spectroscopic examination of the secondary structure in the dried-state conformation of interleukin-2 was used to determine the pH conditions and stabilizers that provide optimal storage stability for the lyophilized product (126). At pH 7, the biopolymer unfolds extensively upon lyophilization, whereas at pH values less than 5 it remained essentially in its native form. A strong correlation was observed between the retention of the native structure during freeze-drying and enhanced stability, as interleukin-2 prepared at pH 5 is approximately an order of magnitude more stable with regard to the formation of soluble and insoluble aggregate protein prepared at pH 7. A similar pH stability profile was observed in the presence of excipients, although the excipients alter the overall stability profile. It was concluded that excipients with the capacity to substitute for water upon dehydration better preserve the native structure, and this, in turn, results in enhanced stability.

Chemical Mapping and Imaging

Probably one of the most exciting new applications of IR spectroscopy is that of chemical mapping and imaging, where investigators generate a chemical image of a two-dimensional area of a sample (127). The spectroscopic chemical imaging technique relies on the interface of an optical microscope, equipped with a motorized stage, to an IR spectrophotometer. The operator is able to visually focus upon a sample of interest and make optical observations regarding the sample, such as morphology, separation of layers, and the like. Subsequently, the same two-dimensional visual area is defined for spectroscopic analysis. Utilizing a raster pattern (i.e., the stepwise movement of the motorized stage along the x- and y-axes), individual spectra are acquired for each spatial location within the two-dimensional area. Spatial resolution is typically defined by the technique (approximately 5–10 μm for IR work). Once the individual spectra are obtained for the two-dimensional area, the intensity of a specific spectral feature within each spectrum can be plotted as a function of the spatial position. In this manner, one is able to obtain a contour plot showing the spatial position of a chemical entity.

One example of the use of IR mapping is the physical characterization of multicomponent pharmaceutical formulations. In order to provide immediate or controlled release rates of the active pharmaceutical ingredient, pharmaceutical formulations are becoming more complex. In addition, the mode of action for a drug substance may require specific dissolution of the compound within the intestinal tract (typically via an enteric coating) or within the stomach. To facilitate these types of drug delivery, multiple-layered solid dose formulations have become commonplace and are ideally suited to the IR mapping methodology.

To facilitate the release of the drug substance within the intestinal tract, a formulation was produced where the active pharmaceutical ingredient was spray coated onto a 90–100-μm sugar sphere. This coated core was overlaid with an intermediate excipient layer, and then a final enteric coating was applied to provide protection to the microsphere from acidic dissolution within the stomach. Chemical and physical characterization of the microsphere was required to determine the thickness of the various layers, their integrity, and their degree of uniformity. In order to obtain this information, a number of microspheres were embedded within an epoxy matrix such that the block could be mounted into a microtome. The epoxy block was then sectioned in controlled cuts so that a half-sphere microsphere was exposed on itssurface. The block was then mounted on a leveling stage positioned on the stage of an IR microscope. It was required that the surface of analysis be kept orthogonal to the IR beam so that a constant focus could be maintained for the entire two-dimensional area sampled by the mapping experiment.

After obtaining an optical microscopy picture of the area to be sampled (Fig. 10A), a two-dimensional area of analysis was defined by the IR mapping software in conjunction with the operator's view of the sample. For this particular example, a line map was performed where a line of analysis was drawn by the

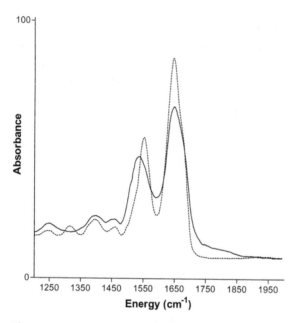

Figure 10A Comparison of the infrared absorption spectra of α-chymotrypsin in its aqueous (*dashed trace*) and dried solid (*solid trace*) states. *Source*: Adapted from Ref. 113 and the absorbance scale is in arbitrary units

mapping software, and the operator designated that IR spectra be collected along this line in 20-μm increments. In actuality, a rectangular sampling area having dimensions of 20 × 70 μm was designated. After the set-up of the sampling parameters, an IR spectrum was acquired at each spatial position. As one can imagine, thousands of spectra can easily be accumulated for very large sampling areas that were designated with small spatial increments. Fortunately, automated processing software is available to subsequently produce the spectral images in accord with a diagnostic spectral feature.

In this case, Figure 10B shows the entire IR spectrum for each spatial position sampled by the experiment. The *x*-axis represents traditional IR energy (in units of wavenumbers), and the *y*-axis represents a spatial position along the line of analysis. The *z*-axis in this case represents the intensity of each spectral feature, and in the figure, blue signifies strong intensity and red signifies weak intensity). By examining Figure 10B, one can clearly discern unique spatial areas due to the envelope of common spectral features from one spatial position to another. Starting at the bottom of Figure 10B, the first 20 μm represents one chemical entity based upon its unique spectral features. After the spectral interpretation based on consultation with IR libraries (128), this layer was identified as a particular enteric coating excipient. A specific polymorphic form of the drug substance was determined to reside approximately 20–80 μm further in,

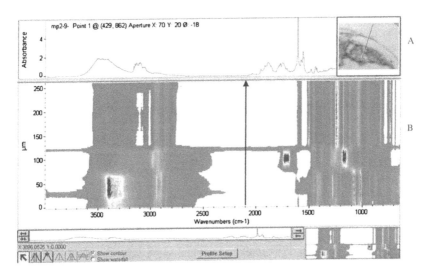

Figure 10B The infrared mapping results for sampling across a microtomed beadlet containing a drug substance. An optical image of the half-sphere beadlet is displayed in the *upper right corner* of the figure. The spectrum presented in Section **A** represents one spatial position from the total sampling area. Section **B** of the figure presents all of the spectra collected from the total sample area. Within Section **B** of the figure, the *y*-axis represents distance along the sampling area (*top of the arrow* pointing toward the core of the beadlet), the *x*-axis is the infrared absorption band (in cm^{-1}), and the *black and gray colors* represent the intensity of spectral features (*black* being most intense, and *gray* being least intense).

and a relatively small intermediate layer was then identified from approximately 80–110 μm. Finally, the sugar core was identified from approximately the 110-μm position and into the core of the microsphere. From this type of analysis, one can easily identify different chemical species (i.e., drug substances or formulated excipients), different physical forms of the drug substance, and the spatial locations of different chemical species. This technique has been widely applied to drug products (tablets, microspheres, etc.) and drug delivery devices (transdermal patches, stents, etc.).

REFERENCES

1. Markovich RJ, Pidgeon C. Pharm Res 1991; 8:663.
2. Bugay DE, Williams AC. Vibrational spectroscopy. In: Brittain HG, ed. Physical Characterization of Pharmaceutical Solids. New York: Marcel Dekker, 1995:59–91.
3. Bugay D. Pharm Res 1993; 10:317; Threlfall TL. Analyst 1995; 120:2435.
4. Brittain HG. J Pharm Sci 1997; 86:405.
5. Brittain HG. Methods for the characterization of polymorphs and solvates. In: Brittain HG, ed. Polymorphism in Pharmaceutical Solids. New York: Marcel Dekker, 1999:227–278.

6. Griffiths PR, Lange AJ. J Chromatogr Sci 1992; 30:93.
7. Johnson DJ. Compton DAC Spectroscopy. 1988; 3:47.
8. Rodriguez C, Bugay DE. J Pharm Sci 1997; 86:263.
9. Kim Y, Rose CA, Liu Y, Ozaki Y, Datta G, Tu AT. J Pharm Sci 1994; 83:1175.
10. Lynch J, Riseman S, Laswell W, Tschaen D, Volante R, Smith G, Shinkai I. J Org Chem 1989; 54:3792.
11. Dempster MA, MacDonald BF, Gemperline PF, Boyer NR. Anal Chim Acta 1995; 310:43.
12. Herzberg G. Infrared and Raman Spectra of Polyatomic Molecules. 6th edn. New York, NY: D. Van Nostrand Reinhold Co., 1954.
13. Rao CNR. Chemical Applications of Infrared Spectroscopy. New York: Academic Press, 1963.
14. Kendall DN. Applied Infrared Spectroscopy. New York: Reinhold Pub., 1966.
15. Conley RT. Infrared Spectroscopy. Boston, MA: Allyn and Bacon, 1966.
16. Colthup NB, Daly LH, Wiberley SE. Introduction to Infrared and Raman Spectroscopy, 2nd edn. London: Academic Press, 1975.
17. Nakanishi K, Solomon PH. Infrared Absorption Spectroscopy. San Francisco, CA: Holden-Day, 1977.
18. Anderson DH, Woodall NB, Infrared spectroscopy. In: Weissberger A, Rossiter BW, eds. Physical Methods of Chemistry. Vol. I, New York: Wiley-Interscience, 1972: 1–84.
19. Griffiths PR, de Haseth JA. Fourier Transform Infrared Spectrometry. New York: John Wiley & Sons, 1986.
20. Stewart JE. Infrared Spectroscopy: Experimental Methods and Techniques. New York: Marcel Dekker, 1970.
21. Spectrophotometric Identification Tests, Infrared Absorption, General Test <197>, United States Pharmacopeia, 28th edn. United States Pharmacopeial Convention, Rockville, MD, 2005:2295.
22. Bell VA, Citro VR, Hodge GD. Clays and Clay Minerals 1991; 39:290.
23. Mutha SC, Ludemann WB. J Pharm Sci 1976; 65:1400.
24. Hlavay J, Inczédy J. Spectrochim Acta 1985; 41A:783.
25. Fuller MP, Griffiths PR. Anal Chem 1978; 50:1906.
26. The Complete Guide to FT-IR, Spectra-Tech, Inc., Stamford CT, 1993.
27. Bugay DE, Newman AW, Findlay WP. J Pharm Biomed Anal 1996; 15:49.
28. Culler SR. Diffuse reflectance spectroscopy: sampling techniques for qualitative/ quantitative analysis of solids. In: Coleman PB, ed. Practical Sampling Techniques for Infrared Analysis. Boca Raton, FL: CRC Press, 1993:93–105.
29. Compton SV, Compton DAC. Optimization of data by internal reflectance spectroscopy. In: Coleman PB, ed. Practical Sampling Techniques for Infrared Analysis. Boca Raton, FL: CRC Press, 1993:55–92.
30. Urban MW. Attenuated Total Reflectance Spectroscopy of Polymers. Washington, DC: American Chemical Society, 1996.
31. Barer R, Cole ARH. Nature 1949; 63:198.
32. Reffner JA, Coates JP, Messerschmidt RG. Amer Lab 1987; 19:5.
33. Bell AG. Amer Assoc Proc 1881; 29:115.
34. Vidrine DW. Appl Spectrosc 1980; 34:314.
35. Rockley MG. Appl Spectrosc 1980; 34:405.
36. McClelland JF, Luo S, Jones RW, Seaverson LM. In: Bićanić D, ed. Photoacoustic and Photothermal Phenomena III. Berlin: Springer-Verlag, 1992:113–124.

37. MacLaurin P, Crabb NC, Wells I, Worsfold PJ, Coombs D. Anal Chem 1996; 68:1116.
38. Brittain HG. Unpublished Results for infrared spectra obtained at a resolution of $2 \, cm^{-1}$, using a Shimadzu model 8400 Fourier-transform infrared spectrometer, and sampled against the ZnSe crystal of a Pike MIRacle^{TM} single reflection horizontal ATR sampling accessory.
39. Raghavan K, Dwivedi A, Campbell GC, Nemeth G, Hussain MA. J Pharm Biomed Anal 1994; 12:777.
40. Brittain HG, Morris KR, Bugay DE, Thakur DE, Serajuddin ATM. J Pharm Biomed Anal 1994; 11:1063.
41. Stephenson GA, Borchardt TB, Byrn SR, Bowyer J, Bunnell CA, Snorek SV, Yu L. J Pharm Sci 1995; 84:1385.
42. Dash AK, Tyle P. J Pharm Sci 1996; 85:1123.
43. McMahon LE, Timmins P, Williams AC, York P. J Pharm Sci 1996; 85:1064.
44. Suryanarayanan R, Wiedmann TS. Pharm Res 1990; 7:184.
45. Matsuda Y, Akazawa R, Teraoka R, Otsuka M. J Pharm Pharmacol 1994; 46:162.
46. Nguyen NAT, Ghosh S, Gatlin LA, Grant DJW. J Pharm Sci 1994; 83:1116.
47. Kalinkova GN, Dimitrova L. Vib Spectros 1995; 10:41.
48. Byrn SR, Gray G, Pfeiffer RR, Frye J. J Pharm Sci 1985; 74:565.
49. Kamat MS, Osawa T, DeAngelis RJ, Koyama Y, DeLuca PP. Pharm Res 1988; 5:426.
50. Langkilde FW, Sjöblom J, Tekenbergs-Hjelte L, Mrak J. J Pharm Biomed Anal 1997; 15:687.
51. Harris RK, Kenwright AM, Say BJ, Yeung RR, Fletton RA, Lancaster RW, Hardgrove GL. Spectrochim Acta 1990; 46A:927.
52. Deeley CM, Spragg RA, Threlfall TL. Spectrochim Acta 1991; 47A:1217.
53. Christopher EA, Harris RK, Fletton RA. Solid State Nucl Magn Reson 1992; 1:93.
54. Gao P. Pharm Res 1998; 15:1425.
55. Gao P. Pharm Res 1996; 13:1095.
56. Sarver RW, Meulman PA, Bowerman DK, Havens JL. Int J Pharm 1998; 167:105.
57. Martínez-Ohárriz MC, Martín C, Goñi MM, Rodríguez-Espinosa MC, Tros de Ilarduya-Apaolaza, Sánchez M. J Pharm Sci 1994; 83:174.
58. Chao RS, Vail KC. Pharm Res 1987; 4:429.
59. Harris RK, Kenwright AM, Fletton RA, Lancaster RW. Spectrochim Acta 1998; 54A:1837.
60. Stephenson GA, Stowell JG, Toma PH, Dorman DE, Greene JR, Byrn SR. J Am Chem Soc 1994; 116:5766.
61. Variankaval NE, Jacol KI, Dinh SM. J Biomed Mat Res 1999; 44:397.
62. Bartolomei M, Ramusino MC, Ghetti P. J Pharm Biomed Anal 1997; 15:1813.
63. Doherty C, York P. Int J Pharm 1988; 47:141.
64. Lin S-Y. J Pharm Sci 1992; 81:572.
65. Brittain HG, Bogdanowich SJ, Bugay DE, DeVincentis J, Lewen G, Newman AW. Pharm Res 1991; 8:963.
66. Raghavan K, Dwivedi A, Campbell GC, Johnston E, Levorse D, McCauley J, Hussain M. Pharm Res 1993; 10:900.
67. Kiss A, Répási J, Salamon Z, Novák C, Pokol G, Tomor K. J Pharm Biomed Anal 1994; 12:889.
68. Fletton RA, Lancaster RW, Harris RK, Kenwright AM, Packer KJ, Waters DN, Yeadon A. J Chem Soc Perkin Trans II 1986; 1705.

69. Zhu H, Khankari RK, Padden BE, Munson EJ, Gleason WB, Grant DJW. J Pharm Sci 1996; 85:1026.
70. Yan J, Giunchedi P. Boll Chim Farmaceutico 1990; 129:276.
71. Grunenberg A, Keil B, Henck J-O. Int J Pharm 1995; 118:11.
72. Lynch IR, Buxton PC, Roe JM. Anal Proc 1988; 25:305.
73. Agatonovic-Kustrin S, Tucker IG, Schmierer D. Pharm Res 1999; 16:1477.
74. Madan T, Kakkar AP. Drug Dev Ind Pharm 1994; 20:1571.
75. Carlton RA, Difeo TJ, Powner TH, Santos I, Thompson MD. J Pharm Sci 1996; 85:461.
76. Roston DA, Walters MC, Rhinebarger RR, Ferro LJ. J Pharm Biomed Anal 1993; 11:293.
77. Neville GA, Beckstead HD, Shurvell HF. J Pharm Sci 1992; 81:1141.
78. Morris KR, Newman AW, Bugay DE, Ranadive SA, Singh AK, Szyper M, Varia SA, Brittain HG, Serajuddin ATM. Int J Pharm 1994; 108:195.
79. Hartauer KJ, Miller ES, Guillory JK. Int J Pharm 1992; 85:163.
80. Pitrè D, Stradi R. Arch Pharm (Weinheim) 1991; 324:57.
81. Fletton RA, Harris RK, Kenwright AM, Lancaster RW, Packer KJ, Sheppard N. Spectrochim Acta 1987; 43A:1111.
82. Brand HV, Rabie RL, Funk DJ, Diaz-Acousta I, Pulay P, Lippert TK. J Phys Chem B 2002; 106:10594.
83. Yamamura S, Gotoh H, Sakamoto Y, Momose Y. Int J Pharm 2002; 241:213.
84. Alkhamis KA, Obaidat AA, Nuseirat AF. Pharm Dev Tech 2002; 7:491.
85. Caira MR, Alkhamis KA, Obaidat RM. J Pharm Sci 2004; 93:601.
86. van Tonder EC, Maleka TSP, Liebenberg W, Song M, Wurster DE, de Villiers MM. Int J Pharm 2004; 269:417.
87. Cantera RG, Leza MG, Bachiller CM. J Pharm Sci 2002; 91:2240.
88. Caira MR, Bettinetti G, Sorrenti M. J Pharm Sci 2002; 91:467.
89. Hu T-C, Wang S-L, Chen T-F, Lin S-Y. J Pharm Sci 2002; 91:1351.
90. Ferrari ES, Davey RJ, Cross WI, Gillon AL, Towler CS. Cryst Growth Design 2003; 3:53.
91. Fujimoto D, Tamura R, Lepp Z, Takahashi H, Ushio T. Cryst Growth Design 2003; 3:973.
92. Wang S-L, Lin S-Y, Wei Y-S. Chem Pharm Bull 2002; 50:153.
93. Lin S-Y, Chien J-L. Pharm Res 2003; 20:1926.
94. Févotte G. Int J Pharm 2002; 241:263.
95. Feng L, Berglund KA. Cryst Growth Design 2002; 2:449.
96. Profir VM, Furusjo E, Danielsson L-G, Rasmuson AC. Cryst Growth Design 2002; 2:273.
97. Fujiwara M, Chow PS, Ma DL, Braatz RD. Cryst Growth Design 2002; 2:363.
98. Doki N, Seki H, Takano K, Asatani H, Yokota M, Kubota N. Cryst Growth Design 2004; 4:949.
99. Brittain HG. J Pharm Sci 2002; 91:1573.
100. Moribe K, Yonemochi E, Oguchi T, Nakai Y, Yamamoto K. Chem Pharm Bull 1995; 43:666.
101. Mura P, Cirri M, Faucci MT, Gines-Dorago JM, Bettinetti GP. J Pharm Biomed Anal 2002; 30:227.
102. Chawla G, Bansal AK. Pharm Dev Tech 2004; 9:419.

103. Teraoka R, Otsuka M, Matsuda Y. Int J Pharm 2004; 286:1.
104. Max J-J, Chapados C. J Phys Chem A 2002; 106:6452.
105. Meade EM. Sodium Hydrogen Divalproate Oligomer, United States Patent 4,988,731, issued January 29, 1991; United States Patent 5,212,326, issued May 18, 1993.
106. Guillory JK, Hwang SC, Lach JL. J Pharm Sci 1969; 58:301.
107. Serajuddin ATM, Thakur AB, Ghoshal RN, Fakes MG, Ranadive SA, Morris KR, Varia SA. J Pharm Sci 1999; 88:696.
108. Yang R. Schulman, Zavala PJ. Anal Chim Acta 2003; 481:155.
109. Wu W-H, Chin T-F, Lach JL. J Pharm Sci 1970; 59:1234.
110. Blaug SM, Huang W-T. J Pharm Sci 1972; 61:1770.
111. Hartauer KJ, Guillory JK. Drug Dev Indust Pharm 1991; 17:617.
112. Yan B, Kumaravel G. Tetrahedron 1996; 52:843.
113. Yan B, Kumaravel G, Anjaria H, Wu A, Petter RC, Jewell CF, Wareing JR. J Org Chem 1995; 60:5736.
114. Manning MC, Patel K, Borchardt RT. Pharm Res 1989; 6:903.
115. Lai MC, Topp EM. J Pharm Sci 1999; 88:489.
116. Costantino HR, Langer R, Klibanov AM. J Pharm Sci 1994; 83:1662.
117. Byler DM, Susi H. Biopolymers 1986; 25:469.
118. Sarver RW, Krueger WC. Anal Biochem 1991; 199:61.
119. Harris PI, Chapman D. Biopolymers (Peptide Science) 1995; 37:251.
120. Venyaminov S, Kalnin NN. Biopolymers 1990; 30:1259.
121. Dong A, Prestrelski SJ, Allison SD, Carpenter JF. J Pharm Sci 1995; 84:415.
122. Huang P, Dong A, Caughey WS. J Pharm Sci 1995; 84:387.
123. Prestrelski SJ, Arakawa T, Carpenter JF. Arch Biochem Biophys 1993; 303:465.
124. Carpenter JF, Crowe JH. Biochem 1989; 28:3916.
125. Souillac PO, Middaugh CR, Rytting JH. Int J Pharm 2002; 235:207.
126. Prestrelski SJ, Pikal KA, Arakawa T. Pharm Res 1995; 12:1250.
127. Krishnan K, Powell JR, Hill SL. Infrared microimaging. In: Humecki H, ed. Practical Guide to Infrared Microspectroscopy. New York: Marcel Dekker, 1995: 85–110.
128. Bugay DE, Findlay WP. Pharmaceutical Excipients: Characterization by IR, Raman, and NMR Spectroscopy. New York: Marcel Dekker, 1999.

9

Raman Spectroscopy

David E. Bugay
SSCI, Inc., West Lafayette, Indiana, U.S.A.

Harry G. Brittain
Center for Pharmaceutical Physics, Milford, New Jersey, U.S.A.

INTRODUCTION

Raman spectroscopy is a form of vibrational spectroscopy that is finding an ever-increasing degree of use in pharmaceutical investigations. Applications include chemical structure elucidation (1), routine chemical identification (2), and characterization of materials in the solid (3–5). Raman spectroscopy is also applicable to drug product characterization within drug delivery devices (6), contaminant analysis, drug–excipient interaction, and problem solving. Analysis can be performed on virtually any type of sample, such as single crystals, bulk materials, slurries, creams, particulates, films, solutions (aqueous and organic), oils, gas-phase samples, and process streams (the latter through the use of fiber optic probes). Furthermore, Raman spectroscopy is usual typically nondestructive in nature, enabling the analyst to recover the sample for further characterization.

When a compound is irradiated with monochromatic radiation, a certain amount of the incident light will be scattered by the molecule. Of the scattered radiation, the majority of the photons are scattered at the same frequency as the frequency of the incident radiation frequency. As described in the section, "Raman Spectroscopy of the Fundamental Modes of Polyatomic Molecules" of chapter 7, this form of scattering has been termed elastic, or Rayleigh, scattering.

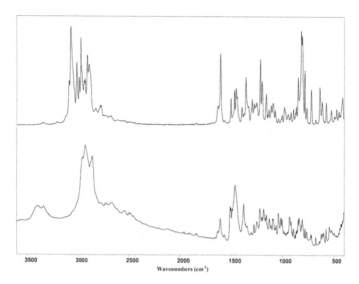

Figure 1   Comparison of the diffuse reflectance infrared absorption spectrum (*lower trace*) and the FT-Raman spectrum (*upper trace*) of an active pharmaceutical ingredient.

If the scattered radiation is passed into a spectrometer, a strong Rayleigh line is detected at the unmodified frequency of radiation used to excite the sample.

As described previously, a very small percentage of photons will be scattered at frequencies higher and lower than the frequency of the Rayleigh line, with the differences between the incident frequency of radiation and the shifted frequencies corresponding to the frequency of molecular vibrations of the molecules in the sample. As both infrared (IR) and Raman spectroscopies are based on vibrational motions, it is not surprising that a Raman spectrum and an IR absorption spectrum for a given compound would bear a strong degree of resemblance. Owing to the different origins of each process, differences in the character of functional groups cause the techniques to be complementary and not duplicative. This is illustrated in Figure 1 where the IR and Raman spectra of an active pharmaceutical ingredient (API) are shown.

SOME FUNDAMENTAL PRINCIPLES OF THE RAMAN EFFECT

In the previous discussion (chap. 7) a purely classical picture of vibrational Raman scattering was developed, in which the oscillating dipole moment induced by the electric field of the incident light was modulated by molecular vibrational motion, resulting in scattering at frequencies higher and lower than the frequency of the incident energy. Nevertheless, accurate descriptions of any type of molecular spectroscopy must originate in the quantum theory, because the states involved in such processes are derived from the quantization of energy levels (7,8).

Nonresonant Raman Scattering

Placzek (9) developed a quantum mechanical model for the Raman effect based on the classical polarizability model, assuming that under appropriate conditions a Raman transition can be treated in a manner analogous to an electric dipole absorption. For those molecular systems having nondegenerate ground electronic states (the usual instance for organic molecules of pharmaceutical interest) that are irradiated with electromagnetic radiation incapable of causing a transition to an excited state, the vibrational Raman intensities are given approximately by the vibrational matrix elements of the electronic polarizability. Through the use of the polarizability theory, one may understand the intensity of vibrational Raman scattering through a correlation with molecular polarizability in a manner that parallels that of IR spectroscopy (where the intensity of absorption is related to the vibrational-coordinate dependence of the dipole moment).

Because in quantum theory the frequency shifts of Raman bands are determined by the energy differences of the vibrational states between which the transitions take place, it is therefore necessary to calculate the transition probabilities of these transitions. As long as the atoms in a polyatomic molecule vibrate with low degrees of amplitude, the motion as a whole can be taken as being a superimposition of numerous harmonic oscillators. As long as the harmonic oscillator model applies, the selection rule of $\Delta v \pm 1$ applies, and therefore only molecular transitions characterized by a change in polarizability will exhibit an allowed Raman transition.

As developed previously in chapter 7 (section on "Raman Spectroscopy of the Fundamental Modes of Polyatomic Molecules"), the family of scattering bands observed at frequencies of ($v_0 - v_v$) are known as the *Stokes* lines, and the family of scattering bands observed at frequencies of ($v_0 + v_v$) are known as the *anti-Stokes* lines. According to the classical theory, the Stokes lines should have the same intensity as the anti-Stokes lines, but in fact the anti-Stokes lines are much weaker in intensity than the Stokes lines. This situation arises because the number of molecules occupying an initial state of $v_i = 1$ can be only $e^{-hcv/kT}$ times the number of molecules in the ground $v_i = 0$ state. The intensity ratio of anti-Stokes to the corresponding Stokes likes is found to be in agreement with the Boltzmann factor.

The intensity of scattered light depends on the magnitude of the induced dipole moment, P, which for a transition between ψ_N and ψ_M is given by the matrix formed from the integrals:

$$\int \psi_N P \psi_M \, d\tau \tag{1}$$

The time-independent part of Equation (1) is given by:

$$[P^0]^{NM} = \int \psi_N P^0 \psi_M \, d\tau \tag{2}$$

where P^0 is the amplitude of P, and the intensity of the $\psi_N \leftrightarrow \psi_M$ transition is proportional to the square of $[P^0]^{NM}$.

Knowing that the magnitude of the induced dipole moment equals the product of the electric field strength and the polarizability, the components of $[P^0]^{NM}$ are given by:

$$[P_X^0]^{NM} = E_X^0 \int \psi_N \alpha_{XX} \psi_M \, d\tau + E_Y^0 \int \psi_N \alpha_{XY} \psi_M \, d\tau + E_Z^0 \int \psi_N \alpha_{XZ} \psi_M \, d\tau \quad (3)$$

$$[P_Y^0]^{NM} = E_X^0 \int \psi_N \alpha_{YX} \psi_M \, d\tau + E_Y^0 \int \psi_N \alpha_{YY} \psi_M \, d\tau + E_Z^0 \int \psi_N \alpha_{YZ} \psi_M \, d\tau \quad (4)$$

$$[P_Z^0]^{NM} = E_X^0 \int \psi_N \alpha_{ZX} \psi_M \, d\tau + E_Y^0 \int \psi_N \alpha_{ZY} \psi_M \, d\tau + E_Z^0 \int \psi_N \alpha_{ZZ} \psi_M \, d\tau \quad (5)$$

In Equations (3), (4), and (5), E_X^0, E_Y^0, and E_Z^0 are the components of the amplitudes of the incident electromagnetic radiations along each Cartesian direction, and the various integrals

$$[\alpha_{XX}]^{NM} = \int \psi_N \alpha_{XX} \psi_M \, d\tau \quad (6)$$

$$[\alpha_{XY}]^{NM} = \int \psi_N \alpha_{XY} \psi_M \, d\tau \quad (7)$$

$$[\alpha_{XZ}]^{NM} = \int \psi_N \alpha_{XZ} \psi_M \, d\tau \quad (8)$$

are the matrix elements of the polarizability tensor.

The diagonal matrix elements ($[\alpha_{XX}]^{NM}$, $[\alpha_{YY}]^{NM}$, and $[\alpha_{ZZ}]^{NM}$) correspond to Rayleigh scattering, whereas the off diagonal elements ($[\alpha_{XY}]^{NM}$, $[\alpha_{XZ}]^{NM}$, $[\alpha_{YZ}]^{NM}$, $[\alpha_{YX}]^{NM}$, $[\alpha_{ZX}]^{NM}$, and $[\alpha_{ZY}]^{NM}$) correspond to Raman scattering associated with the $\psi_N \leftrightarrow \psi_M$ transition. The $\psi_N \leftrightarrow \psi_M$ transition will be allowed Raman scattering if at least one of the six off-diagonal elements does not equal zero.

As would be anticipated, molecular symmetry-based group theory can be used to determine whether one of the off-diagonal elements will have a non-zero magnitude. It has been established before that as long as the ground state of a molecule is not degenerate, then that state will be characterized by all vibrational wave functions being in their $v = 0$ level, and this state must transform as the totally symmetric representation in the molecular point group. To achieve a symmetric end result (and a nonzero magnitude for an off-diagonal element), the direct product of the representation for the excited state with the representation of one of the components of the polarizability must contain the totally symmetric representation of the point group. In other words, the representation of the excited state wave function must be the same as that of the representation for one of the elements of the polarizability operator (10–13). As discussed in a previous chapter, the elements of the polarizability tensor will transform as the squares and binary products of coordinates listed in area IV of the point group character table.

Detailed consideration of the various character tables leads to the *Exclusivity Rule*. For molecules containing a center of symmetry, transitions that are allowed in the IR absorption spectrum will be forbidden in the Raman spectrum, and transitions that are allowed in the Raman spectrum will be forbidden in the IR absorption spectrum. This rule is of great importance for the study of molecules characterized by high symmetry (such as inorganic coordination complexes), but rarely is of consequence in the low-symmetry world of organic molecules having pharmaceutical interest.

Resonance Raman Scattering

In the nonresonant condition, the energy of the incident photons is not equal to that of an allowed electronic transition, and therefore the wavelength of the incident light does not fall within the wavelength range of an electronic absorption band of the molecule. A completely different situation arises when the wavelength of the incident photons lies within the absorption band of an electric dipole-allowed electronic transition.

What has come to be termed the *resonance Raman effect* results from the transition from the ground state into an excited vibronic state, which is subsequently accompanied by immediate relaxation into a vibrational level of the ground state. What makes this resonance Raman process different from that of ordinary photoluminescence is that the transition back to a vibronic level of the ground state is not preceded by prior relaxation of the system to the lowest vibrational level of the excited state. This distinction is illustrated in the Jablonski-type diagram shown in Figure 2. The resonance Raman emission process is essentially instantaneous, and the resulting spectra consist of narrow bands whose energy difference from the exciting light is equal to the energy of a ground state vibrational mode of the molecule.

The consequence of resonance scattering is that the contribution to the scattering from those electronic state(s) with transition energies equal to that of the incident photons (i.e., the states with which the photons are in resonance) becomes very large and dominates all others. This results in a selective enhancement in the intensities of vibrational transitions involving motions of those atoms associated with the molecular orbitals that define the ground and excited states of the electronic transition. The particular vibrational modes enhanced are those that are coupled to the electronic transition responsible for the absorption band. In practical terms, this means that two classes of vibrational modes will produce intense resonance-enhanced spectra, namely totally symmetric vibrations, and those nontotally symmetric vibrations that vibronically couple two electronic states.

Because the molecular orbitals defining an electronic transition are often somewhat localized to a specific functional grouping in a molecule, the resonance Raman effect can provide a way to selectively enhance the Raman bands due to

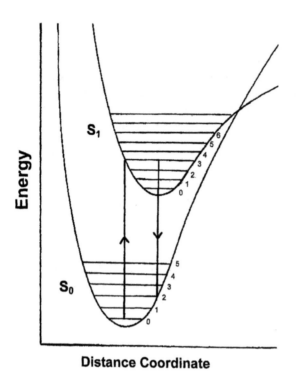

Distance Coordinate

Figure 2 Jablonski-type energy level diagram illustrating the resonance Raman process. Excitation is drawn as the $S_0(v = 0) \rightarrow S_1(v = 3)$ transition, and the resonance Raman transition is the $S_1(v = 3) \rightarrow S_0(v = 2)$ transition.

vibrations of the chromophore associated with the states of the functional group. For instance, in heme-proteins, it has been shown that the resonance Raman bands are solely due to vibrational modes of the tetrapyrrole chromophore and that none of the vibrational bands associated with the protein gain resonance enhancement. The utilities of ultraviolet (UV) resonance Raman spectroscopy for analytical, physical, and biophysical applications have been described (14).

It has been pointed out that the sensitivity of resonance Raman spectroscopy to only chromophore vibrational modes may be considered as being either strength or weakness (15). The strength of the method is that the observed spectra are greatly simplified, and a series of molecules containing slightly different chromophores will yield spectra that are easily distinguished. The main limitation is that if a series of molecules contain the same chromophore but with a variety of aliphatic side chains, the resonance Raman spectra would be nearly identical.

Surface-Enhanced Raman Scattering

In 1974, Fleischmann (16) observed intense Raman scattering from pyridine that had been absorbed onto a roughened silver electrode. It was recognized that the large Raman intensities could not be merely a function of the number of scattering units and that the intensity enhancement had to have its origins in an interaction between the surface and the absorbed scattering molecules (17,18). Subsequently it was found that the same degree of increase in the cross-section of Raman scattering could be obtained if the molecules were allowed to absorb onto the surfaces of colloidally dispersed silver particles (19). The methodology has been reviewed (20), and questions related to scattering from single particles, fractal clusters, and surfaces have been discussed (21).

As discussed earlier, the intensity of Raman scattering is determined by the magnitude of the induced dipole moment, which equals the product of the electric field strength and the polarizability. It follows that the enhancement of scattering must be associated either with an increase in the polarizability of the molecule or with an increase in the electric field experienced by the molecule. The most significant mechanism that could produce an increase in the polarizability would entail some type of charge transfer interaction between the metal substrate and the adsorbed molecule, or even formation of a chemical bond between the two. However, it is now generally agreed that most of the surface enhancement is associated with an amplification of the strength of the local electromagnetic field.

The most effective surface-enhancing systems consist of small metal particles or of rough surfaces of conductive materials. Light impinging on these types of surfaces can excite conduction electrons in the metal, generating polarization in the substrate, and thus the electromagnetic field at the interior of the particle becomes significantly larger than that of the applied field. The magnitude of the enhancement is given by Lorenz-Mie theory and depends on the wavelength-dependent refractive index of the substrate, the size and shape of the rough features or small particles, and the incident wavelength.

The theory predicts two key requirements in order to achieve a large degree of surface enhancement of the Raman scattering. First, the imaginary component of the refractive index of the metal should be small, and the real component should be approximately equal to twice the wavelength of the incident light source. Second, the frequency of the incident light should match the plasma resonance frequency of the substrate particle or the rough feature. For visible wavelength excitation sources, the first requirement leads to the conclusion that only the noble metals would be good surface-enhancing substrates. The ability to exploit the surface enhancement effect for these substrates is limited only by instrumental considerations and by any technical difficulties associated with the preparation of appropriate substrate morphologies.

The selection rules for conventional Raman spectroscopy also are applicable to surface-enhanced Raman scattering. Because the local electric field at the surface is highest in the direction perpendicular to the surface, the vibrational modes that involve changes in the polarizability of the absorbed molecule that are perpendicular to the surface will be preferentially enhanced (22). In addition to the fact that the electromagnetic field amplitude falls off rapidly with distance from the surface, the polarizability dependence allows an investigator to determine the orientation of the adsorbed molecule with respect to the surface and to make some estimate of the distance between a given functional group of the molecule to the surface.

INSTRUMENTATION FOR MEASUREMENT OF RAMAN SPECTRA IN THE SOLID STATE

The type of systems used for the acquisition of Raman spectra will be briefly discussed, and readers seeking more detail are referred to the numerous reference texts available (23–29).

Spectrometer Types

Although the Raman effect was discovered in 1928, the first commercial Raman instruments did not start to appear until the early 1950s. As the laser had not yet been invented, these instruments used elemental sources and arc lamps as the irradiation source. In 1962, laser sources started to become available for Raman instruments, and the first commercial laser Raman instruments appeared in the mid-1960s. The first commercial FT-Raman instruments became available in 1988, which was followed soon by the FT-Raman microscopy. A detailed history of the development of Raman instrumentation is available (30).

The basic configuration and components of a dispersive Raman spectrometer are shown in Figure 3. The source of monochromatic radiation is a laser, which could be helium-cadmium (325, 354, or 442 nm), air-cooled argon-ion (488 or 514 nm), doubled continuous wave neodymium yttrium aluminum garnet (Nd:YAG or $Nd:Y_3Al_5O_{12}$) (532 nm), helium-neon (633 nm), or stabilized diode (785 nm). The stability of the laser radiation is a key attribute of a good spectrometer, and good stability is essential for good function. Frequency stabilization of the laser under standard laboratory conditions (slight temperature fluctuations, vibrational effects, etc.) is required. Laser lifetimes and cost are also considerations of choice to use the laser.

One additional consideration associated with laser selection in dispersive Raman systems concerns the use of wavelengths that could potentially generate molecular fluorescence. As noted previously, the intensity of the Raman effect is fairly weak, and photoluminescence can be so intense as to mask the scattered Raman photons. If fluorescence does not pose a problem for a given sample, lower frequency lasers can be used (532 or 514 nm) for enhanced sensitivity, as the efficiency of Raman scatter is proportional to $1/\lambda^4$ (where λ is wavelength). If fluorescence is a problem when using these high-energy sources, then lower energy sources, such as those used in Fourier transform (FT)-based Raman spectroscopy, can be used to minimize the fluorescence effects.

In a dispersive Raman spectrometer, the sample is positioned in the laser beam, and the scattering radiation is collected either in a 180° (the backscattering method) or a 90° (the right-angle method) scattering configuration. Subsequently, a laser-line rejection filter is placed in the scattered beam path to filter out

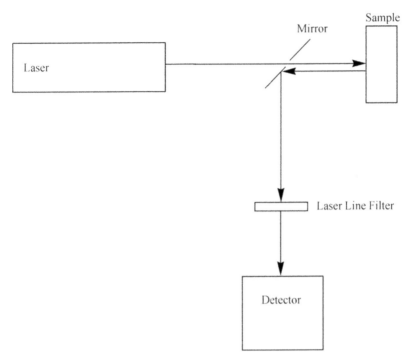

Figure 3   Schematic representation of a dispersive-type Raman spectrometer.

Rayleigh scattering, and the intensity of scattered light measured by a detector positioned in the spectrometer. For dispersive systems, a charge-coupled device (CCD) is typically utilized. Silicon CCD detectors are normally used for Raman spectrometers in which visible wavelength lasers are used. At one time, photomultiplier tubes were used for detection, but since the advent of CCD detectors and their inherently better performance, photomultiplier tubes are not normally used today. All commercial spectrometers are digitally controlled by a computer system.

The basic configuration and components of an FT-based Raman spectrometer is found in Figure 4. Advantages of an FT-Raman spectrometer are wavelength accuracy, and the use of a near-infrared (NIR) laser that usually eliminates spurious effects due to fluorescence. In the FT-based system, a $Nd:YVO_4$ laser (1064 nm output) is used to irradiate the sample; the sample is positioned in the laser beam, and the scattering radiation is collected either in the 180° back-scattering or the 90° right-angle scattering configuration. By utilizing the longer wavelength of the $Nd:YVO_4$ laser in a FT-Raman spectrometer, fluorescence is minimized because it is unlikely that a change in electronic state could be promoted through the use of a low-frequency laser.

The scattered photons are then passed into an interferometer with laser line filtering. Detection of the scattered photons from systems that utilize

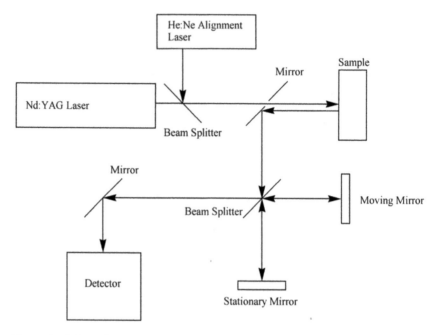

Figure 4   Schematic representation of a Fourier-transform type Raman spectrometer.

lasers emitting light with wavelengths greater than 1000 nm are of the single-element type, being either high purity p-type germanium or indium/gallium/arsenic (InGaAs) detectors. Such detectors are noisier than the CCD or PMT detectors, but they do exhibit high quantum efficiencies. By cooling the Germanium detector to liquid nitrogen temperature (77 K), the frequency response can be extended to 3400 cm^{-1}. Unfortunately, germanium detectors are subject to interference by cosmic rays, and artifacts in their output may be generated. However, corrections to these artifacts have been incorporated into hardware and/or software methods.

Sampling and Data Acquisition

Sampling techniques for Raman spectroscopy are relatively simple because the only requirements are that the monochromatic laser beam irradiates the sample of interest and that the scattered radiation is focused upon the detector. Raman spectroscopy may be performed on very small samples (samples as small as a few nanograms). Powders do not need to be pressed into discs or diluted with KBr as would be the case for most IR spectroscopic sampling procedures, but instead the material merely needs to be irradiated by the laser beam. Solid samples are often examined in stainless steel or glass sample holders that hold approximately 25–50 mg of the sample.

Liquid samples are typically analyzed in quartz or glass cuvettes, which may have mirrored rear surfaces to improve the signal intensity. Water is a good solvent for Raman studies, because the Raman spectrum of water consists of essentially one broad, weak band at 3500 cm^{-1}. Glass does not scatter Raman bands very efficiently, and therefore many samples (liquid and solid) can be simply analyzed in situ in a bottle, or for convenience in a nuclear magnetic resonance (NMR) sample tube. These spectral advantages have been used to develop a method for the noninvasive identification of substances inside USP vials using spectra acquired through the glass and a Raman spectral library (31).

One sample type that may pose a problem is any material that is darkly colored. Often, these samples absorb excessive heat and burn, causing sample and spectral degradation. Amorphous materials also have a tendency to absorb heat in the laser beam. In order to avoid sample burning, it may be necessary to dissipate the heat, which can be accomplished by reducing the laser power or by using an accessory that spins the sample to avoid the irradiation of a single point in the sample. Further reduction of the laser power can also be accomplished through the use of a neutral density filter. Sometimes, the sample can be diluted in KBr to aid in a reduction of sample burning.

The complete Stokes Raman spectrum spans in the range of 100–3500 cm^{-1} and can be obtained under conditions where the intensity of Raman scattering is directly proportional to the concentration of the scattering species (an essential factor for quantitative analysis). Unlike IR absorption spectroscopy, where the strong absorptivity of the bands enables analytes to be detected down to concentration of approximately 0.01%, the weakness of the Raman effect leads to the requirement that an analyte should be present at a concentration of at least 1% for accurate assessments.

As discussed previously, the presence of fluorescence, typically due to additives in the glass sample tubes or impurities within the sample of interest, is problematic for Raman studies. Data massaging techniques can sometimes be used to blank out fluorescence or alternatively one may use photobleaching as another means of suppressing fluorescence. The latter technique involves irradiating the sample for a prolonged period of time (seconds to hours) with the laser. During this time, the fluorescence may decrease due to the destruction of the fluorescing component due to the prolonged exposure to the laser irradiation, and then the spectrum is acquired after photobleaching is complete.

In any Raman study, the possibility that the laser radiation may change the component of interest must always be considered. For example, a solvated crystalline material may desolvate upon exposure to laser radiation. It is appropriate and prudent to establish the integrity of the sample by comparing spectra acquired with short data acquisition times to those acquired with long acquisition times. Additionally, ancillary techniques, such as X-ray powder diffraction, thermal analysis, or IR absorption spectroscopy, could be used to learn if the laser irradiation caused changes in the quality of the sample as a result of the conduct of the Raman spectroscopic analysis.

Variable temperature studies in Raman spectroscopy can be used to study the mechanisms of thermally induced reactions. Because a Raman spectrum typically covers a wavelength range that extends beyond the usual mid-IR region down to the far-IR region, information about the lattice vibrations of crystalline organic compounds can be readily acquired. By varying the temperature of a sample, the lattice energies of the compound become changed, which then facilitates interpretation of the nature of the crystal lattice. In addition, information on crystal form changes and the nature of solvatomorph association, usually obtained with other methods of thermal analysis or thermospectroscopic studies, can be obtained through Raman investigations at variable temperatures.

Use of the Raman microprobe to measure spectra from small amounts of material is of considerable interest. Utilizing the microscope, the Raman scattered photons are collected in a 180° backscattering configuration that allows the operator to optically view the sample, focus the incident radiation, and subsequently collect the Raman spectrum. Most commercial Raman microscope systems utilize confocal microscopy to increase axial resolution (along the z-axis). Confocal points are defined as the point source, the in-focus sample location, and the focused image of the sample point. Axial resolution, defined as the distance away from the focal plane in which the Raman intensity from the sample decreases to 50% of the in-focus intensity, can be approximated from the numerical aperture used in the microscope. When utilizing a 0.95 numerical aperture (NA) objective on a confocal microscope system, the axial resolution is proportional to the square of the NA, or in this case 0.9025 μm.

An additional advantage of Raman microscopy is that of spatial resolution, as opposed to axial resolution. The spatial resolution (in the xy-plane) is dependent upon the NA of the collecting objective and the wavelength of the laser radiation. Larger NA values and shorter wavelengths provide higher degrees of spatial resolution, which can be as low as 1 μm. One consideration is that because the high intensity, monochromatic, laser radiation is focused upon a small area of the sample, any sample degradation by the laser must be monitored. With this caveat in mind, the Raman microprobe has been found to be ideal for investigating polymorphism in single crystal samples, particulate contamination, and small quantities of analytes. Using an apparatus similar to that used for IR microscopy, variable temperature studies can be performed with a Raman microprobe.

Fiber optics has been used in Raman spectroscopy since the early 1980s, facilitating the analysis of solids and liquids by means of an arrangement of optical fibers on the end of a probe. Today, much of the research being conducted in the use of fiber optics in FT-Raman spectroscopy centers around fiber type and fiber-bundle design. The number, type, and arrangement of the fibers in a fiber bundle are key factors that are varied to produce fiber bundles for different applications. Fiber systems include single fibers (where the laser excitation and collected scattered radiation travel along the same fiber) and multifibers (where laser excitation is transmitted along one or multiple fibers and the scattered radiation is transmitted to the detector along different fibers). The arrangements of the fibers in a multifiber system can also vary. For example, one arrangement is where one excitation fiber is surrounded by several collection fibers, and another example is where several excitation and collection fibers are randomly mixed in a bundle. The greatest single advantage of the use of fiber optics in Raman spectroscopy is the ability to sample analytes in a remote location, where fiber optics can link the spectrometer to the sample, typically in a distances of tens of meters. Common applications include monitoring process streams or hazardous reactions.

GROUP FREQUENCY CORRELATIONS IN RAMAN SPECTROSCOPY AND EFFECTS OF THE SOLID STATE

As discussed previously, the concept that a given functional group would consistently exhibit vibrational frequencies within a narrow range is due to the fact that the frequency of a vibrational mode in a molecule is determined primarily by the force constant of the bond or bonds and by the reduced masses of the nuclei involved in the motion. In a complex molecule, the characteristic frequency range is modified by the interplay of many other factors, the resultant of which determines the exact energy of the absorption band. Some of these factors are associated with intramolecular origins, relating to changes in the molecular geometry, the masses of substituent groups, the occurrence of mechanical coupling between one vibration and another, the effects of steric strain, and the electrical influences of the substituent groups operating either along the bonds or across space. Other factors that are of great interest to solid-state spectroscopy concern the external environment of the vibrating group, including such effects as details associated with the crystal form or the influence of hydrogen bonding.

Despite these limitations, there is no doubt that correlation tables of group frequencies continue to be of great value in vibrational spectroscopy. A number of such compilations are available that summarize the frequencies associated with the IR absorption of various functional groups (32–35), but it goes without saying that the same principles of vibrational group frequencies must apply equally to the scattering bands in a Raman spectrum. In fact, such a compilation has been published specific to group frequencies in Raman spectroscopy (36).

Although excitation of a given vibrational mode must give rise to IR absorption and Raman scattering bands having identical frequencies, the fact that each effect originates from fundamentally different principles causes the observation of large differences in relative intensities. This is illustrated in Figure 5, where the intensities of the C=O stretching, methyl deformation, and C—Cl stretching bands in the IR absorption and Raman spectra of 2,5-dichloroacetophenone are seen to exhibit very different intensities (37). Most significant is the aromatic C=C stretching mode that appears prominently in the Raman spectrum but which exhibits only moderate intensity in the IR absorption spectrum.

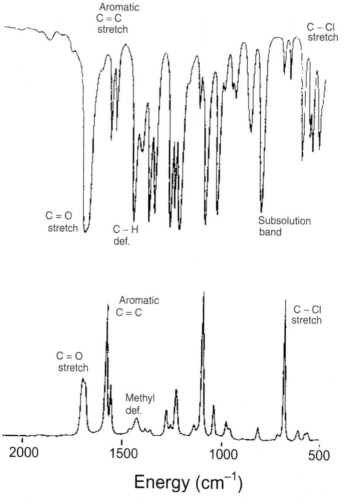

Figure 5 Comparison of the Raman (*lower trace*) and infrared absorption (*upper trace*) spectra in the fingerprint region of 2,5-dichloroacetophenone, illustrating the difference in relative intensities for each spectral type. *Source*: Adapted from Ref. 37.

As previously discussed, the vibrational modes that exhibit the strongest intensities in a Raman spectrum are those that are associated with functional groups that are characterized by high degrees of polarizability, because such groups will be able to exhibit significant changes in the induced dipole moment upon irradiation. Such functional groups include, for example, the C—S, S—S, C—C, N=N, and C≡C groups. Grasselli et al. have provided a useful table summarizing the characteristic wavenumbers for a variety of organic functional groups and have summarized the Raman and IR absorption intensities of these (38); a selection of these can be found in Table 1.

Table 1 Characteristic Frequencies of Some Organic Functional Groups and the Corresponding Raman and Infrared Absorption Intensities

Functional group	Frequency region (cm^{-1})	Raman intensity	Infrared absorption intensity
—O—H	3000–3650	Weak	Strong
—N—H	3300–3500	Medium	Medium
≡C—H	3300	Weak	Strong
=C—H	3000–3100	Strong	Medium
—C—H	2800–3000	Strong	Strong
—S—H	2550–2600	Strong	Weak
—C=O	1680–1820	Strong–weak	Very strong
—C=C—	1500–1900	Very strong– medium	Weak–absent
—C=N—	1610–1680	Strong	Medium
—C=S	1000–1250	Strong	Weak
—C—O—C—	1060–1150	Weak	Strong
—O—O—	845–900	Strong	Absent
—S—S—	430–550	Strong	Absent
—C—Cl	550–800	Strong	Strong
—C—Br	500–700	Strong	Strong

Source: Adapted from Ref. 38.

    Because of the polarizable nature of sulfur, Raman spectroscopy is particularly well suited for studies of the nature of sulfur bonding. Although the S—H stretching mode around 2580 cm^{-1} is weak in the IR absorption spectrum, it is strongly observed in the Raman spectrum. Also prominent in the Raman spectra of mercaptans are the C—S stretching modes (observed between 570 and 785 cm^{-1}) and the symmetric and antisymmetric C—S—C modes (also observed between 570 and 800 cm^{-1}).

    For most molecules in the gas state, a molecular vibration studied by Raman (or IR absorption, for that matter) spectroscopy can be regarded as being essentially free from the influence of other molecules. In the liquid state, however, other molecules and the bulk solvent, which may influence its frequency either through the change they have produced in the dielectric constant of the medium or through molecular association, surround the vibrating group. For example, for carboxylic acids dissolved in a solvent, formation of dimers through hydrogen bonding leads to shifts in carbonyl and hydroxyl frequencies by as much as 50 cm^{-1} and 500 cm^{-1} relative to the vapor state, respectively. The concept of group frequencies still has validity, however, but the range for a given vibrational type becomes wider in the condensed phase.

    The situation existing in a crystalline solid is different relative to either the vapor or liquid phases. Frequency shifts can occur due to a further increase in intermolecular forces, but these are usually relatively small unless the vibrations are involved in hydrogen bonding. Owing to the increased order of a crystalline structure, it is frequently observed that absorption bands associated with some bonds will not be observed in the spectrum. This effect arises from the fact that rotational isomerism cannot take place in an oriented crystal, which causes absorption bands associated with any other conformer to disappear.

In other instances, additional vibrational bands will appear. Within the rigid crystal environment, strong intermolecular forces exist so that the individual group vibrations become strongly influenced by the nature of the unit cell. In suitable circumstances, a process analogous to linear combination takes place, where in-phase and out-of-phase vibrations of the same groups in different molecules are established, leading to splitting of the original vibrational band into two components. Alterations in the crystal lattice, associated with the existence of multiple polymorphic forms, can result in spectral changes due to the alterations, which take place in the immediate environments of the vibrating groups. This phenomenon is particularly wellsuited for Raman spectroscopic investigations, because Raman can provide information in the low-frequency region where the differences between the different polymorphs are appreciable.

APPLICATIONS OF RAMAN SPECTROSCOPY TO AREAS OF PHARMACEUTICAL INTEREST

The widespread use of Raman spectroscopy as a tool to solve problems of pharmaceutical interest required the advent of the FT method, because prior to that the technique was viewed as being too much of a research method and not practical for use on a more routine basis. Part of the problem was that too many compounds exhibited fluorescence when irradiated in the visible region of the spectrum (i.e., argon-ion lasers operating at 488 or 514 nm). However, because the technique is based on light scattering, samples can be investigated in a noninvasive and nondestructive manner, and the analytes do not have to be diluted with inert filler before being studied. A number of detailed reviews are available regarding the use of Raman spectroscopy in areas of pharmaceutical interest (4,5,39–41).

Qualitative Identification of Pharmaceutical Compounds

The identification of a substance on the basis of its Raman spectrum is completely analogous to the identification of a compound on the basis of its characteristic IR absorption spectrum. For example, the fingerprint region of the Fourier-transform Raman spectrum of paracetamol is shown in Figure 6, with its diagnostic peaks being marked. In practice, an analyst would choose five to six of the most intense lines and enter these in a search library. Then, when a new spectrum of a compound purported to be paracetamol was obtained, one could match the observed peaks with the library peaks to make the identification. Because the number of Raman spectral libraries lags considerably behind the number of IR spectral libraries, many workers compile their own libraries for identification purposes.

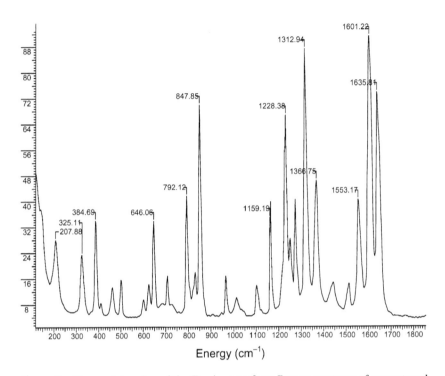

Figure 6 Fingerprint region of the Fourier-transform Raman spectrum of paracetamol, with the diagnostic peaks marked.

However, a compilation of spectra obtained on a wide variety of pharmaceutical excipients is available (42).

The Raman and IR absorption spectra of a wide range of active pharmaceutical ingredients have been reported, and the corresponding spectra critically compared (1). In all the compounds studied, ring breathing modes of monosubstituted phenyl groups, the —C—C—C— symmetric skeletal mode of tricyclic fused ring systems, the N=O symmetric stretch of aromatic nitro groups, and the symmetric stretching vibrations of various double bonds were found to be most prominent and useful for identification purposes. It was noted that although the polar regions of drug substances could be easily studied by IR absorption spectroscopy, the regions that are less polar and have little dipole character (aliphatic chains, aromatic carboxylic groups, and heterocyclic groups) are more amenable to study by Raman spectroscopy.

The complementarity between IR absorption and Raman spectroscopies was demonstrated through a study of benzocaine hydrochloride (43). In addition to the static spectra, measurements were also made at a variety of temperatures and the results of all the spectral studies correlated with the results of semiempirical calculations. Because the amino group was known to play an important role in the reactivity of the molecule, a detailed analysis of its torsional and wagging modes was carried out. The band assignments thus obtained turned out to be the most definitive of their type.

The IR absorption and Raman spectra of four compounds structurally related to diazepam have been reported (44). A number of spectral characteristics emerged that facilitated the distinction between diazepam and delorazepam, fludiazepam, flurazepam, and tetrazepam. All compounds exhibited a strong C=N stretching mode near $1610 \, \text{cm}^{-1}$, and the 1,4-benzo-diazepine nucleus was characterized by a strong band at approximately $1160 \, \text{cm}^{-1}$. It was concluded that owing to the close similarities in their structures, a combination of both IR absorption and Raman spectroscopies was required to obtain unequivocal differentiation of individual benzodiazepines.

The spectral sensitization associated with surface-enhanced Raman scattering has been used to develop an extremely sensitive identification method for papaverine (45). Employing a sequential injection system, an automated apparatus was developed that enabled the reliable and rapid recording of data. The colloidal silver substrate (with or without analyte) was prepared in a cuvette by the reduction of silver nitrate by hydroxylamine. It was shown that the Raman spectrum of papaverine, obtained by either surface-enhanced or conventional means, was the same.

A FT Raman method has been developed to identify the two active components in formulated capsules (2). A spectral region in the Raman spectrum of the formulated product was found where Raman bands unique to the active ingredients, uracil and tegafur, were observed, and the presence of these bands enabled a confirmation to be made that both components were present in a given formulated product.

Raman microspectroscopy is well suited for in situ analysis of contaminants that may be found in pharmaceutical process materials. Because the technique is nondestructive in its nature, the analytes may be subjected to additional confirmatory experiments, such as energy dispersive X-ray analysis or IR microspectroscopy. A positive aspect for contaminant analysis by Raman spectroscopy is the favorable axial and spatial resolutions of the technique as compared with IR microspectroscopy. In general, IR microspectroscopy is diffraction-limited for investigating samples that are typically larger in size than 5 μm. When working with a 0.95 NA objective, spatial resolutions as small as 1 μm can be achieved with a Raman microscope, thus permitting the analysis of very small contaminating particulates.

Raman Spectral Studies of Polymorphic Systems

When the differing crystal structures of polymorphs or solvatomorphs translates into a perturbation of the pattern of molecular vibrations, then the techniques of vibrational spectroscopy (such as Raman scattering) can be used to study the solids. As a result, Raman spectroscopy has found widespread use for the qualitative and quantitative characterization of different solid-state forms of compounds having pharmaceutical interest. For example, the two polymorphs of famotidine exhibit substantially different Raman spectra (46) in the fingerprint region (Fig. 7) and can be readily differentiated from each other on the basis of their characteristic spectra.

An advantage of using Raman spectroscopy for polymorphic investigations is the ease of the technique and the ability to measure low-frequency vibrations $(500–50 \text{ cm}^{-1})$ that are associated with lattice vibrational modes. In many cases, two different solid-state forms of a pharmaceutical entity will display spectral differences in the low-frequency region of the Raman spectrum. These spectral differences will also be observed when comparing the spectra of crystalline and amorphous states of the same substance, which can be used to develop a method to determine the degree of crystallinity in a sample.

Raman spectroscopic studies were found to play a valuable role during studies of the two polymorphs formed by losartan (47), and a consideration of the spectral trends serves to illustrate the magnitude of a vibrational shift due to crystal structure effects. The C—H out-of-plane motion in the biphenyl ring of Form-I is observed at 763 cm^{-1}, whereas in Form-II the band is split into two bands at 710 and 760 cm^{-1}. The ring-breathing mode associated with the imidazole ring was observed at 803 cm^{-1} in Form-II but is split into bands at

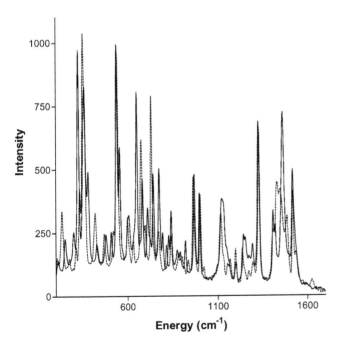

Figure 7 Raman spectra obtained within the fingerprint region for famotidine Form-A (*solid trace*) and Form-B (*dashed trace*). *Source*: Adapted from Ref. 46.

807 and 819 cm^{-1} in Form-I. Significant differences in the bands associated with lattice modes were also reported.

The Raman spectra of two polymorphs of fluconazole were found to permit the ready differentiation between the crystal forms (48). Numerous differences in vibrational band energies are evident in the fingerprint and lattice regions, and even the high-frequency spectral region contains well-resolved spectral features that permit an easy characterization of the structural differences between the two systems. Thirteen samples of spironolactone, obtained from different sources, were evaluated by Raman spectroscopy, and four polymorphic forms of the drug substance were identified in these (49). When IR absorption spectra were obtained for samples in compressed KBr pellets, no differences attributable to polymorphism could be detected, indicating interconversion during processing of the sample. However, when the samples were processed using a non-destructive method, the existence of one (or sometimes more) of the polymorphs could be readily detected.

It should not be automatically assumed that the Raman spectra of different polymorphs will automatically exhibit significant differences. Stavudine is known to exist in two anhydrous polymorphs and one hydrated form (50), and the Raman spectra of these different forms are shown in Figure 8. Even

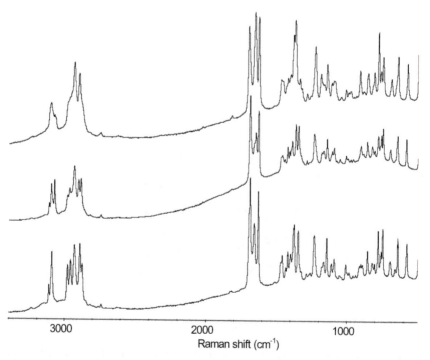

Figure 8   Raman spectra of the three forms of stavudine.

through the crystal structures of the three forms (as evidenced by their X-ray powder diffraction patterns) were found to be quite different, virtually no spectral differences were noted among the Raman spectra, except for some bands in the C—H stretching region. A similar finding was reached regarding the Raman spectra of the polymorphs of roxifiban (51).

The ability of Raman spectroscopy to interrogate the surface of a sample in a nondestructive manner was used to study the phase transformation of cilostazol in its solid dose form (52). Exposure to dissolution media cause the metastable forms of the drug substance to be converted into the stable phase, and these changes can be spectroscopically followed without perturbing the sample.

In those instances where the structural differences between the members of a polymorphic system are sufficiently different so that the corresponding Raman spectra are diagnostic, the degree of spectral simplification associated with Raman data permits a more rapid evaluation of new chemical entities. This feature has been successfully exploited in the cases of 2-{[4-(4-fluoro-phenoxy)phenyl]-methylene}-hydrazine carboxamide (53), *rac*-5,6-diisobuty-ryloxy-2-methylamine-1,2,3,4-tetrahydro-naphthalene hydrochloride (54), 3-(*p*-thioanisoyl)-1,2,2-trimethyl-cyclopentanecarboxylic acid (55), and *trans*-3,4-dichloro-*N*-methyl-*N*-[1,2,3,4-tetrahydro-5-methoxy-2-(pyrrolidin-1-yl)]-naphth-1-ylbenzeneacetamide (56).

Raman spectroscopy has been shown to be a suitable technique for crystal product triage in the high throughput method for polymorph screening (57). In this work, Raman spectroscopy was used to measure the amount of dissolved carbamazepine and then to determine whether Form-I or Form-III crystallized upon thermal cooling of the solutions. The advantage inherent in use of the Raman technique is that the work can be performed in 35- or 100-μL fluid samples, where mixing was effected by diffusion. The advantage of miniaturized techniques over the conventional macrotechniques is that equilibrium can be attained much faster using less drug substance, facilitating higher degrees of throughput.

Raman Spectral Studies of Solvatomorphic Systems

The characterization of solvates and hydrates by Raman spectroscopy also requires that the differing crystal structures of the crystal forms cause a pertur-bation of the pattern of molecular vibrations. In many cases, the degree of crystal-lographic difference is sufficiently large, but in other instances the solvent of crystallization only affects a small number of vibrations. Naproxen sodium has been shown to crystallize in an anhydrate, a monohydrate, and a dihydrate crystal form, and methods have been published regarding a number of their phys-ical characteristics (58). The anhydrate and monohydrate forms of this compound have been prepared (59), and these two solvatomorphs were found to exhibit very similar Raman spectra (46). However, as evident in Figure 9, the two similar

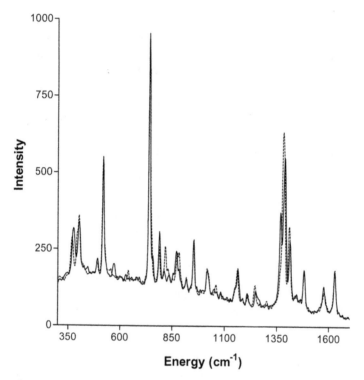

Figure 9 Raman spectra obtained within the fingerprint region for the anhydrate form of naproxen sodium (*solid trace*), and its monohydrate phase (*dashed trace*). *Source*: Adapted from Ref. 46.

forms can still be distinguished on the basis of a few characteristic Raman peaks (such as the $1368/1388/1414 \, cm^{-1}$ triplet of the anhydrate phase, compared with the $1382/1410 \, cm^{-1}$ doublet of the monohydrate phase).

The 1:1 methanol solvate of lasalocid has been found to be capable of crystallizing in two polymorphic forms, and vibrational spectroscopy has been used to study the differences between these (60). In particular, the stretching frequency of the carbonyl moiety in the ketone functional group (observable in both IR absorption and Raman spectroscopies) was found to be a sensitive indicator of the crystal form. The data were interpreted to demonstrate that the observed spectroscopic differences did not depend on the presence or absence of bound solvent, and instead reflected the existence of small conformational changes of the lasalocid molecule.

Raman spectroscopy has been used to study the lattice and intermolecular vibrations of griseofulvin and various solvates (61). Only weak van der Waals interactions between the substance and the solvent were identified for the benzene solvate, whereas for the bromoform and chloroform solvates, weak hydrogen bonding existed between the solvent and the griseofulvin carbonyl group. Using variable temperature methods, it was found that no intermediate structures formed during the desolvation of the solvatomorphs and that the solvatomorph structure simply transformed into the structure of the nonsolvated substance.

Torosemide is known to crystallize in two nonsolvated forms in addition to a channel solvate form that can be formed by water or alcohol (62). Although all three forms were readily distinguishable on the basis of their X-ray powder diffraction patterns, the Raman spectra of the solvate and Form-II were barely distinguishable. Both of these forms were found to be metastable with respect to Form-I, and the phase transformations could be initiated by differing storage conditions. The solvate structure was formed through a framework of hydrogen bonds that established continuous channels of varying sizes along the *b*-axis of the crystal but was found to be unstable when exposed to extreme relative humidity conditions.

Three nonsolvated polymorphs of tranilast have been reported, in addition to the chloroform and methylene chloride solvatomorphs (63). The Raman spectra contained strong bands derived from the methylene stretching modes, the aromatic ring vibrations, and various low-frequency lattice modes. It was reported that the intensity of the amide band in the Raman spectrum of Form-II (observed at 1643 cm^{-1}) was particularly strong relative to the intensities of the analogous band in the other forms. Owing to the ease of its measurement, Raman spectroscopy was developed as an in-process method to evaluate any contamination by nonsolvated Forms II or III in bulk nonsolvated Form-I. The pattern of lattice vibrational modes noted for nonsolvated Form-I and the chloroform solvate indicated the existence of a structural similarity, and a similar conclusion was deduced for the nonsolvated Form-III and the methylene chloride solvatomorph.

The Raman spectra of four different hydrate forms of risedronate were found to be dominated by vibrations associated with the substituted pyridine ring (64). Both the anhydrate and the hemipentahydrate forms yielded two intense peaks derived from an in-plane pyridine ring deformation mode that was similar to that of neat pyridine. On the other hand, the monohydrate and variable hydrate forms were dominated by a single peak derived from this pyridine ring deformation. The spectroscopic differences appeared to originate from differing types of intermolecular interactions originating from hydrogen bonding between the pyridine ring and an adjacent phosphate group. Additional differences in Raman spectra between the different hydrates were observed in the high-frequency symmetric and antisymmetric methylene C—H stretching modes.

Raman Spectroscopy Studies of Phase Transformations

Owing to its ease of acquisition, and noninvasive nature of sample handling, Raman spectroscopy has proven to be an ideal technique for the in situ study of chemical and physical reactions and has found particular use in the study of phase transformations. For example, different sieve fractions of trehalose dihydrate were subjected to isothermal heating at 80°C, and the thermally induced effects were monitored through Raman spectra obtained during the heating process (65). It was found that after heating a fine particle size for 210 minutes, the substance appeared to convert into an amorphous phase that was equivalent to the material obtained after lyophilization. On the other hand, coarse trehalose dihydrate appeared to convert into the crystalline anhydrate at the end of the heating process. Analysis of the Raman spectra permitted an evaluation of the kinetics involved.

In an analogous study, the kinetics of the thermally induced solid-state transformation of carbamazepine Form-III into Form-I was studied by Raman spectroscopy (66). The rate of transformation was monitored through changes in the relative intensities of two C—H bending modes, as this approach was chosen to avoid experimental artifacts. The initial stages of the temperature dependence of the phase conversion are illustrated in Figure 10, and the activation of the process was determined to be 344–368 kJ/mol using four kinetic models that provided good fits to the data. The authors concluded that FT-Raman spectroscopy would be a useful probe technique for in-process control technique, especially for manufacturing steps that entailed a change in the crystal form of a substance.

Temperature-dependent Raman spectroscopy has been used to study the low-temperature phase transitions in a biphenyl-fullerene single crystal (67). It was found that three biphenyl vibrational modes and the C—H stretching mode were indicative of the phase transition, and changes in the planarity of the biphenyl molecule appeared to accompany the phase change. The behavior of most of the C_{60}-fullerene vibrational modes could be correlated with either of the transitions, indicating changes in its site symmetry. In contrast, Raman spectroscopy was used to demonstrate that single crystals of different polymorphs of pentacene did not undergo any phase transformation reactions over the temperature range of 79 to 300 K (68).

The melting behavior of dipalmitoyl-phosphatidylcholine has been studied by Raman spectroscopy, with the C–H and C=O stretching modes being found to be particularly sensitive to lateral expansions of the lipid matrix and increased motion of the *sn*-2 chain within the bilayer interface region (69). The order–disorder phase transition in three crystal modifications of oleic acid has also been studied, where it was found that the phase transformations also involved changes in molecular conformation (70). In addition, the Raman spectra were used to deduce structural information about the nature of the β-phase, for which no crystal structure was known at the time of the work.

Like many amino acids, glutamic acid is capable of existing in two polymorphic forms, and the production of a desired form in a batch crystallization process requires in-process control. Raman spectroscopy has been used to measure the polymorphic composition of glutamic acid suspensions, and good

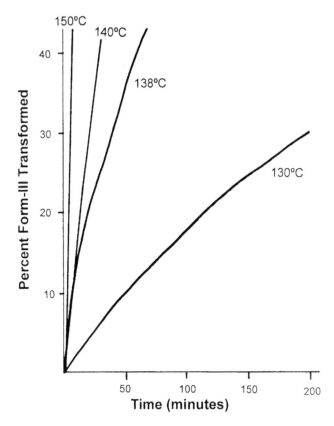

Figure 10 Early-stage time evolution of the phase transformation of carbamazepine Form-III into Form-I as a function of the isothermal heating temperature. *Source*: Adapted from Ref. 66.

calibration curves were obtained for the quantitative analysis of suspensions containing both the α- and β-forms (71). For example, it was shown that for crystallizations conducted at 25°C, the α-form nucleated and completed the crystallization but that the β-form grew subsequently through a solution-mediated phase transformation mechanism. These processes were successfully modeled in a later work that required the appropriate combination of experiment and simulation to adequately understand and predict the crystallization behavior (72).

Using a diamond anvil cell and an appropriate spectroscopic interface, it is possible to study the effect of pressure on the physical state of a substance. For example, different Raman spectra were obtained for 2-adamatanone at pressures of 1 atm and 21 kbar, with the spectral pattern of the high-pressure phase strongly resembling that measured for the ordered phase, which can be produced by cooling the substance to less than 150 K (73). The lattice vibrations of crystalline tetrafluoro-1,4-benzoquinone have been found to be particularly sensitive to the effect of pressure (74), as have the lattice modes of crystalline hexamethylbenzene (75). In the latter study, the Raman spectra provided evidence that the intramolecular methyl torsional mode played an important role in the solid-state phase transformation.

Raman Spectroscopic Studies of Manufacture and Processing

The real-time monitoring of pharmaceutically relevant processes continues to represent a growth area for Raman spectroscopy. On-line monitoring by Raman spectroscopy has been utilized to examine synthetic organic reaction schemes to investigate kinetics, and to identify nonisolated reaction intermediates. Distinct advantages for Raman spectroscopy in this area are: (*i*) the ability to work with aqueous-based systems with little spectral interference from water, (*ii*) the utilization of a fiber optic probe for direct and/or remote sampling, (*iii*) collection of the Raman spectrum directly through the reaction vessel with little or no spectral interference and, (*iv*) the ability to quantitatively analyze the spectral results.

Obviously, if information about the chemical or physical identity can be measured rapidly in the presence of the entire product matrix, in situ process monitoring is the next logical step. This is almost exclusively the realm of fiber optic sampling (76,77), where the probe is inserted either into the process stream so that it is in intimate contact with the products and reactants or where it is mounted in a sight glass window for monitoring in a noncontact mode. Several considerations dictate the instrument configuration, and often the process area imposes restrictions on the physical state of the instrumentation. Fiber optic sampling allows positioning of the laser, launch optics, spectrograph, detector, and any conditioning systems in remote locations from the process.

Although a number of fiber optic probe designs are commercially available that address temperature, pressure, and interface integrity in various ways, most share a common geometry. Fibers are typically composed of low-hydroxide silica to reduce the silicon oxide contributions from the incident and collection fibers. Additional holographic filtering optics are housed in the probe head to reject nuisance scatter from the incident laser and to prevent the Rayleigh scatter from entering the collection fibers. Utilizing fiber optics, a number of processes have been either analyzed or controlled based upon data derived from Raman spectrometers.

The degree of control exerted over a crystallization process and the avoidance of forming undesirable metastable by-products are important principles in the manufacturing of crystalline substances (78). For example, polycarboxylic acids are known to affect calcium carbonate crystallization, and Raman spectroscopy has been used to monitor the levels of different polymorphs (calcite, vaterite, and aragonite) as a function of additive identity and concentration in a crystallization study (79).

The antisolvent crystallization of cortisone acetate from acetone or mixed acetone solvents by water have been monitored using in situ Raman spectroscopy (80). Once the Raman features of the solutes, solvent systems, and antisolvents were profiled, characteristic scattering bands were subsequently used to follow the levels of these compounds during conduct of the process. Given that Raman spectroscopy could provide information on dissolved and crystalline substances, it was judged to be superior to IR absorption spectroscopy (obtained using the attenuated total reflectance sampling mode) because the latter method could only provide information on dissolved substances.

Some of the most interesting applications of real-time process monitoring include the interconversion of polymorphic solids under slurry conditions, and Raman spectroscopy has been an important tool in the characterization arsenal (81,82). For example, the spectra in Figure 11 show the migration of a specific peak of progesterone during a temperature-controlled solid-state conversion, and a plot illustrating the disappearance of Form II and the appearance of Form I as measured in real time for process optimization.

In another study of this type, Raman spectroscopy was used to determine the rate of polymorph turnover for a multipolymorphic research compound, known to exist in four anhydrate forms, two hydrates, and numerous solvates (83). The penultimate and purification steps of the manufacturing process involved a coupling reaction that generated a mixture of crystal forms, which was followed by phase conversion to the desired form. Eventually, after the performance of empirical and simulation work, the use of thermodynamic data to establish process boundaries, and the use of kinetic data to establish process time cycles, a highly robust process that yielded the desired crystal form was developed.

A variety of physical changes can take place in materials as they are manufactured in the dosage form, and Raman spectroscopy has been shown to be useful for the monitoring of these processing steps (84). For example, the effect of excipients on the formation of theophylline monohydrate from the anhydrate in wet massed products has been studied, and Raman spectroscopy was used to evaluate the phase composition (85). It was found that lactose monohydrate could not prevent the phase conversion, and silicified microcrystalline cellulose (a material known to be able to adsorb large quantities of water) provided protection from the phase conversion only at low-moisture levels. The selectivity of the Raman spectra enabled evaluation of the phase composition to be made even in the presence of large amounts of excipients.

The ability to use Raman spectroscopy as an in situ probe for wet granulation processes has been demonstrated. In one work, anhydrous theophylline and caffeine were processed using wet granulation, and the formation of the corresponding hydrate phases monitored through the coupled method of Raman scattering (CCD detector) and NIR absorption (86). In another study, Raman spectroscopy was used to follow the properties of the drug substance during the conduct of a high-shear wet granulation process (87). Knowledge of the

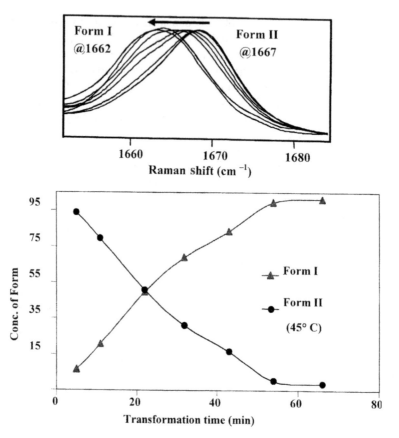

Figure 11 Interconversion of the two solid-state forms of progesterone, as monitored by Raman spectroscopy. *Source*: Courtesy of Kaiser Optical Systems, Inc.

real-time transformation kinetics facilitated the modeling of the processing step, which was used to deduce time scales for hydrate formation during the process.

Quantitative Analytical Applications

The ability to use Raman spectroscopy for the performance of quantitative analysis is a significant advantage of the technique. It has been shown that the Raman scattering intensity is proportional to the number of molecules being irradiated, and the intensity of the scattered radiation is also proportional to the intensity of the incident radiation and to the fourth power of the difference in frequencies between the laser frequency and the molecular vibrational frequency. It follows that increased degrees of Raman scattering intensity and potentially lower limits of detection can be achieved by increasing the intensity of the laser radiation, and/or the frequency of the laser irradiation. This quantitative relationship between Raman scattered intensity and concentration can be expressed as:

$$I_R = (I_L \sigma K)PC \qquad (9)$$

where I_R is the measured Raman intensity (units of photons/sec), I_L is the laser intensity (units of photons/sec), σ is the absolute Raman cross-section (units of cm^2/molecule), K is a constant composed of measurement parameters, P is the sample path length (units of cm), and C is the concentration (units of molecules/cm^3). The constant K represents measurement parameters, such as utilizing the same spectrometer (collection optics efficiency), sample positioning, and the overall efficiency of the Raman spectrometer.

In the past, various problems inherent with dispersive systems precluded the use of Raman spectroscopy for quantitative analysis. With the advent of Raman spectrometers based on the FT approach, the feasibility of quantitative applications greatly improved (88). Some of the advantages associated with the FT approach that are conducive toward the conduct of quantitative applications are Jacquinot's advantage, Fellgett's advantage, and Connes' advantage. Jacquinot's advantage (also known as the throughput advantage) depends on the fact that the entrance aperture of the interferometer is large, making the sampling geometry less sensitive to absolute repositioning of the sample cell. Fellgett's advantage (otherwise known as the multiplex advantage) concerns the increased speed at which spectra can be collected, because all wavelengths of the spectrum are collected simultaneously. This rapid collection implies that FT-Raman spectroscopy might be less sensitive to instrument drift. Finally, Connes' advantage (also called the precision advantage) depends on the fact that there is only one moving part in an FT system (one mirror in the interferometer).

The consequence of these advantages is that an FT system is very rugged, allowing for excellent day-to-day wavelength precision. Although there is not presently an approved ASTM material for calibrating Raman spectrometers, the very strong Raman peak at 520 cm^{-1} of silicon has become the industry standard for calibrating intensity and wavelength.

Although FT-Raman spectroscopy is more applicable to quantitative applications than is dispersive Raman spectroscopy, there are still some drawbacks concerning the optimization of sampling conditions. In FT-Raman spectroscopy, one typically samples a relatively small area (typically a 1–2 mm spot size) of the total sample. As such, it is important that any spectra collected be truly representative of the bulk sample. For solution-phase studies, homogeneity of multicomponent samples is not an issue, and one need not worry about whether the measurement is representative of the bulk or not.

On the other hand, the typical inhomogeneous composition of a solid phase can create significant difficulties in the conduct of quantitative analysis. One way to improve sample uniformity is to prepare all solid mixtures by geometric mixing. Slurry mixing is another good technique, if the sample can withstand contact with the slurry solvent. For data acquisition, sample-spinning accessories should be used to spin the sample cup containing the mixture during spectral acquisition. Using this type of accessory results in the collection of a more representative sample spectrum (a ring of data is collected, instead of just at a single point). To effectively minimize any sample heterogeneity concerns, different rings of data can be collected for each sample cup. The individual spectra from each sampled area can then be coaveraged into groups to create representative spectra for the sample in the cup.

Another factor to consider when using Raman spectroscopy for quantitative analysis is that the power output of the Raman laser can vary from day-to-day, affecting the intensity of spectral peaks. It is advisable to normalize any data to a standard before using these for quantitative purposes. One possible method for normalizing spectra is to ratio the spectral response of the analyte against a peak response for a non-changing component (such as an excipient in a drug product or an internal standard). Alternatively, a ratio can be measured using a peak response of a component changing in an opposite direction to that of the component being monitored.

The surface-enhanced Raman scattering of drug adsorbed onto silver colloids has been used for quantitative purposes. Using this approach, trace quantities of nitrogen-containing drug substances have been detected, where a detection limit of 35 ng was determined for p-amino-benzoic acid and 7 ng for 2-aminofluorene (89). In another work, after sulfadiazine, sulfamerazine, or sulfamethazine were adsorbed onto colloidal silver dispersions, the compounds could be detected in the low nanograms per milliliter level (90).

Quantitative Raman spectral analysis has been used to determine the amorphous-to-crystalline phase content in indomethacin samples (91). Some of the highlighted aspects in this paper concerned the need for producing homogeneous calibration/validation samples and the difficulties associated with collecting a Raman spectrum that is truly representative of the concentration. A linear correlation curve was constructed, which could be used to detect and predict low concentrations of both amorphous and crystalline material in mixtures. It was concluded that the largest source of error in the measurements arose from inhomogeneous mixing of the amorphous and crystalline components in the blends.

Raman spectroscopy was used to quantitatively determine the level of amorphous material in lactose samples, and excellent accuracy was obtained when the amorphous phase content was less than 10% (92). In a related study, amorphous lactose was prepared by lyophilization, crystalline lactose monohydrate by crystallization, and the Raman spectra of blended standards were used to develop calibration curves (93). The best discrimination between the two forms was obtained through measurements of depolarization ratios in the prominent Raman bands observed at 865 and 1082 cm^{-1}.

With suitable calibration, the same Raman spectra that were used to identify the crystal form existing in a polymorphic mixture can also be used to develop quantitative methods for the phase composition. For example, binary mixtures of chlorpropamide forms A and B were analyzed for both qualitative and quantitative purposes, even though the Raman spectra of both crystal forms strongly resembled each other (94). In another study, a combination of FTIR and Raman spectroscopies was used for the identification and quantitation of the orthorhombic and monoclinic polymorphs of paracetamol (95). The precision of the latter method was found to be less than 5%, and the limit of detection for the monoclinic form was estimated to be 0.012 in units of mole fraction.

A quantitative Raman method has been used to analyze various mixtures of β- and δ-mannitol, where it was found that levels of β-mannitol as low as 2% could be accurately quantified (96). As illustrated in Figure 12, the C—C—O stretching mode of β-mannitol was observed at 1037 cm^{-1}, the analogous peak for δ-mannitol was observed at 1052 cm^{-1}, and mixtures of the two polymorphs exhibit peaks due to both forms. The effect of particle size was investigated, and it was reported that when the particle size range of the sample was controlled to be less than 125 μm, spectra more representative of the overall sample mixture were obtained. In addition, it was also demonstrated that experimental artifacts could be further reduced through the use of a rotating sample holder.

As long as one can differentiate the most intense Raman peaks of a drug substance from those of its associated excipients, a quantitative method for determination of the API in its formulation can be developed. For example, the determination of acetylsalicylic acid in tablets containing corn starch has been developed, where the calibration curve was constructed using intensity ratios of aspirin and starch scattering peaks (97). In another study, the level of sulfasalazine in microspheres produced from Eudragit RS has been determined using the characteristic Raman bands of the drug substance (98). In this latter study, it was found that a quantitative relationship existed between the quantity of drug substance in the microsphere and the ratio between the area of a sulfasalazine peak and the area of a Eudragit peak.

A chemometric Raman spectroscopic method has been reported for the precise evaluation of the microcrystallinity of indomethacin in a model tablet matrix (99). The calibration model was constructed by the partial least-squares analysis based on the multiplicative scatter correction and second derivative Raman spectra. The method was validated down to 2% amorphous or crystalline material in the tablet, which in turn represented only 0.2% of the total mass of the tablet. Not only could the amorphous content in a compressed tablet be evaluated, but differences in microcrystallinity at the surface and interior of the tablet could also be determined.

The relative amounts of the three polymorphic forms of benzimidazole in a 5% formulation premix for animal use were determined by FT-Raman spectroscopy (100). To achieve this end, after comparison to the reference spectra of the three pure polymorphic forms, the raw intensities of 78 selected

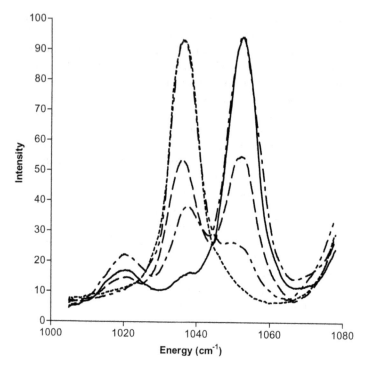

Figure 12 Raman spectra for the C—C—O stretching modes of β-mannitol (*dashed trace*) and δ-mannitol (*dashed trace*), and mixtures of these having a β-phase content of 20% (— · — · — -), 40% (— — — — — —), and 80% (— · · — · · — · -). *Source*: Adapted from Ref. 96.

peaks were vector-normalized, and stepwise linear regression models were then used to estimate the relative quantities of forms A, B, and C in the different samples.

Raman Spectroscopic Studies of the Character of Drug Substances in Formulations

One possibility that must always be considered during the development of an active pharmaceutical ingredient is that even through the physical form of the drug substance may have been carefully controlled during its manufacture, it may change as a result of secondary processing when the dosage form is formulated. The potential of Raman spectroscopy as a technique to establish the solid-state form of an API in a tablet matrix has been evaluated, and it was found that in many cases the drug substance could be studied even when it constituted less than 1% of the tablet mass (101). For example, as illustrated in Figure 13, the Raman spectrum of ibuprofen contains scattering peaks at energies where lactose or

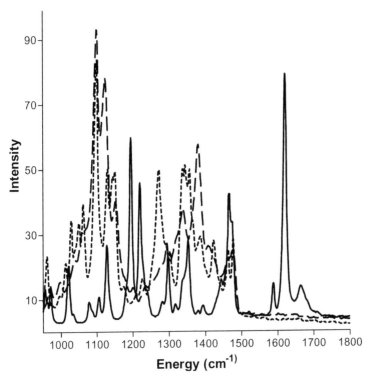

Figure 13   Carbonyl region of the Raman spectrum of ibuprofen (*solid trace*), illustrating the lack of scattering peaks compared with those for lactose monohydrate (*short dashed trace*) or microcrystalline cellulose (*long dashed trace*). *Source*: Adapted from Ref. 101.

microcrystalline cellulose have no peaks, and therefore a method to monitor the state of ibuprofen in tablet formulations is possible.

The determination of ambroxol content in tablets has been accomplished using Raman spectroscopy, with a calibration being established in the range of 8.3–16.26% (102). The samples were rotated to eliminate effects associated with compositional variation, baseline and noise effects were minimized by pre-processing the spectra using wavelet transformation, and then the spectra were normalized prior to being subjected to partial least-squares regression analysis. Four different geometric laser irradiation patterns were examined to study the effect of tablet subsampling (103). Not surprisingly, it was found that the best result in terms of prediction error was obtained by irradiating a large tablet area.

The potential of Raman spectroscopy as a means to characterize the relationship between the hydration state of a drug substance and the physical stability of its tablet has been studied (104). Risedronate sodium was chosen as the drug substance for the study, because its water of hydration is contained within channels in the crystal structure and therefore, potentially mobile. The substance was wet granulated and dried down to granulated moisture contents in the range of 1–7%. Raman spectra were obtained during the course of the drying step, providing information on the contraction of the crystal lattice that accompanied the dehydration. The changes induced in the drug substance crystal structure were found to be reflected somewhat in the stability of the dosage forms.

As discussed earlier, the favorable sampling and spectroscopic characteristics of Raman spectroscopy make it an ideal technique for monitoring pharmaceutical processing. For example, the on-line technique has been used to monitor the blend uniformity of azimilide dihydrochloride formulations and to assist in the optimization of the blending process (105). The methodology has also been used to measure variabilities in tablet coatings, because Raman spectral changes could be correlated with tablet exposure times in a pan coater (106). It was found that the results for coating thicknesses could be improved by preprocessing the spectral data with multiplicative scatter correction, standard normal variate transformation, and Savitzky-Golay second derivative smoothing.

Water that becomes associated with amorphous polymers is known to affect their chemical and physical properties, and Raman spectra of polymers under different environmental conditions have been used to study these effects. In one study, the nature of the polymer-water hydrogen bond interactions in two molecular weight grades of poly(vinylpyrrolidone) (PVP) and also in poly (vinylacetate) and poly(vinylpyrrolidone-*co*-vinylacetate) (107). To assist in interpretation of the Raman spectra, the vapor absorption isotherms and glass transition temperatures of the polymers were also obtained, and it was determined that the pyrrolidone group tended to interact more with water than did the acetate group. Using principal components analysis, differentiation between polymers in the rubbery state and those in the glassy state could be made. In a related study, Raman spectroscopy was used in conjunction with electron microscopy to study the interaction of water with different cellulose ethers and to monitor the swelling and drying behavior of these materials when they went in and out of the gel phase (108).

Raman spectroscopy has been used to evaluate the interactions between drug substances and polymeric excipients. The molecular structure existing in solid dispersions of indomethacin and PVP has been studied, with the carbonyl stretching mode of the drug substance being used as the spectroscopic probe (109). Addition of PVP to amorphous indomethacin increased the intensity of the nonhydrogen bonded carbonyl groups, while the carbonyl-stretching band of the PVP polymer decreased to lower energies. These findings led to the conclusion that indomethacin interacted with PVP through hydrogen bonds formed between the drug hydroxyl and polymer carbonyl groups, resulting in disruption of the indomethacin dimers that existed in the bulk drug substance.

In another study, Raman spectra of promethazine, diclofenac, theophylline, and indomethacin in polymeric diluents based on polyethylene oxide, sodium alginate, and hydroxypropylmethyl cellulose were obtained (110). It was determined that the characteristic Raman spectra of each drug substance could be used to quantitate its concentration in the formulations and also to provide information on the character of any drug–excipient interactions and incompatibilities that might exist in the formulation matrix. The latter aspect, and the ease of obtaining Raman spectral results, clearly demonstrates the utility of this technique as one of the tools in the arsenal of the preformulation scientist.

The use of supercritical fluids in pharmaceutical manufacturing is becoming more widespread, and Raman spectroscopy has been used to study formulations obtained by this processing route. For example, the impregnation of ibuprofen into PVP in supercritical carbon dioxide has been monitored using in situ Raman and IR absorption spectroscopies (111). The spectral results led to the conclusion that the ibuprofen became dispersed as molecules in the polymer matrix through site-specific associations of the drug substance with the carbonyl groups of PVP. It was also established that the incorporation of the drug substance into the PVP perturbed the degree of water sorption by the polymer.

Raman spectroscopy has been used to evaluate the physical state of chlorhexidine, when it was formulated into viscoelastic, bioadhesive, semisolids (112). The formulations were prepared by dispersing the drug-free base into buffered polymer matrices consisting of hydroxyethylcellulose, PVP, and polycarbophil, and the Raman spectra of the drug substance were acquired within a variety of formulation conditions. The existence of an acid–base reaction between chlorhexidine and polycarbophil was demonstrated, forming the dictation of the drug substance that was solubilized by the hydroxyethylcellulose component. The utility of Raman spectroscopy as a process analytical technique during formulation of topical gels and emulsions has been demonstrated, with changes in the spectral characteristics of the thickening agent carbopol and the emulsifying agent tefose being monitored after performance of major processing steps (113).

Chemical Mapping and Imaging

One of the more exciting new applications of Raman spectroscopy concerns chemical mapping. By incorporating a programmable, xyz-movement stage into a Raman microscope, it is now possible to generate a chemical image of a two-dimensional area of a sample. Through the use of a mapping stage, a sample, such as a sliced tablet, can be moved in the x- and y- directions, obtaining spectra at each step. If the sample requires refocusing at different locations, the z-direction can be automated as well. The distance, which the stage moves in the x- and y-directions is called the step size and usually can be as small as 1 μm.

There are three types of maps, and these are identified as point, line, and area. A point map provides several different areas of a sample to be analyzed consecutively, but the spectra are not related to each other spatially. A point map can be considered as a type of auto-sampler. An example of a point map application is a 96-well plate that contains combinatorial beads in each well. The second map type (the line map) defines a series of spectra to be obtained along one dimension. In line maps, chemical changes that occur along this dimension are investigated. An example of a line map sample is a cross-sectioned pharmaceutical beadlet that has several different layers exposed. One of the more practical reasons that line maps are so popular is that they can provide detailed information regarding the chemical changes that occur across a sample without the need to collect as many spectra as is required for area maps.

The final map type (the area map) defines a series of spectra to be collected in two dimensions (i.e., over an entire region). This type of map provides a Raman image that can be directly compared with the visual image, often allowing nonvisible (to the eye) features to be identified. A common area map sample is a tablet. The tablet can be mapped and the various ingredients can be monitored for content uniformity. A series of images demonstrating the dispersion of several excipients in a tablet is displayed in Figure 14. This particular area map was obtained with a step size of 4 μm, a sampling spot size of approximately 1 μm, and a total sampling area of 87 × 52 μm. The time required to collect this map was approximately 26 hours, but it should be noted that not all Raman mapping experiments require this amount of time. For example, if a map is being performed to search for a particular component, the step size

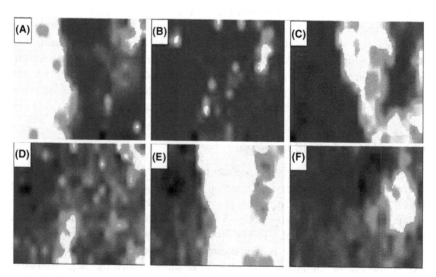

Figure 14 Peak area profiles (images) representing (**A**) mannitol, (**B**) aspartame, (**C**) cellulose, (**D**) magnesium stearate, (**E**) corn starch, and (**F**) monoammonium glycyrrhizinate. A high concentration of material is represented by the *gray color*, whereas a low concentration is represented by the *black color*.

need only be in the order of the particle size of that substance. A good approach would be to first obtain a larger area map on the sample with larger step sizes and shorter sampling times per point. Once an area of interest is defined by analyzing the data from the first map, a smaller, higher resolution map can then be defined.

Area maps and imaging have experienced a recent surge in popularity, mainly due to the use of CCD array cameras and liquid crystal tunable filter (LCTF) technology, respectively. These detectors greatly reduce the amount of time required to collect an area map (114). In one case, the time required to collect an area map decreased by approximately four hours upon switching from a CCD camera (longer than five hours for the experiment) to LCTF technology (less than one hour for the experiment).

Once line or area mapping experiments have been performed, profiles are created that enable certain spectral features to be monitored spatially on the sample. The data sets generated by mapping can consist of hundreds or even thousands of spectra, and viewing all of these spectra concurrently is unrealistic. Profiles can aid in this task by reducing the data set into a more easily viewed format (an image). A profile is a representation of map data in which a measurement of spectral intensity or some other characteristic is shown for each sample point. A profile will verify the presence, location, and extent of a defined spectroscopic feature in the sample. There are many types of profiles. Some profiles (such as chemigram or component profiles) compare an entire reference spectrum to every spectrum in the map, with the resulting image showing spectral similarity across the mapping region. Other images can be created based on profiling a specific peak area, peak height, peak area ratio, or peak height ratio. Profiles can also be performed based on a group of peaks specific for certain functional groups (e.g., alkanes) or even on a quantitative method.

In recent Raman applications of chemical imaging, the content uniformity of an active drug substance within a drug product was investigated (115,116). A pharmaceutical tablet containing cyclobenzaprine hydrochloride was shaved, such that a chemical image was obtained for each sampled depth within the tablet (115). Because the Raman spectrum of cyclobenzaprine hydrochloride exhibited spectral features different from those of the excipients, a chemical image specific to the drug substance could be obtained, or, for that matter, for each excipient. Figure 15 displays the chemical image for one plane (depth) of the tablet. It is clearly observed that the drug substance is detected at the 0,0 spatial position within the sampled plane of the tablet. The corresponding excipient chemical image also shows the lack of excipient concentration (signal intensity) at the same spatial position. These results demonstrate the excellent sensitivity and selectivity (spectrally and spatially) for the spectroscopic, chemical imaging technique. In the second example of Raman imaging of a drug product (116), a map of the ratio of ibuprofen/PVP concentration for a 45×25 μm area was generated. The results showed that ibuprofen was homogeneously distributed throughout the polymer PVP.

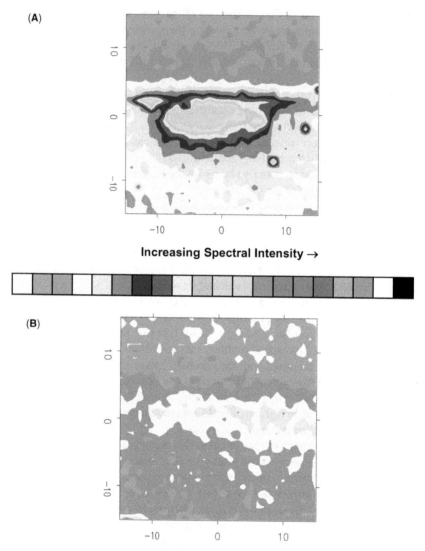

Figure 15 Raman spectral chemical images of a shaved surface of a tablet containing cyclobenzaprine hydrochloride. Image (**A**) maps the intensity of the drug substance, whereas (**B**) is an image of the same spatial area, mapping one of the excipients. *Source*: Adapted from Ref. 115.

Determination that ibuprofen was equally distributed throughout the polymer suggests high stability of the formulation (no localization of drug or polymer) leading to desired dissolution characteristics.

Other works concerning mapping applications of interest to the pharmaceutical industry include crystal formation in hormone replacement therapy patches [Bug5], particle size analysis in mixtures (117), and pharmaceutical matrix determination of dosage formulations (118).

REFERENCES

1. Cutmore EA, Skett PW. Spectrochim Acta 1993; 48A:809.
2. Petty CJ, Bugay DE, Findlay WP, Rodriguez C. Spectroscopy 1996; 11:41.
3. Threfall TL. Analyst 1995; 120:2435.
4. Bugay DE, Williams AC. Vibrational spectroscopy, In: Brittain HG, ed. Physical Characterization of Pharmaceutical Solids. New York: Marcel Dekker, 1995:59–91.
5. Findlay WP, Bugay DE. J Pharm Biomed Anal 1998; 16:921.
6. Armstrong CL, Edward HGM, Farwell DW, Williams AC. Vibrat Spect 1996; 11:105.
7. Eyring H, Walter J, Kimball GE. Quantum Chemistry. New York: John Wiley & Sons, 1944:121–123.
8. Herzberg G. Infrared and Raman Spectra of Polyatomic Molecules. New York, NY: Van Nostrand Reinhold Co., 1945:239–271.
9. Placzek G. Handbuch der Radiologie. Vol. 6. Leipzig: Akademische Verlagsgesellschaft, 1934:208.
10. Hochstrasser RM. Molecular Aspects of Symmetry. New York: W.A. Benjamin Inc., 1966.
11. Orchin M, Jaffe HH. Symmetry, Orbitals, and Spectra. New York: Wiley-Interscience, 1971.
12. Cotton FA. Chemical Applications of Group Theory. 2d ed. New York: Wiley-Interscience, 1971.
13. Harris DC, Bertolucci MD. Symmetry and Spectroscopy. New York: Oxford University Press, 1978.
14. Asher SA. Anal Chem 1993; 65:59A; ibid 1993; 65:201A.
15. Morris MD, Wallan DJ. Anal Chem 1979; 51:182A.
16. Fleischmann M, Hendra PJ, McQuillan AJ. Chem Phys Lett 1974; 26:163.
17. Jeanmaire DL, van Duyne RP. J Electroanal Chem 1977; 84:1.
18. Albrecht MG, Creighton JA. J Am Chem Soc 1977; 99:5215.
19. Creighton JA, Blatchford CG, Albrecht MG. J Chem Soc Faraday Trans II 1979; 75:790.
20. Garrell RL. Anal Chem 1989; 61:401A.
21. Campion A, Kambhampati P. Chem Soc Rev 1998; 27:241.
22. Moskovits M, DiLella DP, Maynard K. Langmuir 1988; 4:67.
23. Durig JR, Harris WC. Raman spectroscopy. In: Weissberger A, Rossiter BW, eds. Physical Methods of Chemistry, Part IIIB. Vol. 1. New York: Wiley-Interscience, 1972:85–205.
24. Grasselli JG, Snavely MK, Bulkin BJ. Chemical Applications of Raman Spectroscopy. New York: Wiley-Interscience, 1981.
25. Strommen DP, Nakamoto K. Laboratory Raman Spectroscopy. New York: Wiley-Interscience, 1984.
26. Hendra P, Jones C, Warnes G. Fourier Transform Raman Spectroscopy. Chichester: Ellis Horwood Ltd., 1991.
27. Schrader B. Infrared and Raman Spectroscopy. Weinheim: VCH Publishers, 1995.

28. Mcreery RL. Raman Spectroscopy for Chemical Analysis. New York: Wiley-Interscience, 2000.
29. Lewis IR, Edwards HGM. Handbook of Raman Spectroscopy. New York: Marcel Dekker, 2001.
30. Adar F. Evolution and revolution of Raman instrumentation, chapter 2. In: Lewis IR, Edwards HGM, eds. Handbook of Raman Spectroscopy. New York: Marcel Dekker, 2001:11–40.
31. McCreery RL, Horn AJ, Spencer J, Jefferson E. J Pharm Sci 1998; 87:1.
32. Flett MC. Characteristic Frequencies of Groups in the Infra-Red. Amsterdam: Elsevier Pub., 1963.
33. Bellamy LJ. Advances in Infrared Group Frequencies. London: Methuen & Co., 1968.
34. Colthup NB, Daly LH, Wiberley SE. Introduction to Infrared and Raman Spectroscopy, 2d ed. London: Academic Press, 1975.
35. Socrates G. Infrared and Raman Characteristic Group Frequencies – Tables and Charts. 3rd ed. Chichester: John Wiley & Sons, 2001.
36. Dollish FR, Fateley WG, Bentley FF. Characteristic Raman Frequencies of Organic Compounds. New York: John Wiley & Sons, 1974.
37. Hendra PJ. American Laboratory, December issue, 17, 1996.
38. Reference 24, pp. 38–39.
39. Huong PV. J Pharm Biomed Anal 1986; 4:811.
40. Frank CJ. Pharmaceutical applications of Raman spectroscopy. In: Pelletier MJ, ed. Analytical Applications of Raman Spectroscopy. Oxford: Blackwell Science, 1999:224–275.
41. Bugay DE. Adv Drug Del Rev 2001; 48:43.
42. Bugay DE, Findlay WP. Pharmaceutical Excipients: Characterization by IR, Raman, and NMR Spectroscopy. New York: Marcel Dekker, 1999.
43. Cutmore EA, Skett PW. Spect Lett 1993; 26:1395.
44. Neville GA, Beckstead HD, Shurvell HF. J Pharm Sci 1994; 83:143.
45. Leopold N, Baena JR, Bolboaca M, Cozar O, Kiefer W, Lendl B. Vibrat Spect 2004; 36:47.
46. Brittain HG. Unpublished Results for Raman spectra obtained at a resolution of 5 cm^{-1}, using an Ocean Optics model R-300-785 Raman spectrometer. Bulk powder samples were contained in aluminum sample cups, and then excited at 785 nm to obtain the spectrum.
47. Raghavan K, Dwivedi A, Campbell GC, Johnston E, Levorse D, McCauley J, Hussain M. Pharm Res 1993; 10:900.
48. Gu XJ, Jiang W. J Pharm Sci 1995; 84:1438.
49. Neville GA, Beckstead HD, Shurvell HF. J Pharm Sci 1992; 81:1141.
50. Gandhi RB, Bogardus JB, Bugay DE, Perrone RK, Kaplan MA. Int J Pharm 2000; 201:221.
51. Maurin MD, Vickery RD, Rabel SR, Rowe SM, Everlof JG, Nemeth GA, Campbell GC, Foris CM. J Pharm Sci 2002; 91:2599.
52. Stowell GW, Behme RJ, Denton SM, Pfeiffer I, Sancilio FD, Whittall LB, Whittle RR. J Pharm Sci 2002; 91:2481.
53. Cheung EY, Harris KDM, Johnson RL, Hadden KL, Zakrzewski M. J Pharm Sci 2003; 92:2017.

54. Giordano F, Rossi A, Moyano JR, Gassaniga A, Massarotti V, Bini M, Capsoni D, Peveri T, Redenti E, Carima L, Alberi MD, Zanol M. J Pharm Sci 2001; 90:1154.
55. Terol A, Cassanas G, Nurit J, Pauvert B, Bouassab A, Rambaud J, Chevallet P. J Pharm Sci 1994; 83:1437.
56. Raghavan K, Dwivedi A, Campbell GC, Nemeth G, Hussain M. J Pharm Biomed Anal 1994; 12:777.
57. Anquetil PA, Brenan CJH, Marcolli C, Hunter IW. J Pharm Sci 2003; 92:149.
58. Kim T-S, Rousseau RW. Cryst Growth Design 2004; 4:1211.
59. The solvatomorphs of naproxen were supplied by Drs. Alan Salerno and Paul Isbester, Albany Molecular Research, Albany, NY.
60. Friedman JM, Fousseau DL, Shen C, Chiang CC, Duesler EN, Paul IC. J. Chem. Soc. Perkin II 1979; 835.
61. Bolton BA, Prasad PN. J Pharm Sci 1981; 70:789.
62. Rollinger JM, Gstrein EM, Burger A. Eur J Pharm Biopharm 2002; 53:75.
63. Vogt FG, Cohen DE, Bowman JD, Spoors GP, Zuber GE, Trescher GA, Dell'Orco PC, Katrincic LM, Debrosse CW, Haltiwanger RC. J Pharm Sci 2005; 94:651.
64. Redman-Furey N, Dicks M, Bigalow-Kern A, Cambron RT, Lubey G, Lester C, Vaughn D. J Pharm Sci 2005; 94:893.
65. Taylor LS, Williams AC, York P. Pharm Res 1998; 15:1207.
66. O'Brien LE, Timmins P, Williams AC, York P. J Pharm Biomed Anal 2004; 36:335.
67. Palles D, Marucci A, Penicaud A, Ruani G. J Phys Chem B 2003; 107:4904.
68. della Valle RG, Venuti E, Farina L, Brillante A, Masino M, Girlando A. J Phys Chem B 2004; 108:1822.
69. O'Leary TJ, Levin IW. J Phys Chem 1984; 88:1790.
70. Kobayashi M, Kaneko F, Sato K, Suzuki M. J Phys Chem 1986; 90:6371.
71. Ono T, ter Horst JH, Jansens PJ. Cryst Growth Design 2004; 4:465.
72. Ono T, Kramer HJM, ter Horst JH, Jansens PJ. Cryst Growth Design 2004; 4:1161.
73. Harvey PD, Butler IS, Gilson DFR, Wong PTT. J Phys Chem 1986; 90:4546.
74. Ikuta A, Suzuki Y, Nibu Y, Shimada H, Shimada R. Bull Chem Soc Jap 1999; 72:963.
75. Takeshita H, Suzuki Y, Nibu Y, Shimada H, Shimada R. Bull Chem Soc Jap 1999; 72:381.
76. Wang F, Wachter JA, Antosz FJ, Berglund KA. Org Proc Res Dev 2000; 4:391.
77. Vergote GJ, DeBeer TRM, Vervaet C, Remon JP, Baeyens WRG, Diericx N, Verpoort F. Eur J Pharm Sci 2004; 21:479.
78. Rodriguez-Hornedo N, Murphy D. J Pharm Sci 1999; 88:651.
79. Agerwal P, Berglund KA. Cryst Growth Design 2003; 3:941.
80. Falcon JA, Berglund KA. Cryst Growth Design 2003; 3:947.
81. Brittain HG. J Pharm Sci 1997; 86:405.
82. Brittain HG. Solid-state analysis, chapter 3. In: Ahuja S, Scypinski S, eds. Handbook of Pharmaceutical Analysis. New York: Marcel Dekker, 2001:57–84.
83. Starbuck C, Spartalis A, Wai L, Wang J, Fernandez P, Lindemann CM, Zhou GX, Ge Z. Cryst Growth Design 2002; 2:515.
84. Brittain HG. J Pharm Sci 2002; 91:1573.
85. Airaksinen S, Luukkonen P, Jorgensen A, Karjalainen M, Rantanen J, Yliruusi J. J Pharm Sci 2003; 92:516.
86. Jorgensen A, Rantanen J, Karjalainen M, Khriachtchev L, Rasanen E, Yliruusi J. Pharm Res 2002; 19:1285.

87. Wikstrom H, Marsac PJ, Taylor LS. J Pharm Sci 2005; 94:209.
88. Walder FT, Smith MJ. Spectrochim Acta 1991; 47A:1202.
89. Torres EL, Winefordner JD. Anal Chem 1987; 59:1626.
90. Sutherland WS, Laserna JJ, Angebranndt MJ, Winefordner JD. Anal Chem 1990; 62:689.
91. Taylor LS, Zografi G. Pharm Res 1998; 15:755.
92. Niemela P, Paallysaho M, Harjunen P, Koivisto M, Lehto V-P, Suhonen J, Jarvinen K. J Pharm Biomed Anal 2005; 37:907.
93. Murphy BM, Prescott SW, Larson I. J Pharm Biomed Anal 2005; 38:186.
94. Tudor AM, Church SJ, Hendra PJ, Davies MC, Melia CD. Pharm Res 1993; 10:1772.
95. Al-Zoubi N, Koundourellis JE, Malamataris S. J Pharm Biomed Anal 2002; 29:459.
96. Roberts SNC, Williams AC, Grimsey IM, Booth SW. J Pharm Biomed Anal 2002; 28:1135.
97. Kontoyannis GG, Orkoula M. Talanta 1994; 41:1981.
98. Watts PJ, Tudor A, Church SJ, Hendra PJ, Turner P, Melia CD, Davies MC. Pharm Res 1991; 8:1323.
99. Okumuar T, Otsuka M. Pharm Res 2005; 22:1350.
100. de Spiegeleer B, Seghers D, Wieme R, Schaubroeck J, Verpoort F, Slegers G, van Vooren L. J Pharm Biomed Anal 2005; 39:275.
101. Taylor LS, Langlilde FW. J Pharm Sci 2000; 89:1342.
102. Hwang M-S, Cho S, Chung H, Woo Y-A. J Pharm Biomed Anal 2005; 38:210.
103. Johansson J, Pettersson S, Folestad S. J Pharm Biomed Anal 2005; 39:510.
104. Hausman DS, Cambron RT, Sakr A. Int J Pharm 2005; 299:19.
105. Hausman DS, Cambron RT, Sakr A. Int J Pharm 2005; 298:80.
106. Romero-Torres S, Perez-Ramos JD, Morris KR, Grant ER. J Pharm Biomed Anal 2005; 38:270.
107. Taylor LS, Langlilde FW, Zografi G. J Pharm Sci 2001; 90:888.
108. Fechner PM, Wartewig S, Kiesow A, Heilmann A, Kleinebudde P, Neubert RHH. J Pharm Pharmacol 2005; 57:689.
109. Taylor LS, Zografi G. Pharm Res 1997; 14:1691.
110. Davies MC, Binns JS, Melia CD, Hendra PJ, Bourgeois D, Church SP, Stephenson PJ. Int J Pharm 1990; 66:223.
111. Kazarian SG, Martirosyan GG. Int J Pharm 2002; 232:81.
112. Dias CS, Mitra AK. J Pharm Sci 2000; 89:572.
113. Islam MT, Rodriguez-Hornedo N, Ciotti S, Ackermann C. Pharm Res 2004; 21:1844.
114. Zugates CT, Treado PJ. Int J Vib Spec 1999; 2:59.
115. Houghtaling MA, Bugay DE. Raman Spectroscopic Imaging: A Potential New Technique for Content Uniformity Testing. Paper presented at the American Association of Pharmaceutical Scientists Annual Meeting, New Orleans, LA, Nov 14–18, 1999.
116. Breitenbach J, Schrof W, Neumann J. Pharm Res 1999; 16:1109.
117. Theophilus A, Lancaster R. Particle Size Analysis of Binary or Tertiary Mixtures Using Raman Image Analysis. Paper presented at the Federation of Applied Spectroscopy Conference, Nashville, TN, Sep 24–28, 2000.
118. Clarke F, Famieson M, Hammond S, Clard K. Chemical Images — The Key to Pharmaceutical Matrix Determination. Paper presented at the Federation of Applied Spectroscopy Conference, Nashville, TN, Sep 24–28, 2000.

10

Near-Infrared Spectroscopy

Robert P. Cogdill and James K. Drennen, III

School of Pharmacy, Duquesne University,
Pittsburgh, Pennsylvania, U.S.A.

INTRODUCTION

The primary analysis methods, such as infrared (IR) absorption spectroscopy, are most often applied within the controlled confines of analytical laboratories. Over the years their performance frontiers have been expanded, giving researchers unprecedented capabilities in the analysis of materials.

A high sensitivity is generally achieved, however, at the expense of robustness and flexibility. Only in select instances when sufficient demand is present to justify the expense of technology transfer have primary analytical methods been adapted for field application. In contrast, the development of near-infrared (NIR) spectroscopy has followed a unique path relative to other spectroscopic methods for analyzing pharmaceutical solids. As discussed in the following paragraphs, the current level of knowledge in NIR spectroscopy can be mainly attributed to its roots as a rugged, field-tested technique. This chapter is devoted to illustrating the intricacies and advantages of NIR spectroscopy for the analysis of pharmaceutical solids, beginning with some perspectives on the historical development of NIR technology.

In his famous work, "Experiments on the Refrangibility of the Invisible Rays of the Sun," presented to the Royal Society in 1800 (1), Sir William F. Herschel first demonstrated the existence of optical radiation beyond the limits of the visible spectrum. In a series of experiments probing the relationship between color and heat, Herschel observed an increase in temperature from violet to red. In a serendipitous moment, however, he discovered that the hottest temperature was actually beyond the red limit of visible light. The mysterious invisible radiation, which he initially termed as "calorific rays," is actually what has come to be known as short-wave near-infrared (SW-NIR), or the Herschel near-infrared.

As a number of texts dedicated to NIR spectroscopy have chronicled (2–6), the NIR spectral region is largely ignored as an analytical asset for nearly 150 years following Herschel's discovery until Karl Norris, an agricultural engineer working for the U.S. Department of Agriculture (USDA), saw its potential for rapid, quantitative analysis of complex biological samples. Norris's pioneering work eventually spawned the development of an NIR industry that produced analyzers capable of rapidly determining the concentration of constituents in whole foodstuffs with very little sample preparation. From the mid-1960s until 1986 the rate of publication for articles mentioning NIR spectroscopy increased at a dizzying pace, with most articles concerned primarily with the rapid analysis of food and agricultural samples. From 1800 to 1970 only about 50 papers concerning NIR analysis were contributed (7). In 1987 Phil Williams and Karl Norris edited a comprehensive text with nearly 1000 references on the subject of NIR technology in the food and agricultural industries (5); their book, which is now in second edition (4), should be considered a staple text for anyone (regardless of their industry) interested in applied NIR spectroscopy for the analysis of solid materials.

Unlike primary analytical methods, which have pushed innovation in new sensors and exotic optical and magnetic materials on their own behalf, NIR technology has generally advanced by riding the wake of technology developments in seemingly unrelated industries. The interest in the NIR region of the electromagnetic spectrum, generally accepted as the range from 780 to 2500 nm, or 12,820 to 4000 cm^{-1}, lagged behind the ultraviolet (UV), visible (VIS), and mid-IR for most of the twentieth century. Unlike the sharp absorbance bands of the mid-IR fingerprint region (frequencies below 1500 cm^{-1}), which are utilized for qualitative analyses in synthetic organic chemistry, the absorbance bands in the NIR, which consist primarily of overtones and combinations of the C—H, N—H, and O—H fundamental vibrations, are very broad, overlapping, and orders of magnitude weaker, thereby precluding its use as a primary qualitative analysis method. Furthermore, because of the propensity for matrix and solvent interactions, as well as the great complexity in absorbance bands, early chemists had difficulty specifying baseline effects in the NIR region, and they viewed NIR spectroscopy as being less reliable for quantitative assays relative to UV/VIS spectrophotometry.

The outlook for NIR spectroscopy eventually brightened, however, as enabling technologies began to appear following World War II. Karl Norris's early work, which was carried out using the Beltsville Universal Computerized Spectrophotometer (BUCS), was revolutionary in that it utilized dual Cary model 14 grating/prism monochromaters and new broad-band PbS detectors (commercialized from military efforts to build heat-seeking weaponry), and powerful digital computing capabilities (Karl's computer had an immense memory capacity of 256 kb) (8). The popularity of NIR spectroscopy accelerated during the 1970s and 1980s on the heels of the development of low-cost personal computers (PCs), consumer imaging devices based on silicon detectors, and re-writable digital memory. Since the 1990s, NIR spectroscopy has benefited from the significant public investments in the telecom industry, which brought economies of scale by mass-introduction of InGaAs diode array detectors, tunable NIR lasers, fiber optics, and advances in dielectric optical coatings.

Over the past decade, NIR spectroscopy has earned its reputation as a powerful analytical technology in its own right. The aforementioned technological developments and the knowledge and understanding gained by the early adopters of NIR have turned the negative aspects of the technology into positive features. The NIR spectroscopy was ultimately found to be useful because the relatively weak absorbance bands in the region allow for interrogation of sample matrices to greater depth than is possible in the IR and UV/VIS regions. Furthermore, in contrast to the IR, NIR spectroscopy has the unique ability to quantify constituents in the presence of significant amounts of water. The complex nature of NIR absorbance bands and baseline effects were initially viewed as a puzzle; the study of chemometrics (one of the few scientific fields owing much of its early development to NIR spectroscopy) has yielded the ability to resolve NIR spectra for simultaneous multicomponent analyses.

With respect to its evolution from a nondestructive method for analyzing whole grain samples, it could be said that the development of NIR spectroscopy has been chiefly concerned with the analysis of solids from the very beginning. As a mature technology, a number of books (2–4,6), reviews (9–11), and a peer-reviewed journal dedicated to NIR technology (12) that discuss the application of NIR spectroscopy in specific industries are available. The following sections of this chapter provide a synopsis of the unique aspects of the fundamental science of NIR spectroscopy, an overview of the equipment and methods employed, and a discussion of the specific areas where NIR spectroscopy is being used for the analysis of pharmaceutical solids.

FUNDAMENTAL PRINCIPLES

At this point it should be clear that the NIR is a unique locale on the continuum of the electromagnetic spectrum. Some potential practitioners of NIR spectroscopy have failed in the past by treating the NIR as an extension of the IR or UV/VIS during method development or application. Indeed, the unique aspect of the interaction between solid materials and NIR radiation at the molecular and particle scales set NIR spectroscopy apart from other methods.

Molecular-Scale Phenomena

Most comprehensive texts on applied NIR spectroscopy begin with a discussion of the anharmonic oscillator model of the fundamental molecular vibrations (and their overtones and combinations) induced by photonic absorption, which have already been covered in earlier chapters of this book. More important to the current discussion, however, is how the complexity of the NIR spectrum contrasts with the IR, which is more readily described by fundamental models and quantum theory.

The previous chapters have discussed the harmonic and anharmonic oscillator models of molecular vibration. The concept of anharmonic oscillation holds significant implications for the existence and complexity of absorption bands in the NIR (13,14):

- Overtone transitions, which involve an increase in a vibrational quantum number of greater than one, are allowed to occur,
- Combination modes, which involve a simultaneous increase in the vibrational quantum numbers of two or more different vibrations from absorption of a single photon, are allowed to occur, and
- The separations of the vibrational energy states of a given vibration are not equal in the NIR and are offset depending on the degree of anharmonicity.

Thus, overtone and combination absorption bands would not exist, and the NIR region would be virtually useless for analytical purposes, if the anharmonic model did not apply. Anharmonic oscillation adds significant complexity to NIR spectra; combinations of fundamental bands and overtones, as well as those of fundamental and overtone bands, are possible. However, vibrations must involve the same functional group and have the same symmetry. The most common bands in the NIR involve C—H, N—H, and O—H stretching; C—H, N—H, and O—H bending, as well as C—O, C—N, and N—O stretching that are represented to some extent. It is well known that, given a molecule with N atoms, up to $(3N - 6)$ overtone absorption bands are possible. A simple methyl group, for example, will have up to six absorption bands. Hence, a typical small-molecule active pharmaceutical ingredient (API) will have many possible overtone bands, with significant opportunities for combination bands. Correlation charts indicating the expected location of absorbance bands in the NIR spectrum for common organic bonds and functional groups are available in many classic spectroscopy texts (15–18).

The degree of anharmonicity determines the amount by which the frequency of overtone bands are displaced from exact integer multiples of the fundamental vibrations and their intensity. As the mass of hydrogen in C—H, N—H, and O—H bonds is so low, such vibrations tend to be very anharmonic and will tend to shift further from the integer multiples of their fundamental frequencies. The degree of anharmonicity for a molecule is further dependent on the level of hydrogen bonding, temperature, and interaction with other molecules in the sample matrix.

Resonance Effects

The NIR spectrum is further complicated by the existence of Fermi and Darling-Dennison resonances. Fermi resonance is attributable to the cubic terms of the anharmonic oscillator potential energy functions and may occur in cases when two vibrational modes of the same functional group have the same symmetry. If a fundamental transition is close to the energy of an overtone or combination transition in the same functional group, the overtone or combination mode "borrows" energy from the fundamental mode. This causes the intensity of the overtone and combination bands to increase relative to the intensity of the fundamental band and the differential frequency shift between the bands will increase as well.

Darling-Dennison resonances involve different overtone transitions within a molecule and are attributable to quartic and higher terms in the potential energy function. Such resonances are pronounced when interacting energy levels are very close to one another and when the degree of anharmonicity is large. Darling-Dennison resonances tend to increase along with the level of excitation. Thus, as the level of excitation increases, resonance bands will tend to be more pronounced and will shift even farther from the position described by the normal mode transition. In terms of practical applications, however, the major implication of resonance is that, in some cases, two absorbance bands might exist where only one band is expected.

Third Overtone Decoupling

The NIR spectra tend to be relatively more featureless in the second and third overtone regions. The reasons are: (*i*) the intensity of overtone absorption bands decrease by orders of magnitude as the cumulative change in quantum numbers increases, and (*ii*) the vibration of multiple bonds with greater degrees of anharmonicity, especially bands involving hydrogen, begin to decouple in the third overtone region (780–1300 nm). As the energy level increases, vibrational modes begin to decouple, and the bonds begin to move more independently of one another, resulting in much broader, less intense absorbance bands. Furthermore, symmetry will be reduced, leading to a reduction in the occurrence of resonance effects.

Functional Group Effects

The important attributes of functional groups that affect the position and intensities of NIR bands include: atomic mass, bond strength, dipole moment, symmetry, and anharmonicity; the latter three have the greatest effect on band intensity. Although it is not a factor for fundamental modes, the effect of anharmonicity is greater with asymmetric stretching vibrations than for symmetric stretching vibrations of the same group (18,19).

Hydrogen Bonding

The interaction of a hydrogen bond acceptor with an O—H group makes the O—H stretching vibration more harmonic, thus greatly reducing the intensity of stretching overtone bands in the NIR spectrum. Hydrogen bonding is the most significant effect in NIR spectroscopy that does not involve a change in composition. The most common hydrogen bond donor groups are —OH and —NH; there are many possible hydrogen bond acceptor groups, including: O—H, N—H, C—O—C, C=O and —N< groups (13). A summary of the effects of hydrogen bonding on band location and shape follows:

- Stretching bands of groups involving the acceptor atom decrease in frequency,
- Donor-hydrogen bending bands increase in frequency, and
- Donor-hydrogen stretching bands decrease in frequency and become broader.

Hydrogen bonding causes a decrease in the intensity of overtone and combination bands involving the donor-hydrogen stretching vibration, which depend greatly on anharmonicity to exist. Hence, overtone and combination bands for O—H and N—H groups are more intense for the "free" (nonhydrogen-bonded) forms than for the hydrogen-bonded forms (references).

Neighboring Group Effect

Even identical functional groups can produce very different spectra depending on the composition and structure of the sample matrix. The exact position of NIR absorbance bands in specific materials can be influenced by the composition of the neighboring functional groups. This is especially important if the neighboring functional groups are either strongly electron-withdrawing or electron-donating, both of which greatly affect bond strength and dipole moment.

Water

Water is the most important and misunderstood absorber in the NIR region. It is unique due to its low reduced-mass, strong dipole moment, and the fact that it is both a hydrogen-bond donor and acceptor. Thus, water has great propensity for hydrogen bonding. As a very anharmonic molecule, water absorption bands commonly exhibit Darling-Dennison resonance effects. Water has intense absorption bands at approximately 1940-, 1420-, 1200-, and 970 nm. Indeed, determination of moisture in solids was one of the first applications of NIR spectroscopy and has provided many published references across diverse fields of application. Pharmaceutical process analytical applications of NIR for moisture monitoring are discussed at length in subsequent sections.

The varying presence and character of water molecules in pharmaceutical solids will generate significant and complicated effects on their observed NIR absorbance bands and baselines. Water absorption bands can be quite broad and shifted by a considerable amount from their anticipated location because of matrix interactions. Water is present in pharmaceutical solids in two principal forms: (*i*) free water adhered to the surface and interstitial spaces of the particle matrix by London dispersion forces, and (*ii*) bound water held tightly inside individual crystals. The level of free and bound water in pharmaceutical solids has implications for drug dosage form stability, solubility, and bioavailability.

Free and bound water exhibit different levels of hydrogen bonding and unique intermolecular distances. Recent works have demonstrated that the spectral signatures of the two forms can be determined via NIR spectroscopy (20–23). Hogan and Buckton (22) demonstrated that the mono-, di-, tri-, tetra-, and pentahydrate states of crystalline and amorphous raffinose could be distinguished from NIR spectra (Fig. 1). Other groups (21,23) have utilized NIR spectroscopy, along with multiple chemometric methods, to develop process analytical techniques for effectively discerning surface and bound water in various drugs and drug products. Each of these groups showed that the presence of free surface water causes increased NIR absorbance intensity near 1905 nm, and increasing levels of bound water lead to relatively higher absorbance at lower frequency bands. For example, Zhou et al. (23), studying an unnamed model drug compound, reported decreasing absorbance intensity near 1936 nm as the relative levels of bound water (in the form of crystalline hydrates) decreased (Fig. 2). Additionally, in two separate studies (20,21), researchers from the Rantanen group showed that bound water solvated in

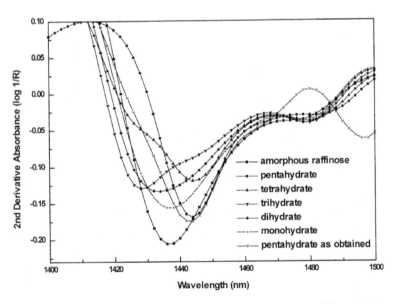

Figure 1 The NIR second derivative spectra corresponding to different hydrate levels. *Abbreviation*: NIR, near-infrared. *Source*: Courtesy of Elsevier, 2001.

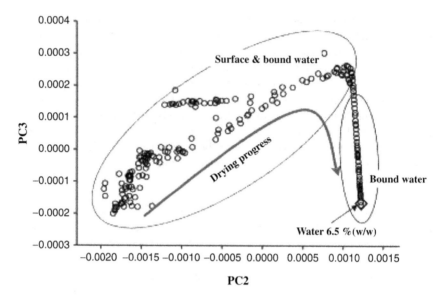

Figure 2 Principle component scores plot of PC2 versus PC3 during the drying process of a wet cake of model drug by in-line NIR monitoring. *Abbreviation*: NIR, near-infrared. *Source*: Courtesy of Wiley-Liss, Inc. and the American Pharmaceutical Association, 2003.

caffeine is detected near 1960 nm, whereas bound water in theophylline is detected near 1970 nm.

In 1998 Buckton et al. (24) published the results of their work investigating changes in the form of crystalline and amorphous lactose using NIR spectroscopy. The researchers were able to produce samples of amorphous, form-α and form-β lactose by exposing spray-dried amorphous lactose to conditions of elevated relative humidity for time periods of up to three months. The crystalline form of the lactose samples was confirmed using differential scanning calorimetry (DSC) and thermal gravimetric analysis (TGA). In contrast to earlier studies on crystallinity, the NIR reflectance analyses of the samples suggested that the major source of sensitivity was due to changes in the character of water in the lactose matrix. As the amorphous lactose crystallized into the α form and eventually converted to the β form, intense changes in the band shapes near 1440- and 1920 nm were observed, which agree with the observations of other researchers analyzing different materials.

In more recent work, Räsänen et al. (25), utilized NIR spectroscopy along with a novel multichamber microscale fluidized bed dryer to investigate the effect of temperature and moisture content of the drying air on the dehydration of disodium hydrogen phosphates with wet theophylline granules and three different levels of hydrates. The removal of bound crystalline water was monitored by in-line Fourier transform near-infrared (FT-NIR) spectroscopy and off-line X-ray powder diffraction. The real-time drying curves illustrated the stepwise change in moisture content as hydrates were liberated from the solids.

Temperature Effect

The sample temperature variability produces a complex effect on the NIR spectra of pharmaceutical solids, especially for solids with high moisture content, as is typical during granulation and drying operations. As the temperature increases, the degree of anharmonicity of absorbing groups will tend to increase, leading to an upward shift in the frequency center and a broadening of absorbance bands. Because of their reduced mass, molecules with significant hydrogen bonding are most affected by temperature variation. Thus, the temperature effect on NIR absorbance is most relevant for pharmaceutical solids when it is varying in conjunction with high moisture concentration or when critical analytical bands for a weakly absorbing constituent are heavily overlapped by water absorbance bands.

Maeda et al. (26), utilized multiple chemometric and curve fitting tools to investigate the relationship between water temperature and spectral changes as a result of the temperature effect on hydrogen bonding. They observed a strong, multifactor effect (Fig. 3) that could be used to accurately predict water temperature, which has been verified in more recent work by other groups (27). During operations that cause substantial fluctuation of temperature and moisture, such as high-shear granulation or drying, the evolution of the water spectrum with shifting temperature will often induce significant nonlinearity in the NIR

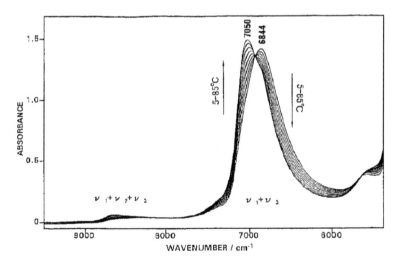

Figure 3 FT-NIR spectra of water measured over a temperature range of 5 to 85°C in increments of 5°C. *Abbreviation*: FI-NIR, Fourier transform near-infrared. *Source*: Courtesy of NIR Publications, 1995.

response curve. If the NIR analyses are to be conducted over wide temperature and/or moisture concentration ranges, it will generally be necessary to increase the complexity of the mathematical method for resolving the spectra by either adding model factors, or by estimating nonlinear model terms.

Particle-Scale Phenomena

The discussions of molecular-scale phenomena in the previous section dealt with the mechanisms of absorption and the confounding factors associated with optical absorption in the NIR. Such phenomena are general to most sample types analyzed by NIR spectroscopy. This section on particle-scale phenomena will be focused on the unique interactions between NIR radiation and solid particulate samples.

NIR Transmittance in the Solid State

One of the most important aspects of the interaction between NIR radiation and particulate solids is the unique combination of relatively weak absorbance intensity with high scattering efficiency. This combination allows the NIR radiation to probe the interior of many solid samples with little or no sample preparation. The spectroscopic analyses of solids in the NIR region are performed using two primary modes of sampling geometry: transmittance and diffuse reflectance (DR).

The transmittance measurement through particulate solids is a distinctive capability of spectroscopy in the NIR relative to UV/VIS and IR. Sample transmittance, T, is estimated as the ratio of intensities for light transmitted through an empty path (e.g., cuvette), I_0, and light transmitted through an equal distance of a particulate sample, I_s. Transmittance data are most often reported in terms of Beer-Lambert absorbance:

$$A = \log\left(\frac{1}{T}\right) = \log\left(\frac{I_s}{I_0}\right) = abc \tag{1}$$

where, for a single wavelength, λ:

A = Beer-Lambert optical absorbance, AU
T = transmittance ratio
a = absorption coefficient, cm^{-1}
b = pathlength (or sample thickness), cm
c = concentration of absorbing species.

In contrast to UV/VIS (transmission) absorption spectrophotometry, which is typically performed in solution, the transmittance of NIR radiation through a sample matrix comprised of tightly packed solid particles is expected to deviate nonlinearly from the Beer-Lambert law of absorption. The source of these deviations becomes apparent when the derivation and simplifying assumptions of the Beer-Lambert law, or simply Beer's law, are considered.

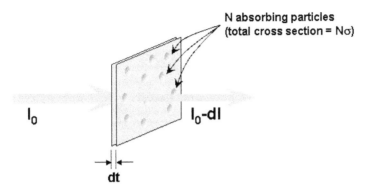

Figure 4 Illustration of optical absorption from an infinitesimally thick layer of a particle matrix.

The derivation of Beer's law is best visualized by considering a two-dimensional plane of a sample material containing a homogeneous distribution of an absorbing analyte (Fig. 4). For a collimated beam of light directed normal to a sample plane of infinitesimal thickness, dt, there is a finite probability that a fraction of the photons, dI, will be absorbed by the analyte, thereby reducing the intensity of the transmitted radiation. This probability is proportional to the intensity of the photons, I_0, the number of analyte molecules encountered, N, and the effective absorption cross section, σ, where:

$$-dI = \sigma N I_0 \, dt \tag{2}$$

The relationship can be extended to estimate the intensity of the beam remaining after transmittance through an optically thick layer, b, by dividing each side by the initial intensity, I_0, and integrating from $t = 0 \rightarrow b$, generating the following expression:

$$\ln(I_0) - \ln(I_s) = \sigma N b \quad \text{or} \quad \ln\left(\frac{I_0}{I_s}\right) = \ln\left(\frac{1}{T}\right) = \sigma N b \tag{3}$$

This result [Eq. (3)] is converted to the more user-friendly version [Eq. (1)] by replacing the effective absorption cross-section, σ, with the catch-all absorption coefficient term, α, according to:

$$\sigma = \frac{\alpha}{2.303 \, N} \tag{4}$$

The most notable deviations from Beer's law for NIR transmittance through clear liquids are due to changes in absorption coefficients across wide ranges of concentration. Otherwise the response between optical absorption, A, and concentration c, is expected to be linear. The transmittance through particulate solids is less accurately described by the Beer-Lambert law, however, because it assumes that all radiation encountering an absorbing particle will

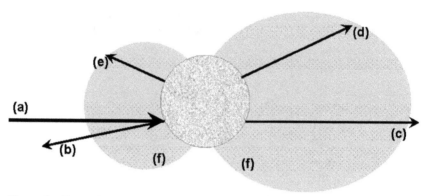

Figure 5   The optical interactions, of which the three fluxes are comprised of: (**A**) incident beam absorption, (**B**) specular reflectance, (**C**) transmittance, (**D**) forward scatter, (**E**) back scatter, (**F**) anisotropic scattering intensity fields.

either be transmitted or absorbed. There is, however, finite probability that incident photons will be scattered (inelastic or elastic), or reflected either forward or backward relative to the direction of propagation of the incident beam, I_0 (Fig. 5).

Thus, the incident photons are not transmitted directly through particulate samples, rather, they take a tortuous path through the material with multiple opportunities for scattering and backward reflection (Fig. 6). Because the intensity of scattered radiation decreases along with absorptivity, the transmittance through a solid sample will become less diffuse as the absorptivity decreases. These effects tend to increase the path a photon will travel before being emitted from the sample. Thus, the *effective* pathlength for transmittance

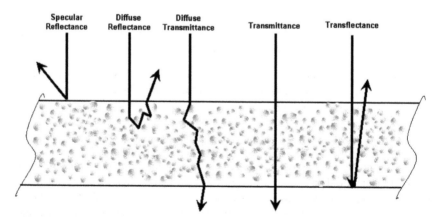

Figure 6   Schematic depiction of the most common modes of sample interface for NIR spectroscopy of solids. *Abbreviation*: NIR, near-infrared.

through a plane of compressed particulates of thickness b will be some nonzero, normally distributed amount greater than b. Furthermore, as b increases, the transmitted radiation will become more diffuse until the *diffuse thickness* (28) is achieved, at which point the maximum amount of incident radiation reflected back toward the source occurs. Hence, solid-state transmittance in the NIR region might more aptly be termed *diffuse* transmittance spectroscopy.

The absorptivity, reflectance, and scattering characteristics for a particular sample are dependent on multiple material qualities, such as particle size and morphology, packing density, and index of refraction. Hence, the effective pathelength and the level of nonlinearity relative to Beer's law are difficult to predict in practical situations. When taken over narrow concentration ranges, however, the simple $\log(1/T)$ approximation works fine for the majority of applications for NIR analysis of pharmaceutical solids.

The transmittance NIR spectroscopy is usually performed in the third overtone region between 780–1100 nm, where absorption bands are orders of magnitude less intense than the lower-order overtones. Although scatter coefficients do increase nominally in the NIR with increasing frequency, the scattering efficiency in the third overtone region is low because the amount of radiation loss to backward reflection is reduced along due to the decreasing absorption coefficient, as described by the complex Fresnel equations (29).

DR Spectroscopy

The DR sampling geometry is arguably the defining quality of NIR spectroscopy for process control applications. The DR sampling enables true noncontact, nondestructive analysis of solid materials. The DR analysis poses fewer constraints on sample positioning for such applications as high-speed tablet analysis (30). Remote-sensing via DR permits quantitative analysis of material qualities for granular or particulate samples in fluidized beds, mixing vessels, or on moving belts.

The sample reflectance, R, is most often estimated as the ratio of intensities for light reflected from a nonabsorbing, diffusely reflecting surface (e.g., white ceramic or inner surface of an integrating sphere), I_0, and light reflected from a sample, I_s. Even though reflectance measurements bear little resemblance to the scenario depicted in the derivation of Beer's law (shown earlier), reflectance data are most often reported in terms of absorbance units, $\log(1/R)$. Furthermore, as is typical for absorbance spectra collected in transmittance, the absorbance spectra collected in reflectance are treated as being linearly correlated with the concentration according to Beer's law. For most practical applications, nonlinearities are either ignored over a narrow concentration range or are approximated by additional empirical model factors or nonlinear terms.

Scientists have been attempting to develop rigorous theoretical models of DR for more than 200 years. These early development efforts are summarized in Birth & Hecht (28). The earliest efforts to explain the mechanism of DR treated it as strictly a surface phenomenon. In 1760, Bouguer posited that DR could be considered as the summation of mirror-type reflections from an assembly of microcrystalline faces statistically distributed over all possible angles. Lambert published a mathematical relationship describing the link between the intensity of an incident beam of light, angle of incidence, angle of observation (known today as the Lambert cosine law) in the same year. Although the Lambert cosine law has been shown to be theoretically sound, it fails in practice because an ideal diffuse reflector, or true *Lambertian* reflector, has never been found.

It wasn't until more than 100 years after the theories of Bouguer and Lambert were published that scientists began to consider that optical remission occurred through interaction with the surface and the interior of samples placed within an optical path (28,29). Seeliger described radiation as penetrating the surface of materials, where it is either absorbed or returned to the surface via reflection, refraction, or diffraction from the surfaces of the internal microstructure (28,29). Diffusely reflected radiation is now considered to be the superposition of two components: (*i*) *reflection* from external surfaces, and (*ii*) elastic *scatter*. Numerous theories of elastic scattering phenomenon have been developed to explain the relationship between the frequency of radiation and the scattering intensity. The major difference between the theories is their treatment of particle size and morphology.

Mie theory, which relates primarily to the scattering of radiation by isolated, spherical particles, is suggested to be the most relevant to NIR analysis of solids, because the Mie equation does not include a limitation on particle size (31,32). According to Mie's theory, scattered radiation is not distributed isotropically; rather, a complex scattering intensity pattern is produced, with forward scatter preferred over scattering in the reverse direction (similar to Fig. 5):

$$\frac{I_{\theta scat}}{I_0} = \frac{\lambda^2}{8\pi^2 R^2}(i_1 + i_2) \tag{5}$$

where, for a single wavelength, λ:

$I_{\theta scat}$ = Intensity of scattered radiation at distance R and angle θ from the center of the scattering particle.

i_1, i_2 = Complex functions of the angle of the scattered radiation, the spherical harmonics, or their derivatives with respect to the cosine of the angle of scattered radiation, the refractive index of both the sphere and surrounding medium, and the ratio of the particle circumference to wavelength.

For practical NIR applications, however, particles are not isolated but are generally in intimate contact with one another. Furthermore, Mie's theory does not address the consequences of multiple forward- or back-scattering events as individual particles propagate through a solid matrix. An intuitive assessment of the overall relationship between wavelength and scattering intensity can be

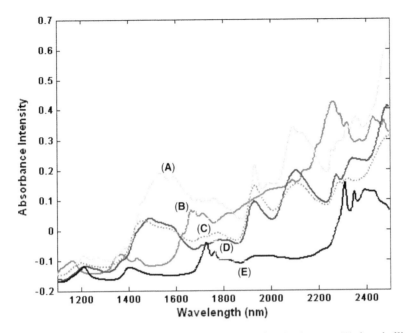

Figure 7   Pure-component NIR reflectance spectra for: (A) lactose, (B) theophylline, (C) starch, (D) MCC PH 101, (E) magnesium stearate. *Abbreviations*: NIR, near-infrared; MCC, microcrystalline cellulose.

gained from Equation (5): As the wavelength decreases, the intensity of scattered radiation will increase, thereby decreasing the apparent absorbance in the third overtone region. This phenomenon is one of the factors contributing to the shape of the upward-sloping baseline that is typically observed in the NIR reflectance spectra of solids (Fig. 7).

Experiments by Theissing, which concur with Birth's concept of diffuse thickness described earlier, showed that as the number of times a photon is scattered increases the distribution of scattered radiation will deviate further from Mie's theory, becoming more isotropic and with greater proportion of the radiation being scattered in the reverse direction (33,34). In general, a comprehensive theory of elastic radiation scattering is lacking for most NIR applications; hence, complete coverage of the various theories of elastic scattering are beyond the scope of this chapter.

Theoretical Models of DR

Besides the theoretically flawed $\log(1/R)$ format in which most NIR absorbance data are reported, the absence of a quantitative theoretical relationship explaining multiple-scattering phenomena has led to the development of phenomenological models from a generalized radiation transfer equation in some ways similar to

Equation (2). A modification of the radiation transfer equation to accommodate forward and reverse reflection, as well as anisotropic scattering and absorption, leads to a rather messy integration problem that is difficult to employ in practice. In 1905, Shuster reported a simplified solution of the radiation transfer equation for the case of reflectance, summarized here from a later reference (33,34). By assuming the total radiation flux is comprised of two components, one flux traveling in the forward direction, I, and a second flux traveling in the reverse direction, J, Shuster derived the following two differential equations:

$$\frac{-\mathrm{d}I}{\mathrm{d}t} = (k+s)I - sJ \tag{6}$$

$$\frac{\mathrm{d}J}{\mathrm{d}t} = (k+s)J - sI \tag{7}$$

where, for a single wavelength, λ:

$$k = \frac{2\alpha}{\alpha + \sigma_{\mathrm{scat}}} = \text{Absorption coefficient for a single-particle scattering event}$$

$$s = \frac{\sigma_{\mathrm{scat}}}{\alpha + \sigma_{\mathrm{scat}}} = \text{Scatter coefficient for a single-particle scattering event}$$

$$\sigma_{\mathrm{scat}} = \text{Effective absorbance cross-section due to scattering}$$
$$\alpha = \text{Effective molecular absorbance coefficient.}$$

Schuster made the additional simplifying assumption that only single-particle scattering occurs. By setting boundary conditions such that, at infinite thickness ($t \to \infty$), there is no further change in flux in the reverse direction with increasing depth: $\Delta J / \Delta t \to 0$, Equations (6) and (7) can be solved for reflectance at infinite depth, R_∞:

$$R_\infty = \frac{1 - (k/(k+2s))^{1/2}}{1 + (k/(k+2s))^{1/2}} \tag{8}$$

which can be rewritten as:

$$\frac{(1 - R_\infty)^2}{2R_\infty} = \frac{k}{s} = \frac{2\alpha}{\sigma_{\mathrm{scat}}} \tag{9}$$

Schuster's result is virtually identical to a later, more well-known expression derived by Kubelka and Munk, published in 1931 (35). Whereas Schuster derived his expression using k and s, which are applicable to single-particle scattering events, Kubelka and Munk employed the more general versions, K and S, which are defined as the absorption and scattering coefficients, respectively, for a densely packed sample layer as a whole. Although this change in parameters leads to an entirely different path for derivation (33,34), the result is (operationally) identical and is in a form that can be correlated with concentration:

$$\frac{(1 - R_\infty)^2}{2R_\infty} = \frac{K}{S} = \frac{2.303ac}{S} \tag{10}$$

where a is the absorptivity at wavelength λ; the 2.303 scaling factor corresponds to a conversion to the more-traditional base-10 representation of absorptivity. Because the K/S is generally related to concentration in practice via linear regression, the scaling factor can be omitted, and the absorptivity be treated as an unknown.

Kubelka and Munk made several assumptions in deriving their solution to the radiation transfer equation (some of which are shared by Schuster's derivation):

1. The total radiation flux is comprised of two components, one flux traveling in the forward direction, I, and a second flux traveling in the reverse direction, J.
2. The sample is illuminated with monochromatic radiation of intensity I_0 at wavelength λ; otherwise the next point (3) applies.
3. Particles must be assumed to be much larger than the wavelengths of illumination (so the scattering coefficient will be independent of wavelength).
4. Scattered radiation is distributed isotropically (implying that the specular component of reflection is zero).
5. Particles are distributed heterogeneously (randomly) throughout the sample as a whole.
6. Particles have much smaller diameter than the overall thickness of the sample, t.
7. The sample is subject only to diffuse illumination.
8. The area of the sample layer is large compared to the sample thickness, t (so as to avoid edge effects).

In light of the discussion from the preceding paragraphs, it is apparent that many of the assumptions are in conflict with physical reality, resulting in deviation from linear response to concentration. Some early practitioners posited that observed nonlinearity was due to the nonzero specular component of first-surface reflection. Even after removing the specular component of reflectance (via crossed polarizers), however, the Kubelka-Munk absorbance coefficient, K, will not be a linear function of concentration. Other researchers have developed augmentations or complete reformations of the theory of DR in an effort to relax some of the assumptions of the Kubelka-Munk model, some of which have seen significant application (36). It has been shown (34), however, that their refinements result in little more than a rescaling of the Kubelka-Munk function stemming from variations in the assumed lighting conditions. The theoretical shortcomings notwithstanding the Kubelka-Munk reflectance transformation, often denoted by $f(R_\infty)$, has been used extensively because it works reasonably well when applied correctly and has relatively few parameters such that it can easily be employed in practical situations.

Dahm and Dahm have shown in their recent work that both $\log(1/R)$ and the Kubelka-Munk transformation are inherently nonlinear functions of concentration. The $\log(1/R)$ transformation is inherently nonlinear: (*i*) because it treats the absorbance coefficient of a sample as being an additive sum of the absorbance coefficients of all absorbing species in the sample, which is true only for weakly absorbing constituents, and (*ii*) because it fails to account for the effects of scatter, which results in nonconstant effective pathlength (29). They demonstrate that the inherent nonlinearity of the Kubelka-Munk relationship is due to the erroneous assumption that the scattering coefficient, S, is generally assumed in practice to be constant, and independent of the level of absorption. Their work (37) is an extension of the mathematical theories of plane-parallel layers introduced by Stokes in 1860 and subsequently refined by Benford in 1946, then Wendlandt and Hecht in 1966 [references found in Dahm & Dahm (29)].

Their work contrasts with Kubelka-Munk theory in that it treats samples not as a homogeneous mass of scatterers but rather as a series of parallel infinite planes. The total radiation flux (from a source normal to the plane) encountered by any layer is accounted for as the sum of fluxes transmitted through the layer, T, remitted (reflected back toward the source), R, and absorbed, A. Hence, the optical flux incident upon a sample comprised of two *identical* planes (or layers), n and $n + 1$, can be accounted for by the Benford equations:

$$T_{n+1} = \frac{T_n T_1}{(1 - R_n R_1)} \qquad R_{n+1} = R_n + \frac{T_n^2 R_1}{(1 - R_n R_1)}$$
$$A_{n+1} = 1 - T_{n+1} - R_{n+1} \tag{11}$$

By considering the effect of doubling or halving the thickness of an individual layer on each of the flux components, Dahm and Dahm arrived at an iterative relationship for the estimation of absorbance, which is constant for samples of arbitrary thickness. Their postulated theorem is known as the absorbtion/remission function, $A(R, T)$, or, simply, the Dahm equation:

$$A(R, T) = \frac{[(1 - R)^2 - T^2]}{R} = \frac{(1 + T - R)A}{R} = \frac{(2 - A - 2R)A}{R} \tag{12}$$

The Dahm equation is proportional to the ratio of the linear absorption, K_{AR}, and remission, B_{AR} coefficients, as well as the fraction of light absorbed, A_0, and remitted, R_0, by a sample of infinitesimal thickness:

$$A(R, T) \propto \frac{2K_{AR}}{B_{AR}} \propto \frac{2A_0}{R_0} \tag{13}$$

Hence, in practical applications, it is necessary to know two of the three flux components (A, R, or T) to evaluate the Dahm equation for an individual sample. By doing so, the relationship between absorbance and scatter for a sample of infinite thickness is implicitly defined. The Dahm equation relates to the Kubelka-Munk relationship by:

$$2A(R_\infty, 0) \propto F(R_\infty) = \frac{K}{S} \tag{14}$$

Utilizing the Dahm equation to estimate the Kubelka-Munk absorbance and scattering coefficients, K and S, respectively, is theoretically advantageous because it accounts more effectively for the wavelength-dependent relationship between absorbance and scatter.

Although the Dahm equation and the Kubelka-Munk transformation (to a lesser extent) can be shown to be superior to the $\log(1/R)$ transformation in terms of linearity, the bulk of the DR data is likely to continue to be dealt with using the Beer's law representation for quantitative applications. This can be attributed to the simple fact that most NIR methods are ultimately empirical in nature and, in the case of NIR analysis of pharmaceutical solids, are expected to cover a relatively narrow range centered about some nominal condition. As the understanding of DR (and diffuse transmittance) phenomena continues to mature, however, these advancements may become essential parts of routine method development by enabling calibration using fewer training samples or by extending the limit of detection (LOD) of very subtle material qualities.

Physical Sample Factors Affecting the NIR Spectroscopy of Solids

Based on the discussion in the preceding paragraphs, it should be clear that the NIR spectra are the convolution of chemical and physical interactions between the sample and NIR radiation. Although the value of the chemical information in NIR spectra is obvious, the physical factors, which affect NIR spectra may also yield useful information related to the quality (state) of the sample material. In other cases, the presence of strong variations in concomitant physical parameters may overwhelm the analytical signal of interest.

Particle Size

The particle size distribution (PSD) and/or mass-mean particle size (MMPS) are critical quality control factors for raw pharmaceutical solids or for process control during unit operations, such as milling and granulation. Drug and excipient PSD variability can affect dosage form performance by altering dissolution rate or bioavailability, or it can affect drug product quality in manufacturing by altering the flow properties of the materials. It is a well-known fact that the angle of monochromatic optical diffraction for a single-particle event is inversely proportional to particle size. The actual angle of diffraction will depend not only on particle size and wavelength of incident light but also on particle morphology and refractive index, as described by Mie and Fraunhofer theories of optical scattering (15). Thus, as particle size varies, the relative level of forward and backward scattering events will vary as well, causing a change in apparent optical absorption. Larger particles, for example, will have a greater proportion of forward scatter, which reduces the probability that a diffusely scattered photon from an incident beam will be returned to the surface of a sample in DR (Fig. 8). The particle size variation will also affect particle packing characteristics, leading to more complex spectral effects. Hence, there has been great interest in using NIR spectroscopy for rapidly determining MMPS or PSD of pharmaceutical solids and granules.

Because much of the first widespread application of NIR spectroscopy was in the area of rapid content analysis of ground grain samples, the effect of particle size on NIR spectra was for a long time considered to be a detrimental factor affecting model performance. Possibly the earliest study (38) of the effect of particle size on NIR spectra was published in 1981 by Osborne, et al. While their study was intended to develop an NIR calibration for wheat kernel texture, their reference measurement of texture was actually a particle size fraction of the flour from the ground wheat kernels used for NIR analysis.

Norris and Williams (8) published some of the first work directly investigating the effect of sample particle size on calibration model performance. Their work, published in 1984, describes a method of optimal treatment of NIR spectra to reduce the effect of MMPS variation on NIR analysis of hard

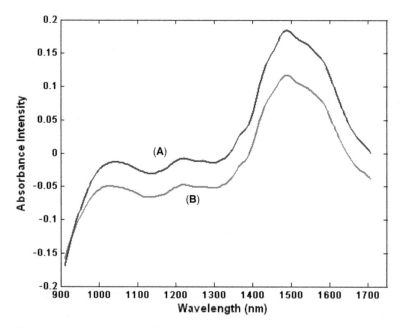

Figure 8 Comparison of NIR process spectra for two particle sizes of MCC: (A) PH 200~180 μm, (B) PH 101~50 μm. *Abbreviations*: NIR, near-infrared; MCC, microcrystalline cellulose. *Source*: Courtesy of Control Developments, Inc.

red spring (HRS) wheat for the prediction of Kjeldahl protein content. They observed that increasing particle size tended to increase the slope and baseline of NIR reflectance spectra of HRS wheat so that the absorbance values at longer wavelengths were more inflated by an increase in particle size than was observed at shorter wavelengths. Although they did not provide a detailed explanation for their observations, they deduced (via regression analysis) that the primary relationship between particle size and the magnitude of the effect on NIR spectra was due to the level of apparent absorption $[\log(1/R)]$ rather than wavelength. Thus, for solids with many strong NIR absorbance bands, it should be expected that the PSD variation will have a complex, nonlinear effect on the $\log(1/R)$ spectrum. Although this suggests that Kubelka-Munk transformation [Eq. (10)] of the reflectance spectra should be more effective in reducing the effect of particle size variation on NIR analysis, Norris and Williams went on to show that a ratio of two selected wavelengths in second derivative spectra was superior.

Bull (39,40) continued to explore the observations of Norris and Williams utilizing Kubelka-Munk theory (and the same dataset as Norris and Williams) in her 1990 and 1991 publications. Bull ultimately proposed a ratio of Kubelka-Munk functions at two wavelengths to mitigate the effect of particle size on the measurement of protein and water concentration in milled samples of HRS wheat (40). Although her result was not empirically superior in predictive performance, Bull suggested the solution would be less sensitive to particle size variation over a wider range of experimental parameters.

Ciurczak (41) published some of the first pharmaceutical work relating the particle size of pharmaceutical raw materials to variation in NIR spectra. Ciurczak demonstrated a linear relationship between the absorbance at any wavelength and the reciprocal of particle size as measured using low-angle laser light scattering (LALLS). Subsequent references on the use of NIR for particle size analysis include the application of NIR for real-time median particle size determination during granulation (42) and that of MMPS of an API in a binary mixture (43). Pasikatan et al. (38), published a detailed review of NIR for particle size analysis of powders and ground materials covering pharmaceutical and agricultural applications. Their review includes a theoretical discussion of the basics of NIR spectroscopy related to particle size determination and the aspects of sampling, reference methods, calibration development, and method validation.

The effect of varying particle size on the NIR spectra of solids is often described as a sloping, multiplicative baseline offset. Although this is qualitatively correct, the true effect of particle size variation is convolved with absorbance, morphology, density, and other sample factors. Thus, various chemometric treatments have been applied to extract MMPS information from NIR spectra of pharmaceutical solids, including Kubelka-Munk transformation (44), multiple linear regression (MLR) (45,46), principal component regression (PCR) and partial least-squares (PLS), as well as artificial neural networks (ANN) (47). Some form of laser diffraction or sieve analysis was used for reference testing in all cases. Although Frake et al., did not observe significant advantages by using more complex modeling algorithms, they succeeded in demonstrating that the ability to predict MMPS is not diminished when the NIR spectra are preprocessed to remove the effects of scatter. Although they did not discuss this aspect at length, this is an important concept because certain aspects of instrument and interface stability may necessitate some preprocessing before the analysis of NIR spectra. Chemometric topics, such as preprocessing and model algorithm selection, will be discussed in later sections.

Beyond prediction of MMPS and single-parameter particle size indicators, some researchers have explored methods of predicting entire cumulative particle size distributions from NIR spectra (46). Their method involved correlating the NIR reflectance spectra of a series of samples against 11 quantiles of their respective cumulative PSDs measured using forward angle laser light scattering (FALLS). Thus, 11 calibration equations were estimated to predict the 5, 10, 20, 30, 40, 50, 60, 70, 80, 90, and 95% cumulative size fractions using both MLR and PCR. They observed that, for both algorithms, the lowest calibration error was achieved near the MMPS. Although their results demonstrated an ability to predict cumulative PSD from NIR spectra using a multitude of calibration equations, a more detailed investigation of the correlation among PSD quantiles might have yielded a simpler solution along with some sense of the specificity (and the basis for specificity) of prediction at individual quantiles. Higgins et al. (48), recently demonstrated the use of NIR spectroscopy and PLS regression to accurately predict the volume-weighted particle size of nanoparticles in a high solid-content colloidal dispersion.

Hardness and Other Physical Parameters

Similar to particle size, variation in sample hardness has been observed to have a measurable effect on NIR spectra. Tablet hardness is a common product release parameter and is important for process control, because varying hardness might affect product friability, disintegration, dissolution, and ultimately, bioavailability. The hardness and density (or porosity, solid fraction) are common parameters for process control during roller compaction.

Drennen (49) published the first examples of using NIR spectroscopy and chemometrics for nondestructive tablet hardness in 1991. Subsequent publications by Ciurczak and Drennen in 1992 (50,51) and Drennen and Lodder in 1993 (52) further explored the application using both quantitative and qualitative methods. It was observed in all of these experiments that changes in dosage form hardness are indicated by a sloping baseline shift of the apparent absorption spectrum measured in reflectance (Fig. 9). As the tablet (or compact) hardness increases, the apparent absorption baseline increases. Although the root cause of the spectral effect is not conclusively established, some references do posit hypotheses. The results of some of the work recently conducted in the Drennen laboratory suggests that, as tablet hardness increases (and the total

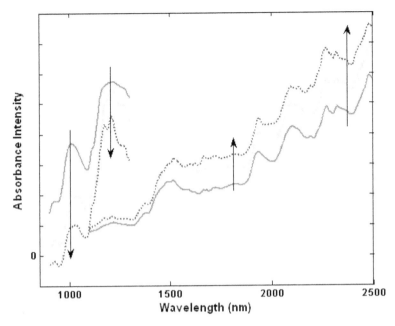

Figure 9 Effect of increasing hardness on NIR Transmittance and reflectance spectra (*arrows* indicate direction of increasing hardness). The transmittance spectra have been scaled for comparison. *Abbreviation*: NIR, near-infrared.

surface area of interparticle contact increases), a lesser fraction of the radiation is backscattered due to the air–particle interface, reducing the magnitude of reverse flux. This concurs with similar observations in earlier work by Kirsch and Drennen (53). At the same time, a greater fraction of the radiation propagates through points of interparticle intimate contact, increasing the magnitude of forward flux. As fewer scattered photons reach the reflectance detector via reverse flux, the apparent absorbance increases. This hypothesis is supported by the observed relationship between the tablet hardness and NIR transmittance, whereby increasing hardness reduces the apparent tablet absorbance (Fig. 9). This intuitively suggests that as powders are compressed more firmly, the resulting matrix becomes more "glasslike." As is for the effect of particle size, the rate of baseline increase is more pronounced near strong absorption bands. Although baseline shifting is the most pronounced spectral change observed with varying tablet hardness, other spectral changes (e.g., peak shifting) have been observed (54).

The development of NIR calibration models for hardness have utilized MLR and PCA (55), PLS (30,56,57), and more complicated algorithms, such as ANN (58). A survey of the literature suggests that there is little accuracy to be gained by using very complex chemometric methods, such as ANN, for prediction of hardness. Indeed, it has been shown that the hardness can be predicted with reasonable accuracy by simply fitting linear or nonlinear functions to the NIR baseline (53,57,59) and then correlating hardness to selected coefficients of the baseline fit. Because the NIR spectral baseline can be influenced by tablet alignment variation (55) and certain instrumental drift factors (57), a more robust calibration may be developed when preprocessing is used to reduce unspecific baseline variations. Despite the subtlety of the wavelength-specific nature of the hardness effect on NIR spectra, hardness calibrations have demonstrated specificity in the presence of concomitant variation in coating thickness (55), drug content (53,57), lubricant content (60), and both the type and level of filler material (59).

Donoso et al. (61) recently published the results of experiments involving the creation of calibrations for porosity and hardness of theophylline and placebo tablets. Although the validation results suggest that some nonlinearity was not accounted for in the model, the data indicate NIR sensitivity for porosity. Because Donoso et al. neither supplied many details of the hardness and porosity calibrations, nor did they present the regression coefficients or model parameters, the specificity of the predictions is suspected. It is because the porosity variation used for calibration was generated by varying compression force, similar to the methods used for hardness calibration.

Otsuka et al. (62) published similar findings in applying NIR spectroscopy for the evaluation of pharmaceutically relevant properties of antipyrine granules, such as MMPS, PSD, true density, angle of repose, and compressibility, as well as the porosity and hardness of the tablets produced from the granules. They observed NIR sensitivity to each parameter, and the collinearity among the physical parameters were tested. This is not surprising because the variation induced in all of the granulations was developed by varying the amount of water added during granulation. Indeed, spectral decomposition of the physical parameter data might have revealed that fewer independent factors were present in the measurements. More detailed multivariate experiments with variation in multiple production factors are likely to improve the level of understanding and specificity of the NIR response.

Effective Mass Sampled During NIR Analysis

For any new measurement technology or application to be deployed with confidence (and validated), there must be at least a mechanistic understanding of the measurement scale involved. This concept is critical for composition uniformity measurements (such as blend and content uniformity release testing) because, as is well known, variability between sample units increases as the size of the sample unit decreases (63). Thus, a sample considered to be homogeneous at the scale of a few hundred grams but would be quite heterogeneous at the level of tens of grams.

The safety and efficacy of most drug therapies is dependent on a specific level of uniformity among individual dosage forms. For this reason the uniformity of finished and in-process pharmaceutical materials is determined at a scale determined by the magnitude of a unit dose, sometimes referred to as the "scale of scrutiny" (64,65). For in-process measurements, such as blend uniformity analysis, precedent (66) suggests that the effective sampling mass should be no more than three times the mass of a single dosage form. Thus, while utilizing larger collection optics may increase signal levels, smaller sensors may be desirable for quantifying blend uniformity to ensure the appropriate scale has been assayed. The case is reversed for the analysis of individual dosage forms, however, where it is desirable to include as much of the product mass in the analysis as possible. Indeed, there has been some sentiment among early practitioners that NIR transmittance analysis of intact dosage forms would be superior to reflectance measurements based on the supposition that more of the dosage form would be effectively sampled in transmittance.

Besides being a qualification or validation issue, the effective mass sampled during NIR analysis has a complex, nonlinear effect on the shape of reflectance or transmittance spectra. Whereas the independent sample parameters affecting a spectrum (composition, PSD, hardness, etc.) can be adjusted relatively easily, the effective mass sampled is largely a convolution of all of such effects on the NIR spectrum. Hence, effective mass sampled must be inferred from indirect experiments, rather than measured directly. In her 1990 experiments investigating the effect of particle size variation on NIR spectra, Bull (39) developed the following relationship to quantify the observed relationship between multiple optical properties and the depth of penetration, d_λ, of the measurement into a sample at a given wavelength (for a particular measurement geometry). This relationship is directly related to effective mass sampled:

$$F(R_\infty) = \frac{K}{S} = \frac{x}{2d_\lambda r_\lambda} \tag{15}$$

where K/S is the Kubelka-Munk function of reflectance at infinite depth, x is the mean particle size of the sample material, and r_λ is the reflectance at a single air-particle discontinuity. The depth of penetration, d_λ, is specified as the $1/e$ information depth; in other words, the sample depth at which the fractional change in reflectance is less than $1/e$ ($e = 2.718$) (40). This depth of penetration is consistent with the depth of infinite thickness defined in earlier paragraphs.

Olinger and Griffiths (33,34) demonstrated a procedure for determining information depth of DR NIR spectroscopy using mixtures of carbazole, graphite, and sodium chloride. Using the standard absorbance values, known particle size and absorptivity values for the mixtures, they were able to calculate the number of particles that the NIR radiation reaching the detector had encountered while passing through the matrix. Although they estimated an information depth of $\sim$1 mm, the researchers suggested that the true depth might be much less due to the effect of scattering. Furthermore, they observed a strong relationship between absorbance, particle size, and information depth.

In perhaps the first work investigating the information depth of penetration for pharmaceutical applications of NIR reflectance spectroscopy, Cho et al. (63), conducted a series of experiments to estimate the effective mass sampled by NIR fiber-optic probes for blend uniformity analysis (63). Their work was based on theoretical models relating sampling variance to effective sample mass. By making certain assumptions regarding d_λ and true blend homogeneity, they were able to estimate two parallel regression models relating observed spectral variance and actual sample mass for a series of samples (of known mass) analyzed using a laboratory NIR reflectance spectrophotometer:

$$\log(\sigma_s^2) = b_0 - b_1 \log(M_s) + e \qquad (16)$$

$$\sigma_s^2 = b_0 - b_1 \frac{1}{M_s} + e \qquad (17)$$

where σ_s^2 and M_s are observed spectral variance and true sample mass for a set of seven known mixtures (of the same nominal composition), respectively, and e is assumed to be normally distributed indeterminate error. Following the estimation of the linear slope and intercept coefficients, b_1 and b_0, effective mass sampled could be estimated for other NIR measurement systems by evaluating spectral variance at individual wavelengths for repeat measurements at different locations in the same sample. To simulate monitoring of blend uniformity, Cho et al. collected 10 replicate spectra each for a series of blends (of the same nominal composition), which were mixed for an increasing number of rotations. The spectra were collected by carefully inserting a thin fiber-optic probe into multiple locations within each of the blends at each time point.

Both models [Eqs. (16) and (17)] were quite consistent with one another in the prediction of effective mass sampled for the fiber-optic probe system. The experimenters observed effective mass sampled to be wavelength-specific, ranging from 0.154 to 0.858 g with standard deviation ranging from 0.0379 to 0.1581 g. They further noticed that the effective mass sampled was significantly lower for wavelengths where the major constituents of the mixture contribute most to the spectral variance, though the experimenters were unable to determine the root cause. Given this disparity and the fact that the wavelength dependency of effective mass sampled was not monotonic, it seems likely that effective mass sampled is dependent on the apparent absorption of the material rather than wavelength, which is in agreement with the relationship posited by Bull [Eq. (15)]. Thus, for reflectance, as absorbance increases (or scattering decreases) the effective mass sampled is reduced. This hypothesis suggests that, for NIR blend uniformity analyses relying on spectral RSD, individual wavelengths might be selected or weighted by NIR absorptivity to evaluate homogeneity at the most representative scale.

A second group of researchers published results (64,65) from a series of experiments aimed at estimating effective sample size for general applications of NIR spectroscopy for the analysis of pharmaceutical solids. The studies explored the wavelength-dependent effective sample mass curve for multiple instruments and materials of diverse particle size and composition. Two models were employed to estimate effective mass sampled: (*i*) variable layer thickness (VLT) model, and (*ii*) equation of radiative transfer (ERT) model; each model necessitated a unique experimental strategy.

The VLT experiments (65) were conducted using a modified reflectance sample cup consisting of a strongly absorbing polyamide back plate upon which a series of spacer rings were stacked. A uniform column of powder was contained within the cavity at the center of the rings; the thickness of the powder layer was adjusted by adding or removing spacer rings to vary the height of the powder slug. Thus, as the powder layer thickness increased, a larger fraction of incident radiation was diffusely reflected by the powder while a smaller fraction of the photons reached the back plate to be absorbed. By fitting a curve to the depth versus absorbance data at every wavelength, the wavelength-specific information depth could be estimated via interpolation (Fig. 10). The results were presented in relative units (mg/cm^2); because the effective measurement area was never provided, correlation of the authors' results with those of Cho et al., is not straightforward. The experiment was repeated for four instruments (two dispersive, two FT-NIR) using each of

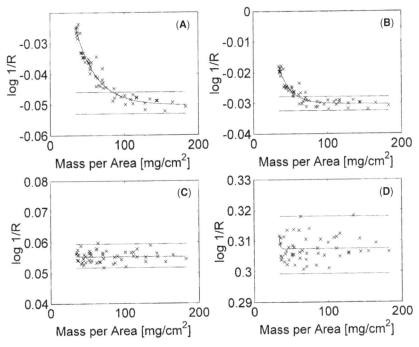

Figure 10   The effect of powder mass per unit area in sample cell on absorbance (log $1/R$) for DR measurements of MCC at: (**A**) 900, (**B**) 1300, (**C**) 1700, and (**D**) 2100 nm. *Abbreviations*: DR, diffuse reflectance; MCC, microcrystalline cellulose. *Source*: Courtesy of Elsevier, 1998.

seven samples of varying composition and particle size (microcrystalline cellulose (MCC), film-coated pellets, polystyrene microspheres—two sizes, nonpareil pellets—three sizes). It was observed for MCC and the film-coated pellets that at wavelengths longer than 1350 and 1600 nm, respectively, the effective sample information depth was too shallow for the quantification using the VLT method due to the higher absorbtivity. Hence, the expected relationship where increasing absorptivity reduces effective sample mass was observed. The researchers noted that, across all the samples tested, increasing the particle size resulted in greater effective mass sampled. This concurs with the hypothesis related to the particle size effect on NIR spectra stated earlier. Interestingly, bias shifts in the effective mass sampled profiles were observed among the instruments tested. The authors suggest that the differences in lighting intensity and geometry may be the root cause.

The ERT experiments (64) were based on a three-flux approximation of the equation of radiative transfer described by Chandrasekhar (36), which is one of the augmented theories on DR cited in earlier paragraphs discussing Kubelka-Munk theory. The solution of the ERT model yields a set of functions relating the effective depth of penetration and the effective mass sampled to the observed reflectance and mass-specific scattering and absorption coefficients. As it is quite involved, the derivation and solution of the ERT model is left to the references cited by Berntsson et al. In order to determine the mass-specific absorption and scattering coefficient profiles for the series of test samples, the authors utilized directional-hemispherical NIR transmittance and reflectance measurements. By applying the experimental data to the ERT model solution, an effective mass sampled could be predicted for reflectance measurements from samples of the same materials. Thus, relative to the VLT method and the methods employed by Cho et al., the ERT method relies to a greater extent on the mechanistic understanding of the phenomena than empirical observation. Despite this distinction, the VLT and ERT methods were in very close agreement (Fig. 11) in predicting effective mass sampled per area.

The only areas of noticeable disagreement between the two methods were in spectral regions of high absorbance, where the VLT method was not sufficiently sensitive. One noted advantage of the ERT method is its ability to estimate the fraction of the total reflected radiation contributed by each progressively deeper layer of powder. As would be expected, the nearest layers to the surface of the powder contributed the majority of reflected radiation. The authors emphasized that there is a distinction between the fraction of reflected light and the fraction of total information contributed by a particular layer. This point suggests that the actual effective mass sampled is probably somewhat smaller than estimated by any of the methods discussed.

Clarke et al. (67), utilized a method similar to the VLT procedure of Berntsson et al., to estimate the effective mass sampled during NIR microscopic imaging. To accurately determine the spatial distribution of chemical and physical effects in a dosage form or powder mixture, it is imperative that the

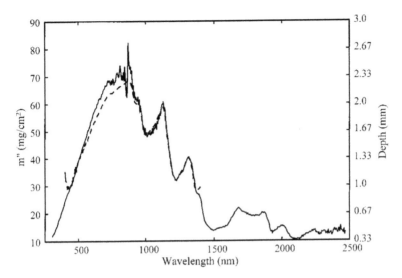

Figure 11 Effective sample size for film-coated pellets MCC calculated using the ERT method (*solid line*) and the VLT method (*dashed line*). The left ordinate gives the sample mass per unit area and the right ordinate the corresponding effective sample thickness calculated using tapped density of 0.30 g/cm³. *Abbreviations*: MCC, microcrystalline cellulose; ERT, equation of radiative transfer; VLT, variable layer thickness. *Source*: Courtesy of American Chemical Society, 1999.

depth of penetration be controlled to a limited range of surface features, otherwise stronger absorbers located deeper within the matrix may overwhelm the signal of an analyte of interest on the surface (or vice versa).

Because it was anticipated that the sampling depth and volume for microscopy measurements would be much smaller (due to the very small detector area for individual pixel measurements), the investigators devised a novel solution using layers of paper to achieve precise control over layer thickness. Because paper is composed primarily of cellulose, they were able to show that their results were applicable to systems using granular cellulose excipients, such as MCC. Clarke et al., concluded that, depending on the wavelength region for analysis, the effective mass sampled for individual pixel measurements varied between .03 and 418 µg; they went on to suggest that, along with the judicious choice of wavelength, NIR microscopy data could be used reliably in conjunction with Raman imaging, which is known to have a very limited effective mass sampled.

Whereas the efforts discussed in the preceding paragraphs have focused on the effective mass sampled by NIR for the analysis of free-flowing powders (during blend uniformity analysis), effective mass sampled for single-tablet analysis poses a somewhat different problem. Unlike blend uniformity analysis, the ideal nondestructive tablet analysis would sample as much of the tablet (or capsule) mass as possible. Furthermore, whereas the earlier works have focused on effective mass sampled during NIR reflectance analysis, many pharmaceutical tablets can be analyzed in both DR and transmittance modes. Thus, early NIR tablet analysis efforts were confronted with the following questions: (*i*) is enough of the dosage form sampled for the measurement to accurately capture the sample characteristics, and (*ii*) what is the relative difference in effective mass sampled between single-tablet transmittance and reflectance NIR analyses? Indeed, there was some early concern regarding the effect of compositional heterogeneity within individual tablets on the accuracy of NIR reflectance measurements (64,65). The initial conclusion was that the transmission measurement would be less sensitive to such sample inhomogeneities and would therefore be more suitable for tablet analysis. Efforts to rigorously answer these questions were undertaken by Iyer et al. (68), by performing two series of physical pharmaceutical experiments at Duquesne University that were published in 2002. The first series of experiments sought to determine the effective mass sampled for both NIR transmittance and reflectance measurements.

In order to estimate effective mass sampled, a series of specially designed model tablets were constructed to determine the effective information depth of NIR reflectance and transmittance measurements and the effective optical path diameter of transmittance measurements. Effective information depth was determined by analyzing a series of tablets of increasing thickness; the experiment was repeated using four strengths of cimetidine (other constituents were lactose, MCC PH101, and magnesium stearate). A polynomial function was fit for each wavelength studied to describe the change in spectral signature as a function of tablet thickness (Fig. 12); effective information depth of reflectance was determined by locating the first zero of the derivative of the polynomial function fit to the absorbance data.

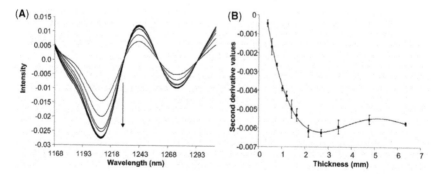

Figure 12   Effective sampling depth for reflectance measurements: (**A**) Second derivative spectra at different thickness for the 4% Cimetidine tablets at the Cimetidine and lactose bands (*arrows* indicate increasing tablet thickness). Tablet thickness ranged from 0.36 to 6.44 mm. (**B**) Second derivative values at 1186 nm plotted against tablet thickness.

The limiting tablet thickness for transmittance measurements was determined by monitoring the transmittance spectrum as tablet thickness is increased until the transmittance signal showed signs of distortion due to stray light around the tablets. The limiting tablet thickness for transmittance measurement was observed to range between 3.5 and 4.8 mm. The effective information depth of reflectance measurement varied according to wavelength, ranging from 1.9 to 2.7 mm. Effective information depth of reflectance at a wavelength of high cimetidine absorbance was observed to decrease as the concentration of cimetidine increased; information depth at wavelengths of high lactose absorbance was shown to increase with increasing lactose concentration. These results conflict with all previous studies, where the depth of penetration (or effective mass sampled) was directly related to the level of absorbance. The researchers suggest that these findings are the result of the effect of particle size on the NIR spectra because lactose has a much larger equivalent diameter than cimetidine. Hence, it was suggested that as the level of lactose increased (and cimetidine decreased), the level of scatter decreased, reducing the level of observed absorbance. The tablet composition had no observed effect on the limiting tablet thickness for transmittance measurement. The results for the depth of penetration and limiting tablet thickness were confirmed by analyzing a set of double-layer tablets with varying layer thickness. Although the observed results were in general agreement with those for the single-layer studies, a more linear spectral effect of tablet thickness was observed for the transmittance measurements.

In order to determine the effective optical path diameter, a series of cavities were drilled 2 mm into the tablets at three distances from the center of the tablet: 0, 2, and 4 mm. After the cavities were drilled, the transmittance spectra of the tablet were acquired and the cavities were subsequently filled with a mixture of ibuprofen and orange dye. The transmittance spectra from the tablets after the cavities were filled were compared with spectra prior to filling the cavities to identify any detected ibuprofen signal in the NIR spectra. Ibuprofen signal was observed in the spectra for the tablets with cavities at 0 and 2 mm but not for the tablets with a cavity at 4 mm. Thus, the estimated effective diameter of transmittance measurement was assumed to be 7 mm (the distance from the tablet center to the edge of the cavity placed at 4 mm was 3.5 mm).

Based on these results, and the measurements of effective information depth of reflectance, it is concluded that: (*i*) neither transmittance nor reflectance measurement sampled the entire tablet mass in any case, and (*ii*) the estimated mass sampled for transmittance and reflectance measurements were comparable based on the assumptions employed. Depending on the wavelength of analysis, the effective mass sampled for both measurements was observed to be near 200 mg.

The second series of experiments was aimed at determining the relative abilities of NIR transmittance and reflectance measurements to detect tablet heterogeneities in controlled locations. The experiments were performed by comparing reflectance and transmittance spectra of the double-layer tablets and the tablets with drilled cavities from the earlier studies. The double-layer tablets simulated heterogeneity along the *y*-axis, while the tablets with drilled cavities simulated heterogeneity along the *x*-axis. Qualitative examination of the data revealed that both transmittance and reflectance measurements were affected similarly by varying tablet heterogeneity in both directions. Thus, transmittance measurement geometry does not necessarily improve the robustness of single-tablet NIR measurement. Indeed, later studies published by Cogdill et al. (30), demonstrate that, for a high-speed process analytical scenario, transmittance measurement may have an inherent disadvantage due to the effect of tablet positioning variation.

The fact that neither transmittance nor reflectance measurements sample the complete volume of individual solid dosage forms will only generally be a minor drawback, especially for process control applications. Rather, because NIR spectroscopy offers so much more speed and flexibility in process analysis, the ability to gather relevant information on orders of magnitude more samples offsets such limitations. Finally, as mentioned earlier, the effective mass sampled is hardly a constant for any given material. The effective mass sampled will change along with packing and hardness parameters, sampling geometry, sample moisture content, and ambient humidity. It is important that for process analytical applications of NIR spectroscopy, where real-time control or release decisions might be based on NIR analyses, such spectral effects should be "built into" the instrument calibration as much as possible.

The preceding sections were meant to provide some of the historical background that differentiates NIR from other optical techniques and to better inform the reader of the true power and complexity of the NIR spectrum. Although the complexity of the NIR spectra from pharmaceutical solids may seem to be drowning in overlapping, competing signals from complex absorption and scattering phenomena, NIR spectroscopy has repeatedly proven to be an indispensable tool when paired with adequate instrumentation, chemometrics, and rigorous method understanding. The subsequent portion of this chapter is devoted to the discussion of NIR instrumentation, mathematics, and details of responsible method development. The third and final section of this chapter will cover broad applications of NIR spectroscopy in the analysis of pharmaceutical solids.

PRACTICE OF NIR SPECTROSCOPY

In the past the NIR spectrum was usually either an option or an afterthought in the marketing of optical instruments. For example, the Cary-14 recording spectrophotometer, with which much of the groundbreaking work of NIR spectroscopy in solids was performed (8), was actually a UV–VIS spectrophotometer with optional capabilities for measurements in the NIR region. The convergence of inexpensive computing power, improved digital detection systems, and new data analysis methods, along with the years of fundamental research, have led to dramatic growth in the demand for NIR instrumentation. The demand for

NIR technologies is growing quickly mainly due to process analytical applications in the pharmaceutical, chemical, and semiconductor industries. This demand is fueling research into more advanced technologies, such as laser-based instrumentation and microspectrometers. This section of the chapter will illustrate aspects of analytical instrumentation, data analysis, and method development relevant to NIR spectroscopic analysis of pharmaceutical solids.

Instrumentation

Because of its favorable position in the electromagnetic spectrum the development of NIR instrumentation has been relatively inexpensive compared to other optical techniques. Unlike the fingerprint region of the IR, simple and inexpensive quartz glass optics, fiber-optics, and lamps have excellent efficiency in the NIR region. Furthermore, the NIR region has been used for many military, aerospace, and telecommunications tasks, such as night vision, remote sensing, and dense-wave division multiplexing (DWDM). Thus, much of the fundamental research in optics and broadband digital detector materials [essential for many process analytical technology (PAT) applications of NIR spectroscopy] are paid for by the needs of other industries.

Modern NIR instrumentation for the analysis of pharmaceutical solids is best classified according to the demands of the application. According to the PAT guidance, at-line, on-line, and in-line measurements are defined as follows (69):

- At-line: Measurement where the sample is removed, isolated from, and analyzed in close proximity to the process stream.
- On-line: Measurement where the sample is diverted from the manufacturing process and may be returned to the process stream.
- In-line: Measurement where the sample is not removed from the process stream and can be invasive or noninvasive.

Within this framework, the demands for instrument speed and robustness increase, beginning with at-line instrumentation and ending with in-line analyzers. The need for flexibility, in terms of the capacity to be used for a diversity of measurement tasks, however, decreases from at-line through in-line measurement.

At-Line and Off-Line Instruments

At-line (and off-line) applications of NIR for the analysis of pharmaceutical solids include raw material identification, investigation of in-process and finished materials, and data collection for method development. The instrumentation for at-line analysis is generally self-contained (including power supply, lamp, optics, and detector), may have an integrated computer platform for data handling, and is designed for either table top or mobile (e.g., pushcart) applications. Many at-line NIR analyzers have automated sample carousels and/or fiber-coupled probes.

As they may be used for a wide variety of tasks, at-line analyzers generally offer the widest spectral range (often covering the entire NIR region), the highest spectral resolution, and internal means for instrument background correction and performance qualification (PQ). At-line NIR analyzers can be both interferometric, usually FT-NIR, or dispersive, typically utilizing a grating monochromator.

The FT-NIR spectrometers are most advantageous when high-resolution capabilities are important or if the spectrometer needs to have many options for sample interaction. The spectral resolution of an FT-NIR analyzer is constant across the spectral range detected and is adjusted by varying the length of the interferogram (cm) (Fig. 13); for a Michelson interferometer, this corresponds to increasing the range of displacement for the moving mirror (Fig. 14) (15). Thus, increasing the spectral resolution of an FT-NIR spectrometer will require longer scan times for equivalent performance in terms of signal-to-noise ratio (SNR). It is useful to keep in mind that the SNR of optical measurements is proportional to the square root of the number of photons collected, which scales linearly with scan time and/or the number of exposures. The FT spectrometers have a combination of three theoretical advantages, which make the technology attractive. They are: (*i*) all frequencies (wavelengths) are detected simultaneously (Fellgett's advantage), (*ii*) wavelength accuracy is constantly maintained by an internal laser line (Connes advantage), (*iii*) the interferometer has much higher optical throughput than other (dispersive) technologies (Jacquinot's advantage).

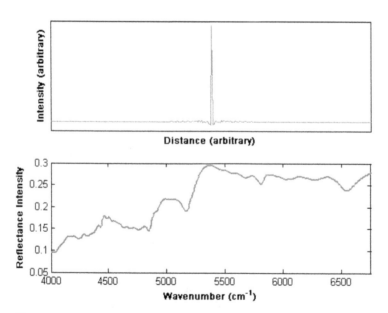

Figure 13   Fourier interferogram, or power spectrum (*upper panel*) and the corresponding reflectance spectrum (*lower panel*).

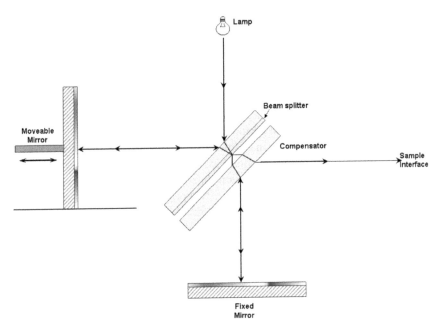

Figure 14 Michelson interferometer.

The FT spectrometer technology is extensively characterized in a number of comprehensive reference texts (70–72).

The typical resolution for FT-NIR applications is 2–8 cm^{-1} for explorative or qualitative analysis, and 8–32 cm^{-1} for routine applications. In contrast to the analyses of gases or liquids, the resolution needs for most applications of NIR for the analysis of pharmaceutical solids could be met with relatively low resolution (8–64 cm^{-1} or 6–12 nm). This is due to the extensive coupling of molecular bond vibrations in pharmaceutical solids, which results in rather wide absorbance bands. Indeed, it is not uncommon for practitioners to overestimate the need for spectral resolution when it would have been more beneficial to increase the SNR. The wavelength accuracy of FT-NIR spectrometers is guaranteed by using a constant laser line reference. The spectral range of FT spectrometers is controlled by adjusting the resolution of the interferogram (e.g., adjusting the step length of the moving mirror). The spectral range of modern FT-NIR spectrometers is mainly limited by the sensitivity of the detector material, rather than the mechanical stability or transmittance efficiency of the optical platform.

There are a variety of detector materials (72,73) suitable for the NIR region. The earliest NIR detectors utilized materials, such as silicon (<1100 nm) or lead sulfide (PbS, ~850–3300 nm), among others. Modern NIR spectrometers often use indium gallium arsenide (InGaAs, 850–1750 nm) or extended-InGaAs (1100–2200 nm). The InGaAs detectors are capable of high performance at room temperatures and work well in digital systems; however, early extended-InGaAs detectors had the reputation for unpredictable lifespan and variable detector linearity. Recently, more stable versions of mercury cadmium telluride (MCT or HgCdTe) and lead sulfide detectors have become available in linear and 2-D diode array formats with excellent sensitivity over the entire NIR range (850–2500 nm); these devices are likely to be useful for future instrumentation designs. The choice of detector material will be determined by the price, desired wavelength range, and performance requirements.

Numerous methods (15) (in addition to the Michelson interferometer) exist to exploit the inherent Fourier relationships of optical–mechanical systems. The Mach-Zender, Sagnac, or polarization interferometers are some examples. The polarizer interferometer relies on a two-piece birefringent crystal wedge polarizer and birefringent compensator to modulate optical energy in the pattern of a Fourier interferogram (15,74). As shown in Figure 15, randomly polarized radiation passes through a linear polarizer and a birefringent crystal, which separates the light into two orthogonally polarized parallel rays with a (wavelength specific) phase shift, thereby slightly rotating the plane of polarization. The phase-shifted rays are transmitted through the two-piece wedge polarizer, which then rotates the plane of polarization in the opposite direction, reducing the phase shift. The angle of polarization rotation is dependent on the effective pathlength through the wedge polarizer, and wavelength. At the initial position, the thickness of the wedge polarizer is set so that the polarization rotation angle (at all wavelengths) perfectly offsets the compensator, resetting the polarization state to that of the first polarizer. In this state, the maximum light energy at all wavelengths will pass through the second polarizer (DC intensity). As the thickness of the wedge polarizer is varied, the polarization angle incident on the second polarizer will vary in a sinusoidal pattern between 0 and 100% transmittance; the frequency of the sinusoidal pattern will vary according to optical frequency. Therefore, the sum of the components incident on the detector will produce an optical interferogram. The change in the thickness of the wedge polarizer (and, therefore, the change in phase shift) is analogous to the distance traveled by the moving mirror of a Michelson interferometer.

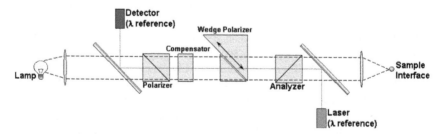

Figure 15 Simplified view of a polarization interferometer.

At first glance, the advantages of the polarization interferometer for process spectrometry are not obvious. The optical efficiency of the device is limited by absorption in the beam path, especially as the thickness of the wedge polarizer reaches a maximum (which will limit maximum resolution). Spectrometers utilizing the polarization interferometer have been available for many years. Although early instrument models suffered from some wavelength stability problems (74), recent upgrades of the technology have yielded a very robust instrument with more than adequate resolution capability (8 cm^{-1}) (75). In fact, the advantage of the polarization interferometer is its ruggedness; there is only a single optical path through the instrument (compared with the dual path of Michelson, Sagnac, or Mach-Zender interferometers). Thus, the polarization interferometer is less affected by small misalignments in the beam path, the sort of which might arise during at-line or in situ operation. Furthermore, the simple design is more cost-effective and can be produced with a smaller footprint.

The grating monochromator is the dominant dispersive technology for at-line NIR analyzers. Although there are many grating monochromator designs, they all rely on mechanical modulation (rotation of a scanning mirror, or rotation of the grating) of the instrument geometry to scan regions of the electromagnetic spectrum (Fig. 16). Although grating monochromators are not subject to the theoretical advantages of FT-NIR (e.g., Jacquinot, Fellgett, Connes), the high practical efficiency and robustness of their optical systems (which often utilize nearly 100% reflective optics), as well as their relatively low cost, have made them a popular choice for many NIR analytical applications. These analyzers have a long history of NIR applications in food and agriculture. Despite their mechanical limitations, they are capable of stable performance in relatively harsh environments (4,5,72,73).

The spectral resolution of a grating monochromator is determined by the slit function and the quality of the grating (15). Similar to FT-NIR, increasing the spectral resolution of a grating monochromator instrument will involve an SNR trade-off because the optical flux is reduced as the slit is narrowed. The spectral range of grating monochromator instruments is limited by the geometry of the grating and optical bench as well as the detector material. Some instruments utilize multiple gratings and detectors to increase spectral range and/or resolution. The wavelength accuracy of rotating grating instruments is limited by the quality of the grating and the mechanical stability of the optical system. Grating monochromator instruments have a proven track record of stability and transferability. In fact, the NIR grain analyzer networks exist where hundreds or thousands of instruments utilizing the same calibration equations have been established in many countries and trading regions around the world.

The stability and transferability of rotating grating instruments are due to the following characteristics:

- Optical standardization utilizing a simple gain and bias adjustment function can sometimes be employed, because each wavelength is

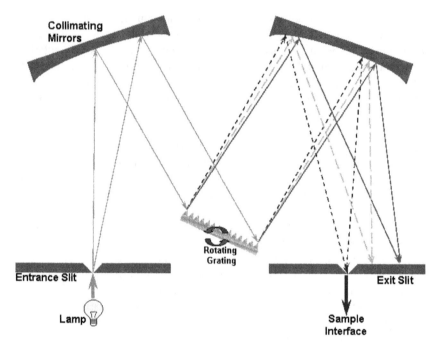

Figure 16   Czerny-Turner rotating grating monochromator.

sampled by the same detector(s). If the standardization function is wavelength dependent, a simple (e.g., quadratic) function is expected, depending on the geometry of the grating.

- Wavelength errors can typically be characterized by a linear shift or stretch of the axis (76), which is related to variability in the position of the detector relative to the monochromator.
- Spectral resolution is generally very stable and will vary quadratically across the spectral range according to the monochromator slit function. Some instrument manufacturers utilize spectral convolution to increase the effective bandpass of the monochromator to ensure a more constant bandpass across the spectrum and between instruments. Because resolution is rarely a limiting factor in NIR analysis, it is an efficient method of increasing transferability.

One instrument manufacturer (77), in particular, undertook a study of the instrumental factors affecting calibration transfer. Besides the factors mentioned, their work showed that the precise control of the geometry between the sample, monochromator, and detector is critical to calibration transferability among at-line or off-line instruments.

On-Line Instruments

The choice of instrumentation for on-line applications will be determined by the performance requirements in terms of sensitivity, robustness, and speed. Because process analyzers are often located in hazardous or controlled environments, their performance must be stable and robust to environmental variation. The expense of frequent instrument standardization and maintenance may be prohibitive due to access or scheduling restrictions. Furthermore, frequent restandardization or calibration updates can become a regulatory burden. Hence, after meeting the demands for sensitivity, the critical factor affecting technology selection (and method development) for PAT implementations of NIR technology should be robustness (78).

If the speed of analysis is not critical, an at-line type of spectrometer can be adapted for on-line use by employing an automated sample handling mechanism. However, speed will often be a critical factor for on-line analyses of in-process materials or finished products. Thus, spectral range, resolution, sensitivity, or SNR may be sacrificed to increase the measurement speed. For example, high-speed FT-NIR instruments have been developed for on-line applications by significantly reducing the length of measured interferograms and utilizing step scanning. Because on-line applications of NIR spectroscopy for solids analysis will generally not require high spectral resolution, relatively few interferogram points must be measured to generate a suitable broadband spectrum (79). Although grating monochromators can be deployed effectively for some on-line applications, two newer dispersive technologies have some advantages in such cases: acousto-optic tunable filter (AOTF) and holographic grating diode array.

The AOTF modulator relies on an interesting quality of optical diffraction in an anisotropic medium. AOTFs utilized for NIR spectroscopy consist of a bire-fringent tellurium dioxide (TeO_2) crystal bonded to an acoustic transducer (72,73,80–83). When the transducer (Fig. 17) is excited by an applied radio frequency (RF) signal, the acoustic pressure waves generated propagate through the crystal lattice, resulting in a periodic modulation of the refractive index of the crystal. Thus, a narrow band of optical frequencies that satisfy the phase matching conditions of the RF signal will be cumulatively diffracted, while other frequencies will be transmitted without diffraction (the zero-order beam). Because the crystal is birefringent, two diffracted beams are produced with orthogonal polarization states. One of the beams is directed to the sample for detection, and the other is either directed to a beam stop, or is detected directly as a reference channel. In the latter case, the instrument is operating as a double beam (in space) spectrometer. The center wavelength of the diffracted beam(s) is varied by modulating the frequency of the RF driver. Because the frequency of RF driver can be modulated very quickly, AOTF devices are capable of extre-mely rapid scanning. Furthermore, since the TeO_2 crystal is mostly transparent in the NIR region, AOTF devices have high optical efficiency. The spectral range of AOTF devices is determined by the transmittance curve of the crystal

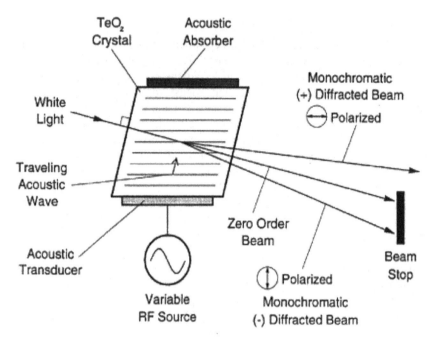

Figure 17 Acousto-optical tunable filter (AOTF) operation.

and the sensitivity of the detector material. The bandpass of commercially available devices in the NIR is similar to that of a grating monochromator (~6–10-nm FWHM). The wavelength accuracy is reported to be less than ~0.5 nm (84), though the variation depends on a number of factors related to SNR, such as tuning speed.

The AOTF spectroscopy (80) was popularized (beginning in the 1980s) because it presented a real breakthrough in simplicity, speed, and efficiency. The devices were considered to be inherently stable because there were no apparent moving parts and were, therefore, more practical for rugged process applications or high-speed reaction monitoring. For spectroscopists unfamiliar with diode array technology, AOTF devices were desirable because, they utilized simple, single-element detectors, like FT-NIR and grating monochromators that were assumed to be more amenable to calibration transfer. Furthermore, because diode array devices were viewed to be limited by the tradeoff between resolution and spectral range, which is set by instrument geometry and the number and size of detector elements, the broadband high-resolution capacity of AOTF devices was an apparent advantage. In long-term applications, however, Cogdill et al., have reported the occurrence of complex, unpredictable patterns of baseline variation between measurements over time and those on similar instruments (85). These baseline shifts are thought to be the result of changes in the character of the bond between the acoustic transducer and the

TeO$_2$ crystal. The time constant of these changes is thus far unknown and have been observed over both short (weeks) and long (months) periods (unpublished findings). Moreover, the complexity of the baseline variations are in stark contrast to the simple (linear or quadratic) baseline variations expected for competitive technologies. Although these phenomena do not necessarily rule out successful implementation of AOTF technology for pharmaceutical applications, their existence does necessitate a comprehensive strategy for instrument performance monitoring, calibration transfer, and calibration update (30,57,85).

The development of multielement detector arrays, such as the linear charge-coupled device (CCD) and photodiode array (PDA, or diode array), more than 30 years ago spurred the development of multichannel spectrophotometers based on imaging polychromators, known as linear diode array spectrometers (72,86,87). A polychromator (Fig. 18) is an augmentation over a monochromator whereby the spectrum is acquired by capturing an image of the holographic grating on a diode array (either CCD or PDA) in a parallel fashion, rather than serial scanning the surface of the grating through mechanical translation. Thus, though diode array spectrometers still utilize dispersive technology, they are subject to Fellgett's (multiplex) advantage and therefore offer high SNR. Depending on the source intensity and the means of optical collection, a typical diode array spectrometer can acquire a broadband NIR spectrum of good SNR in less than 100 ms. Because diode array spectrometers have no physically modulated components and are constructed using a solid block of material, diode array spectrometers are extremely robust to environmental conditions and aging. The diode array spectrometers have only gained significant application in NIR applications with the relatively recent availability of InGaAs photodiode arrays, the development of which is largely attributable to demand by the telecommunications industry.

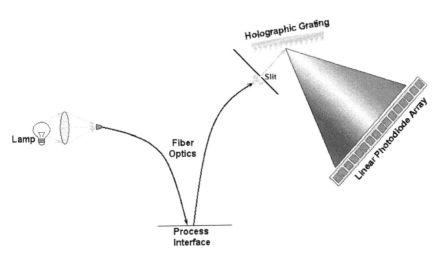

Figure 18   Fiber-coupled process diode array spectrometer.

As mentioned in earlier paragraphs, diode array spectrometers were initially received by some spectroscopists with a bit of skepticism. This is largely attributable to the poor stability of early linear photodiodes, which tended to make calibration transfer and maintenance rather complex. Although simple standardization functions can often be estimated for most other spectrometer technologies (e.g., slope and bias correction), diode array devices require that every detector is treated as an individual spectrometer. This is no longer viewed as a significant limitation of the technology, especially given the extreme robustness of the diode array optical platform.

The housing of a diode array spectrometer is typically constructed using a single piece of aluminum or even glass/ceramic to reduce temperature-induced flexure. Other concerns surrounding diode array spectrometers were related to their limitations in terms of absolute resolution and wavelength range. The spectral resolution and range of a diode array device will typically vary quadratically across the wavelength range, and is determined by the number of detector elements, their size, instrument geometry, and the slit width. Typical diode array devices have nominal spectral resolution of ~6 nm, which is more than adequate for process analysis of most pharmaceutical solids. The NIR diode array spectrometers might have 128-, 256-, or 512-element arrays and will cover the wavelength range from 850–1750 nm for standard InGaAs or 1100–2200 nm for extended-range InGaAs. In some cases the curvature of the polychromator grating is optimized to project a uniform wavelength axis on the linear diode array. During the manufacture of a diode array spectrometer, an internal lookup table is created to determine the actual wavelength sensed by each photodiode. Even if the grating has been designed to have a linear wavelength axis, most diode array spectrometers utilize the lookup table along with some internal preprocessing (interpolation) to correct for slight differences in the wavelength axis between instruments.

The diode array spectrometers most often interact with samples via fiber-optic probes and might not have an integrated light source. Although this might hinder the use of such instruments for at-line applications, it adds significant flexibility for on- and in-line use. Early versions of commercially available diode array spectrometers were produced without internal means of photometric correction (dark current subtraction, normalization). With such instruments, it is necessary to manually collect background correction spectra using standard reference materials, though the mathematical background correction procedure is generally automated. One diode array instrument manufacturer (88) has begun to offer an integrated process interface system with redundant internal light sources, an automated background correction mechanism, and one or more performance monitoring standard reference materials (Fig. 19).

In-Line Instruments

The instrumentation for in-line NIR applications is much the same as for on-line. The main difference is the way the sensor interacts with the process. Whereas

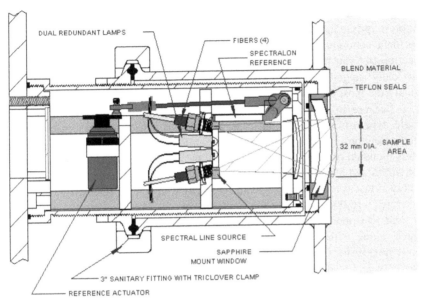

Figure 19 Design of a self-referencing process interface for a fiber-coupled diode array spectrometer. *Source*: Courtesy of Control Developments, Inc. (patents pending).

on-line measurements occur in a controlled side stream, where the sampling geometry can be easily controlled and the flow can be stopped, the in-line measurements are typically performed by noninvasive or noncontact sensing directly through an optical window while the process is running. For example, an NIR spectrometer for fluidized bed powder processing might be positioned to acquire spectra of the fluidized material directly through a window in the processor at the same height as the fluidized layer. Because the material being analyzed by in-line sensors is typically in motion, very high-speed spectrometers, especially diode array devices, are preferred. This is in spite of the fact that slower-scanning instruments will often work surprisingly well for dynamic processes, because total scan time is still likely to be much less than the overall process time. Robust design of the sampling system is a critical aspect of in-line sensor design. Flowing materials might be abrasive, wet, or sticky; thus, the spectrometer window should be made of a hard material (e.g., sapphire) and might have an automated cleaning or antifouling mechanism. Some scientists have indicated that slight heating of the optical probe above the process temperature will reduce the incidence of probe fouling (unpublished conversations).

In-line instrumentation may often be smaller and less integrated than at-line analyzers and is likely to resemble a sensor much more than an analyzer. In-line instruments may be subjected to temperature extremes and vibration, or they may be deployed in remote, inaccessible locations. A fairly recent innovation in in-line instrumentation is RF wireless communication. By using wireless data communication and battery-powered source illumination a process spectrometer can be attached to moving equipment, such as a V-blender. The physically rugged, lightweight devices, such as diode array instruments, are usually the most ideal for in-line use.

Because in-line applications often involve long-term measurement of a single type of material, filter-type instruments (discrete photometers) are another option. Filter instruments often employ a set of 10–20 interference filters mounted on a high-speed rotating filter wheel positioned between collection optics and a single detector. High-speed measurement is possible (approximately as fast as a diode array spectrometer) by spinning the filter wheel at a very high rate (~10,000 rpm). Fast Fourier transform (FFT) of the detector signal can be used to isolate the spectral intensity data from instrumental noise effects, which, along with the simple, efficient design, make filter instruments very robust devices for single-material measurement. Filter instruments have been significantly deployed as in-line moisture gauges, though some have been successfully calibrated for multiconstituent content prediction, especially in agricultural applications (72,73). Because they provide such limited spectral information, however, their applicability has begun to erode as low-cost diode array systems have become widely available.

As the market demand for process spectrometers in the pharmaceutical and other industries has continued to expand, new technologies are constantly evolving. A current trend in sensor development is miniaturization. A new type of NIR "microspectrometer" (89) has been developed, which has the potential to greatly increase the pervasiveness of NIR process monitoring by producing extremely small, relatively inexpensive devices, based on microscopic tunable Fabry-Perot interferometers that are built into a single microchip. Fabry-Perot interferometers (15,90) consist of a stack of partially reflective parallel dielectric plates, or etalons, for wavelength selection. For any given wavelength, the distance between an individual pair of plates, and the refractive index of the space between the plates, will determine whether or not the multiple reflections between the plates will lead to constructive or destructive interference. Thus, by varying the thickness of the etalons, or by modulating the refractive index of the void space [e.g., via liquid crystals (90)], a tunable filter can be constructed. Because each etalon passes harmonics of each optical frequency, a stack of etalons is used to select individual wavelength bands. The Fabry-Perot etalons are capable of very high efficiency and resolution; however, the free spectral range of an etalon is inversely proportional to the resolution. Thus, a relatively narrow spectral range is required to produce a high resolution device. Because the microspectrometer is so small and efficient, however, a bank of multiple devices can be assembled to cover a wide spectral range (if such flexibility is necessary) while maintaining a very small instrument footprint.

Imaging Spectroscopy

Chemical imaging spectroscopy as a micro- or macroscopic analytical tool has recently become available. Commercially available visible, NIR, Raman, IR (FT-IR), and most recently, terahertz (91) imaging spectrometers, capable of analyzing a diversity of pharmaceutical solids, have been developed. The NIR imaging spectroscopy, however, is unique in its combination of speed, analytical capabilities, resolution, and efficiency. A variety of optical systems have been created for high-resolution NIR imaging spectroscopy, including point-mapping systems, push-broom type sensors utilizing holographic gratings or linearly variable filters (LVF, or "wedge") or more often tunable filters, such as the AOTF and the liquid crystal tunable Lyot filter (LCTF) (72,81–83,92–94). The current dominant technology for pharmaceutical imaging spectroscopy is based on the combination of an LCTF and an InGaAs or InSb (83) digital focal plane array (Fig. 20). The LCTF-based systems operate in a manner whereby single-plane images are collected of a sample (typically by reflectance) over a wide range of individual wavelengths. The collection of single plane images forms a three-dimensional matrix of data called a hypercube. Thus, a high-resolution spectrum is collected for every pixel in a single-plane image of the sample (Fig. 21). Clarke (92) and Tran (81) have recently published reviews on the methods, instrumentation, and applications of imaging spectroscopy.

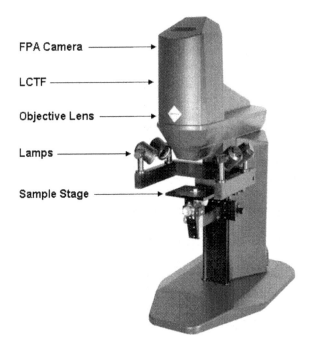

FPA Camera

LCTF

Objective Lens

Lamps

Sample Stage

Figure 20 Sapphire imaging spectrometer from spectral dimensions.

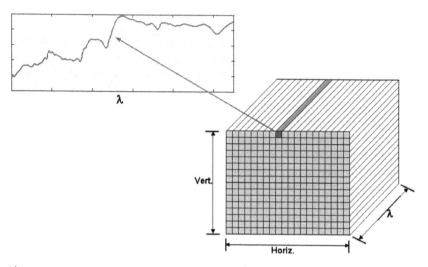

Figure 21 Example of a spectral image data hypercube.

At the heart of the LCTF-based imaging system is the camera focal plane array (FPA) and the tunable filter. The systems based on InGaAs cameras cover the wavelength range from 850 to 1750 nm, whereas the InSb system is sensitive over the range of 1200 to 2450 nm. The NIR FPAs are typically either 0.08 (320 × 256 pixels) or 0.20 (640 × 320 pixel) megapixel arrays. The highest-volume applications of NIR FPAs are either military/emergency (night vision, ballistic detection, heat sensing), or industrial (preventative maintenance, machine vision). Further maturation of the industry is likely to yield larger imaging arrays; however, many pharmaceutical process analysis applications of imaging spectroscopy can be solved with the 0.08 megapixel camera. Imaging spectroscopy FPAs are available in 12-, 14-, and 16-bit versions. Although one might typically expect a 16-bit system to have much higher performance, the true performance limit of an FPA may be limited by the electron well depth, rather than the quantization limit. Because the pixels of an NIR FPA are typically much smaller (~30 μm) than the photodiode of an FT-NIR or even diode array, their electron well capacity and, hence, their individual SNR will be much lower. The SNR performance of an imaging spectrometer can be greatly improved, though, by leveraging the three-dimensional correlation structure of the data by using 3D convolution, 2D image processing, or spectral preprocessing, such as smoothing, normalization, and derivatives. Furthermore, compared to single-point spectra of pharmaceutical materials, which represent the average spectral intensity over a large area of the sample, single-pixel spectra are often dominated by the absorbance of a single constituent (Fig. 22). Thus, by interrogating a much smaller area (volume or mass) of material, concentration variation between pixels is magnified. It is in this way that extremely

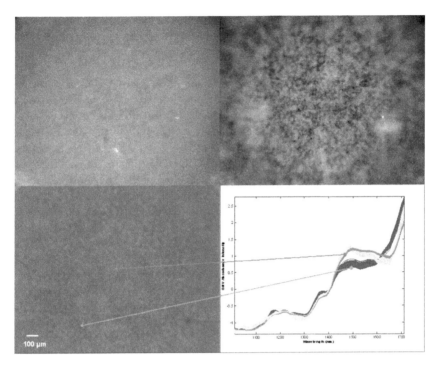

Figure 22 Spectral image of a directly compressed tablet consisting of theophylline, MCC, and magnesium stearate. The *upper left* and *right* images were acquired at 1050 and 1570 nm, respectively. The *lower left* image is a pseudocolor PLS prediction image (*gray*, theophylline; *black*, MCC; *light gray*, other/Mg stearate). The single plane images clearly show the local sensitivity of pixel intensity to composition. The periphery of the images demonstrates the focal problems associated with spectral imaging of samples with irregular surface height (such as a convex tablet). *Abbreviations*: MCC, microcrystalline cellulose; PLS, partial least-squares.

high-quality spectral information about a sample can be gathered, even though the individual pixel-level spectra are of comparatively low SNR. Furthermore, very low-concentration constituents can be detected and quantified by NIR imaging spectroscopy.

The most common LCTF design used for pharmaceutical NIR imaging is based on a tunable Lyot filter (94), which exploits the index of refraction characteristics of birefringent liquid crystals. Birefringent liquid crystals exhibit an index of refraction characteristic that depends on the polarization parameters of incident light. Liquid crystals transmit either the left- or the right-handed polarization state, depending on the helical ordering of the crystals, and reflect the opposite polarization state. Application of an electric field parallel to the optical axis causes the crystals to rotate in the plane perpendicular to the electric field. This rotation changes the index of refraction of the crystal by changing the position of the crystal relative to the optical axis (90). Thus, tunable optical filter devices that employ liquid crystals to vary the index of refraction can be tuned by simply adjusting an applied voltage.

Lyot filters, also known as polarization interference filters, are constructed in a sandwich arrangement of multiple stages. An individual stage consists of a birefringent polarizer and an analyzer (linear polarizer). Similar to the polarization interferometer, the birefringent polarizer effects a wavelength-dependent rotation of the plane of polarization. When the plane-polarized light impinges on the analyzer, a fraction (from 0 to 100%) of the light in the plane of the analyzer passes, while the orthogonal component is blocked (94). Because each stage passes all harmonics of a single frequency band, a complete Lyot filter is constructed by stacking multiple stages which combine to pass only a single optical frequency band. By adding a tunable liquid crystal layer to the Lyot stages, the Lyot filter becomes an LCTF (Fig. 23).

The LCTF holds advantages over similar devices constructed using Fabry-Perot etalons and AOTFs, because it is capable of tuning across very wide regions of the electromagnetic spectrum, while maintaining reasonable optical throughput with no discernible image shift or deterioration. The LCTFs have been constructed with resolutions ranging from 0.25–60 nm, operating in

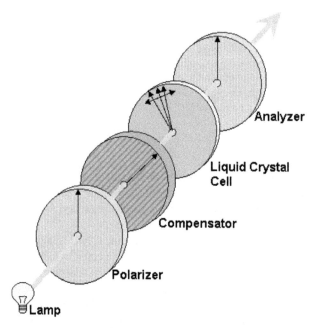

Figure 23   Exploded view schematic of a single Lyot filter stage. A complete filter would be created by stacking multiple stages with increasing fixed compensator thickness.

the visible (400–720 nm), the SW-NIR (700–1100 nm), or the NIR (900–1800 nm) (93), and more recently 1200 to 2450 nm (95). As expected, increasing the spectral resolution of an LCTF decreases the optical throughput. Commercially available NIR imaging spectrometers have nominal spectral resolution of approximately 6 nm, which increases quadratically by wavelength. Although LCTFs are relatively fast, capable of tuning between adjacent wavelengths in ~150 ms, exceeding the maximum tuning speed can produce erratic results as tuning relies on dynamically altering the orientation of the liquid crystals. A major drawback of the LCTF, however, is that by placing the filter between the camera and the sample, the depth of the field of the imaging optics can be greatly limited. Thus, as the level of magnification is increased, the effective aperture is greatly reduced (reducing the SNR). Furthermore, the focal distance will vary greatly across the spectrum due to chromatic aberration (Fig. 24). Depending on the surface characteristics of the sample, however, the focal quality of linear combinations (e.g., principal component score images, concentration maps) of images will be somewhat better than the focal quality at individual wavelengths.

One factor that has limited the use of imaging spectroscopy among diverse industries is the cost of the equipment and skills needed for successful implementation of the technology. Among pharmaceutical applications, however, the NIR imaging spectroscopy has the potential to increase the efficiency of formulation and process development and root cause analysis, as well as in-process testing and PAT method development. A common misconception about applying imaging spectroscopy for routine applications, such as PAT, is that the data flow would be too great to be feasible for on-line or in-line use. Because the magnification factor of imaging spectroscopy greatly increases sensitivity over single-point spectroscopy (as discussed in earlier paragraphs), most beginning

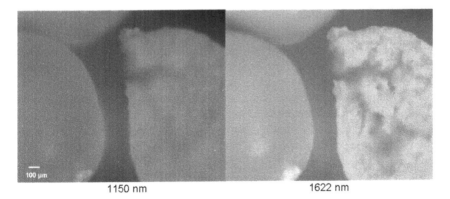

Figure 24 Image of extruded particles at 10× magnification. The difference between the *left* and *right panels* illustrates the effect of chromatic aberration on focal quality. Depth-related blurring can also be seen in areas of major thickness variation.

practitioners greatly overestimate the number of wavelengths that should be sampled. For a typical multicomponent tablet analysis, reflectance imaging (320×256 pixels) at <64 wavelengths is likely to capture most relevant spectral information and allow many opportunities for spectral preprocessing and modeling, yet consume only ~ 10 Mb/hypercube (assuming an unsigned 16-bit integer). Furthermore, the elimination of background pixels, image and spectral data compression, and metadata extraction can reduce the data load even further. Spectral image data do not have an intrinsically high information density because much of the data are correlated in all three dimensions. Thus, as with any instrumental PAT method, rigorous, knowledgeable method development is necessary to determine which data (whether spectral images, NIR spectra, or other sensor data) are relevant to critical material properties. The NIR imaging spectroscopy has been deployed for root-cause investigation, solid dosage form analysis, and blend homogeneity analysis. As imaging spectroscopy continues to be explored, and the equipment, methods, and understanding become more refined, many more applications will be discovered. Because the field continues to evolve quite rapidly, those interested should consult current literature for a more in-depth discussion of imaging spectroscopy (81,92).

Information Technology

Whether deployed in a development laboratory or embedded within a complex process, the efficient use of NIR technology in the future will require careful consideration of the information technology (IT) infrastructure to be employed. The NIR instruments have historically included layers of proprietary systems developed by their manufacturer for data handling and calibration development. For example, most early NIR spectrometers included an internal computer and chemometrics software package and might have used a proprietary data/ model structure. Although such vertical integration is suitable for many at-line analytical applications, on- and in-line measurement applications demand real-time interaction with a larger data system (e.g., management execution system, MES), and may require a common interface for many types and brands of instrumentation. In the ideal scenario, a process NIR spectrometer should be little more than a sensor, leaving data processing demands to remote software applications. The modular approach increases flexibility and transparency and allows more customization of the system to adhere to any current data management architecture involved.

As the demand for multivariate process analytics grows, systems integration hardware and software will play a larger role in managing NIR data and those from other sensor technologies. Eventual linkage of the highest-level planning systems (enterprise resource planning, ERP) with instrumentation and process control systems (components of the MES layer) will increase the importance of rigorous method development. Finally, the pharmaceutical industry could gain from a strategic investment in the development of a comprehensive set of multivariate data/model standards for storage, secure communication, and interoperability. Such an effort would fit into the scope of the ASTM (American Society of Testing and Materials) E55 committee on pharmaceutical application of process analytical technology.

From the onset of technology selection, and continuing through the process of deployment and operation, careful consideration of sensor performance parameters is important (6). All too often it seems that NIR projects begin with the purchase of equipment, followed by significant efforts to use what turns out to be instrumentation that is not suitable for the application. Whether the objective of the application is process analysis or materials research during process development, the chemical and physical constraints imposed by the materials and processes should guide the selection of an analyzer (Fig. 25).

Some questions to be considered during this phase include:

- Is the measurement required to be noncontact or noninvasive?
- What is the required speed of analysis?
- What environmental factors have the potential to affect instrument performance?
- What is the acceptable level of analytical error?
- Is in situ maintenance feasible in a given deployment?
- Is instrument stability suitable such that the maintenance can be performed at feasible intervals (e.g., between production campaigns)?

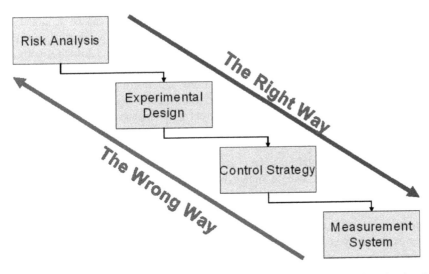

Figure 25 The ideal path of development actions is similar across multiple levels of method development. Costly errors in technology selection, model development, or systems integration that might hinder the progress of deployment will be avoided by following a rigorous, science-based approach from the beginning of any PAT application. *Abbreviation*: PAT, process analytical technology applications.

- What is the sensitivity and selectivity of the analytical device for the analyte(s) of interest?
- From an analytical perspective, does the instrument have a suitably broad range of performance to be expected to serve effectively over the expected composition range of the product(s) being monitored?

The first few items on the list will play a significant role in determining the design of the instrument and data handling system to be chosen, whereas the latter items will guide method development and the determination of performance standards.

The United States Pharmacopeial (USP) convention has established a monograph, USP <1119>, that sets forth rigorous procedures for NIR instrument performance qualification (96). The monograph describes the procedures and materials for testing such instrument performance parameters as photometric linearity, range, high- and low-flux noise, wavelength accuracy, spectral resolution, and wavelength repeatability. The monograph also suggests performance qualification limits for each of the tests, frequency of performance verification, and discusses methods for model transfer and validation. Although the guidance serves as a good framework for qualification (and validation), the proposed performance limits are likely to be excessive for some applications, and insufficient for others (30).

Some of the performance parameters, such as photometric linearity, spectral resolution, and wavelength accuracy, are often less critical in technology selection because they are generally "fixed" parameters that can be accurately described for a particular instrument. The most important parameters would relate to measurement precision and stability (e.g., high- and low-flux noise, wavelength repeatability, baseline stability) because they ultimately dictate the maximum level of confidence in predictions and instrument suitability. For example, depending on the optical system and detector material employed, the NIR instruments would have vastly different performance in terms of spectral noise, even though they are manufactured to cover the same wavelength range (Fig. 26).

If the instrumental noise level is too great in the primary spectral regions of sensitivity, the net analyte signal (NAS) for the component of interest will be diminished, and an otherwise appropriate instrument might be determined to have insufficient SNR. Alternatively, choices made during method development can greatly impact acceptable levels of various types of noise (30,57,85). For example, by utilizing high levels of spectral smoothing, or by building wavelength variation into the analytical model, the influence of wavelength repeatability on prediction stability can often be diminished without degrading the accuracy significantly. Whatever the application may be, in the same way that process understanding reveals the product and process parameters that are critical to quality, *method understanding* will help with the determination of appropriate instrument performance requirements.

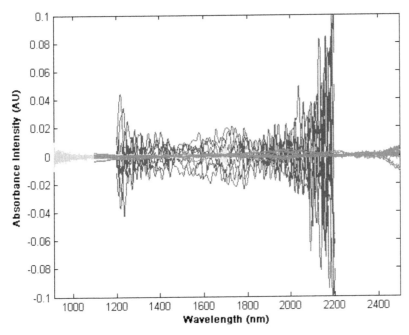

Figure 26 Relative repeat-scan noise profiles for three different instruments in normal operation: *gray*, diffraction grating-based monochromator (at-line); *black*, AOTF (on-line); *light gray*, diode array (in-line). All three sets of profiles were collected by mean-centering of a matrix of repeat scans of a single pharmaceutical solid material. The relative scan time was as follows (from longest to shortest): monochromator > AOTF > diode array. Although sample presentation varied somewhat for each of the instruments, all scans were collected without moving the sample. (The grating-based instrument profile is a composite noise profile collected from three separate tablets, which should not significantly affect the noise profile.) It should be noted that the performance of each of the instruments can be adjusted by altering the exposure time, light intensity, rescans, or sample positioning; furthermore, the scaling of the noise will be affected by the magnitude of the reference scan used in calculating the absorbance spectrum. *Abbreviation*: AOTF, acousto-optic tunable filter.

Data Analysis

Because of the complex chemical and physical factors affecting NIR spectra of pharmaceutical materials, the NIR data analysis is a multidisciplinary exercise, utilizing the aspects of chemistry, physics, digital signal processing, and multivariate statistics. Chemometrics, as defined by the International Chemometrics Society (ICS), is "the science of relating measurements made on a chemical system or process to the state of the system via application of mathematical or statistical methods." Chemometrics is a scientific field of study, and the credit of much of its development can be attributed to the development of algorithms for NIR data analysis and calibration. Whether the goal of data analysis is to interpret some observed patterns in the spectral data or to develop a predictive calibration model, the same toolbox of chemometric methods is employed. Furthermore, almost all chemometric tools are based on statistical and mathematical concepts of correlation and convolution. A *chemometrician*, however, seeks to find a valid chemical or physical interpretation for the relationships estimated.

In recent years, a great deal of excitement and mysticism has been associated with chemometric data analysis. Powerful, highly automated tools, such as the PLS and ANN, have made the development of a predictive equation from a myriad of data types virtually trivial. However, the importance of truly understanding the chemical and physical phenomena underlying the spectral data and the modus operandi of the chemometric algorithms cannot be understated. No combination of chemometric procedures can ever increase the amount of signal within a set of data. In the most abstract sense, chemometrics can only serve to optimally isolate relevant signals from noise and background, it is the duty of the scientist to extract information and knowledge from the signals. The following paragraphs will outline some of the most important chemometric tools for NIR method development and are not meant to be a rigorous treatment of the subject. For more information on chemometrics, numerous reviews of the subject are available (79,97–101). Three well-respected journals are dedicated to the science and application of chemometrics (102,103) and the application of chemometrics and NIR spectroscopy (12).

Data-Analysis Objectives

The selection of chemometric tools to be employed in solving a particular problem is guided, mainly, by the data-analysis objective. There are two general categories of data analyses, *quantitative* and *qualitative*, though the two categories are, at times, indistinct.

Quantitative chemometric data analysis involves multivariate calibration (79,98,104) for prediction and estimation. In a pattern recognition sense, quantitative data analysis is consistent with "supervised learning," where a functional relationship between a set of inputs (NIR spectra) and outputs is sought. Hence, quantitative chemometric tools are typically related to regression (e.g., multiple linear regression, PLS) and function estimation (e.g., ANN). As discussed in section I of this chapter, functions can sometimes be estimated to predict derived properties of pharmaceutical materials (e.g., dissolution rate) and typical chemical and physical properties from the NIR spectra.

Qualitative chemometric methods are a collection of methods to explore patterns observed (or hidden) within a set of NIR spectra. Conversely, qualitative data analysis is comparable to "unsupervised learning." Thus, while the results of qualitative analyses may be applied in very quantitative ways, the primary distinction is that, rather than directing the algorithm toward a particular solution, qualitative methods will reveal whatever relationships happen to be observed in the data (depending on the mathematical treatment). Beyond exploratory data analysis, examples of qualitative methods include curve resolution, discrimi-nate analysis and classification, and conformity testing (105).

Preprocessing

The purpose of spectral preprocessing is to reduce the amount of noise and/or background signal in a set of data so that a more robust model can be developed and the basis of the model developed can be more easily understood. Spectral pre-processing associated with early spectrometer designs, which sampled relatively few wavelengths, such as filter instruments, consisted of time-based noise filter-ing of spectral signals and the optimal selection of spectral filters. Indeed, a great deal of effort was often spent to optimize wavelength selection, typically through the use of stepwise multiple linear regression (MLR) (106,107).

With the advent of scanning spectrometers and full-spectrum regression methods, such as PLS, the need for optimal wavelength selection has diminished. However, for most pharmaceutical applications of NIR, it is a good practice to utilize instrument performance testing and pure-component scanning to eliminate regions of the spectrum where analyte signal is greatly reduced by instrument noise (57) or where model performance may suffer due to concomitant signals (e.g., temperature-induced shifting of water absorption bands). Model robustness can be improved in some cases where limiting the spectral range will remove interfering signals while preserving most of the net analyte signal. Finally, it is often advantageous to retain a contiguous spectrum to ease model interpretation and to facilitate full-spectrum preprocessing operations.

Most preprocessing operations (besides wavelength selection) are actually a form of mathematical bandpass (convolution) filter. The most common NIR preprocessing operations used to mitigate the effects of scatter and instrument baseline variation are similar to a high-pass filter. These operations include multiplicative scatter correction (MSC) (98), standard normal variate (SNV) (108) transformation, and polynomial detrending (108). As its name suggests, polynomial detrending is simply the operation of subtracting a polynomial fit of a spectrum.

The SNV transformation is the simplest of the three, does not require the storage of a reference spectrum, and is calculated according to the following formula:

$$\hat{x} = \frac{x - \bar{x}}{\sigma_X} \tag{18}$$

where:

x = Vector of NIR absorbance intensities (e.g., NIR spectrum)
$\hat{x}$ = SNV transformed vector of NIR absorbance intensities
$\bar{x}$ = Mean absorbance intensity for the current spectrum
σ_X = Standard deviation of absorbance intensity for the current spectrum.

The MSC transformation is calculated by subtracting the linear fit of a spectrum against a reference spectrum (typically, the mean of the data set). Although many experienced spectroscopists would suggest one or the other transformation to be superior, it can easily be shown that the difference between the two algorithms is quite trivial. The MSC transformation can be derived from the SNV transformation by replacing σ_X and $\bar{x}$ from Equation (18) with slope and offset terms from the linear fit, respectively. Only in certain, nontrivial circumstances will there be a significant performance difference between the two algorithms (Fig. 27).

Spectral smoothing derivatives are an example of a common preprocessing operation that resembles a signal bandpass filter; the derivative calculation suppresses low-frequency shapes, while the necessary smoothing calculation suppresses high-frequency shapes in the NIR spectra. Smoothing derivatives are used to enhance the subtle features in the NIR spectra and to reduce the effect of baseline variation (Figs. 28 and 29).

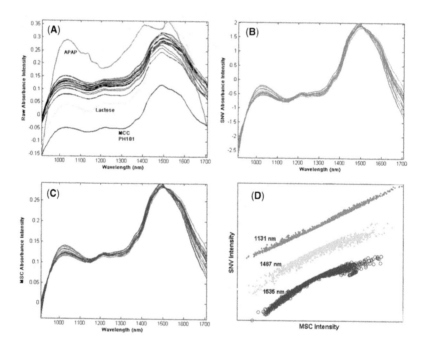

Figure 27 Comparative illustration of the effect of SNV and MSC preprocessing transformations: (**A**) raw spectra and pure-component spectra from a blending experiment, (**B**) SNV transformation of the raw spectra, (**C**) MSC transformation of the raw spectra, (**D**) correlation between the SNV and MSC transformation at single wavelengths (offset to ease viewing). The wavelength-dependent nonlinear relationship between SNV and MSC is related to the absorbance intensity, magnitude of baseline variation, and the source of baseline variation. Both MSC and SNV are expected to induce some small nonlinear effects in practical application. *Abbreviations*: MSC, multiplicative scatter correction; SNV, standard normal variate.

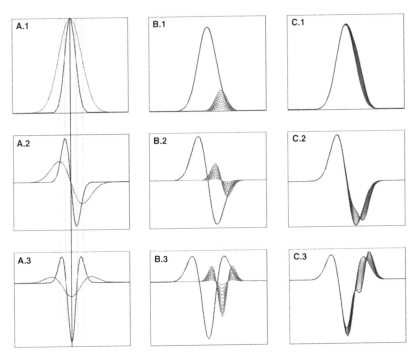

Figure 28 Illustration of the relationship between absorbance peak shape, derivative order, and the resultant derivative spectrum. The first row of panels illustrate shapes in raw data, the second row is the first derivative, and the third row is the second derivative of the raw data. In panels **A.1–3** it can be seen that increasing the width of an absorbance band will tend to reduce the intensity of the derivative spectrum. Furthermore, it can be seen that the zero crossing in the first derivative corresponds to the center of a pure band (**A.2**), and the distance between the zero crossings of the second derivative are related to the band width (**A.3**). Panels **B–C.1–3** illustrate the effect of intensity on derivative shape and the impact of derivatives on the ability to detect minor components. The spectra in panel **C.1** are the superposition of the spectra in **B.1** and are intended to simulate the effect of a minor constituent in a strongly absorbing matrix (e.g., drug in aqueous solution).

A very simple way to calculate a spectral derivative at any given wavelength, λ, is the first-difference approximation:

first derivative: $\dfrac{X_{+G} - X_{-G}}{2G}$ (19)

second derivative: $\dfrac{X_{-G} - 2X_G + X_{+G}}{2G}$ (20)

where:

G = Derivative gap, nm or variables
X_G = Absorbance intensity at λ

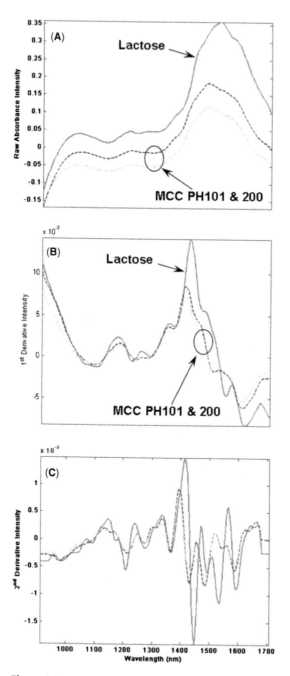

Figure 29 Raw, first- and second-derivative pure component spectra of lactose, MCC PH101, and MCC PH200. *Abbreviation*: MCC, microcrystalline cellulose.

X_{+G} = Absorbance intensity at $\lambda + G$
X_{-G} = Absorbance intensity at $\lambda - G$

Thus, by repeatedly applying the derivative approximation [Eqs. (19) and (20)] to a series of wavelengths from near the beginning of a spectrum to near the end, a full-spectrum derivative can be calculated. The so-called derivative gap parameter, G, can be adjusted to vary the bandpass frequency of the derivative. A narrow setting of G will tend to enhance subtle features in the spectra while mitigating low-frequency effects but also increase the sensitivity of the derivative function to noise. A wide setting of G will attenuate high-frequency features in the spectra while preserving larger-scale features. Hence, the ideal derivative gap is related to the width of the absorbance bands of interest in the spectra. Smoothing is accomplished by using a wider G, and by co-averaging (boxcar smoothing) the absorbance intensity at multiple wavelengths near X_G, X_{+G}, and X_{-G}.

The first-difference derivative is sometimes called "Gap Derivative," or the "Norris Derivative," with respect to Karl Norris, who was the first to widely publish the use of derivatives in NIR spectroscopy. The first-difference formula is especially attractive for use with fixed-filter or imaging spectrometers for which it is either impossible or too time consuming to measure NIR absorbance at many wavelengths. Norris and Williams have demonstrated that the ratio of derivative functions at different wavelengths can be used for the prediction of constituent concentration in the presence of significant baseline variation due to particle size or scatter effects (8).

The derivative function described first by Savitzky and Golay (109) in 1964 is more applicable to continuous, high-resolution NIR spectra and, hence, is used more often in modern NIR methods. Their method utilizes least squares fitting of piecewise polynomial functions to continuous signals to create a smoothed approximation. Thus, the derivative at any given wavelength can be estimated by evaluating the derivative of a piecewise polynomial fit centered at that wavelength. Rather than repeatedly fitting polynomials (and estimating their derivative) at every variable across a spectrum, Savitzky and Golay showed that exact local polynomials can be calculated by convolving the spectrum by a convolution kernel derived using least squares criteria. Convolution is a much faster operation than least squares estimation at every wavelength because only a single multiplication and summation is required at every convolution step across the spectrum. Furthermore, they showed that the exact derivative function can be calculated by convolving the spectrum by the appropriate derivative of the convolution function. Savitzky and Golay include an explicit derivation, tables of integer convolution filters, and FORTRAN code to implement their method in the appendix of their paper.

The application of Savitzky-Golay derivatives to NIR spectra requires proper specification of the polynomial order of the smoothing function, the convolution kernel width, and the derivative. For most full-spectrum NIR applications, stable performance can be achieved by using a quadratic poly-nomial and a smoothing kernel width of 11–23 variables (assuming ∼2-nm increment). As the wavelength increment is decreased, or as the noise level increases, a wider smoothing function will be more appropriate. Increase in the polynomial order of the least-squares approximation should also accompany the increase in filter width. The use of a smoothing filter that is too wide (well beyond the width of the absorbing bands) will actually produce artifacts in the derivative trace. The ideal combination of derivative parameters, however, is generally specific to the instrument and material at hand. Only first- and second-derivative spectra are typically used for NIR spectroscopy, because each successive derivative order has the effect of intensifying more high-frequency noise. The order of baseline effect suppressed by the derivative is one less than the order of derivative. For example, the first derivative transformation suppresses zero-order baseline shift, and the second derivative mitigates the baseline slope. Just as any convolution can be carried out by multiplication in Fourier-transformed space, spectral derivatives can be calculated by FFT, or by discrete wavelet trans-form (DWT) with a derivative-of-Gaussian (DOG) mother wavelet.

It may seem somewhat redundant, yet combinations of multiple prepro-cessing operations are often helpful in developing a sensitive, yet robust NIR method. For example, common combinations of operations include SNV + first derivative or SNV + second derivative. Utilizing normalization *after* derivative transformation can be useful, though it is seen less often in the literature. The use of other convolution kernels, FFT, or wavelet functions (100,110), and a wide variety of linear projection methods, such as orthogonal signal correction (OSC) (111–113), generalized least-squares (GLS) weighting (104,114–118), and direct orthogonalization (DO) (111), along with baseline correction and derivatives, is not uncommon. Because these are more specialized algorithms, they will not be further detailed in this chapter. Finally, even though the selection and optimization of NIR spectral preprocessing seems to be much of an empirical art form, it is important to keep in mind that each operation is targeted to either suppress or enhance particular spectral shapes (or variance components) that define the frequency bandpass of the operation. It is the job of the spectroscopist (or chemometrician) to match the combination of preprocessing operations to chemical, physical, and instrumental demands of the application.

Exploratory Data Analysis

Exploratory data analysis is performed to gain a qualitative sense of the signifi-cant patterns in the variance of the spectral data. The first steps of qualitative data analysis are quite simple and straightforward. Once a matrix of new NIR spectra has been assembled, it is important to begin by simply plotting the spectra (with and without preprocessing). Although the process is not as glamorous, a careful visual examination of the raw data will often reveal relationships that might be obscured by more advanced data analysis methods. Furthermore, it is sometimes important to look at spectra either one at a time or in small groups, because aberrant scans can sometimes be buried within a plot of many traces. Once the raw data have been visually characterized, chemometric tools dedicated to exploratory data analysis, such as bootstrap inference analysis (119–121), principal components analysis, and multivariate curve resolution, can be applied with confidence.

Principal components analysis (PCA) is a form of eigenvalue decomposition [see singular value decomposition-SVD (122)] whereby a matrix, X, of p correlated variables (e.g., NIR spectra) is reduced to a subset of latent variables, P, scores, T, and some residual error, E_x. The PCA algorithm, first described by Harold Hotelling in 1933 (123), has become an essential tool for multivariate data analysis across a diverse range of fields. The power of the PCA model is its automatic data compression and orthogonalization capabilities. As a consequence of the PCA model estimation process, the matrix of principal components, P, describes an orthonormal basis set, which maximizes the amount of information captured in the first few components (123):

$$X = TP^T + E_X \tag{21}$$

where:

$X = [n \times p]$ matrix of absorbance spectra
$T = [n \times r]$ matrix of principal component scores
$P^T = [r \times p]$ matrix of principal component loadings
$E_X = [n \times p]$ matrix of spectral residuals
$r =$ Number of principal components retained
$n =$ Number of samples (spectra) in the matrix.

Each latent variable, P_i, represents the linear combination of the original variables that captures the maximum residual variance. Hence, for highly correlated systems, such as raw NIR spectra, the first principal component, P_1, will often capture >95% of the total spectral variance, and each subsequent latent variable will capture a decreasing fraction of the variance until the total information has been captured. Once all of the viable information has been captured in r principal components, the remaining $p - r$ components will be correlated mostly with random noise and can be omitted (as part of E_x). The scores are simply the projection of each spectrum into the PCA basis, that is, the covariance between the spectrum and each individual loading. When considered as a whole, the scores for the X matrix used to derive the PCA model will be mutually orthogonal. However, the scores of new data (spectra) projected into the PCA model will not necessarily be mutually orthogonal.

For NIR spectra, chemometricians often notice that the addition of preprocessing operations, such as scatter correction and derivatives, will reduce the variance captured by the first few principal components. This is observed because preprocessing suppresses low-frequency components of the spectra, which typically account for a large fraction of the correlated variance. For example, without scatter correction, one of the first principal components will

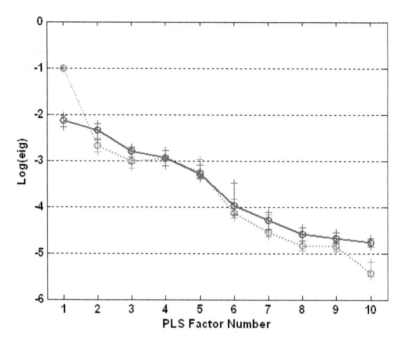

Figure 30 Logarithm of median eigenvalues during PLS regression using SNV + first derivative preprocessing (*solid line*) and raw (*dashed line*) spectra. The "whiskers" extending above and below each point on the lines illustrates the upper (75%) and lower (25%) quartiles of the data. The curves show that preprocessing has removed much of the low-frequency variation that would otherwise dominate the first latent variables. The magnitude of the later eigenvalues (>5) for the preprocessed data are actually larger; this may be because the preprocessing adds some nonlinear components to the spectra that are difficult to approximate using a linear factor structure. *Abbreviations*: PLS, partial least-squares; SNV, standard normal variate.

generally correspond to the shape (or slope) of the spectral baseline. The absolute amount of information contained in the matrix does not change with preprocessing, though, so the eigenvalue plots will eventually converge (Fig. 30). It can be helpful to plot the logarithm of eigenvalues to quickly compare the magnitude of information captured. For factor models of NIR spectra (e.g., PCA, PLS, etc.), logarithmic eigenvalues less than −5 are likely to be correlated primarily to spectral noise and will thus be uninformative or unreliable for modeling (unpublished observations).

The principal component loadings, *P*, are a mathematical representation of the unique correlation structures in the data. Thus, for a single factor (Figs. 31 and 32), regions of large positive magnitude are correlated with one another, as are those of large negative magnitude, and regions of positive magnitude are inversely correlated with those of negative magnitude. Furthermore, regions of low magnitude

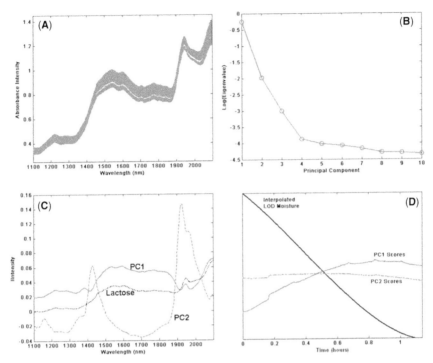

Figure 31 Principal component analysis of NIR spectra (**A**) of moisture loss during tray drying of granulated lactose. Since the two constituents are bound by closure, the first principal component accounts for ∼99% of the spectral variation (**B**). Analysis of the loadings (**C**) and score (**D**) profiles suggests that the first component is correlated to the relative concentration gain of lactose as water is removed. Comparison of the first component loading and the lactose pure-spectrum (**C**) demonstrates that the loading is actually the superposition of the lactose spectrum with the inverse of a moisture spectrum (along with the shape of the baseline variation). The second principal component is related to the nonlinear change in the water spectrum as moisture concentration is reduced. The vertical axis of panel (**D**) is in arbitrary units. *Abbreviation*: NIR, near-infrared.

have little variance in that principal component direction. By observing the shapes in a principal component loading it is sometimes possible to deduce the chemical or physical mechanism(s) responsible for the variance approximated by that principal component. Spectroscopists sometimes attribute a principle component to a single constituent (e.g., "the first PC appears to be attributable to API concentration").

Although a single constituent (or physical factor) can dominate a principal component, the divergent orthogonality and closure constraints of the PCA model and sample matrix, respectively, strictly prevent such a relationship. In other words, it is important to keep in mind that a principal component is nothing

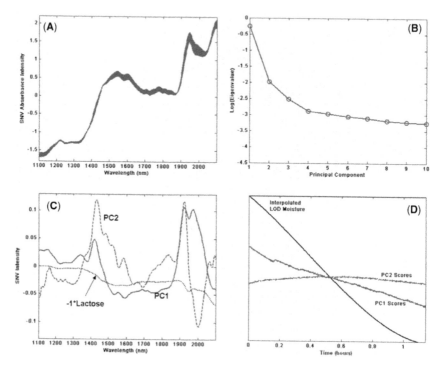

Figure 32 Principal component analysis of SNV preprocessed NIR spectra (**A**) of moisture loss during tray drying of granulated lactose. The magnitude of the first two eigenvalues (**B**) barely changed compared to the model without preprocessing (Fig. 31), suggesting that either the preprocessing did not have a great effect or that it increased the efficiency of information capture in the initial factors. The later eigenvalues remain larger for the SNV model, which may be related to nonlinearity introduced by the transformation. The overall shape of the first principal component is similar to the raw spectra model, but the direction is reversed (**C**). Furthermore, because the baseline variation has been suppressed, the concentration features in the loading vector are more intense, and are centered around zero. The second principal component loading (*dashed line* in **C**) is much sharper after SNV preprocessing, but the score profile is the same, which suggests that the same process is being modeled. The derivative-shaped feature in the second component (centered near 1950 nm) is similar to hydrate phenomena described in the literature and is apparently related to the level of bound and unbound water in the lactose matrix. The vertical axis of panel (**D**) is in arbitrary units. *Abbreviations*: NIR, near-infrared; SNV, standard normal variate.

more than an estimate of the linear combination of variables most correlated to a variance factor in the space spanned by the original variables. It is not guaranteed to have a direct chemical or physical interpretation. Indeed, experimental design (of X) has a major impact on the shape of principal component loadings (124). Finally, it is important to keep in mind that, unlike PLS factors, which are positively correlated with a target (Y) matrix, PCA loadings are nondirectional.

For example, it may be that if a data matrix is changed only slightly (e.g., some samples are deleted), the direction of a loading will be reversed (with a corresponding change in the sign of the scores). In other cases, what appears to be an absorbance band in a principal component loading may actually correspond to a region of low sensitivity lying between two correlated bands. Thus, if it is difficult to interpret the features in a particular principal component loading, it may be helpful to plot the inverse of the vector.

The orthogonal basis sets created during PCA offer a convenient platform for outlier detection, multivariate statistical process control, and matrix analysis by investigating individual sample scores, variable influence and contribution plots (125), spectral residuals, and leverage statistics. A very common application of PCA for qualitative analysis is to determine if there is any substructure or grouping within the matrix. For example, score–score plots are a rapid means of identifying and analyzing subclasses within a set of samples (Fig. 33) (30).

The residual and leverage statistics for samples projected into a PCA model are useful for determining the condition of the model (85) and for multivariate statistical process control (MSPC) (122,125–128). For each 1 x p sample spectrum, x_i, the PCA model residual lack-of-fit, or Q statistic, is calculated as the sum of squared reconstruction error across all wavelengths:

$$Q_{ik} = \sum (x_i - t_{ik}P_k^T)^2 \tag{22}$$

where:

Q_{ik} = Sum of squared reconstruction error for ith sample spectrum using model k
x_i = Sample spectrum
t_{ik} = Latent variable scores for ith sample spectrum, using model k
P_k = Model k loadings, or eigenvectors.

Sample leverage, or distance-to-model, is evaluated by using Hotelling's T^2 statistic, which is calculated as:

$$T_{ik}^2 = t_{ik}(\lambda_k^{-1})t_{ik}^T \tag{23}$$

where:

T_{ik}^2 = Hotelling's T^2 for the ith sample spectrum using model k
t_{ik} = Latent variable scores for ith sample spectrum, using model k
λ_k^{-1} = Diagonal matrix of normalized eigenvalues of the covariance matrix for model k.

The normalized eigenvalue matrix is calculated for model k as:

$$\lambda_k = \frac{(T_k^T T_k)}{n_k} \tag{24}$$

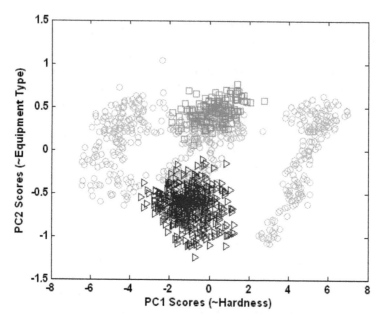

Figure 33 Scores plot for a series of tablets produced at varying levels of compression force, using laboratory-scale single-punch, laboratory-scale rotary, and production-scale rotary tablet presses. The first principal component was correlated with hardness, whereas the second component was correlated to tablet press scale. The plot illustrates the capability of NIR spectroscopy to detect subtle process signatures that would not be detected by other methods. Although the PCA model was estimated using nonproduction samples (*light gray circles*), an independent data set of production-scale tablets projected into the PCA model (*gray squares, black triangles*) showed separation. The separation was attributed to the use of two different tablet presses in the process stream. No quality or performance difference in the dosage forms was observed. *Abbreviations*: NIR, near-infrared; PCA, principal components analysis.

where:

T_k = Matrix of calibration sample model scores for model k
n_k = Number of calibration samples used for the estimation of
      model k.

As demonstrated by Equations (23) and (24), Hotelling's T^2 is a measurement of statistical distance in score space. Hence, T^2 is not influenced by spectral variation out of the plane (hyperspace) of the model (Fig. 34). A large T^2 indicates that the sample has high leverage on the model and may exceed the confidence limits of the model hyperspace, which is consistent with the conditions of extrapolation. Confidence limits for the T^2 statistic are calculated using a chi-square distribution. Hotelling's T^2 is very similar to Mahalanobis distance (105,129–131). Although both are statistically based distance measurements,

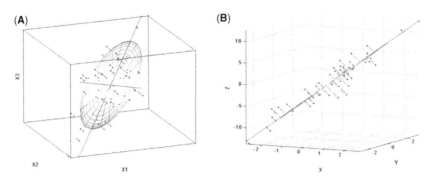

Figure 34 Demonstration of the approximation of a three-dimensional dataset by a two-component PCA model (panel **A**). The ellipse illustrates the 95% confidence limits of the Hotelling's T^2 statistic. The projection shown in panel **B** demonstrates the variation out of the plane of (orthogonal to) the model, which is measured by the Q residual statistic. *Abreviation*: PCA, principal components analysis.

the Hotelling's T^2 distance is used for distance measurement among orthogonal variables (such as principal component scores), whereas Mahalanobis distance takes into account covariance among the variables directly.

In contrast to T^2, Q describes the lack of fit, or spectral variation orthogonal to the plane, which is unexplained by the model. Thus, Q and T^2 are completely independent measurements. A large Q residual indicates that the sample is poorly reconstructed by the model, which is an indication that either a new factor may be present in the sample matrix (contamination, phase change, etc.) or that an instrumental fault has occurred (e.g., sampling error, component failure, etc.).

Soft independent modeling of class analogies (SIMCA) is an extension of principal component analysis, which was first described by Wold (99,132) in 1976. SIMCA can be used for discriminate analysis, multiclass classification or as part of an exploratory cluster analysis routine. The SIMCA procedure, which is in some ways similar to canonical correlation or Fisher's linear discriminate analysis (122), begins by dividing the data matrix (e.g., spectra) into m groups, based either on some sort of clustering routine or by user-supplied reference information on class membership. After splitting the data, a unique PCA decomposition is performed separately for each of the m groups; each PCA model can utilize different numbers of latent variables (or other model parameters), which are typically determined using cross-validation. After suitable models have been estimated for each group, a new sample can be classified by comparing a weighted combination of the T^2 and Q statistics in each of the models. Thus, a sample is either attributed to the group with the lowest score beneath a threshold or is attributed to none of the groups if it does not satisfy the threshold (based on an F-test criterion) for any group.

Principal component analysis is one of the most basic and widely used chemometric tools devoted to finding the number and direction of the relevant sources of variation in a matrix of bilinear data. A major drawback of PCA for the analysis of mixtures and process data, however, is the orthogonality constraint of the principal component loadings. As discussed earlier, since the factors must be orthogonal, it is impossible to isolate the shapes of individual chemical components using PCA when the mixture is constrained by closure. Self-modeling curve resolution (SMCR) (133), or self-modeling mixture analysis, is a rapidly evolving family of algorithms which overcome this drawback through either direct or iterative means. The goal of all resolution methods is to mathematically decompose a set of mixed instrument responses into the pure contributions attributable to each of the active components in a system. Similar to PCA, SMCR analysis does not explicitly require any a priori knowledge of the system or chemical reference values, though such knowledge can be incorporated into the model. For example, given nothing more than a time-series of NIR spectra of a chemical reaction, SMCR can be used to estimate the concentration and identity of each of the constituents at any point in the reaction without developing a calibration (134). Alternatively, given a single NIR spectral image hypercube of a tablet or powder sample, SMCR can be used to resolve the data into a set of pure-component intensity maps that describe the distribution of the chemical constituents in the sample (135).

The self-modeling resolution methods utilize a factor structure similar to PCA, but the factors are allowed to be correlated to one another as shown:

$$X = CS^T + E_X \tag{25}$$

where:

$X = [n \times p]$ matrix of absorbance spectra
$C = [n \times k]$ matrix of pure component concentration scores
$S^T = [k \times p]$ matrix of pure component sensitivity vectors
$E_X = [n \times p]$ matrix of spectral residuals
k = number of resolvable absorbing species in the matrix.

There are two types of ambiguity, rotational and intensity, that confound the process of resolving the true concentration and sensitivity profiles (133). Rotational ambiguity refers to the fact that there are an infinite number of pure-component sensitivity profiles (shapes) that can describe X and the true profiles; intensity ambiguity refers to the fact that there are an infinite number of intensity levels at which C and S^T can exist without changing the magnitude of E_X in Equation (25). Unique solutions to Equation (25) can be found, however, by properly constraining the form of the solution.

The constraints employed during SMCR analysis are mathematical representations of the cumulative knowledge of the chemical, physical, and spectroscopic limitations on the system. For example, the pure component concentration and sensitivity vectors may be constrained so that negative values are not allowed, which reflects the fact that (in normal situations) it is impossible to have negative real optical absorbance or concentration. For mixture systems, a closure constraint can be used because it is known that, when the matrix is considered as a whole, the concentration of all constituents at all times (or pixels, locations, etc.) must sum to 100%.

There are numerous articles and reviews (133) on the growing topic of self-modeling resolution for chemical analysis. Two exciting new areas of self-modeling research are hard–soft mixed model algorithms, which use specific knowledge of the reaction, such as hard stoichiometric models, to add powerful constraints to the soft-model solution (136), and multiway analyses (e.g., parallel factor analysis—PARAFAC), which are proving useful in the analysis of batch systems (125,133). The field of self-modeling resolution analysis is only beginning to be utilized for mainstream analytical tasks and will likely be very useful as an ever-greater volume of process analytical data must be efficiently analyzed.

NIR Calibration and Estimation

Multivariate calibration (79,98,104) is the essence of quantitative NIR data analysis. The NIR calibration models can be used to estimate a model for qualitative purposes (e.g., to indicate regions of spectral sensitivity), to estimate unknown sample qualities within a data set where some samples are of known quality or to generate predictions of quality for "unknown" samples. Early univariate instrumental calibrations utilized so-called classical least-squares (137) (CLS) regression, whereby the model residual error is attributed to the instrument data. The CLS method was extended for use with multivariate systems, such as NIR spectroscopy years later (138–140). The CLS model has some advantages in terms of interpretability and model transfer/update. A serious drawback of the CLS method, however, is that the concentration of all chemical constituents in the calibration matrix, X, must be known, otherwise the estimation fails due to rotational ambiguity.

Although augmentations (27) of the CLS model have been formulated to overcome this limitation, most spectroscopic calibrations are based on a first-order general linear model for inverse least-squares (ILS):

$$Y = XB^T + E_Y \tag{26}$$

where:

$X = [n \times p]$ matrix of absorbance spectra (dependent variable)
$Y = [n \times m]$ matrix of concentration profiles
$B^T = [m \times p]$ matrix of regression coefficients
$E_Y = [n \times p]$ matrix of concentration residuals
m = number of independent variables to be modeled.

It can be seen from Equation (26) that the ILS model is based on the assumption that the spectra are free of error, are that all the errors are attributable to the reference data. In practice, this is actually very close to reality, though there is always some nonzero level of spectral error. Equation (26) is referred to as ILS because, even though spectral absorbance is *dependent* on the composition of the sample matrix, it is treated as being the *independent* estimator. The ILS model is unconstrained by any assumptions related to the pure spectral profiles and is completely determined by the correlation structure of the data. As a result, the ILS model can be estimated for a single chemical constituent without specifying the concentrations of any of the other components in the matrix. Although this added flexibility is convenient, it shifts a great deal of burden toward responsible calibration dataset design. For example, depending on the calibration experimental design, the variance of some of the matrix constituents might be aliased with the component of interest. Thus, although it is assumed that the model is sensitive to gain in the target component, the model may actually be quantifying the loss of a secondary component (124). If the ideal calibration design is infeasible, there is a greater burden on the method developer to analyze the regression equation to determine the specificity of prediction.

The ILS model, also commonly known as multiple least squares regression (MLR), is estimated by solving the normal (106,107) equations for the following:

$$B = (X'X)^{-1}X'Y \tag{27}$$

For full-spectrum calibration problems, as is typical for NIR analyses, Equation (27) suffers during the covariance matrix inversion because of the multicollinearity among the spectral variables. Spectral preprocessing can reduce the level of multicollinearity by reducing the baseline correlation, but the full-spectrum MLR solution may, nevertheless, be unstable. Early efforts to mitigate the effect of collinearity involved combining MLR and one of the several methods of variable selection. More recently a number of regression techniques have been developed to address the problem of collinearity directly, including ridge regression (106,107) and latent variable methods, such as the PCR and PLS regressions.

The PCR (79,98,106,141–143) is an extension of the general linear model whereby the matrix of absorbance spectra, X, is precompressed into a basis of orthogonal factors by using the PCA. The calibration model is then estimated using the principal component scores in lieu of the absorbance spectra as the dependent ILS variable:

$$B = P[(T'T)^{-1}T'Y] \tag{28}$$

The latent variable compression adds complexity to the calibration process by increasing the number of user-selected parameters. In general, it is best to minimize the number of parameters required to optimize a method. As the number of principal components (factors) included in the model is increased, the significance (eigenvalue) of each additional factor would decrease, thereby increasing the likelihood that the model will overfit the data. There are many methodologies suggested to determine the ideal number of latent variables to be included in a predictive model. Some suggestions include adding latent variables up to the point of a sharp change in the screen plot (Figs. 31 and 32), using all latent variables prior to reaching a minimum absolute eigenvalue, retaining all latent variables that capture $>2\%$ of the variance in the Y (or X) matrix, or continuing adding latent variables until the loading vectors begin to appear unstable (noisy). The most defensible tactic, when feasible, is to use independent test sets or independent cross-validation (e.g., batch-wise cross-validation) (57) to determine the maximum number of informative latent variables to include in the model. Unless there is a high degree of nonlinearity present, the ideal number of latent variables to include in a calibration model will typically be within one component more or less than the number of varying constituents in the sample matrix. Because principal components are estimated in order of decreasing spectral variance and are not necessarily associated with the target, Y, it is sometimes advantageous to omit components with larger eigenvalues while retaining smaller eigenvalue components that are more directly correlated with Y.

The PLS regression (79,98,141–146) is quite similar to PCR in that they are both biased regression methods utilizing latent variable compression prior to estimation of a set of regression coefficients using ILS. In many nontrivial situations, PCR and PLS regression analyses yield equivalent solutions (142,143). The difference between the algorithms lies in the method of estimating the basis vector set, P. In contrast to PCR, PLS utilizes an iterative procedure, whereby an intermediate matrix of weight vectors, W, is formed in a manner similar to CLS (141), utilizing information from the dependent variable to estimate the model basis vectors. There are two principal algorithms for the estimation of the PLS regression solution. They are: (1) NIPALS (nonlinear iterative partial least-squares), and (2) the more recent SIMPLS (simple PLS). Although both algorithms will yield identical results when the target matrix, Y, is univariate, NIPALS is the preferred method when Y is multivariate (e.g., PLS-2) (98,141). In contrast to PCR basis vectors, PLS regression produces a basis vector set arranged in order of decreasing covariance between the spectral data, X, and the analytical target, Y. Hence, although PLS and PCR will often produce virtually identical results, PLS will generally arrive at the solution by estimating fewer latent variables (142). Consequently, PLS is a very powerful, user-friendly algorithm for empirical regression analysis and is considered by many practitioners to be the preferred tool for chemometric regression problems associated with NIR method development.

A number of recent publications have demonstrated the benefits of a new hybrid multivariate calibration procedure based on the NAS work of Lorber, Faber, and Kowalski (147,148). Given a linear system, the NAS of a single component, k, in a multicomponent sample matrix can be computed as the part of its spectrum orthogonal to the contribution of other coexisting constituents (147). Thus, NAS processing seeks to reduce the portion of the spectrum used for prediction to a univariate quantity. Although the NAS regression procedure follows a unique course, it has been shown that NAS regression can produce a regression solution identical to PLS or PCR (149).

The NAS regression method has been presented in the literature as a generalization of the spectroscopic calibration problem. The benefit of NAS calibration cited most often is the ability to calculate useful figures of merit (FOM), such as selectivity, SNR, sensitivity, and LOD, which may be beneficial for describing the performance limitations of an NIR method during validation. The NAS processing has also been suggested as a useful means for variable selection (150) and outlier detection (149). Goicoechea and Olivieri (151) have shown NAS to be useful for multivariate calibration using fewer training samples.

Unlike ILS regression methods, the unrestrained addition of calibration training samples with only marginal spectral variation caused by the constituent of interest should, intuitively, benefit the NAS calculation. This is true because NAS projection works to remove portions of the hyperspace spanned by the samples that are subject to spurious correlations. Such a situation is advantageous, because, for typical process analytical applications, production-scale samples are relatively inexpensive, whereas samples having sufficient range for CLS and ILS calibration methods may be expensive, difficult, or even impossible to obtain. As NIR for process analysis becomes more pervasive in pharmaceutical manufacturing, there will be an increased need for efficient methods of calibration. Methods, such as NAS and a new variant of GLS (104,114–118), known as pure-component projection (152) (PCP) seek to address this need.

Although the brief introduction to multivariate calibration and data analysis presented in this section is far from comprehensive, the section would not be complete without some mention of linearizable and inherently nonlinear methods of NIR data analysis. Calibration models for concentration-based sample qualities are expected to follow a nominally linear model, such as Beer's law. Performance characteristics, such as dissolution or disintegration, however, are likely to be inherently nonlinear. For solution of this type of data analysis problem there are entire families of algorithms that may prove useful. The simplest form is the polynomial augmentation of MLR (106,107), which has been extended to work with latent variable regression (e.g., quadratic PLS). When very large data sets are available, and if a static model equation is not required, local-linear methods, such as locally weighted regression and local-PLS (LOCAL) (153), as well as nonparametric boostrapping methods (119–121), have been shown to be very powerful for predictive discriminate analysis of NIR spectra.

The ANN approach has recently been explored for function estimation and classification in many fields including NIR spectroscopy and pharmaceutical science (153–158). An ANN prediction model consists of a collection of nodes (or neurons) and activation functions, each connected by a series of weight coefficients (Fig. 35). The ANN predictions progress, layer by layer, from the input nodes, through the layer(s) of hidden nodes, to the output nodes. At the input layer, the nodes can be raw data points or some other preprocessed form of the data. For NIR spectroscopy, the inputs are often a subset of latent variable scores in order to avoid problems of multicollinearity. At each subsequent layer, the node values are a function of the sum of the connected nodes in the previous layer multiplied by their respective weight coefficients.

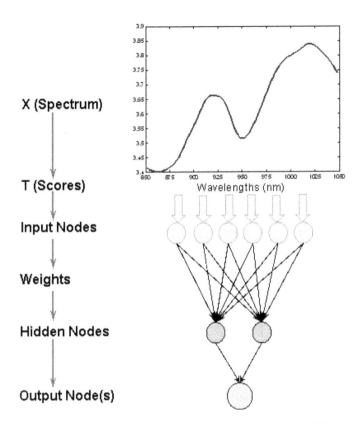

Figure 35 Graphical illustration of the form of an ANN prediction function. *Abbreviation*: ANN, artificial neural network.

The node activation function can be linear or nonlinear, and must be selected prior to the derivation of the ANN model.

An ANN function is trained by tuning its weight coefficients according to the squared error of prediction until some stopping criterion is reached. Unlike linear least squares regressions, such as PLS, which arrive at a unique, explicit solution by making certain assumptions about the distribution of the calibration data, ANN iteratively derives an implicit solution using error-gradient back-propagation (159) while making no assumptions about the distribution of the calibration data. Because the ANN derivation process can be prone to generating only locally minimal solutions, it is necessary to repeatedly train and test the model using random starting weights to create the best possible model.

It is easy to see that the ANN model complexity grows quickly as nodes are added. Due to the high number of trained weights, an ANN could "memorize" the calibration set, leading to poor performance when making predictions from new data. Various preprocessing techniques and training algorithms allow the avoidance of model overfitting the training data. Depending on the application, an enormous amount of training data (e.g., >500 spectra for NIR calibration) may be necessary to train an accurate, reliable ANN prediction model. Because of the vast amount of peripheral knowledge required to competently derive ANN models, they are generally only developed by expert practitioners. A more in-depth discussion of the ANN methodology can be found by consulting ANN literature (156,160).

The statistically based methods of nonlinear function estimation have been developed. These methods may ultimately replace ANN techniques for NIR method development in situations where complex nonlinear modeling is necessary. Known as support vector machines (161,162) (SVM), these methods utilize a so-called nonlinear kernel function to transform a nonlinear regression or classification problem into a higher-order space in which a linear solution is feasible. Indeed, with suitable constraints on the algorithm, it is possible to solve the SVM optimization using a form of ridge regression (LS-SVM) (161) or even PLS (163). Additionally, Cogdill and Dardenne (161) have shown that nonlinear SVM functions can be developed with good predictive ability using far fewer training spectra than for a comparable ANN.

Method Development

The specific details of NIR method development (164) will vary considerably depending on the analytical task and with the performance requirements of the greater system involved. The overall plan, however, is basically the same whether NIR spectroscopy is being used for basic research or process analysis. A multipart series of articles by Cogdill et al. (30,57,85) illustrates NIR method development for a process analytical application. The remaining portion of this section briefly describes the basic steps of the NIR method development: (*i*) risk analysis and feasibility studies, (*ii*) design of experiments, (*iii*) data analysis, (*iv*) validation, and (*v*) maintenance.

The selection of a particular type of instrumentation is determined by the definition of the analytical task, which is one of the results of risk analysis and feasibility studies. Once the analytical task has been defined (whether it is process analytical or for research), the choice to use NIR or some other technology should be dictated by careful consideration of the risk/benefit profile of all feasible methods. If NIR is indeed the technology to be employed, a quantitative risk and performance evaluation of the various instrument types (and vendors) should be undertaken. Some aspects to consider include speed and sensitivity requirements, instrument size and accessability, cost, and the available knowledge base. The technology feasibility study should yield a clear choice of technology, and an estimate of the scope of data that will be required for calibration or model development.

The NIR model development (either quantitative or qualitative) can either proceed through passive sample selection, where calibration samples are drawn from routine operations, or through active experimentation, where calibration samples are created according to a designed experiment, which may or may not exceed the design space (165) of the process. Although passive sample collection is typically useful for many qualitative analyses, well-controlled processes (which are the norm for pharmaceutical solid dosage manufacturing) offer relatively little matrix variation. This lack of leverage can make it difficult to interpret observations from qualitative models and can prevent the development of quantitative models with acceptable robustness and specificity when using ILS regression methods (152). Thus, some portion of the model development and validation datasets must usually be generated from non-production samples.

The specificity of an ILS regression solution is contingent on the calibration of experimental design (DOE). Inadequate designs may produce spurious or unspecific correlations that are ultimately unreliable in a predictive situation. Spiking of the nominal, or production-scale, mixture with pure API is a common practice that is often seen in published applications of NIR for solid dosage content uniformity analysis. Although this method is suitable for feasibility studies, it introduces an aliasing effect with the inactive excipients in the dosage matrix. In order to reduce the potential for such effects, NIR calibration models that are to be deployed in a process analytical environment are often developed using hundreds, even thousands, of calibration samples having significant variation balanced in all controllable factors (e.g., composition, material source, particle size, hardness, etc.). Current research is aimed at reducing the number of nonproduction samples, and the associated cost of reference testing, required for robust NIR method development (152).

Data analysis is one of the most important and engaging aspects of method development. It is during the data analysis phases that hidden processes behind sample variation begin to be understood and performance limits of the method are defined. The NIR data analysis is largely an exercise of prudent data management and optimization through simulation. Problems with NIR calibration methods that ultimately fail upon deployment will often stem from a failure to properly segregate the training, testing, and validation data to adequately preserve independence and diversity in the respective datasets. Specifically, it is absolutely critical that adequate independence is available in the calibration dataset (e.g., multiple distinct batches) to avoid overfitting of the training data. Nearly every qualitative or quantitative NIR method will involve the optimal determination of model parameters, such as the number of PLS factors to include. Truly independent test sets, or batch-wise cross-validation, should be employed to ensure that parameter selections are optimal, if not conservative. The performance limits of the NIR method can be determined by utilizing simulation (e.g., addition of synthetic noise sources and variation to the training data), or by bootstrapping (119–121). Finally, the data analysis portion of method development may ultimately reveal the need to augment the calibration data set with a more extensive DOE, select alternative instrumentation, or revise the analytical goals.

Method validation is a critical part of the development process. However, if the prior steps of method development have been completed in a rigorous manner, validation is simply a formal description of the method and its performance limitations. This augments with the testing of method performance using appropriate independent "validation" samples. Besides the validation of the model, effort is necessary to test the capability of the entire system (e.g., sample presentation, systems integration, etc.). Detailed descriptions of analytical method validation are described in the International Conference on Harmonization (ICH) Q2A and Q2B guidelines, USP monograph <1119>, and FDA guidance documents (69), as well as agricultural guidances (164). Even though the basic validation parameters (accuracy, precision, specificity, linearity, range, and robustness) are common among NIR methods and primary methods, such as UV–VIS spectrophotometry, the definition, description, and means of determining each of these parameters is somewhat unique for each application and for NIR spectroscopy as a whole. As the use of NIR as a critical process analysis tool grows, literature examples describing validation procedures will continue to define the requirements for NIR methods in diverse industrial pharmacy applications.

The final phase of NIR method development involves deploying, maintaining, and propagating the method. Even though some instrument manufacturers have made considerable improvement in the uniformity and stability of NIR analyzers, calibration models that will be utilized over long periods of time, or in numerous installations, will ultimately require transfer or updating (85). Calibration transfer describes either the propagation of the mathematical calibration model to new instrumentation (or installations), or the redeployment of a model after an instrument has been substantially modified. The goal of calibration transfer is to derive a function that changes either the analyzer response or the calibration model so that instrument output after transfer matches the method performance observed during calibration development. Numerous mathematical techniques are developed for the estimation of the calibration transfer function; a more detailed description of these methods is available in the current literature (166). A common problem for NIR calibration transfer is that standard reference materials are typically not sufficient for the development of the transfer function. This occurs because the shapes of the NIR spectra are partially the result of the geometry of the instrument relative to the scattering pattern of the sample. Thus, not only must standard reference materials have suitable absorption characteristics, it is believed that their scattering pattern must match that of the sample material to be useful for calibration transfer. Although research in this area is ongoing, the current solution is to maintain a dynamic supply of calibration standard "rescue" samples for the purpose of maintaining the calibration in the event of instrument failure (85). The composition of the rescue set, in terms of the number of samples and spectral diversity, is anticipated to be very application specific.

In some cases, the performance of an NIR method might be reduced because of a slight change in the product over time as a result of process equipment wear or due to drift in raw material characteristics. Due to the sensitivity of NIR spectroscopy, even though a drug product or material may change very little in quality (even staying well within the space of acceptable product), the performance of a calibration model may vary significantly. To mitigate the effects of such method drift, the model should be periodically updated by adding new sample data as part of a comprehensive parallel testing regime. The addition of new calibration samples will alter the distribution of data in the model, as a result of which the model performance might change. However, as suggested in ICH Q2A (167), a new validation exercise should not be required as long as the specifications for calibration update are set a priori (the same applies for calibration transfer). Hence, it is important to adequately describe calibration transfer, update, and maintenance routines as they fit into the overall analytical procedure prior to validation and deployment. If calibration update or transfer significantly alters the performance of the method or is a deviation from a validated procedure, it should be viewed as re-calibration; in these cases it is most prudent to undertake a new validation study. Indeed, knowledgeable use of chemometrics, and some method understanding, can build robustness into the model, greatly reducing the extent and frequency of effort in calibration update or transfer.

APPLICATIONS IN PHARMACEUTICAL SOLIDS ANALYSIS

The previous sections of this chapter have described the theory, equipment, and methods for NIR analysis of pharmaceutical solid materials. The objective of this section is to provide a brief summary of how NIR spectroscopy is applied in the pharmaceutical sciences, highlighting the areas that are sometimes unique relative to other fields where NIR spectroscopy is used. Although this section is devoted to covering a broad range of pharmaceutical applications and references will be provided where appropriate and available, it is not meant to be a comprehensive review of the literature. Thorough reviews are presented elsewhere (9,11,168). For the purpose of organization, the NIR applications discussed in this section will be divided into three categories: (*i*) basic materials analysis, (*ii*) in-process analysis, and (*iii*) analysis of finished products.

Basic Materials Analysis

For a number of reasons, NIR spectroscopy is well-suited for many fundamental analyses of the performance of pharmaceutical materials. Such analyses, typically performed using at- or off-line instruments, are performed during formulation and process development stages. Many applications of the NIR for basic material analysis are focused on the content and character of moisture in pharmaceutical materials. Additionally, the micrometric properties of pharmaceutical materials are important relative to product performance characteristics. The details of such analyses are already covered in the first section of this chapter.

Beyond these applications, the following paragraphs will discuss the application of NIR for the analysis of pharmaceutical polymer materials and the quantitative and qualitative determination of crystalline morphology.

Analysis of Pharmaceutical Polymers

As modern pharmaceutical drugs and therapeutic regimens continue to grow in complexity, the share of advanced solid dosage forms and delivery systems utilizing natural and synthetic polymers will increase. Polymeric materials, especially cellulose derivatives, have served an important role in the manufacture of immediate- and controlled-release tablets for many years and are nearly ubiquitous in modern solid dosage forms as either fillers or coating agents. Hydrophilic polymers, such as poly(ethylene glycol) (PEG), poly(lactic acid) (PLA), or poly(DL-lactic-co-glycolide) (PLGA), are increasingly utilized for sustained-release implant dosage forms (169) and for the delivery of proteins, peptides, or DNA in the form of nano- or microspheres. For every application of polymers in pharmaceutical drug formulation, there are associated polymer quality parameters which have a significant impact on dosage form performance or manufacturability. Furthermore, as is the case of compressed matrix tablets, the quality, content, and distribution of individual polymers can have a significant impact on drug release. For all of these reasons, there is great need for rapid, accurate and nondestructive means for measuring polymer quality and concentration in pharmaceutical raw materials and dosage forms.

As a class of materials, polymers are excellent subjects for NIR analysis. In addition to the fact that most polymer residues are rich in —CH, —NH, and —OH intramolecular bonds, the critical performance parameters of polymers, such as viscosity, yield strength, or glass transition, can be correlated to the long-range bonding structure of the material (e.g., chain length, cross-linking, crystallinity), residual monomer concentration, or the degree of substitution, which are readily detected using NIR spectroscopy. Consequently, the examples of NIR applications for polymer analysis in the chemical, petroleum, and textile industries abound. Lachenal published a review paper discussing many aspects of NIR spectroscopy for the analysis of polymers, including instrumentation, process quality control, monitoring water sorption characteristics, morphology, and crystallinity (170). Other general references on NIR spectroscopy of polymers include a chapter in the general NIR text by Siesler et al. (171) and a book on modern polymer spectroscopy by Zerbi et al. (172).

One of the first published applications of NIR spectroscopy for the direct analysis of pharmaceutical polymers, undertaken by the Lodder group in 1994, involved the determination of moisture content of intact gelatin capsules (173). Studies indicated that repeated or prolonged exposure of gelatin capsules to high humidity conditions resulted in decreased drug release upon dosing. It was noted that gelatin stored in high-humidity conditions experienced changes in conformation and cross-linking, which led to the formation of a swollen, water-insoluble gelatin (pellicle), thereby reducing the in vitro dissolution rate.

Thus, NIR spectroscopy was explored as a rapid, nondestructive means of determining the water content and, hence, the likelihood of decreased dissolution rate. Similar studies were performed much later by Berntsson et al. (174). Later studies by the Lodder group (175) utilized NIR spectroscopy to rapidly detect formaldehyde-induced cross-linking in soft gelatin capsules.

As discussed in earlier paragraphs, it is well known that the quality of polymeric excipients has an important impact on the physical and performance attributes of compressed tablets. The NIR spectroscopy has become a valuable tool for rapid identification and qualification of pharmaceutical excipients. Early studies by Gemperline et al. (129–131) concerning raw material identification and qualification demonstrated the successful application of NIR spectroscopy and chemometrics (SIMCA) for correctly identifying cellulose derivatives with varying levels of quality or impurities. Indeed, they showed that NIR spectroscopy was able to distinguish between individual batches of polymeric excipients from single and multiple suppliers. Svensson et al. (176), later demonstrated the combination of NIR spectroscopy and similar chemometric methods to classify 11 cellulose derivatives among more than 400 batches of pharmaceutical excipients with a high degree of accuracy.

Beyond simply identifying and qualifying polymer excipients based on a qualitative NIR test, Gustaffson et al. (177) recently published their work utilizing NIR and FT-IR spectroscopy and chemometrics to accurately predict tablet performance quality parameters from the spectra of raw hydroxypropyl methylcellulose (HPMC). Their work, which is a continuation of earlier work which explored the relationship between HPMC quality parameters and tablet performance, analyzed 12 grades of HPMC in terms of methoxy- and hydroxylpropyl concentrations, total degree of substitution, methoxy/hydroxypropyl ratio, Heywood's shape factor, apparent density, and specific surface area. Tablet performance parameters analyzed included radial tensile strength, axial tensile strength, specific surface area, pore radius, elastic recovery, yield pressure, and drug release after one hour. They utilized principal component analysis of the raw material and tablet parameters to identifiy correlations. In contrast to earlier studies by other researchers, Gustafsson, et al. observed that particle shape and powder surface area were more critical in predicting compaction behavior than the degree of substitution of the HPMC powders. Both NIR and FT-IR were useful for predicting excipient quality parameters and tablet performance. Although both techniques demonstrated high correlations, the predictive ability of NIR was superior to FT-IR in all cases.

Although the opportunity to measure the physical and chemical characteristics of polymers with NIR is well established, the ability to optimize a pharmaceutical manufacturing process based on the quality of incoming raw materials will require additional study. Such studies will ultimately allow pharmaceutical production processes to be controlled rapidly, and with flexibility, to variable quality of incoming raw materials via real-time process analytical measurements and feed-forward controls. Alternatively, such functional modeling might be used to set evidence-based limits on the quality of incoming raw materials, thereby replacing more arbitrary qualitative methods, which rely on pattern recognition algorithms, such as SIMCA. The majority of published process analytical applications of NIR spectroscopy involve the adaptation of NIR spectroscopy to existing processes. However, by deploying NIR spectroscopy for exploratory studies much earlier during formulation and process development, the establishment of a functional product or process design space [as described in ICH Q8 (165)] might be significantly augmented by increasing both the speed and value of data collection.

Tablet coating analysis is yet another polymer analysis application for which NIR spectroscopy is particularly well-suited. Coating thickness and uniformity impact dosage form performance, especially in terms of drug release. A direct correlation can be observed between the thickness of the coating layer and the NIR signal (Fig. 36). Precedence suggests, however, that analysis below the coating surface is possible through judicious application of chemometrics (e.g., content analysis of coated tablets, quantification of multiple coating layers). Kirsch and Drennen (55) demonstrated the effectiveness of NIR spectroscopy for the prediction of coating thickness, dissolution rate ($t_{50\%}$), and core hardness of theophylline tablets coated with ethylcellulose. Their results proved that NIR spectroscopy can be used to quantify tablet quality parameters through a polymer coating. Recently, Pérez-Ramos et al. (178), building upon the early work of Kirsch and Drennen, utilized NIR spectroscopy to estimate the rate constants of a semiempirical model for tablet

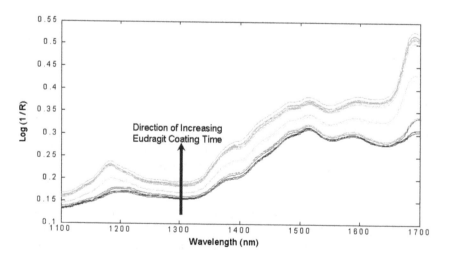

Figure 36 NIR spectra of coated solid dosage tablets. Tablets were coated for four different times: 0, 20, 40, and 44 minutes. The effect of increased coating time can be observed in the increasing baseline offset, and the sharp features near 1180 and 1680 nm. *Abbreviation*: NIR, near-infrared.

coating thickness. In developing a calibration model for the prediction of coating thickness, they observed that the calibration regression vector utilized both the positive correlation with the increase in coating thickness, and the negative correlation with the decreasing intensity of absorbance bands related to tablet constituents.

Functional polymer coatings, such as controlled-release films, multilayer films, and drug-laden polymer films are becoming more common means for dosage form design. The spatial uniformity of functional coatings is a critical quality parameter for process development or control. The NIR imaging spectroscopy is a powerful means for rapidly assessing coating thickness and uniformity during process development. By correlating NIR absorbance with coating time, a sensitive indicator of coating thickness can be estimated without complicated reference testing. By projecting NIR image spectra onto the regression vector, a pseudocolor map of coating uniformity can be created for visual interpretation of coating uniformity (Fig. 37). Indeed, a similar qualitative prediction vector might be estimated by subtracting the spectrum of the uncoated tablet core from the pure coating spectrum.

Arguably, the future of drug delivery technology will involve sustained-release or programmable-release implants and dosage forms, as well as advanced methods for delivering protein and peptide drugs parenterally and, eventually, orally. The performance of these systems will rely heavily on the drug diffusion characteristics of polymer matrices, membranes, or hydrogels. For example, the uniformity of the programmable pulsatile release profiles of modular Accudep^{TM}

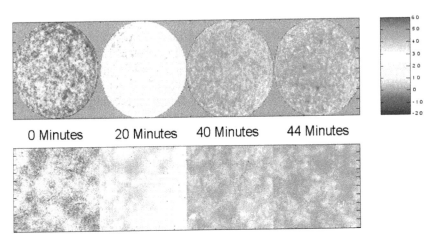

0 Minutes   20 Minutes   40 Minutes   44 Minutes

Figure 37 Sample images of coated tablets at $2\times$ (*upper image*) and $10\times$ (*lower image*) magnification. The coating thickness calibration was developed by regressing single-point NIR spectra against the total time of coating. Tablets were precoated with Surelease before being coated with Eudragit for 0, 20, 40, and 44 minutes. The images clearly show the coating thickness variability for early time points. *Abbreviation*: NIR, near-infrared.

dosage forms (179,180) relies on reliable characterization of the diffusivity of polymer membranes. It is easy to imagine how NIR imaging spectroscopy could be used to qualify the uniformity of the physical characteristics of the film sheets used to create the diffusive films or to investigate the physicochemical properties related to film performance. For polymer implants, the rate of drug release can be controlled by dispersing inert, water-soluble excipients throughout the polymer matrix. The dissolution of these excipients in vivo creates a pore structure for drug diffusion and release. Hence, control of the concentration and distribution of these excipients is an important quality control parameter in the manufacture of such dosage forms. Brashear et al. (169), demonstrated the use of NIR spectroscopy to quantify the content of PEG and loading of Lomeflox-acin in poly(ε-caprolactone) (PCL) matrix implants. They utilized ratios of second derivative functions at two wavelengths, as described by Norris and Williams (8), to develop calibration models with very high correlation for both constituents.

Polymer hydrogels are increasingly being considered for sustained release of injectable biotechnological therapeutics, scaffolds for release of growth factors in tissue engineering, and oral delivery of gastric labile protein or peptide thera-peutics. Continual concerns with these systems include the quantification of drug loading and characterization of drug release potential via diffusion. Early work in this area by Nerella and Drennen (158) utilized dispersive NIR spectroscopy with a novel controlled aperture to develop a method for depth-resolved determination of drug content in a polymer hydrogel. They utilized the NIR method to solve the Fickian diffusion constants of salicylic acid through a hydrogel matrix, which served as a practical in vitro model for a transdermal drug delivery system. In much later work, Blanco and Romero (181) demonstrated the use of NIR trans-flectance spectroscopy for the determination of dexketoprofen in a hydrogel. They noted that because hydrogels have poor flow characteristics and a highly aqueous structure, they are difficult to characterize using either transmittance or reflectance methods; however, transflectance, via a cuvette modified with a gold-plated reflector allowed for repeatable sample presentation and for the development of a robust, quantitative calibration model. As multicomponent hydrogels see more routine use in critical drug delivery and tissue engineering applications, NIR spectroscopy may enable in situ feedback control of drug loading operations, perhaps to compensate for variability in hydrogel structure, thereby stabilizing the delivery characteristics of the system in vivo.

Crystalline Morphology

The NIR spectroscopy has been applied extensively for detection and quantifi-cation of changes in crystalline forms of numerous pharmaceutical solids. Although variation in polymorphic form might not be expected to have a sig-nificant NIR signature, because there is no change in composition, per se, the realignment of crystal lattices associated with unique polymorphs does often cause a noticeable, although sometimes subtle, variation in the NIR spectra.

This may often be due to variations in intermolecular hydrogen bonding. The analysis of crystalline hydrates will not be discussed at length in this section because they are covered in the fundamentals section related to the spectral characteristics of water.

The first application of NIR spectroscopy for the analysis of crystalline morphology was published by Gimet and Luong in 1987 (182). Their work was concerned with the quantitative determination of two polymorphic forms (α and β) of an unnamed drug compound in a multicomponent formulation matrix. The more stable form, β, exists as fine acicular crystals (2–50 μm). Certain manufacturing operations, however, cause enantiotropic conversion to form α, which exists as larger rectangular plates (60–230 μm); thus, a rapid method of accurately determining the concentration of the two forms in a drug product matrix was needed. The baseline shapes of pure-component spectra of the two forms were basically the same; however, the spectrum of form α had many more sharp spectral features. Although the authors chose not to venture a qualitative explanation for the differences in absorption between the two forms, they did develop a simple quantitative calibration model for the prediction of the content of both forms in a drug matrix.

Buchanan et al. (183) published their work investigating NIR spectroscopy for the determination of enantiomeric purity of valine in 1988. The determination of optical isomers is typically performed in solution using circular dichroism or optical rotatory dispersion spectroscopy. The authors noted that an DR NIR spectroscopy of solids is inherently insensitive to optical rotation. (DR efficiently randomizes the state of polarization.) However, absorbance differences between the *d* or *l* and the racemic *dl* forms of valine arise due to differences in the orientation (packing) of groups in the crystalline matrix which affect hydrogen bonding.

Because the *d* and *l* forms of valine are structurally indistinguishable, a calibration model was developed to predict the ratio of the two forms according to the following equation:

$$R = \frac{x}{1-x}$$

where x is the mole fraction of the minor component. Thus, for a racemic mixture R would be equal to one, and for pure *d* or *l*, R would be equal to zero. The fact that the major and minor components cannot be determined spectroscopically is of no consequence, because it would typically be known a priori. A series of 14 physical mixtures of the *d* and *l* isomers, which spanned the range of R values from 0 to 0.99, were prepared and analyzed by NIR reflectance. The calibration model, which was optimized using wavelength selection and MLR, used three wavelengths for the prediction of R with correlation greater than 0.99. A spectral reconstruction algorithm was used to interpret the calibration model by comparing the relative shape of band intensities. The first two wavelengths, 1562 and 1576 nm, were not located in conjunction with an observed absorbance peak and were suggested to function as baseline correction factors in the calibration model. The third wavelength (1660 nm) was selected just off of an inflection point of the first overtone C—H stretch absorbance band. The authors suggested that because the major regression coefficient was located just near an absorbance intensity maximum, the spectral phenomenon being detected was likely a band shift. However, no further investigation of the spectral signature was discussed.

In 1989, Miller and Honigs (184) published what was, at that time, the most detailed analysis of changes in NIR spectra due to differences in crystalline form. They presented a detailed analysis of the spectral differences between the α and γ crystalline forms of glycine. Rather than developing a calibration model for predicting the content of one or the other forms, Miller and Honigs focused on comparing the relative spectral shapes with expected bands based on mathematical combinations of the fundamental frequencies observed for the two forms. Their results supported the earlier works, which suggested that the basis for NIR discrimination of polymorphic forms was related to differences in hydrogen bonding. Specifically, the authors noted intensity changes in O—H and N—H absorbance bands and shifts of 4–5 nm in the location of C—H vibrations. Their results correlated well with earlier observations using Raman spectroscopy and with band locations predicted using mathematical combinations of the fundamental absorption bands.

Three studies of NIR spectroscopy for the quantification of crystalline form for a variety of pharmaceutical substances were published in 2000 by the Luner group of the University of Iowa in conjunction with Foss NIR Systems. Binary mixtures of two forms of six different compounds were analyzed to create a series of calibration models. They are as follows: Form-I and Form-III sulfathiazole (185,186), Form-I and Form-II sulfamethoxazole (185), ampicillin anhydrate and trihydrate (185), β-lactose and α-lactose monohydrate (185), amorphous and crystalline indomethacin (187), and amorphous and crystalline sucrose (187). As described in the three papers, DSC, TGA, and X-ray powder diffraction (XRPD) were used to verify the different crystalline forms. Although some qualitative interpretations of the spectroscopic results were given, the overall objective of the experiments was to determine the suitability of NIR spectroscopy for quantitative prediction of the content of particular crystalline forms. Good predictive performance was observed in all cases, with concentration limit of quantification (LOQ) as low as 0.3% (186). These results should not be surprising, though, given that the calibrations and predictions were generated using simple binary mixtures. Other work by the Luner group published in 2000 demonstrated the ability of the NIR to predict the relative concentration of Forms-I and II sulfamethoxazole in mixtures diluted with up to 90% of a pure absorbing excipient; LOQ was observed to increase marginally to 1–2% (188). Similar experiments using binary physical mixtures of Forms-I and III carbamazepine were performed (with similar results achieved) in more recent work by Braga and Poppi (189).

Otsuka et al. (190) published the first assessment of NIR spectroscopy for the determination of crystalline form of drug compounds in finished dosage forms in 2003. The objective of their work was to develop an NIR calibration model to predict the relative concentrations of α- and γ-indomethacin in tablets. The researchers developed a series of calibration samples by compressing (at constant pressure) mixtures consisting of a 50:50-w:w combination of a multicomponent excipient matrix and indomethacin. The composition of the indomethacin in the mixture ranged from 0–100% γ-indomethacin. The performance of the calibration was evaluated by analyzing a series of samples with unknown relative levels of α and γ indomethacin obtained by a metastable transformation process; the reference test in all cases was XRPD.

In analyzing the calibration samples, Otsuka et al. attributed most of the spectral features used to discriminate between the two forms to C—H stretch vibrational absorbance bands. The researchers observed no loss in accuracy or precision in the prediction of polymorph content in tablets compared with powder mixtures. Furthermore, good correlation was observed between the NIR and XRPD predicted values for the unknown samples, though NIR demonstrated lower standard deviation of repeatability. Thus, the authors suggested that NIR should be useful (if not superior to XRPD) for the analysis of polymorphic purity in finished pharmaceutical products.

Process Development and Scale-Up

The majority of the NIR applications referenced throughout this chapter have been concerned with demonstrating the sensitivity of NIR to, or the development of an NIR calibration for, certain drug product components or quality characteristics. Without detracting from the significance of these efforts, it should be noted that the development of NIR methods and calibration models are small steps toward a larger goal of utilizing NIR spectroscopy along with other tools to generate a greater understanding and control of pharmaceutical manufacturing processes. Thus, as the application of NIR spectroscopy for the analysis of pharmaceutical solids continues to mature, NIR sensors will increasingly be utilized as a basis for mechanistic and control models of the processes being monitored. For example, Davis et al. recently demonstrated the use of real-time NIR process spectroscopy in developing a time-based model of polymorphic phase transformation during the granulation of γ-glycine (191).

The advantage of this sort of methodology is that, because NIR analyses can be performed independent of the scale of the operation, process controls can be created using NIR spectroscopy during the process development phase and applied during scale-up to production levels. Furthermore, the complex patterns in the time profiles of predicted values during in-process analyses will be useful in determining the kinetics of certain operations. Methods and examples of NIR spectroscopy for in-process analysis will be discussed in the following paragraphs.

In-Process NIR Analysis

Among analytical technologies, NIR spectroscopy is unique in its combination of sensitivity, robustness, flexibility, and economy. Thus, it is not surprising that the number of NIR process applications increased rapidly long before the PAT guidance was drafted. Process analytical installations of NIR spectroscopic instrumentation are expected to grow rapidly. Many industry observers have noted that PAT is not about sensors, rather, it relates to continuous management of quality through the reduction of variability. Sensors, such as NIR, however, are necessary to gather the relevant quality data in a timely fashion and, hence, are integral to PAT. On-line NIR instrumentation, such as diode array spectrometers, typically gather spectroscopic data at rates sufficient to continuously characterize both the instantaneous mean and variance of many process operations. By analyzing the mean and variance in quality parameters for a particular process over time, estimates of process risk and capability [e.g., CpK (192)] can be monitored for trends. Additionally, because NIR spectroscopy offers multivariate detection, the capability of multiple individual aspects of operations can be assessed simultaneously. In addition to quality monitoring, the most compelling applications of in-process NIR spectroscopy will be to enable the sort of feed-forward control anticipated for future batch and continuous manufacturing operations. Without real-time product quality assessment at each critical-to-quality (CTQ) operation, there is little information to be fed forward through process transfer functions (192).

Although in-process NIR spectroscopy is applied in fundamentally the same manner as for basic materials analyses, there are many unique aspects of process-spectroscopy worth noting. The most significant differences are related to the type of instrumentation used and the characteristics of the data analysis methods. In-process NIR analyses utilize the types of on- or in-line instrumentation described in earlier sections. Furthermore, in-process calibration models are often optimized for robustness, efficiency, and simplicity, sometimes at the expense of sensitivity. Calibration models to be deployed for in-process testing may be qualitative or quantitative and should be developed in a way that accounts for a sufficient portion of the expected process variability (ref. PAT II). The primary characteristic by which in-process NIR applications can be classified is the method of sample interaction. The basic means of sampling pharmaceutical operations by in-line NIR spectroscopy can be classified as either interactive or remote sensing.

Interactive Sampling Systems

For our purposes, interactive sampling refers to the means and methods used to acquire spectra where some part of the instrument is in contact with the sample material. Interactive transmittance measurements of in-process pharmaceutical solids are less common relative to other methods of interactive sampling, due largely to the fact that it can be difficult to reliably transport solids through a transmittance cell in a repeatable manner. Though absorptivity in transmittance is sometimes more linear than DR over a wide range of concentrations (unpublished observations), the increased complexity of sample handling erodes most tangible benefits. Thus, in-process analysis of pharmaceutical solids using interactive NIR transmittance will not be further discussed herein.

Interactive DR, on the other hand, is a very common technique for in-process NIR analysis. For interactive DR, the in-process sample material is typically in contact with some sort of window or lens. For example, fluidized bed processes (e.g., granulation or drying) are monitored using NIR spectroscopy by inserting a quartz or sapphire window in the wall of, or protruding into, the processor at the static bed height. As the materials inside the processor tumble in the airstream, a continuous fraction of the solids will be in dynamic contact with the window, which allows a reflectance spectrum to be acquired by an NIR spectrometer on the outside of the processor. Rantanen et al., have published the details of many experiments demonstrating through-wall NIR spectroscopy for monitoring and control of fluidized bed drying and granulation (21,25,42,193–200). More recent NIR blend monitoring systems, based on RF-coupled wireless diode arrays, use a fixed window for both illumination and data collection. As the blender rotates, the spectra are acquired during the periods when powder is flowing over the window (88).

One of the advantages of interactive DR (relative to remote sensing, discussed later) is that the sampling geometry is stable because the distance between the spectrometer and sample is controlled by the window. There will be, however, a certain amount of unavoidable spectral variability related to the packing density of the material pressing against the window. There are also a number of potential drawbacks to NIR measurement through a window into the process. An obvious consideration is the path of illumination; for instruments where the lamp is separated from the spectrometer, it can be difficult to align the illumination source and instrument detector through the same window. For instruments using a fiber-optic probe facing the window, reflection from the window surface can dramatically reduce the dynamic range of the measurement unless an antireflective (AR) coating is applied. Furthermore, variation in the illumination and detector geometry relative to the window can have complicated effects on the spectral baseline (Fig. 38). Thus, if the through-window NIR sampling system is not rigidly attached to the processor, some instrument positioning variation will be present in the sample spectra.

Another difficulty encountered during interactive DR (and other interactive-type sampling methods) is the interaction of the sampling window with the in-process material. For example, quartz glass windows or AR coatings can be degraded by abrasive, reactive, or high-temperature flowing solids. With respect to NIR reflectance analysis of typical pharmaceutical materials, a more significant problem is the buildup of a layer of sample materials on, or in front of the window, also known as window *fouling*. One manufacturer of filter-based NIR sensors for process analysis has developed a system whereby an

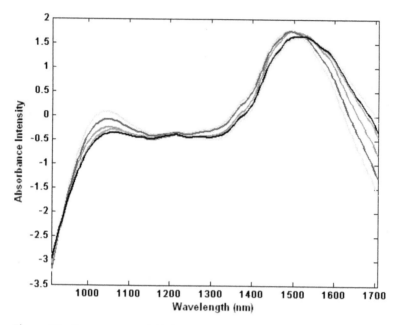

Figure 38   Scatter-corrected (SNV) NIR spectra of a mixture of acetaminophen, lactose, and MCC in a laboratory scale blender. The spectra were acquired using a fiber-optic probe pressed against an antireflective coated window in the side of the blender. Each spectrum is the mean of 50 spectra gathered from the same blend after repositioning the probe. As shown, there is variability in the spectral baseline for the wavelength region beyond 1500 nm. Although the observed effect may be related to the composition balance between lactose and MCC, the effect has been observed to be repeatable. Supposing the spectral variation to be the result of probe alignment with the coated window illustrates the importance of considering all sources of error when developing NIR methods. (The apparent variation near 1020 nm is probably the result of SNV preprocessing because, as the baseline is reduced beyond 1500 nm, the baseline must increase in other areas of the spectrum to compensate.) *Abbreviations*: SNV, standard normal variate; NIR, near-infrared; MCC, microcrystalline cellulose.

automated wiping mechanism periodically clears the window of residue (201). Although the ability to automatically clean in place (CIP) is attractive, there must still be a determination of the ideal cleaning frequency and some means of dealing with the level of residue between cleanings. Hence, it would be a better to devise a means of avoiding the buildup of residue altogether. Rantanen et al. (195) prevented fouling during fluidized-bed wet granulation by continuously heating the optical window to a temperature greater than that of the in-process material. As other researchers have corroborated (unpublished correspondence), heating NIR windows or probes beyond the temperature of the process is an effective means of reducing fouling in the presence of wet materials.

Figure 39 Spray ring for use with an NIR process window. The fluid spray ports located around the periphery of the inner surface are designed to continuously wash the optical window with an inert gas or solvent. *Abbreviation*: NIR, near-infrared.

Another alternative would be to continuously wash the surface of the window with a layer of air or inert gas. For example, J.M. Canty Process Technology (202) has developed an integrated spray ring system for use with their fused silica process windows. By spraying a constant volume of air through a series of ports around the periphery of the window bezel (Fig. 39), wet or dry materials can be deflected from the window.

For obvious reasons, fiber-optic probes are one of the most common means for interactive sampling and are an advantage that sets NIR (and Raman) spectroscopy apart from other analytical technologies. Fiber-optic probes significantly reduce the effort required for sampling materials while increasing the repeatability of some measurements. Fiber probes consist of a set, or bundle, of fibers to deliver illumination from a lamp source to the sample and one or more fibers to collect and transmit the light reflected from the sample. Transflectance probes, used for NIR monitoring of liquid operations, such as reaction monitoring or fermentation, have a nonabsorbing reflector positioned normal to the tip of the probe, separated by a fixed pathlength. Ball lenses or other collimating optics can be used to increase the solid angle of light captured by the fiber probe. For example, if only the bare end of a single fiber is used to collect reflected light, the effective sampling volume will be very small (the acceptance angle of bare fibers is approximately 40–55°). Some fiber-probe manufacturers have claimed that effective sample volume can be increased and that the detection of residue on a process window can be avoided, by using lenses that set the "focal point" of analysis at a depth further beyond the surface of the process window (unpublished communications). Such claims have yet to be substantiated in the literature, however, and seem to be in conflict with the accepted concepts of NIR absorption.

Some probes utilize a single collection fiber, whereas others use a bundle of multiple fibers. Besides their reduced cost, it has been supposed that single fiber probes are advantageous because they have more uniform delivery of light to the diffraction grating (because it is difficult to align the beams emanating from multiple fibers). Whether or not this is true is debatable; on the other hand, single-fiber probes will have lower efficiency and will be more likely to experience complete failure during operation if the collection fiber is damaged. Besides the number and configuration of the optical fibers, the quality of fibers and connections are important in determining the performance of the system (203). The NIR process spectrometers are generally paired with low-OH optical fibers, which have better transmittance characteristics than telecommunication-grade fibers in the NIR region (Fig. 40). The attenuation of transmitted light through optical fibers is mainly related to absorption by the glass and losses to connections and transmission through fiber walls. Thus, the spectral range of fiber-coupled process NIR spectrometers is typically limited to less than 2200 nm.

The NIR fiber-optic probes are typically applied in reflectance or interactance modes (or some combination of the two). For example, NIR fiber probes are often employed using optical interactance for raw material qualification and identification (204), whereby the probe tip is pressed against the wall of a container (e.g. polyethylene bags) (177). In the interactance mode, the light collected by the detector has either been reflected from beneath the surface of the

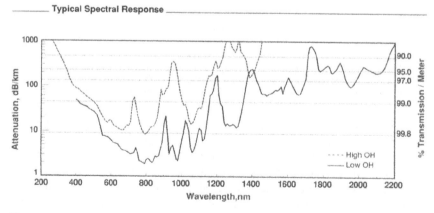

Figure 40   Typical spectral transmittance of standard and low-OH optical fibers in the NIR region. *Abbreviation*: NIR, near-infrared. *Source*: Courtesy of Romack Fiber Optics Inc.

sample or has been transmitted laterally from the point of incidence of the source illumination. Fiber probes can also be inserted directly into a dry powder bed, though care must be taken to avoid fouling of the probe tip. Many of the early studies using NIR for blend uniformity testing were based on the premise of replacing a sample thief probe with a spectroscopic fiber probe (205–207). When used for monitoring chaotic operations, such as particle coating or granulation, fiber probes can be inserted directly into the material stream to gather spectra in the reflectance mode (208); again, probe fouling and contamination is a concern. Furthermore, it can be difficult to deliver sufficient illumination for reliable reflectance measurements unless the probe is in close interaction with the sample (i.e., remote sensing measurements are limited with fiber probes).

Remote Sensing

The capability for noncontact, or remote-sensing, sampling is an advantage unique mainly to NIR and Raman spectroscopy. For our purposes (in contrast to military and agricultural definitions), remote sensing refers to methods of spectroscopic sampling that do not require any contact or interaction between sample materials and sensing components. Remote sensing is more convenient and is less disruptive to the sample material because there is no window or probe interaction. Because remote sensing spectrometers are typically positioned at some distance away from in-process sample materials, ambient light contamination and sample misalignment can add significant noise factors to the spectral data.

Remote sensing NIR measurement can be accomplished in both DR and transmittance modes. As mentioned earlier, transmittance measurements of pharmaceutical solids are becoming relatively less common because of difficulties in sample alignment and limitations on the spectral range. Remote NIR transmittance might be an ideal means for sampling certain manufacturing operations where the materials are uniformly thin and presented in a manner amenable to transmittance, such as thin-film or web processing of diffusion membranes and transdermal dosage forms or granulation via roller compaction of powdered solids. Remote NIR transmittance has been used for rapid analysis of solid dosage tablets by using a two-piece conveyor belt with a channel down the center to hold tablets and allow transmitted NIR radiation to pass. The NIR transmittance analysis of solid dosage forms was initially suspected to be more robust, because, as was theorized, a greater fraction of the tablet mass is sampled. Recent research comparing transmittance and reflectance for in-process tablet release testing via NIR spectroscopy demonstrated, however, that the effect of sample alignment on tablet spectra is more pronounced for transmittance measurements (30).

Remote NIR reflectance is a practical means for in-process spectral data acquisition. Depending on the geometry of the illumination and detector optics (relative to the sample stream), the effect of sample alignment on NIR spectra will be limited and easily mitigated by spectral preprocessing methods. For example, scatter correction and spectral derivatives have been found to be useful for mitigating sample alignment effects during NIR reflectance analysis of tablets for real-time release (RTR) testing (30,57). Remote reflectance sampling has been found to be especially advantageous for measuring the in-process quality of wet or sticky products during wet granulation (20,21,191,193,194,200) and pan coating (178). By positioning the NIR sampling port in an appropriate location of the processing chamber, spectra can be acquired real-time with no concern for sample window fouling. One disadvantage, however, is that the access to all or part of the NIR instrumentation for maintenance might be restricted if it is positioned inside a containment barrier. For certain operations where the in-process material is immobile, such as tray drying, remote reflectance analysis can be carried out through a process window, with no contact between the sample material and the optical window (209).

Remote sensing NIR reflectance spectrometers have been positioned above conveyor belts for continuous moisture and quality measurement in agricultural and other industries for decades. Even so, continuous conveying operations are relatively scarce in pharmaceutical solid dosage production. As pharmaceutical manufacturing begins to transit from exclusive batch production to a combination of batch and continuous operations, however, remote sensing of conveyed materials will be critical for control of operations.

Analysis of Finished Products

The analysis of intact, finished solid dosage forms was, at one time, considered to be a very important goal for NIR analysis of pharmaceutical solids. The PAT initiative has caused some debate regarding the value of developing methods for finished dosage form analysis. The argument is that, rather than simply using NIR (and other PAT methods) as a new tool for inspection, process analytics should be deployed further upstream in the processing chain as a tool for process control. Although this is a valid point, the importance of rapid finished product analysis in accelerating product release should not be diminished. Furthermore, in order to develop and verify the many interacting transfer functions that will ultimately be required for feed-forward control of processes, real-time, comprehensive measurement of product quality must be in place as feedback. Thus, RTR methods for testing finished dosage form quality parameters, such as identification, potency, content uniformity, and hardness, continue to be explored (30,57,85).

The NIR spectroscopy is also demonstrated as a tool for predicting dosage form performance (e.g., bioavailability) in terms of dissolution and disintegration time (55,210–212). Although this is an intriguing and worthwhile concept, there has been too little discussion regarding the specificity of the dissolution prediction. In general, published examples of NIR calibrations for the prediction of dissolution or disintegration are developed by varying specific formulation factors, such as hardness, composition, or raw material quality. In other words, although it is obvious that the researchers have demonstrated an ability to correlate dosage form performance to composition, it has not been rigorously demonstrated that the NIR signal reveals information related to dissolution not related directly to (more tangible) composition factors. Ultimately, as there is currently some debate regarding the overall value of dissolution release testing for certain dosage forms, it remains to be seen if NIR spectroscopy (or any other nondestructive test) will be uniquely useful in predicting dosage form performance in truly relevant terms (e.g., bioavailability and permeability).

Beyond the development lab and manufacturing floor, the combination of chemometrics and NIR spectroscopy is increasingly being considered a high-tech tool for postproduction analytical tasks. Thousands of solid dosage forms are gathered from the marketplace for analysis every year, either through product recall, regulatory investigation, or surveillance. The NIR spectroscopy and imaging can be valuable tools for nondestructively assessing the quality of individual dosage forms before more complicated or destructive tests are performed. One of the earliest applications of NIR spectroscopy of intact dosage forms, published in 1987, was to detect adulteration and tampering of solid gelatin capsules. Lodder et al. (213), showed that NIR spectroscopy and bootstrapping pattern recognition could detect various adulterants with concentrations down to 0.4% of capsule weight. In a later example, Bauer et al. (214), utilized NIR spectroscopy of raw materials to investigate dissolution failures for erythromycin dihydrate tablet formulation. The spectral data indicated that a dehydrated dihydrate of erythromycin is produced during formulation, which gradually binds with $Mg(OH)_2$, delaying the process of dissolution. Furthermore, the researchers were able to successfully predict that humidifying the tablets would reverse the unfavorable binding and therefore increase the dissolution rate.

Finally, as the economic value of the pharmaceutical drug market continues to expand, and as more consumers purchase their personal drug supply through international trade via the internet, the counterfeit drug trade will increasingly grow in sophistication. In response, NIR spectroscopy is currently being explored as a high-speed method of validating drug product identification and integrity throughout the supply chain, whether as part of an automated dispensing system in a high-tech pharmacy or carried as a mobile investigation tool by regulators. Indeed, the future of NIR technology in pharmaceutical science is bright.

ACKNOWLEDGMENTS

The authors would like to thank Hua Ma, Zhenqi (Pete) Shiz, and Steve Short for their efforts in locating and organizing current references and Ms. Ryanne Forcht for her skillful scientific and grammatical editing.

REFERENCES

1. Herschel SWF. Phil Trans Royal Soc 1800; 90:225.
2. Burns DA, Ciurczak EW, eds. Handbook of Near-Infrared Analysis. 1st ed. New York, NY: Marcel Dekker, Inc., 1992. Practical Spectroscopy; No. 13.
3. Burns DA, Ciurczak EW, eds. Handbook of Near-Infrared Analysis. 2nd ed. New York, NY: Marcel Dekker, Inc., 2001. Practical Spectroscopy; No. 27.
4. Williams P, Norris K, eds. Near-Infrared Technology in the Agricultural and Food Industries. 2nd ed. St. Paul, MN: American Association of Cereal Chemists, Inc., 2001.
5. Williams PC, Norris KH. Near-Infrared Technology in the Agricultural and Food Industries. 1st ed. St. Paul: American Assoc. Cereal Chemists, 1987.
6. Ciurczak EW, Drennen JK. Pharmaceutical and Medical Applications of Near-Infrared Spectroscopy. 1st ed. New York, NY: Marcel Dekker, 2002.
7. Hindle PH. Historical development. In: Burns DA, Ciurczak EW, eds. Handbook of Near-Infrared Analysis. 2nd ed. New York, NY: Marcel Dekker, Inc., 2001:1–6.
8. Norris KH, Williams PC. Cereal Chem 1984; 61(2):158–165.
9. Kirsch JD, Drennen JK. Appl Spectrosco Rev 1995; 30(3):139–174.
10. Workman JJJ. Appl Spectrosco Rev 1996; 31(3):251–320.
11. Ciurczak EW. Appl Spectrosco Rev 1987; 23(1&2):147–163.
12. Journal of Near Infrared Spectroscopy.
13. Miller CE. Chemical principles of near-infrared technology. In: Williams P, Norris K, eds. Near-Infrared Technology in the Agricultural and Food Industries. 2nd ed. St. Paul, MN: American Association of Cereal Chemists, Inc., 2001:19–38.
14. Murray I, Williams PC. Chemical principles of near-infrared technology. In: Williams PC, Norris KH, eds. Near-Infrared Technology in the Agricultural and Food Industries. 1st ed. St. Paul: American Assoc. Cereal Chemists, 1987:17–34.
15. Ingle JDJ, Crouch SR. Spectrochemical Analysis. Upper Saddle River, NJ: Prentice Hall, 1988.
16. Colthup NG, Daly LH, Wiberly SE. Introduction to Infrared and Raman Spectroscopy. 3rd ed. New York: Academic Press, 1990.
17. Hirschfeld T, Zeev-Hed A. The Atlas of Near-Infrared Spectra. Philadelphia, PA: Sadtler Research Laboratories, 1981.
18. Hollas JM. Modern Spectroscopy. Chichester, UK: John Wiley & Sons, 1987.
19. Bonanno AS, Olinger JM, Griffiths PR. The origin of band positions and widths in near-infrared spectroscopy. In: Hildrum KI, Isaksson T, Næs T, Tandberg A, eds. Near-Infrared Spectroscopy, Bridging the Gap Between Data Analysis and NIR Applications. New York, NY: Ellis-Horwood, 1992:19–28.
20. Jorgensen A, Rantanen J, Karjalainen M, Khriachtchev L, Räsänen E, Yliruusi J. Pharm Res 2002; 19(9):1285–1291.
21. Räsänen E, Rantanen J, Jorgensen A, Karjalainen M, Paakkari T, Yliruusi J. J Pharm Sci 2001; 90(3):389–396.
22. Hogan SE, Buckton G. Int J Pharm 2001; 227:57–69.
23. Zhou GX, Ge Z, Dorwart J, et al. J Pharm Sci 2003; 92(5):1058–1065.
24. Buckton G, Yonemochi E, Hammond J, Moffat A. Int J Pharm 1998; 168:231–241.
25. Räsänen E, Rantanen J, Mannermaa J-P, Yliruusi J, Vuorela H. J Pharm Sci 2003; 92(10):2074–2081.
26. Maeda H, Ozaki Y. J Near IR Spectrosco 1995; 3:191–201.

27. Haaland DM, Melgaard DK. Appl Spectrosco 2000; 54(9):1303–1312.
28. Birth GS, Hecht HG. The Physics of near-infrared reflectance. In: Williams PC, Norris KH, eds. Near-Infrared Technology in the Agricultural and Food Industries. 1st ed. St. Paul: American Association Cereal Chemists, 1987:1–15.
29. Dahm DJ, Dahm KD. The Physics of near-infrared scattering. In: Williams P, Norris K, eds. Near-Infrared Technology in the Agricultural and Food Industries. 2nd ed. St. Paul, MN: American Association of Cereal Chemists, Inc., 2001:1–18.
30. Cogdill RP, Anderson CA, Delgado-Lopez M, et al. AAPS Pharm Sci Tech 2005; In Press.
31. Ciurczak EW. Principles of near-infrared spectroscopy. In: Burns DA, Ciurczak EW, eds. Handbook of Near-Infrared Analysis. 2nd ed. New York, NY: Marcel Dekker, Inc., 2001:7–18.
32. Ciurczak EW. Principles of near-infrared spectroscopy. In: Burns DA, Ciurczak EW, eds. Handbook of Near-Infrared Analysis. New York, NY: Marcel Dekker, Inc., 1992:7–12.
33. Olinger JM, Griffiths PR. Theory of diffuse reflectance in the NIR region. In: Burns DA, Ciurczak EW, eds. Handbook of Near-Infrared Analysis. New York, NY: Marcel Dekker, Inc., 1992:13–35.
34. Olinger JM, Griffiths PR, Burger T. Theory of diffuse reflection in the NIR region. In: Burns DA, Ciurczak EW, eds. Handbook of Near-Infrared Analysis. New York, NY: Marcel Dekker, Inc., 2001:19–51.
35. Kubelka P, Munk F. Z Tech Phys 1931; 12:593–601.
36. Chandrasekhar S. Radiative Transfer. Oxford: Clarendon Press (reprinted by Dover Publications, 1960), 1950.
37. Dahm DJ, Dahm KD. Appl Spectrosco 1999; 53:647–654.
38. Pasikatan MC, Steele JL, Spillman CK, Haque E. J Near IR Spectrosco 2001; 9: 153–164.
39. Bull CR. J Mod Optics 1990; 37(12):1955–1964.
40. Bull CR. Analyst 1991; 116:781–786.
41. Ciurczak EW. Spectroscopy 1986; 1(7):36.
42. Rantanen J, Yliruusi J. Pharm Pharmacol Commun 1998; 4:73–75.
43. Frake P, Luscombe CN, Rudd DR, Waterhouse J, Jayasooriya A. Anal Commun 1998; 35:133–134.
44. Otsuka M. Powder Technology 2004; 141:244–250.
45. O'Neil AJ, Jee RD, Moffat AC. Analyst 1998; 123:2297–2302.
46. O'Neil AJ, Jee RD, Moffat AC. Analyst 1999; 124:33–36.
47. Frake P, Gill I, Luscombe CN, Rudd DR, Waterhouse J, Jayasooriya UA. Analyst 1998; 123:2043–2046.
48. Higgins JP, Arrivo SM, Thurau G, et al. Anal Chem 2003; 75:1777–1785.
49. Drennen JK. A Noise in Pharmaceutical Analysis [Ph.D.]: College of Pharmacy, University of Kentucky, 1991.
50. Ciurczak EW, Drennen JK. Near-infrared spectroscopy in pharmaceutical applications. In: Burns DA, Ciurczak EW, eds. Handbook of Near-Infrared Analysis. 2nd ed. New York, NY: Marcel Dekker, Inc., 2002:609–632.
51. Ciurczak EW, Drennen JK. Near-infrared spectroscopy in pharmaceutical applications. In: Burns DA, Ciurczak EW, eds. Handbook of Near-Infrared Analysis. 1st ed. New York, NY: Marcel Dekker, Inc., 1992.

52. Drennen JK, Lodder RA. In: Patonay G, ed. Advances in Near-Infrared Measurements. Greenwich, CT: JAI Press, 1993:93–112.
53. Kirsch JD, Drennen JK. J Pharm Biomed Anal 1999; 19:351–362.
54. Kirsch JD, Drennen JK. Paper APQ 1177. Paper presented at: AAPS National Meeting, 1996; Seattle, WA.
55. Kirsch JD, Drennen JK. J Pharm Biomed Anal 1995; 13:1273–1281.
56. Morisseau KM, Rhodes CT. Pharm Res 1997; 14(1):108–111.
57. Cogdill RP, Anderson CA, Delgado-Lopez M, et al. AAPS Pharm Sci Tech 2005; In Press.
58. Chen Y, Thosar SS, Forbess RA, Kemper MS, Rubinovitz RL, Shukla AJ. Drug Dev Ind Pharm 2001; 27(7):623–631.
59. Guo J-H, Skinner GW, Harcum WW, Malone JP, Weyer LG. Drug Dev Ind Pharm 1999; 25(12):1267–1270.
60. Ebube NK, Thosar SS, Roberts R, et al. Pharm Dev Technol 1999; 4(1):19–26.
61. Donoso M, Kildsig DO, Ghaly ES. Pharm Dev Technol 2003; 8(4):357–366.
62. Otsuka M, Mouri Y, Matsuda Y. AAPS Pharm Sci Tech 2003; 4(3):1–7.
63. Cho JH, Gemperline PJ, Aldridge PK, Sekulic SS. Anal Chim Acta 1997; 348:303–310.
64. Berntsson O, Burger T, Folestad S, Danielsson L-G, Kuhn J, Fricke J. Anal Chem 1999; 71(3):617–623.
65. Berntsson O, Danielsson L-G, Folestad S. Anal Chim Acta 1998; 364:243–251.
66. FDA's Summary of Judge Wolin's Interpretation of GMP Issues Contained in the Court's Ruling in USA vs. Barr Laboratories. GMP Institute, ISPE. Available at: www.gmp1st.com/barrsum.htm.
67. Clarke FC, Hammond SV, Jee RD, Moffat AC. Appl Spectrosco 2002; 56(11):1475–1483.
68. Iyer M, Morris HR, Drennen JKI. J Near IR Spectrosco 2002; 10:233–245.
69. FDA. PAT—A Framework for Innovative Manufacturing and Quality Assurance, 2004.
70. Ferraro JR, Basile LJ. Fourier Transform Infrared Spectroscopy: Applications to Chemical Systems. Vol. 1–3. New York: Academic, 1978.
71. Griffiths PR, de Haseth JA. Fourier Transform Infrared Spectrometry. New York: Wiley-Interscience, 1986.
72. Wetzel DL. Contemporary near-infrared instrumentation. In: Williams PC, Norris KH, eds. Near-Infrared Technology in the Agricultural and Food Industries. 2nd ed. St. Paul, MN: AACC, 2001:129–144.
73. McClure WF. Near-infrared instrumentation. In: Williams PC, Norris KH, eds. Near-Infrared Technology in the Agricultural and Food Industries. 2nd ed. St. Paul, MN: AACC, 2001:109–128.
74. Ciurczak EW. Spectroscopy 2005; 20(2):68–75.
75. www.Buchi.com. Buchi FT-NIR NIRFlex Technology.
76. Funk DB. J Near IR Spectrosco 1996; 4:101–106.
77. www.Foss.dk. XDS Series NIR Analyser.
78. Zeaiter M, Roger JM, Bellon-Maurel V, Rutledge DN. Trends Anal Chem 2004; 23(2):157–170.
79. Martens H, Næs T. Multivariate calibration by data compression. In: Williams PC, Norris KH, eds. Near-IR Technol Agr Food Ind. 2nd ed. St. Paul, MN: AACC, 2001:59–100.
80. Chang IC. Optical Eng 1981; 20(6):824–829.

81. Tran CD. Anal Letters 2005; 38(5):735–752.
82. Tran CD. J Near IR Spectrosco 2000; 8:87–99.
83. Treado PJ, Levin IW, Lewis EN. Appl Spectrosco 1994; 48(5):607–615.
84. Brimrose L-M, personal communication. Baltimore, MD: Brimrose Corp.
85. Cogdill RP, Anderson CA, Drennen JK. AAPS Pharm Sci Tech 2005; 6:E284–E297.
86. Milano MJ, Kim K-Y. Anal Chem 1977; 49(4):555–559.
87. Yates DA, Kuwana T. Anal Chem 1976; 48(3):510–514.
88. www.ControlDevelopment.com. CDI, Inc. Diode Array Technology.
89. www.Axsun.com. Axsun Microspectrometer Technology.
90. Schneller W, Noto J, Kerr R, Doe R. Proc SPIE 1998; 3355:877–883.
91. Taday PF, Bradley IV, Arnone DD, Pepper M. J Pharm Sci 2002; 92(4):831–838.
92. Clarke FC. Vib Spectrosco 2004; 34:25–35.
93. Miller PJ. Paper No. 95–3216: Liquid Crystal Tunable Filters (LCTF) for Multispectral Imaging. St. Joseph, MI: ASAE, 1995.
94. Miller PJ, Hoyt CC. Proc SPIE 1995; 2345:354–365.
95. www.SpectralDimensions.com. Spectral Dimensions Imaging Spectroscopy Instruments and Research.
96. United States Pharmacopeial Convention I. Near-infrared spectrophotometry. USP-NF, Second Supplement. Rockville, MD USA: United States Pharmacopeial Convention, Inc., 2003:3337–3344.
97. Brown SD, Sum ST, Despagne F, Lavine BK. Anal Chem 1996; 68(12):21–61.
98. Martens H, Næs T. Multivariate Calibration. New York, NY: John Wiley and Sons, 1989.
99. Massart DL, Vandeginste BGM, Buydens LMC, de Jong S, Lewi PJ, Smeyers-Verbeke J. Handbooks of Chemometrics and Qualimetrics. Vol. Parts A and B. New York: Elsevier, 1998.
100. Mobley PR, Kowalski BR, Workman JJ, Jr, Bro R. Appl Spectrosco Rev 1996; 31(4):347–368.
101. Workman JJ, Jr, Mobley PR, Kowalski BR, Bro R. Appl Spectrosco Rev 1996; 31(1 & 2):73–124.
102. Chemometrics and Intelligent Laboratory Systems: Elsevier.
103. Journal of Chemometrics: Wiley-Interscience.
104. Brown PJ. J Roy Stat Soc B 1982; 44:287–308.
105. Ritchie GE, Mark H, Ciurczak EW. AAPS Pharm Sci Tech 2003; 4(2):Article 24.
106. Draper NR, Smith H. Applied Regression Analysis. 2nd ed. New York: John Wiley & Sons, Inc., 1981.
107. Draper NR, Smith H. Applied Regression Analysis. 3rd ed. New York: John Wiley & Sons, Inc., 1998.
108. Barnes RJ, Dhanoa MS, Lister SJ. Appl Spectrosco 1989; 43(5): 772–777.
109. Savitzky A, Golay MJE. Anal Chem 1964; 36(8):1627–1639.
110. Teppola P, Minkkinen P. J Chemometrics 2000; 14:383–399.
111. Fearn T. Chemometrics Intell Lab Sys 2000; 50:47–52.
112. Goicoechea HC, Olivieri AC. Chemometrics Intell Lab Sys 2001; 56:73–81.
113. Wold S, Antti H, Lindgren F, Ohman J. Chemometrics Intell Lab Sys 1998; 44:175–185.
114. Næs T. Technometrics 1985; 27:301–311.
115. Næs T. Biometrical J 1985; 27:265–275.
116. Næs T. Biomet J 1986; 28:99–107.
117. Skrede G, Næs T, Martens M. J Food Sci 1983; 48:1745–1749.
118. Spjøtvoll E, Martens H, Volden R. Technometrics 1982; 24(3):173–180.

119. Drennen JK, Lodder RA. Spectroscopy 1991; 6(8):34–39.
120. Lodder RA, Hieftje GM. Appl Spectrosco 1988; 42(8):1351–1365.
121. Lodder RA, Hieftje GM. Appl Spectrosco 1988; 42(8):1500–1512.
122. Johnson RA, Wichern DW. Applied Multivariate Statistical Analysis. 4th ed. Upper Saddle River, JN: Prentice Hall, 1998.
123. Hotelling H. J Educ Psychol 1933; 24:417–441, 489–520.
124. Anderson CA, Cogdill RP, Drennen JKI. NIR News 2005; 16(3):7–8.
125. Westerhuis JA, Gurden SP, Smilde AK. Chemometrics Intell Lab Sys 2000; 51: 95–114.
126. Box GEP, Jenkins GM, Reinsel GC. Time Series Anal. Englewood Cliffs, N.J., USA: Prentice Hall, 1994.
127. Gallagher NB, Wise BM, Butler SW, White D, Barna GG. Development and benchmarking of multivariate statistical process control tools for a semiconductor etch process: improving robustness through model updating. Paper presented at: IFAC ADCHEM '97, 1997; Banff, Canada.
128. Jackson JE, Mudholkar GS. Technometrics 1979; 21(3):341–349.
129. Gemperline PJ, Boyer N. Anal Chem 1995; 67:160–166.
130. Gemperline PJ, Webber LD, Cox FO. Anal Chem 1989; 61(2):138–144.
131. Shah NK, Gemperline PJ. Anal Chem 1990; 62:465–470.
132. Wold S. Pattern Recog 1976; 8:127–139.
133. de Juan A, Tauler R. Anal Chim Acta 2003; 500:195–210.
134. Garrido M, Larrechi MS, Rius FX, Tauler R. Chemometrics Intell Lab Sys 2005; 76:111–120.
135. de Juan A, Tauler R, Dyson R, Marcolli C, Rault M, Maeder M. Trends Analy Chem 2004; 23(1):70–79.
136. Ma B, Gemperline PJ, Cash E, Bosserman M, Comas E. J Chemometrics 2003; 17:470–479.
137. Krutchkoff RG. Technometrics 1967; 9(3):425–439.
138. Haaland DM, Easterling RG. Appl Spectrosco 1980; 34(5):539–548.
139. Haaland DM, Easterling RG. Appl Spectrosco 1982; 36(6):665–673.
140. Haaland DM, Easterling RG, Vopicka DA. Appl Spectrosco 1985; 39(1):73–84.
141. Haaland DM, Thomas EV. Anal Chem 1988; 60:1193–1202.
142. Helland IS. Commun Stat B Simulation Comput 1988; 17(2):581–607.
143. Helland IS. Scandinavian J Stat 1990; 17:97–114.
144. De Jong S. Chemometrics Intell Lab Sys 1993; 18:251–263.
145. Wold S, Ruhe A, Wold H, Dunn WJ, III. SIAM J Sci Stat Comput 1984; 5(3):735–743.
146. Geladi P, Kowalski BR. Anal Chim Acta 1986; 185:1–17.
147. Lorber A. Anal Chem 1986; 58:1167–1172.
148. Lorber A, Faber K, Kowalski BR. Anal Chem 1997; 69:1620–1626.
149. Ferre J, Brown SD, Rius FX. J Chemometrics 2001; 15:537–553.
150. Skibsted ETS, Boelens HFM, Westerhuis JA, Witte DT, Smilde AK. Appl Spectrosco 2004; 58(3):264–271.
151. Goicoechea HC, Olivieri AC. Anal Chim Acta 2002; 453:289–300.
152. Cogdill RP, Anderson CA. J Near IR Spectrosco 2005; 13:119–132.
153. Roussel S, Hardy CL, Hurburgh CRJ, Rippke GR. Appl Spectrosco 2001; 55(10):1425–1430.
154. Agatonovic-Kustrin S, Beresford R. J Pharm Biomed Anal 2000; 22:717–727.

155. Borggaard C. Neural networks in near-infrared spectroscopy. In: Williams PC, Norris KH, eds. Near-IR Technology in the Agricultural and Food Industries. 2nd ed. St. Paul, MN: AACC, 2001:101–108.
156. Despagne F, Massart DL. The Analyst 1998; 123:157R–178R.
157. Hussain AS, Zuanqiang Y, Johnson RD. Pharm Res 1991; 8(10):1248–1252.
158. Nerella NG, Drennen JKI. Appl Spectrosco 1996; 50(2):285–291.
159. Rumelhart DE, Hinton GE, Williams RJ. Learning internal representations by error propagation. In: Rumelhart DE, ed. Parallel Distributed Processing. Cambridge, UK: MIT Press, 1986:318–362.
160. Fiesler E, Beale R. Handbook of Neural Computation. Bristol and Oxford, UK: IOP Publishing, Ltd. and Oxford University Press, Inc., 1997.
161. Cogdill RP, Dardenne P. J Near IR Spectrosco 2004; 12:93–100.
162. Vapnik V, Golowich S, Smola A. Support vector method for function approximation, regression estimation, and signal processing. In: Mozer M, Jordan M, Petsche T, eds. Advances in Neural Information Processing Systems. Cambridge, MA: MIT Press, 1997:281–287.
163. Walczak B, Massart DL. Anal Chim Acta 1996; 331:177–185.
164. AACC. "Near-infrared methods: guidlines for model development and maintenance," AACC Method 39–00. Approved Methods of the American Association of Cereal Chemists. 10th ed. St. Paul, MN USA: AACC Press, 1999.
165. ICH. Harmonized Tripartite Guideline Q8: Pharmaceutical Development: International Conference on Harmonization, 2004.
166. Fearn T. J Near IR Spectrosco 2001; 9:229–244.
167. ICH. Harmonized Tripartite Guideline Q2A: Text on Validation of Analytical Procedures: International Conference on Harmonization, 1994.
168. Reich G. Adv Drug Deliv Rev 2005; 57:1109–1143.
169. Brashear RL, Flanagan DR, Luner PE, Seyer JJ, Kemper MS. J Pharm Sci 1999; 88(12):1348–1353.
170. Lachenal G. Vib Spectrosco 1995; 9:93–100.
171. Siesler HW, Ozaki Y, Kawata S, Heise HM. Near-infrared spectroscopy: principles, instruments, applications. Weinheim, Germany: Wiley-VCH GmbH, 2002: 213–246.
172. Zerbi G, ed. Modern Polymer Spectroscopy. Weinheim: Wiley-VCH GmbH, 1999.
173. Buice RGJ, Gold TB, Lodder RA, Digenis GA. Pharm Res 1995; 12(1):161–163.
174. Berntsson O, Zackrisson G, Östling G. J Pharm Biomed Anal 1997; 15:895–900.
175. Gold TB, Buice RGJ, Lodder RA, Digenis GA. Pharm Dev Technol 1998; 3(2):209–214.
176. Svensson O, Josefson M, Langkilde FW. Appl Spectrosco 1997; 51(12):1826–1835.
177. Gustafsson C, Nyström C, Lennholm H, Bonferoni MC, Caramella CM. J Pharm Sci 2003; 92(3):460–470.
178. Pérez-Ramos JD, Findlay WP, Peck GE, Morris KR. AAPS Pharm Sci Tech 2005; 6(1):E127–E136.
179. Friend DR. Pharm News 2002; 9:375–380.
180. Friend DR, Chrai SS, Kupperblatt G, et al. Accudep technology for oral modified drug release. In: Rathbone M, ed. Modified-Release Drug Delivery Technology. New York, NY: Marcel Dekker, Inc., 2002:89–100.
181. Blanco M, Romero MA. J Pharm Biomed Anal 2002; 30:467–472.
182. Gimet R, Luong AT. J Pharm Biomed Anal 1987; 5(3):205–211.

183. Buchanan BR, Ciurczak EW, Grunke AQ, Honigs DE. Spectroscopy 1988; 3(5):54–56.

184. Miller CE, Honigs DE. Spectroscopy 1989; 4(3):44–48.

185. Luner PE, Majuru S, Seyer JJ, Kemper MS. Pharm Dev Technol 2000; 5(2):231–246.

186. Patel AD, Luner PE, Kemper MS. J Pharm Sci 2001; 90(3):360–370.

187. Seyer JJ, Luner PE, Kemper MS. J Pharm Sci 2000; 89(10):1305–1316.

188. Patel AD, Luner PE, Kemper MS. Int J Pharm 2000; 206:63–74.

189. Braga JWB, Poppi RJ. J Pharm Sci 2004; 93(8):2124–2134.

190. Otsuka M, Kato F, Matsuda Y, Ozaki Y. AAPS Pharm Sci Tech 2003; 4(2):1–12.

191. Davis TD, Peck GE, Stowell JG, Morris KR, Byrn SR. Pharm Res 2004; 21(5):860–866.

192. NIST/SEMATECH. e-Handbook of Statistical Methods. http://www.itl.nist.gov/div898/handbook. Available at: http://www.itl.nist.gov/div898/handbook. Accessed October 22, 2005.

193. Jørgensen A, Rantanen J, Karjalainen M, Khriachtchev L, Räsänen E, Yliruusi J. Pharm Res 2002; 19(9):1285–1291.

194. Jørgensen AC, Luukkonen P, Rantanen J, Schæfer T, Juppo AM, Yliruusi J. J Pharm Sci 2004; 93(9):2232–2243.

195. Rantanen J, Antikainen O, Mannermaa J-P, Yliruusi J. Pharm Dev Technol 2000; 5(2):209–217.

196. Rantanen J, Jørgensen A, Räsänen E, et al. AAPS Pharm Sci Tech 2001; 2(4):21.

197. Rantanen J, Känsäkoski M, Suhonen J, et al. AAPS Pharm Sci Tech 2000; 1(2):10.

198. Rantanen J, Lehtola S, Rämet P, Mannermaa J-P, Yliruusi J. Powder Technol 1998; 99:163–170.

199. Rantanen J, Räsänen E, Tenhunen J, Känsäkoski M, Mannermaa J-P, Yliruusi J. European J Pharm Biopharm 2000; 50:271–276.

200. Rantanen J, Wikström H, Turner R, Taylor LS. Anal Chem 2005; 77:556–563.

201. www.NDCInfrared.com. NDC Infrared Engineering.

202. www.JMCanty.com. J.M. Canty Process Technology.

203. www.romackfiberoptics.com. Romack Fiber Optics.

204. Kemper MS, Luchetta LM. J Near IR Spectrosco 2003; 11:155–174.

205. El-Hagrasy AS, Delgado-Lopez M, Drennen JKI. J Pharm Sci. 2006; 95:407–421.

206. El-Hagrasy AS, Drennen JKI. J Pharm Sci 2006; 95:422–434.

207. El-Hagrasy AS, Morris HR, D'Amico F, Lodder RA, Drennen JKI. J Pharm Sci 2001; 90(9):298–1307.

208. Frake P, Greenhalgh D, Grierson SM, Hempenstall JM, Rudd DR. Int J Pharm 1997; 151:75–80.

209. Parris J, Airiau C, Escott R, Rydzak J, Crocombe R. Spectrosco 2005; 20(2):34–41.

210. Donoso M, Ghaly ES. Pharm Dev Technol 2005; 10:211–217.

211. Donoso M, Ghaly ES. Pharm Dev Technol 2004; 9(3): 247–263.

212. Freitas MP, Sabadin A, Silva LM, et al. J Pharm Biomed Anal 2005; 39:17–21.

213. Lodder RA, Selby M, Hieftje GM. Anal Chem 1987; 59:1921–1930.

214. Bauer JF, Dziki W, Quick JE. J Pharm Sci 1999; 88:1222–1227.

11

Solid-State Nuclear Magnetic Resonance Spectrometry

Ales Medek

Pfizer Global R&D, Groton, Connecticut, U.S.A.

FUNDAMENTAL PRINCIPLES OF NMR

Introduction

Nuclear magnetic resonance (NMR) spectroscopy has many similarities with the molecular spectroscopic techniques involving transitions between electron energy levels. However, NMR stands apart in many details, which will be briefly discussed in this introduction section. The obvious distinction is that NMR involves transitions between nuclear energy levels. In the absence of external magnetic or electric fields, the energy levels of the ground state of an isolated nucleus are degenerate. It is the external magnetic field that brings the distinction between the nuclear energy levels. If we apply external magnetic field B_0 on a sample containing nuclei with nonzero spin angular momentum I (i.e., magnetically active spins), the $2I + 1$ nuclear energy levels will be

$$E = -\gamma I_z B_0 \tag{1}$$

in which the proportionality factor, referred to as gyromagnetic ratio, γ is the characteristic constant for a given isotope. Because the values of angular momentum are quantized with the magnetic quantum number m ($I_Z = \hbar m$, where $\hbar$ is the Planck's constant), so are the nuclear energy levels known as Zeeman levels. Following the selection rule governing any magnetic dipole transitions ($\Delta m = \pm 1$), the differences between the Zeeman energy levels are

$$\Delta E = \hbar \gamma B_0 = \hbar \omega_0 \tag{2}$$

where ω_0 is the frequency (in units of radian per second) of the electromagnetic radiation, corresponding to the difference between the spin energy levels, known as Larmor frequency. Equation (2) also predicts linear dependence of the energy level difference on the external magnetic field B_0. For the currently available magnetic fields, the energy difference falls into the radio frequency (rf) portion of the electromagnetic radiation. When the system is at thermal equilibrium, the population of the given energy level is given by the Boltzmann distribution

$$N_i = \exp\left(\frac{-E_i}{k_B T}\right) \tag{3}$$

where N_i is the number of nuclei occupying energy level E_i, k_B is the Boltzmann constant, and T is the absolute temperature. In contrast to the electronic transitions, at temperatures close to the room temperature, the product of $k_B T$ is several orders of magnitude larger than the difference between the Zeeman levels ΔE. Consequently, even random thermal fluctuations can very efficiently induce the nuclear spin transitions, nearly equalizing the number of nuclei occupying each Zeeman energy level. Being directly proportional to the population difference, the NMR signal strength is significantly weaker compared with other atomic or molecular spectroscopies involving electron levels transitions. Out of 200,000 ^1H spins placed into the external magnetic field of 11.7 T (500 MHz ^1H frequency), approximately 100,001 spins will occupy the lower energy level and 99,999 the higher one. Effectively, only 1/200,000 of the sample gives rise to the observable NMR signal. When compared with other spectroscopies, NMR is a relatively insensitive technique.

Another property of nuclear transitions stands apart from spectroscopies involving electronic transitions. It is the relative isolation of the nuclear system from its surroundings. This simplifies the NMR response of the system under the study and, typically, one can understand and simulate the outcome of any NMR experiment based on first principles considerations. When out of thermal equilibrium, nuclear spins tend to spontaneously return to their equilibrium states. This relaxation process is driven by random fluctuations of variables affecting the nuclear transitions. In many cases, encountered especially in solid-state NMR (SSNMR), the relaxation might be prohibitively long. It is not uncommon for the return to equilibrium to take in the order of hours or longer, prohibiting the practical observation of the signal. On the other hand, unlike the electronic spectroscopies, the nuclear spins can form coherent states for long periods of seconds or longer. Modern NMR multidimensional experiments take excellent advantage of this phenomenon. The slow decay of coherent states allows one to design rather complex pulse sequences manipulating the coherent states of nuclear spins to obtain the targeted information about the system under the study.

Nuclear Spin Interactions

To understand the applications of the NMR spectroscopy in both the solution and solid phases, it is useful first to review the various interactions affecting the nuclear spins. Nuclear spins can interact with each other or with internal or external electric and magnetic fields (1). The internal electric and magnetic fields arise due to the effect of electrons surrounding the nuclei of interest in atoms or molecules. The possibility to selectively exploit the nuclear interactions makes NMR an extremely powerful tool, finding its applications across many scientific disciplines.

In quantum mechanics, all the interactions of a system of interest are represented by corresponding Hamiltonians. The state of the system is given by its wave function and all the measurable quantities are represented by their operators. The most basic NMR interaction, the Zeeman interaction, was already introduced in the previous section. Zeeman interaction causes the nuclear spins to interact with the external magnetic field. This interaction is responsible for observing the NMR phenomenon. Based on the Correspondence Principle, the classical description from Equation (1) can be rewritten using the quantum mechanical Zeeman Hamiltonian $\mathcal{H}_Z$

$$\mathcal{H}_Z = -\gamma \hat{I}_Z B_0 \tag{4}$$

where $\hat{I}_Z$ is the spin angular momentum operator. The energy levels of the nuclei are directly proportional to the applied external field B_0. It turns out that the external magnetic field has to be extremely homogeneous to resolve the fine features of NMR spectra. A great deal of attention is paid to this issue. Every modern NMR spectrometer is equipped with a set of shimming coils correcting the local imperfections of the field.

NMR would not be a very chemically useful technique if only Zeeman interaction existed, because all spins from any compound would resonate at the same frequency. The useful NMR properties arise from other nuclear interactions. All the interactions exist simultaneously and can be described by the total Hamiltonian (2–5)

$$\mathcal{H}_T = \mathcal{H}_Z + \mathcal{H}_{CS} + \mathcal{H}_J + \mathcal{H}_D + \mathcal{H}_Q + \mathcal{H}_{rf} \tag{5}$$

where the subscripts CS, J, D, Q, and rf denote the chemical shift interaction, indirect spin-spin (J-coupling), dipolar coupling, quadrupolar coupling, and rf irradiation, respectively. It should be noted that these nuclear interactions do not represent all the interactions that nature provides. They are, however, sufficient to facilitate the majority of discussions of NMR applications in this review. In terms of magnitude, the Zeeman interaction is typically, but not always, the strongest, followed by quadrupolar, dipolar, chemical shift, and J-coupling. This is just a general order, which may change strongly depending on the type and shape of molecules. For example, in many asymmetric molecules with quadrupolar nuclei, the quadrupolar coupling can be larger than the Zeeman interaction, especially when working in weaker magnetic fields. Such nuclei may be difficult to observe by NMR but are a great subject to nuclear quadrupolar resonance (NQR) studies, a spectroscopic technique, which does not require a presence of an external magnetic field, because it deals only with the quadrupolar states. On the other hand, highly symmetrical molecules containing quadrupolar nuclei can show vanishing quadrupolar contribution.

In general, in the laboratory frame, the Hamiltonian representing a particular interaction $\mathcal{H}_\lambda$ can be characterized as a product of two spin operators with a Cartesian tensor $\boldsymbol{A}^\lambda$ of rank two, described by the (3×3) matrix:

$$\mathcal{H}_\lambda = C^\lambda(\hat{I}_x, \hat{I}_y, \hat{I}_z) \cdot \begin{pmatrix} A_{xx}^\lambda & A_{xy}^\lambda & A_{xz}^\lambda \\ A_{yx}^\lambda & A_{yy}^\lambda & A_{yz}^\lambda \\ A_{zx}^\lambda & A_{zy}^\lambda & A_{zz}^\lambda \end{pmatrix} \cdot \begin{pmatrix} \hat{S}_x^\lambda \\ \hat{S}_y^\lambda \\ \hat{S}_z^\lambda \end{pmatrix} \qquad (6)$$

where C^λ is a constant dependent on the interaction λ, $\hat{\boldsymbol{I}} = (\hat{I}_x, \hat{I}_y, \hat{I}_z)$ is the nuclear spin angular momentum operator of spin I, and $\hat{\boldsymbol{S}} = (\hat{S}_x, \hat{S}_y, \hat{S}_z)$ can be either the nuclear spin operator of spin S or B_0 in the special case of $\lambda = CS$ (the chemical shift interaction). From this notation, it is evident that the nuclear interactions depend on properties that exhibit a dependence on a particular orientation in the physical space and have to be described by tensors rather than scalars (numbers). For example, the A_{zx}^λ element of the tensor $\boldsymbol{A}$, describes the reaction of the system along the z direction when a "force" represented by the operator $\hat{S}_x^\lambda$ is applied in the x direction. In a special frame of reference called the principal axis system (PAS), the symmetrical part of the second rank tensor takes a diagonal form, that is, the only nonzero components lie along the diagonal:

$$A^\lambda(\text{PAS}) = \begin{pmatrix} A_{11}^\lambda & 0 & 0 \\ 0 & A_{22}^\lambda & 0 \\ 0 & 0 & A_{33}^\lambda \end{pmatrix} = A_0^\lambda(\text{PAS}) + A_2^\lambda(\text{PAS})$$

$$A_0^\lambda(\text{PAS}) = A_{\text{ISO}}^\lambda \begin{pmatrix} 1 & 0 & 0 \\ 0 & 1 & 0 \\ 0 & 0 & 1 \end{pmatrix}$$

$$A_2^\lambda(\text{PAS}) = \begin{pmatrix} \delta_{11}^\lambda & 0 & 0 \\ 0 & \delta_{22}^\lambda & 0 \\ 0 & 0 & \delta_{33}^\lambda \end{pmatrix}$$

$$\delta_{ii}^\lambda = A_{ii}^\lambda - A_{\text{ISO}}^\lambda \qquad (7)$$

The orientation of the principal axis system of a tensor A^λ depends on the local molecular symmetry and even for the same nucleus can, in general, change depending on the type of interaction λ. The PAS orientation for given coupling λ is fixed with respect to the molecular frame of the compound. Because $\text{Tr}(A_2^\lambda) = 0$, only two parameters are necessary for a complete description of A_2^λ:

$$\delta_\lambda = \delta_{33}^\lambda$$

$$\eta_\lambda = \frac{\delta_{22}^\lambda - \delta_{11}^\lambda}{\delta_\lambda} \tag{8}$$

where η_λ is usually referred to as the asymmetry parameter with values limited to $0 \leq \eta_\lambda \leq 1$. An arbitrary second rank tensor is then defined in its respective PAS by its isotropic part A_{iso}^λ, by its anisotropic contribution δ_λ, and the asymmetry parameter η_λ.

The chemical shift interaction is a coupling of the nuclear spins I with the external magnetic field B_0 mediated through the electronic environment. Diamagnetic currents in the electron orbitals and partial unquenching of their paramagnetism generate local field B_{ind} proportional to B_0, $B_{\text{ind}} = -A^{CS}B_0$. The chemical shift Hamiltonian is defined as

$$\mathcal{H}_{\text{CS}} = -\gamma \hat{I} B_{\text{ind}} = \gamma \hat{I} A^{\text{CS}} B_0 \tag{9}$$

By comparison of Equation (9) and Equation (6), the proportionality constant C^{CS} is simply equal to γ. The values of the components of chemical shift tensor A^{CS} depend, in general, on the orientation and local symmetry of the molecule. The remarkable property of the chemical shift tensor is that the sum of its diagonal elements (known as its trace) does not vanish, thus giving rise to the isotropic shift

$$\delta_{\text{iso}} = \frac{1}{3} \text{Tr}(A^{\text{CS}}) \neq 0 \tag{10}$$

This is why even the fast isotropic tumbling of molecules in solution does not average out the chemical shift to zero. Different nuclei in the same molecule experience different local fields. Based on the chemical shifts alone, many molecules can be uniquely identified.

At the atomic level, a given nucleus affects, and is also affected by, the local dipolar fields of its neighbors. The indirect spin–spin J-coupling can be described by Hamiltonian in Equation (11).

$$\mathcal{H}_{\text{J}} = \hat{I} \cdot A^{\text{J}} \cdot \hat{S} \tag{11}$$

It is called indirect, because it acts through the electronic environment, rather than directly though space. Similar to the chemical shift, the trace of the J-coupling tensor is not zero. Therefore we can observe and measure finite J-coupling values even in solution. Moreover, for majority of practical applications, the J-coupling tensor A^J can be approximated as being symmetrical with vanishing anisotropy and as such can be regarded as scalar (a number) rather than tensor.

Nuclear spins can also interact directly through space. Such an interaction is referred to as dipolar coupling (or dipole–dipole coupling) and can be described by the dipolar Hamiltonian:

$$\mathcal{H}_D = D\left[\hat{I}_1\hat{I}_2 - 3\frac{(\hat{I}_1\mathbf{r})(\hat{I}_2\mathbf{r})}{\mathbf{r}^2}\right] \tag{12}$$

where $\mathbf{r}$ is the internuclear vector and D is the dipolar coupling constant in frequency units, defined as

$$D = \frac{\mu_0}{4\pi}\frac{\gamma_1\gamma_2\hbar}{2\pi r^3} \tag{13}$$

μ_0 is the permeability of vacuum constant. The trace of the dipolar tensor vanishes, which is why dipolar coupling has no direct effect on isotropic solution NMR spectra apart from relaxation-induced processes.

Spin $I = 1/2$ nuclei possess only magnetic dipole moments, because all the higher magnetic and electric multipole moments vanish. Therefore, all their Hamiltonians are linear in $\hat{I}$. However, apart from the magnetic couplings encountered in spin $I = 1/2$ isotopes, nuclei with $I > 1/2$ are distinguished by a nonzero nuclear quadrupole moment Q, which gives origin to the interaction with the surrounding electric field gradients (EFG). The quadrupolar coupling between the quadrupole moment and the electric field gradient is described by the quadrupolar Hamiltonian:

$$\mathcal{H}_Q = \frac{eQ}{2I(2I-1)\hbar}\hat{I}A^Q\hat{I} \tag{14}$$

The symbol e designates the elementary charge. Similar to the dipolar coupling, the trace of EFG tensor vanishes having no effect on solution NMR signals except through the relaxation processes. The majority (72%) of all the magnetically active nuclides in the Periodic Table possess half-integer spin higher than 1 (Fig. 1) (6). This clearly underlines the importance as to why we should study these spins (7–9).

The chemical shift interaction, J, dipolar, and quadrupolar couplings are called internal interactions, because they represent couplings between the various fields present internally within a molecule. On the other hand, Zeeman interaction and the rf irradiation are external interactions, because they represent the interactions of the nuclear spins with the external magnetic fields.

In order to disturb the spin system from its thermodynamic equilibrium, an oscillating rf field B_1 has to be applied in the direction perpendicular to that of B_0. If the frequency of this oscillating field is ω and its phase in the plane

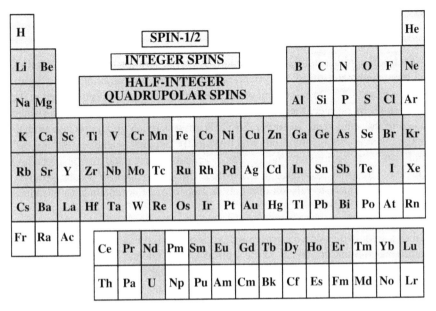

Figure 1   Magnetically active isotopes in the Periodic Table. The majority (72%) of the magnetically active nuclides are half-integer quadrupolar nuclei. If one plans to use solid state NMR for studying ceramics, semi-conductors, catalysts, glasses, superconductors or metal binding processes it is necessary to deal with the spectroscopy of these spins. For elements with several different nuclides, only those with the highest natural abundance are considered.

perpendicular to B_0 is φ, then the B_1 vector can be fully described by:

$$B_1(t) = 2B_1 \begin{pmatrix} \cos\varphi\cos(\omega t) \\ \sin\varphi\cos(\omega t) \\ 0 \end{pmatrix} \tag{15}$$

The irradiation Hamiltonian representing the coupling between the oscillating rf field and the particular components of the magnetic dipole is in the laboratory frame given by:

$$\mathcal{H}_{rf} = -\omega_1[\hat{I}_x \cos(\omega t + \varphi) + \hat{I}_y \sin(\omega t + \varphi)] \tag{16}$$

where $\omega_1 = \gamma B_1$ is the nutational frequency directly proportional to the amplitude of the irradiation field. This value, typically given in kHz, is used frequently in the NMR literature to describe the strength of irradiation. The irradiation field is typically calibrated based on the experimentally observed duration of $90°$ pulse (τ_{90}), which is related to ω_1 by $\omega_1 = 1/(4\tau_{90})$. In practice, its value depends on the available power delivered from the rf amplifier (ω_1 is directly proportional to the rf voltage amplitude) and on the internal characteristics of the probe, such as coil geometry. Of a great interest is the homogeneity of the B_1 field throughout the coil, strongly influencing performance of many pulse sequences.

SSNMR OF POWDER SOLIDS

Introduction

Most NMR applications have traditionally involved solution samples. In solution, all the valuable anisotropic information is lost by the fast tumbling of the liquid. Unlike in solution, the full extent of the anisotropic interactions is observed in the solid state. In essence, there is more NMR information available on the system when in the solid state. However, SSNMR possess much greater technical challenge to mine this information, which typically requires selective "turning on" and "off" the various strong nuclear interactions discussed in the preceding chapter. The great advantage of SSNMR is the "concentration" of the samples. One does not have to rely on dissolution properties of compounds to introduce them into the NMR coil.

The NMR spectra of stationary (not spinning) polycrystalline or disordered solids are known as powder patterns. It is possible to predict their line shapes by numerical simulations based on the a priori known nuclear coupling parameters. Figure 2 and Figure 3 show the typical line shapes of chemical shift and quadrupolar interactions. In simulating these spectra, only a single spin contribution was considered. Because more than one magnetically inequivalent nucleus is typically present in a realistic system, the static spectra can quickly become overcrowded and difficult or impossible to interpret. The available

Chemical Shielding

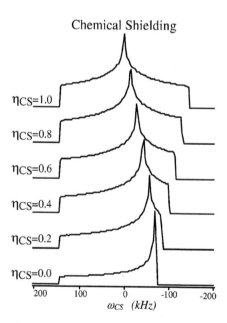

$\eta_{CS}=1.0$

$\eta_{CS}=0.8$

$\eta_{CS}=0.6$

$\eta_{CS}=0.4$

$\eta_{CS}=0.2$

$\eta_{CS}=0.0$

200 100 0 -100 -200

ω_{CS} *(kHz)*

Figure 2 Examples of static powder patterns for chemical shielding as a function of asymmetry parameter η_{CS}. The magnitude of chemical shielding δ_{CS} was arbitrarily set to 143 kHz, and the isotropic shift $\delta_{iso} = 0$.

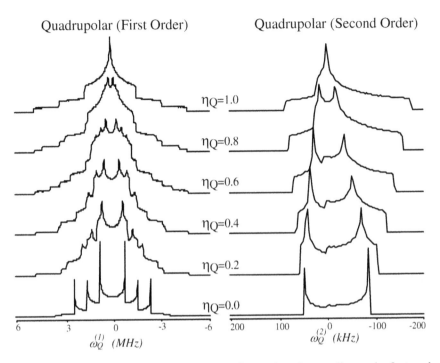

Quadrupolar (First Order) Quadrupolar (Second Order)

η_Q=1.0

η_Q=0.8

η_Q=0.6

η_Q=0.4

η_Q=0.2

η_Q=0.0

6 3 0 -3 -6 200 100 0 -100 -200

$\omega_Q^{(1)}$ (MHz) $\omega_Q^{(2)}$ (kHz)

Figure 3 Examples of static powder patterns for quadrupolar coupling to the first- and second-order as a function of asymmetry parameter η_Q for $I = 7/2$. The quadrupolar coupling constant e^2qQ/h was arbitrarily set to 22 MHz, the Larmor frequency (important for second-order spectra) used in these simulations was $\omega_0 = 47.7$ MHz (^{59}Co at 4.7 T), and no chemical shift was considered.

signal is also spread over broad range of frequencies, which diminishes the achieved sensitivity in terms of the signal-to-noise ratio (S/N). Therefore, a technique improving the resolution of the SSNMR spectra of crystalline solids has long been sought. The two key factors precluding widespread use of SSNMR in the past were sensitivity and resolution. A boost in solid-state applications appeared after a combination of the line-narrowing technique called magic angle spinning (MAS) (10–12) and sensitivity-enhancing technique called cross-polarization (CP) (13) was introduced. Cross-polarization magic angle spinning (CPMAS) (14) brought about the necessary sensitivity enhancement to make SSNMR spectroscopy of natural abundance samples practicable. Spinning around the "magic" angle, which is the root of second-order Legendre polynomial, $\chi = 54.74°$ (Fig. 4), removes completely the second rank anisotropies if these are much smaller than the rotational frequency or may otherwise lead to a spinning sideband manifold. Example of this behavior is shown in Figure 5 on carbon spectra DL-alanine. At spinning speed of 1.0 kHz, all three alanine carbons show significant spinning sidebands. At 2.5 kHz, no appreciable

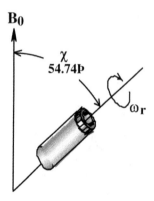

Figure 4 The most conventional example of spinning about a single axis, the magic angle spinning. The time average of $\langle P2(\cos \beta) \rangle$ over integer multiples of rotor period vanishes.

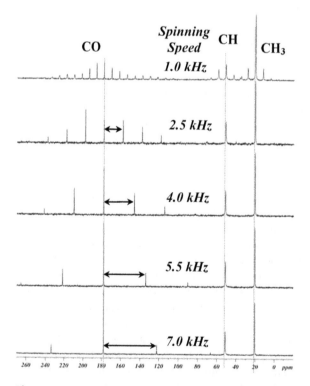

Figure 5 The 11.7 T carbon CPMAS spectra of DL-alanine at variable spinning speeds of 1.0, 2.5, 4.0, 5.5, and 7.0 kHz. The spinning sidebands are spaced in integer multiples of the spinning speed (highlighted by the *arrowhead lines*). *Abbreviation*: CPMAS, cross-polarization magic angle spinning.

spinning sideband intensities are observed for the methyl group. The methine carbon still shows signs of spinning sidebands, albeit with greatly reduced intensities. Even at the maximum spinning speed of 7.0 kHz (the maximum specified spinning speed of standard wall 7.0 mm Bruker-Biospin rotors), the carboxylic carbon signal is still spread between spinning sidebands. The spinning sidebands in this example arise due to the modulation of the carbon chemical shift anisotropies by the MAS spinning. Clearly, the anisotropic contribution of methyl group is smaller than that of methine, which is still smaller than that of carboxylic carbon. In general, the magnitude of the chemical shift anisotropy for different carbons approximately follows the rank order of $CH_3 < CH_2 < CH <$ aromatic-carbonyl, which happens to coincide with the typical rank order of their isotropic shifts. The contribution of methyl groups to spinning sidebands is further reduced due the fact that they exhibit fast wheel-like rotation around the $C-C$ axis, even in the solid state at temperatures close to room temperature.

Spins higher than $\frac{1}{2}$ show additional quadrupolar coupling contribution to their spectral line shapes. Two cases should be distinguished. For half-integer spins, the central transition $(-1/2 \leftrightarrow +1/2)$ is void of quadrupolar coupling to the first order. For compounds with high molecular symmetry (small nuclear quadrupole moment), MAS is likely to remove most of the quadrupolar anisotropies. To get high resolution for compounds with moderately strong quadrupolar coupling, such techniques as double rotation (15,16), dynamic angle spinning (17), or multiple quantum MAS (18,19) have to be applied. There is no central transition for nuclei with integer spins. The full effect of the quadrupolar coupling shows in their spectra. There are at least two pharmaceutically relevant spin 1 nuclei: ^2H and ^{14}N. Deuterium quadrupole moment tends to be small and spinning around magic angle is typically sufficient to average out the anisotropies, leaving only the extensive spinning sideband manifold. Due to the low ^2H natural abundance, isotope enrichment is required for any deuterium studies. The easiest way to introduce ^2H labels is through deuterium exchange of the exchangeable protons. Unlike ^2H, ^{14}N is a nitrogen isotope with almost 100% natural abundance. However, its widespread applications have been largely hindered by the presence of rather sizeable nuclear quadrupole moment. This is why NQR technique has been successfully applied to detect nitrogen-containing explosives. Several NMR approaches to get better ^{14}N resolution have been tried, but no simple method has been found yet (20–27).

Proton SSNMR

In contrast to ^1H spectroscopy in solutions, proton NMR in the solid state has faced significant technical challenges, mainly due to the extensive line broadening originating from strong, through-space dipole–dipole coupling. This interaction is, to a very good approximation, completely averaged out in isotropic solution as molecules tumble fast and completely randomly on the NMR timescale. Recently, it was shown that the dipolar coupling may not get completely averaged out over long distances, if the interacting spins do not share the same diffusion sphere (28–31). This contribution is, however, very small and for most applications of practical interest can be disregarded. The only remaining effect that the dipolar coupling has on isotropic solutions is through the incoherent relaxation. The distance measurements in solution rely on dipolar relaxation-induced nuclear Overhauser effect (NOE) (32). NOE-derived structural constraints are the basis for protein structural determination by NMR. Due to the type and magnitude of motions involved, NOE effect plays much smaller role in the solid state.

In the crystalline solids, molecules are locked in their crystal lattices. Equation (12) predicts the dipolar coupling to be directly proportional to the product of the gyromagnetic ratios of the two involved nuclei (two protons in this case) and inversely proportional to the cube of the distance between the nuclei (r^{-3}). The gyromagnetic ratio of protons is among the biggest of all nuclei. Moreover, the typical organic solids contain many protons in the close vicinity of each other. Both of these factors translate into very strong proton dipolar coupling. In the absence of extensive molecular motions, the dipole–dipole interaction is usually significantly larger than the chemical shielding effect. When out of the equilibrium state, the through-space dipole–dipole interaction induces fast exchange of magnetization between the protons, effective over many molecules. This magnetization exchange is called a spin diffusion (2,4). The term of diffusion is very appropriate as it refers to the diffusion of the NMR signal over the net of nuclei, albeit without any physical movement of the molecules. The signal transfer due to spin diffusion can be described by conventional equations for diffusive motion [Equation (17)], where D is the diffusion coefficient, t is the time during which the diffusion is effective, and $\langle L \rangle$ is the average distance traveled by the NMR signal.

$$\langle L \rangle = \sqrt{6Dt} \tag{17}$$

Thus, domain sizes of phase-separated solids may be estimated from the rates of spin diffusion. Unless the dipolar coupling is averaged out by significant molecular motions or by very fast MAS rotation, the spin diffusion is fast on the NMR timescale. A typical proton static line shape of organic molecules is a single peak, up to 50 kHz wide, precluding resolution of the chemical shifts (2–4) (Fig. 6; bottom trace). As a result of the spin diffusion, the anisotropic proton lineshape broadening is referred to as being homogeneous. In contrast to the inhomogeneous broadening arising from, for example, the chemical shift or quadrupolar coupling (Figs. 2 and 3), selective excitation of arbitrarily narrow part of the proton spectrum leads to saturation (or disappearance) of the whole spectrum. The spin diffusion also counteracts the MAS averaging effect (Fig. 6; second trace from bottom). Because MAS narrows down the proton line shape only marginally, other line-narrowing techniques have been sought. Combined rotation and multiple-pulse spectroscopy (CRAMPS) (33–39) offers a partial

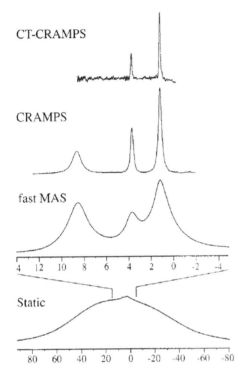

Figure 6 Solid-state proton spectra of L-alanine. Differences between proton spectra acquired in static, fast magic angle spinning, and FSLG-CRAMPS modes. The top is a projection of the indirectly detected domain of 2D constant time $^1H-^1H$ correlation experiment. *Abbreviation*: FSLG-CRAMPS, frequency-switched Lee-Goldburg experiment-combined rotation and multiple-phase spectroscopy. *Source*: Modified from Ref. 38.

solution to this problem (Fig. 6; top two traces). It combines conventional MAS with rotation of proton magnetization around the magic angle in spin space. To avoid interference between the MAS averaging and the rf pulses, successful application of CRAMPS requires the duration of the complete train of pulses (typically 24 or more) to be much shorter with compared one cycle of the MAS rotation. This significantly restricts the applicable MAS rotation speeds. Typically, spinning at speeds in access of 3 kHz is prohibitive. The CRAMPS experiment also puts a large demand on the NMR hardware, especially on NMR probes, because high-power rf pulses are applied in between each acquisition point. High homogeneity of the irradiation field B_1 and careful set-up of experimental variables are required to avoid distortions of the proton peaks. Typical line widths afforded by the CRAMPS experiment are approximately 1 ppm, limiting the application of the CRAMPS experiment to compounds with a small number of well-resolved protons. The more modern variants of CRAMPS are based on frequency-switched Lee-Goldburg (FSLG) experiment

(40–43). The frequency switching removes the 'pedestals' of the proton peaks and better resolution can be afforded, especially when applied in the indirect domain of a 2D ^{1}H-^{1}H experiment (44,45). Both CRAMPS and FSLG work in spite of the MAS spinning. The currently available very fast MAS spinning cannot be applied, because the MAS spatially induced averaging would interfere with the similar averaging in the spin space. Both techniques work under the assumption that the full cycle of pulses is much shorter than one rotor cycle. More details on the homonuclear decoupling can be found in the section describing decoupling techniques. In spite of all the challenges, there is a big incentive to exploit the proton spectroscopy in the solid state, mostly due to the ubiquitous presence of protons in all organic compounds, their 100% natural abundance, and high gyromagnetic ratio, both translating into excellent sensitivity.

Heteronuclear SSNMR

The difficulty to get highly resolved proton spectra of crystalline solids largely precluded widespread applications of proton spectroscopy. The SSNMR spectra of magnetically active nuclei other than protons are typically void of the limitations encountered in proton applications. Carbon, nitrogen, fluorine, and phosphorus are the examples of spin $1/2$ nuclei present in pharmaceutically relevant systems.

Carbon and nitrogen are isotopicaly diluted nuclei, having 1.1 and 0.36% natural abundance, respectively. The spin dilution basically eliminates the homonuclear dipole-dipole coupling. For example, the probability of having two ^{13}C carbons next to each other is $1/10,000$. Even a moderate MAS speed will remove such dipolar coupling completely. Though a strong dipole–dipole coupling of carbons to protons still prevails, it is relatively easy to decouple by a high-power proton irradiation (3,4). Before the development of the signal-enhancing CP technique, the two serious limitations that precluded extensive use of carbon and nitrogen solid-state spectroscopies were low sensitivity and long T_1 relaxation times of these nuclei. Both of these limitations are largely removed by CP of NMR magnetization from protons. CP relies on a strong pool of proton magnetization that is transferred during the CP step by applying spin-locking pulses on both proton and the heteronuclear channels (13,14). The simultaneous spin-locking pulses on the proton and the insensitive nucleus channel must be of the specific rf amplitudes required by the Hartman-Hahn matching condition (13). The theoretical maximum enhancement factor is given by the ratio of gyromagnetic ratios of protons and the insensitive nuclei. The enhancement factor is up to $4\times$ for carbons and up to $10\times$ for nitrogens. At the beginning of the CP experiment the magnetization starts at protons. As such, it is the proton T_1 relaxation that determines the recycle delay between successive scans. Due to the efficient dipole–dipole relaxation mechanism, the proton relaxation (T_{1H}) is typically substantially faster than that of carbon (T_{1C}). The recycle delay between successive scans of CP-based experiment can be shortened significantly when compared with carbon direct polarization acquisitions without the CP signal enhancement. Therefore, the CP experiment improves carbon sensitivity and, at the same time, shortens the necessary recycle delays.

Fluorine and phosphorus are both almost 100% naturally abundant. The obvious advantage is great sensitivity. However, if more than one of these nuclei are present simultaneously in the molecule of interest, the homonuclear dipolar coupling may be significant, calling for CRAMPS-like type of line-narrowing solution. Even when a single fluorine or phosphorus is present, the distances between the two nuclei, each residing on different molecules, may still be short enough to give rise to significant homonuclear dipolar coupling.

Arguably, carbon is the most important of the aforementioned nuclei. Depending on the crystallinity and residual dynamics of the compound, the ^{13}C SSNMR peaks can be very narrow. Compounds with high internal mobility, such as adamantane with its fast rotation of the whole molecule around its center of gravity, narrows down the line widths below 0.01 ppm. Coupled to its favorable relaxation properties, adamantane is frequently used not only as a chemical shift standard but also for optimizing experimental variables and for shimming probes. A typical line width encountered in crystalline organic compounds is in the order of 0.5 ppm, but it is strongly dependent on the compound's mobility and the available proton decoupling field. It is not unusual to resolve the majority of carbon resonances in compounds containing 40 or more carbons.

Oxygen is also present frequently in pharmaceutically relevant materials, but the low natural abundance of magnetically active ^{17}O of 0.038% makes NMR applications on nonisotopically enriched samples impractical. Moreover, ^{17}O is a quadrupolar spin $(S = -5/2)$ bringing additional issues of quadrupolar interactions to cope with.

It is worth noting here that in many cases the nature of the samples studied does not allow for the incorporation of spin labels (e.g., analysis of production lots). Even when dealing with "research" samples, studied for development purposes, it is rather difficult to obtain isotopic enrichment of the small organic molecules in quantities suitable for SSNMR.

In addition to the nuclei discussed previously, other NMR spins may be occasionally present in pharmaceutical materials and may be relevant for NMR studies. Because many of these more "exotic" nuclei, such as chlorine, bromine, sodium, potassium, or manganese, possess spin higher than $\frac{1}{2}$, special experimental arrangements discussed earlier have to be applied.

PHARMACEUTICALLY RELEVANT APPLICATIONS OF SSNMR

Introduction

The goal of this review is to give the readers a balanced overview of all the currently available pharmaceutical applications of SSNMR. Less emphasis is put in the comprehensiveness of reviewing all the original literature. On the other hand, some nonpharmaceutical applications of SSNMR are also included (e.g., SSNMR characterization of polymers, food and cosmetic science, or quantification of natural organic matter). The reason for including these seemingly non-related topics is the potential of these applications for pharmaceutical use. The common thread is that they are applied to natural abundance samples to study crystalline and amorphous phases and, therefore, are directly applicable to characterization of pharmaceutical materials with little or no additional modifications. It is envisioned that the inclusion of these topics will bring new ideas and perspectives to the pharmaceutical SSNMR field. The primary emphasis is given to literature appearing after year 2000, but whenever appropriate, older literature is also included. The pharmaceutical SSNMR applications reviewed (in no particular order) are:

1. Crystal form characterization (polymorphism) and determination of the number of molecules per asymmetric unit.
2. Differentiation of crystalline and disordered (amorphous) phases.
3. Identification and characterization of excipients and nonpolymorphic impurities.
4. Identification of crystal-bound solvents.
5. Identification and quantification of the different phases when in mixtures.
6. Structural characterization based on fitting the experimental chemical shift and lineshape data.
7. Structural characterization based on measurement of interatomic distances.
8. Characterization of domain sizes.
9. Characterization of sample dynamics.
10. Characterization of drug–drug or drug–excipient interactions.
11. Physical and chemical stability and solid-state reactivity studies.
12. Characterization of hydrogen bonds and proton transfer.

Worth noting is that any of these SSNMR applications can be used to generate and claim an intellectual property. In a sense, generating the intellectual property is yet another pharmaceutical application of SSNMR.

Previous Reviews

Only reviews detailing pharmaceutical applications or containing some pharmaceutically relevant aspects are discussed. Brittain et al. (46) published a review on physical characterization of pharmaceutical solids containing a short chapter on SSNMR. Bugay (47) reviewed the theory and pharmaceutical applications of SSNMR. Holzgrabe et al. (48) published review on NMR spectroscopy in pharmacy, mostly dealing with solution application but containing a section on SSNMR. Burn et al. (49) published a chapter devoted to SSNMR in their book *Solid State Chemistry of Drugs*. Duer (50) published a review on the SSNMR studies of molecular motion. Sections on powder line shape analysis, relaxation time measurements, exchange experiments, and applications are included. Smith published a review of SSNMR as applied to various fields, such as organic, organometallic, bio-organic and inorganic materials, liquid crystals and membranes, silicates, microporous materials, and glasses and ceramics (51). This review-contains almost 1000 references to the original literature. Potrzebowski (52) presented a review of SSNMR studies of carbohydrates and their analogs. Many examples of ^{13}C CPMAS, ^{31}P MAS, ^{2}H MAS, 2D correlation experiments and single-crystal NMR studies are shown. Ye et al. (53) presented a concise review of modern SSNMR experiments. Stephenson et al. (54) published a review on quantitative issues of spectroscopic techniques (including SSNMR) when applied to pharmaceutical solids. Bugay (55) published a review of spectroscopic techniques used to characterize pharmaceutical materials in the solid state. A review of SSNMR is included. Harris (56) published a review on recent advances of SSNMR containing a clear and concise overview of important SSNMR experiments. Shapiro and Gounarides (57) published a rivew of high-resolution MAS (HRMAS) applications to combinatory chemistry. Bryce et al. (58) presented a review on practical aspects of modern routine 1D SSNMR spectroscopy. Ando et al. (59) and Kolodziejsi and Klinowski (60) reviewed the quantification aspects of CPMAS technique. Dybowski et al. (61) presented a short review of SSNMR, mainly considering years 2000 and 2001. Auger (62) published a short review on biological and pharmaceutical applications of SSNMR. The review contains mostly biological applications (structure and dynamics of lipid and proteins of biological membranes).

The short review by Reutzel-Edens and Bush (63) described the pharmaceutical applications of SSNMR in characterization of small molecules. The phase identification capability of SSNMR and its suitability for pharmaceutical formulations are stressed. The aspects of intellectual property and regulatory issues are also discussed. Schmidt-Rohr (64) published a review on dynamics in polymer. Although not centered exclusively around pharmaceutical applications, the review by Brown and Emsley (65) brings a concise and easy to-read account of modern SSNMR experiments and applications. Potrzebowski (66) published a short review titled "What high resolution SSNMR can offer to organic chemist." Medek (67) published a review of pharmaceutical applications of SSNMR. Rather than being an exhaustive review of existing literature, this chapter concentrates on review of the possible SSNMR applications to characterize pharmaceutical solids. Tishmack et al. (68) presented a comprehensive and well-balanced review of pharmaceutical applications. Offerdahl and Munson (69) published a short review on SSNMR of pharmaceutical materials. Vinogradov et al. (45) published a review on strategies for high-resolution proton spectroscopy in SSNMR. It compares the available techniques and evaluates their applicability under different conditions (spinning speeds, decoupling power, etc.). Watts (70) published a review on SSNMR in drug design and discovery for membrane-embedded targets. Aliev and Law (71–75) published a series of comprehensive but concise reviews of SSNMR spectroscopy in the years 2000 to 2005.

Review of Basic SSNMR Experiments

NMR can probe various aspects of materials of interest. By exploiting the nuclear interactions discussed in the previous chapter, one can, in general, gain understanding about the chemical and physical nature of the system, the 3D arrangement of its atoms and its dynamics. Before discussing the pharmaceutical applications in detail, it is useful to review the conventional SSNMR experiment arsenal. Only experiments commonly referenced throughout this review are discussed. Pulse sequences designed for specialized, one-of-type of studies, or even more conventional ones, but typically applied to other than pharmaceutically relevant studies are excluded.

The simplest SSNMR experiment is a direct signal observation (one pulse-acquire) on static samples (no sample rotation applied). Because all the nuclear interactions will show their full strength, this is applicable only on systems with a few isolated nuclei. For example, it would be of interest to describe the shielding tensor of a nitrogen in system where the $^{15}N-^{15}N$ homonuclear dipolar coupling can be neglected. Typically the static line shape would be acquired in the spin echo mode (Fig. 7) to eliminate the dead time of the electronics and to keep the pulse voltage ring down. It should be noted that similar information could be obtained with better sensitivity while spinning. This spin echo experiment can be also combined with MAS.

Decoupling Techniques

The most typical SSNMR experiment is the direct polarization (one pulse-acquire) coupled to MAS (Fig. 8). Of the pharmaceutically relevant spins, it is typically applicable only to highly abundant nuclei, such as ^1H, ^{19}F, and ^{31}P. A high-power proton decoupling during acquisition is applied to remove the heteronuclear dipolar coupling. A caution has to be exercized when combining the decoupling with MAS. The MAS can interfere with decoupling if the frequency of spinning is in the order of the decoupling field. An interference with decoupling can also arise if the frequencies of the random internal sample motions approximately match that of the decoupling field (76). Both types of interferences might substantially degrade the decoupling performance.

Figure 7   Spin echo experiment.

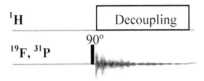

Figure 8 DP experiment, typically coupled with high-power proton decoupling and magic angle spinning. *Abbreviation*: DP, direct polarization.

Two-pulse phase modulation (TPPM) decoupling sequence rather than the simple continuous wave (CW) decoupling is typically used (77). The TPPM scheme uses two pulses of duration corresponding to slightly less than 180° flip angle with a phase difference of 15 to 30°. It shows significantly better off-resonance performance than CW, especially when combined with a moderately fast MAS. More recently, novel heteronuclear decoupling schemes were designed by various groups, outperforming TPPM under special conditions (e.g., very fast MAS) (78–83). Noteworthy are the XiX and SPINAL-64 sequences. The XiX decoupling consisting of windowless rf irradiation with a repeat of two pulses of equal width, phase shifted by 180° (81,84). Gains in peak height up to 29% over TPPM were demonstrated but realized mostly when coupled to high-spinning speeds (>30 kHz) and decoupling fields (>110 kHz). Interestingly, for very high MAS spinning speed exceeding the proton-proton dipolar coupling (i.e., >40 kHz for rigid solids), it has been shown that much lower decoupling power actually achieves almost the same decoupling efficiency as the optimized decoupling at 15 times higher power (Fig. 9) (83,85,86). SPINAL-64 decoupling, originally developed for liquid crystal applications, consists of windowless irradiation with changing of phase in a stepwise manner and then combining the basic cycles into supercycles (79). Better TPPM performance has been shown for crystalline solids, especially for low rf power levels (87). More recently, DePaepe et al. (88) devised a methodology of optimizing the performance of the heteronuclear decoupling. This generic approach, which is referred to as eDUMBO, consists of an iterative optimization of the decoupling performance based on experimental feedback. The eDROOPY decoupling sequence (with a simple cosine modulation of the decoupler RF phase) was designed using the eDUMBO approach and was shown to outperform the optimized TPPM and XiX sequences (Fig. 10) (89,90).

A quite different class of decoupling schemes is used for homonuclear decoupling (i.e., removing the strong proton–proton dipolar interaction while preserving chemical shift or heteronuclear coupling) (34–41,43,91–103). A recent comprehensive account of homonuclear decoupling was given (45). The strong homonuclear dipolar coupling of protons is the direct result of their 100% natural abundance and high gyromagnetic ratio. Due to the homogeneous character of this coupling, MAS narrows down the proton peaks substantially

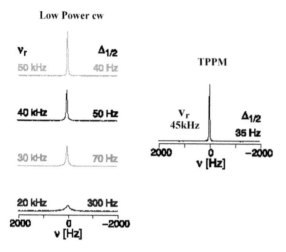

Figure 9 Comparison of low power CW decoupling at various spinning speeds ν_r to optimized TPPM. Relative intensities and line widths of the α-carbon resonance of $[^{13}C\text{-}2]$ glycine are shown. *Abbreviations*: CW, continuous wave; TPPM, two-pulse phase modulation. *Source*: Adapted from Ref. 85.

only at speeds equivalent of exceeding the strength of the dipolar coupling. To efficiently narrow down the proton line widths, MAS must be combined with similar averaging in the spin space. The early pulse sequences to achieve this were Lee-Goldburg (LG) (5,104) and Waugh, Huber, and Hacberlen (WHH4) (5). Because they were applied to static samples, the proton lines were broadened by the chemical shift interaction. CRAMPS (30,32) (combined rotation and MAS) combined both the spin space and real-space averaging techniques. Based on the averaging principle, there are four classes of homonuclear decoupling techniques: (*i*) solid echo-based sequences [WHH4 (5), MREV8 (105,106), BR24 (92), CORY24 (94), BLEW12 (93), and DUMBO (99), (*ii*) magic echo sandwich-based sequences [TREV8 (107) and MSHOT3 (37,108)], (*iii*) Lee-Goldburg–based sequences [LG (104), FSLG (40–42), PLMGn (42,43)], and (*iv*) rotor synchronized sequences [semi windowed WHH4, CN$_\kappa^\chi$

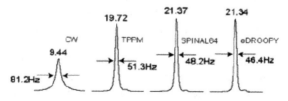

Figure 10 Comparison of the decoupling performance of different decoupling techniques. Relative intensities and line widths of the α-carbon resonance of $[^{13}C\text{-}2]$ glycine are shown. *Source*: Modified from Ref. 89.

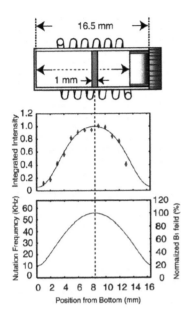

Figure 11 The irradiation field profile of Bruker 7-mm rotor as a function of the rotor z-axis disctance. The profile was obtained by measuring NMR signal of 1-mm thickness rubber disk placed at different z-positions inside the rotor. The irradiation field profile is shown at the *bottom*. Note that this profile is far from the ideal flat response. It can also hamper quantification of multiple components if there is substantial component segregation inside the rotor. *Abbreviation*: NMR, nuclear magnetic resonance. *Source*: Modified from Ref. 173.

(109) and RN^{χ}_{κ} (110)]. PLMG and DUMBO sequences were found to be preferred due to their relative ease of implementation and their adaptability to fast MAS spinning speeds (45). The rf irradiation inhomogeneity (Fig. 11) and the contribution of unaveraged higher order terms remain the main reasons for limiting the attainable resolution.

CP Techniques

Because the majority of the pharmaceutical SSNMR applications involve ^{13}C acquisitions, arguably the most typical pulse sequence is the CP between protons and carbons (Fig. 12), which was already discussed in the previous section. In order to set up the CPMAS experiment parameters, it is useful to know the various relaxation times of the system under study. As discussed next, the CP dynamics is a function of the relaxation times of both nuclei involved (T_{1H}, $T_{1\rho H}$, T_{1C}, and $T_{1\rho C}$) and the CP constant T_{CH}. Due to the long carbon relaxation times, CP dynamics can often be approximated by employing just the proton relaxation. The most sensitive (and fastest) way to determine proton relaxation times is through direct proton detection. However, if a

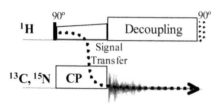

Figure 12 Cross-polarization magic angle spinning (CPMAS) experiment. After the initial 90° proton pulse, spin-locking fields on both channels are applied matched according the Hartmann-Hahn condition, causing a net signal flow from the abundant proton reservoir to rare nucleus (schematically shown as the *dotted arrowhead line*). The rare nucleus signal is detected usually with high-power proton decoupling. The last proton 90° pulse (usually referred to as the proton flip-back pulse) is optional. It returns the remaining proton transversal magnetization into the $+I_z$ direction, thus speeding up the return to equilibrium between the successive scans.

mixture with multiple components is involved, heteronuclear detection may offer advantage in resolving the components. Figure 13 shows the CP-based pulse sequence for measurement of the proton longitudinal relaxation time (T_{1H}). The equilibrium proton magnetization is inverted by the initial 180° pulse, followed by the variable time relaxation delay τ_{relax}, during which the magnetization returns back to equilibrium with the rate corresponding to the inverse of T_{1H}. After the proton 90° pulse, the proton magnetization is transferred to the xy plane, where it can be either directly detected by proton acquisition, or, as actually shown in Figure 13, it is cross-polarized and detected heteronuclearly. Figure 14 shows the pulse sequence for measurement of the proton relaxation time in the rotating frame ($T_{1\rho H}$). The equilibrium magnetization is transferred to the xy plane followed by the spin-locking pulse of variable lengths during which the magnetization decays by the rate corresponding to the inverse of $T_{1\rho H}$. The magnetization can again be detected directly by proton acquisition or, as shown in Figure 14, after cross-polarizing it to heteronuclei. It should be noted that the value of the measured $T_{1\rho H}$, depends strongly on the strength of the applied irradiation field (the stronger the rf, the longer the $T_{1\rho}$).

In many cases, especially at higher magnetic fields, the maximum rotor spinning speed is smaller than (or on the order of) the chemical shift anisotropy,

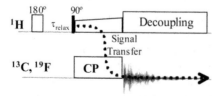

Figure 13 CPMAS-based proton T_{1H} relaxation sequence.

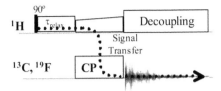

Figure 14 CPMAS-based proton $T_{1\rho H}$ relaxation sequence.

leading to a spinning sideband manifold, which may (*i*) compromise resolution by introducing peak overlap, (*ii*) complicate quantification, and (*iii*) lead to misassignments of peaks in the absence of variable spinning speed repetitions. In 1982, Dixon designed an ingenious and simple way to suppress the spinning sidebands by introducing the total suppression of spinning sidebands (TOSS) sequence (111,112). TOSS cancels the spinning sidebands by destructive interference by adding four (or in general $2n + 2$) of the properly spaced 180 carbon pulses right after the CP step (Fig. 15). An added advantage of TOSS is the long ($\sim$100 μs) delay before the start of acquisition, eliminating the receiver dead time problems. However, if many spinning sidebands are present, TOSS may lead to slight phase distortions, incomplete suppressions, or even negative intensities of the spinning sidebands.

Spectral Editing

A very important class of SSNMR pulse sequences represents the spectra editing techniques, because they simplify spectra according to some factors (usually proton multiplicities) to facilitate the desired peak assignment. As evidenced by many examples of the reviewed literature that follows, the typical 1D SSNMR peak assignment consists of comparison of the chemical shift with solution assignments in combination with spectra editing. In many cases, a close-to-complete peak assignment can be achieved by using this approach. The most simple of the editing techniques is the so-called interrupted decoupling (also referred to as dipolar dephasing or nonquarternary carbon suppression), which consists of switching off the decoupling field for short period of time of

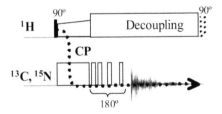

Figure 15 CPMAS sequence with TOSS (total suppression of spinning sidebands) and proton "flip back" pulse.

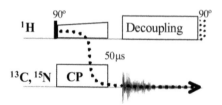

Figure 16 Interrupted decoupling spectral editing.

approximately 40–100 μs after CP and before acquisition (Fig. 16) (113). The decoupling delay dephases the strongly proton-coupled carbons (CH and CH₂ groups) but leaves the qaurternary, and also CH₃, groups largely unchanged. The methyl group is not dephased due to the partial averaging of their ^{1}H–^{13}C dipolar coupling as a result of fast, wheel-like rotation in the solid state. Complementary to this technique is the short CPMAS (also referred to as quaternary carbon suppression), with CP limited to only a short period of approximately 100 μs. During this short CP, unlike the protonated carbons, the quaternary carbons do not have enough time to develop appreciable magnetization and are suppressed. When the CP is combined with short period of polarization inversion with 180° change of phase (CPPI; Fig. 17A), CH signals are cancelled, CH₂ are negative with intensity of 1/3, and methyl and quaternary carbons are attenuated by 62 and 86%, respectively (114). Methylene only spectra can be obtained by the simultaneous phase inversion (SPI) variant of this pulse sequence (Fig. 17B), (115,116). Later, this technique was improved to give better suppression of artifacts and better S/N. It was termed CPDDRCP for CP

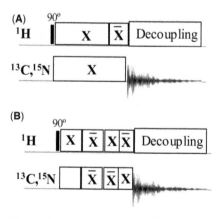

Figure 17 CPPI. The X and $\bar{X}$ symbols denote 0 and 180° phases of the pulses, respectively. (**A**) Standard CPPI affording negative CH₂, positive CH₃ and quaternary and no CH carbons. (**B**) SPI affording only (negative) CH₂ sub-spectra. *Abbreviations*: CPPI, Cross-polarization with polarization inversion; SPI, Simultaneous phase inversion.

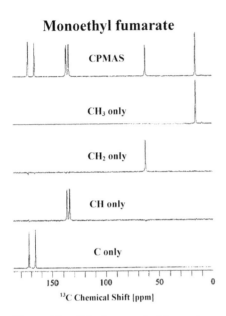

Figure 18   Edited monoethyl fumarate spectra. *Source*: Modified from Ref. 117.

dipolar dephasing re-crosspolarization, and it affords CH and CH_2 only spectra (Fig. 18) (117). Different approaches to select methylene only subspectra were used by Rossi et al. (118). This DQCP approach is based on first generating proton double-quantum coherence, which develops fast for CH_2 groups, filtering it, converting back to single quantum coherence, followed by CP and detection on carbon channel. As an example, an application of these sequences to select side-chain carbons of peptides was shown by Kumashiro et al. (119). De Vita and Frydman (120) utilized the editing power of the amplified dipolar dephasing coupled with the fact that at moderate spinning speeds (>10 kHz) the CH and CH_2 groups behave like isolated spins, that is, they are isolated from the surrounding proton reservoir. The technique distinguishes between CH, CH_2, and quarternary/methyl groups. No homonuclear decoupling has to be applied unlike the windowless isotropic mixing for spectral editing sequence (WIMSE) (121). Various other variants of dipolar coupling-based spectral editing appeared in the literature (122–127). Selection of carbons bound to ^{14}N was shown by Schmidt-Rohr and Moa and named saturation-pulse-induced dipolar exchange with recoupling (SPIDER) (128). For quadrupolar nuclei, editing out the narrow line components with smaller quadrupolar coupling strength was demonstrated (129)

   The editing techniques discussed so far are all based on through-space dipolar coupling. Lesage et al. (130) used hetoronuclear J-coupling editing approach. The result of this through-bond transfer is the solid-state equivalent

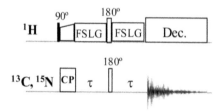

Figure 19 SS-ATP. FSLG is frequency-shifted Lee-Goldburg decoupling, and $\tau = 1/(2J_{CH})$ with typical values ranging from 4 to 6 ms. *Abbreviation*: SS-ATP, Solid-state attached proton test.

of the solution sequence-attached proton test, named SS-ATP (Fig. 19). This sequence affords spectra with CH_2 groups having an opposite phase to the CH and CH_3 groups (quaternary carbons are not observed; Fig. 20). Although excellent editing efficiency is obtained, narrow carbon lines of less than approximately 50 Hz are required to resolve the J-coupling ($^1J_{CH}$, scaled to approximately 70 Hz by the FSLG scaling factor), restricting the use of this technique to highly crystalline samples. The J-coupling editing approach was further developed to afford pure subspectra of carbons according to the proton multiplicity. It is based on multiple quantum filtering of the proton coherences (J-MQF; Fig. 21) (131). Unlike the dipolar coupling-based editing, this method is insensitive of molecular motions and MAS spinning. However, it suffers from rather low sensitivity.

Although it is not the spectral editing technique, a spin selection based on the solution variant of IPAP experiment was shown to enhance resolution of multiple spin-labeled systems, because the contribution of J-coupling to the line width is removed (132,133). A selectivity in coherence transfer for 2D

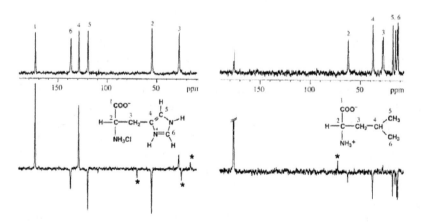

Figure 20 SS-APT spectral editing. *Abbreviation*: SS-APT, solution sequence-attached proton test. *Source*: Modified from Ref. 130.

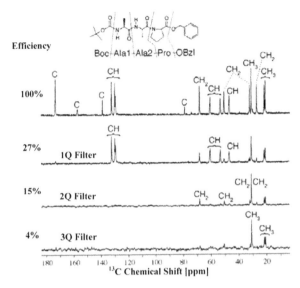

Figure 21 J-MQ-filtered ^{13}C CPMAS spectra of the tripeptide Boc-Ala-Ala-Pro-Obzl. *Abbreviations*: MQ, multiple quantum; CPMAS, cross-polarization and magic angle spinning. *Source*: Modified from Ref. 131.

recoupling experiments, which can lead to enhanced resolution was also reported by Carravetta et al. (134). An 1D technique to separate spinning sidebands according to their order was proposed (135).

Correlation Techniques

To take the full advantage of the wealth of information SSNMR offers, multi-dimensional correlation techniques should be employed. In general, different nuclear spin interactions are selectively switched on and off during the evolution, mixing, and acquisition periods of these experiments. The purpose of using multi-dimensional experiments is to have more information available (e.g., better resolution) by using the higher dimensions. On the other hand, many of the multi-dimensional techniques could be reduced to their 1D variants. In these cases, the mixing period of the originally multidimensional experiment serves as a selective information filter. Some examples of the correlation techniques include chemical shift correlations (tracing through space or through bond atom connectivities and facilitating a site-specific chemical shift assignments and confirming the identity of molecules), recoupling techniques (reintroducing the MAS averaged anisotropic interactions—typically dipolar coupling but also the chemical shift anisotropy or the quadrupolar coupling—to gain structural information, such as distances and torsional angles or also to help assignments), and chemical exchange techniques (studying slow dynamics or chemical reactions). Again, as in other sections, the emphasis is to preferentially review

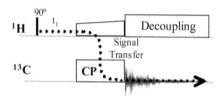

Figure 22 WISE (wideline separation) experiment separates the proton MAS line shapes by the carbon chemical shift of close-by carbons. Because it involves CP transfer, it is a through-space correlation technique. It is typically applied to get information on mobility of the different phases in the sample. *Abbreviations*: MAS, magic angle spinning; CP, cross-polarization.

techniques applicable to nonlabeled materials conducive with most pharmaceutical applications.

WISE (wide line separation; Fig. 22) is a 2D technique separating the proton MAS lines by the carbon chemical shifts (136). The free proton evolution periods follows the initial proton 90° pulse, after which the magnetization is cross-polarized and detected on the carbon channel. Because WISE involves CP-based polarization transfer, the correlation is dipolar coupling driven and, consequently, takes place through space. If multiple phases with limited spin diffusion exist, the mobility of the protons in the different phases can be judged by the line width of the corresponding proton slices (wide lines correspond to rigid domains with low mobility and vise versa). This technique is also frequently applied in its reduced, 1D version, with the proton evolution representing the proton T_2 or "mobility" filter. It is especially suited for polymer science to filter out the rigid crystalline polymer domains. By setting the proton evolution time to longer values, the signals originating from the rigid components decay to zero and only the mobile portion of the sample is detected.

ODESSA (one-dimensional exchange spectroscopy by side-band alteration) is a rotor-synchronized MAS exchange technique, with preparation period fixed to one-half of a rotation period and a mixing time equal to an integer number of rotation periods (Fig. 23) (137). After the half-rotor period and just before the mixing, the spinning sidebands are polarized in the alternate fashion. This creates an initial difference in polarization. Dipolar coupling-driven spin

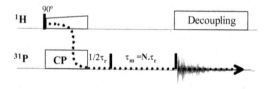

Figure 23 ODESSA sequence. *Abbreviation*: ODESSA, one-dimensional exchange spectroscopy by sideband atteration.

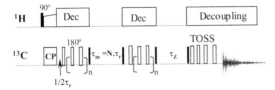

Figure 24 The 1D CODEX characterizes slow motions through chemical shift aniso-tropy modulations. After the cross-polarization step, the magnetization evolves under chemical shift anisotropy (recoupled by the series of 180° pulses), followed by mixing time τ_m and second evolution period under chemical shift anisotropy. At the end of this period, the signal is refocused by stimulated echo. The amplitude of the echo, reduced by the motions, is detected after the TOSS segment as magic angle spinning peak intensity. *Abbreviation*: TOSS, total suppression of spinning sidebands; CODEX, center band only detection of exchange.

exchange then results in a redistribution of the polarization and the corresponding decay in the sideband intensities. Internuclear distances can thus be measured. High abundance nuclei (^{19}F or ^{31}P) or isotope enrichment is required.

A technique related to ODESSA is the center band only detection of exchange experiment (CODEX) (138–140). CODEX enables the observation and characterization of slow molecular motions with the high resolution of ^{13}C chemical shifts (Fig. 24). The chemical shift anisotropy serves as the actual probe of the motions. The type of motions can be established from a series of 1D CODEX experiments.

The 2D HETCOR (heteronuclear correlation) experiment correlates abundant nucleus (protons) chemical shift through space to nearby carbons (or other hetreronulei) based on the dipolar coupling mechanism (Figs. 25, 26) (141–143). For short CP contact times, the one bond H–C correlations are preferably selected, whereas at longer contact time this through-space experiment also often includes intermolecular correlations, which, depending on the application, may or may not be desired. Mao et al. (144) applied 2D HETCOR and 2D SUPER (undistorted powder-patterns by effortless recoupling) to characterize

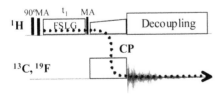

Figure 25 HETCOR (hetoronuclear correlation) correlates proton and carbon chemical shifts by through-space dipolar interaction. The homonuclear frequency-shifted Lee-Goldburg (FSLG) decoupling is used during proton t_1 evolution to decouple protons from each other. MA stands for pulses of magic angle (54.7°) flip angle.

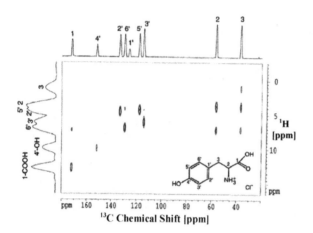

Figure 26 Dipolar coupling-based, trough space 2D $^1H-^{13}C$ FSLG (homonuclear frequency-shifted Lee-Goldburg) heteronuclear correlation on $[U-^{13}C]$ L-tyrosise hydrochloride. *Source*: Modified from Ref. 141.

biocolloidal organic matter (145,146). Sozzani et al. (147) applied HETCOR to study polymer multi-phase (nanocomposite) materials.

DeLacroix et al. (148) developed 2D TOSS-based experiment separating spinning sidebands by their order (Fig. 27). This technique allows for separation of the isotropic and anisotropic chemical shifts along two dimensions, making it possible to analyze the spinning sideband manifolds at moderate spinning speed. Smith et al. (149) applied this experiment to study conformational polymorphism of ROY compounds (Fig. 28). Chemical shift anisotropy of ROY compounds were obtained. Several multidimensional techniques based on magic angle turning and hopping, correlating carbon isotropic shifts to their anisotropic powder patterns, were developed in David Grant's lab. When coupled to ab-initio calculations, FIREMAT (five π-replicated magic angle turning) and its preceding versions, provide an excellent tool to study molecular conformations and polymorphism (150–159). An improvement of FIREMAT enabling faster spinning,

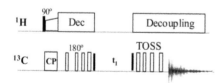

Figure 27 The 2D TOSS: spinning sidebands are separated according to their orders. The t_1 evolution is typically sampled over only one rotor period. The number of t_1 increments must be chosen to correspond to the number of the sidebands. *Abbreviations*: TOSS, total suppression of spinning sidebands.

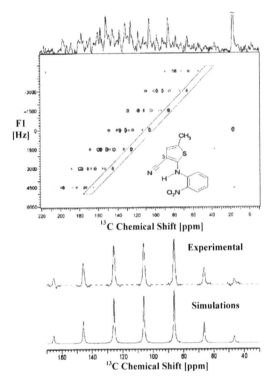

Figure 28 The 2D TOSS spectrum of ROY-orange form (*top*). Simulated spectrum using a best fit to the CSA tensor values (*bottom*). *Abbreviation*: TOSS, total suppression of spinning sidebands. *Source*: Modified from Ref. 149.

and hence avoiding the need for ultrastable slow spinning speeds, was designed at the same lab and is referred to as SPEED MAS (160). The technique uses a series of π-pulses to effectively reduce the apparent spinning speed. Elena et al. (161) developed an experiment separating fast and slow spinning sidebands, dubbed ROSES. The principle of the chemical shift anisotropy amplification was applied to increase the number of sidebands in the indirect domain of the 2D correlation (162).

Gullion and Schaefer (163–174) introduced an important technique to measure internuclear distances, which is referred to as rotational echo double resonance (REDOR, Fig. 29). Inserting several π pulses within the rotor period of MAS interferes with its ability to average out the heteronuclear dipolar coupling (recouples the averaged dipolar coupling), and as such, decreases the amplitude of the rotational spin echoes (Fig. 30). The attenuation of the rotor echoes is proportional to the heteronuclear dipolar coupling and hence to the distance between the nuclei. REDOR was originally devised for measuring $^{13}C–^{15}N$ distances specifically in ^{13}C- and ^{15}N-labeled samples. By judicious choices of

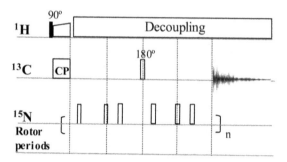

Figure 29 Basic $^{13}C\{^{15}N\}$ REDOR sequence. Inserting the π pulses within the rotor periods of MAS interferes with MAS ability to average out the $^{13}C-^{15}N$ heteronuclear dipolar coupling and, as such, decreases the amplitude of the rotational spin echoes (Fig. 30). The attenuation of the rotor echoes is proportional to the heteronuclear dipolar coupling and hence to the distance between the $^{13}C-^{15}N$ nuclei. *Abbreviations*: REDOR, rotational echo double resonance; MAS, magic angle spinning.

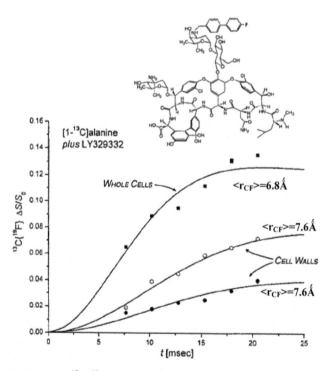

Figure 30 $^{13}C\{^{19}F\}$ REDOR dephasing curves as a function of the dipolar evolution time for complexes of LY329332 (chloroeremomycin) with whole cells and cell walls. The cell walls with LY329332 binding occupancy of 33% (*open circles*) and 16% (*closed circles*) were used (hence the different dephasing corresponding to the same average fluorine–carbon distance). *Source*: Modified from Ref. 174.

the placement of the pairs of labels, a structure of a molecule can be determined one distance at a time. Multiple improvements of the original REDOR scheme appeared in literature (175–178). The notable one, θ-REDOR, enables simulataneous measurements of many distances in multiple-labeled molecules (179–182). The methodology was also extended to measure distances between quadrupolar nuclei (SEDOR, REAPDOR, and TRAPDOR) (183–191). Fu measured $^{15}N-^{1}H$ bond lengths based on REDOR-type experiment (192). FSLG was used during proton evolution. A simple compensation for pulse angle error in REDOR experiments was also devised (193).

Saalwachter and Spiess (194) designed REDOR-based heteronuclear correlation sequences (HMQC, HSQC, HDOR, and RELM), which work optimally at very high MAS spinning speeds (>20 kHz). They use the REDOR-like pulse sequence to excite and re-convert the heteronuclear coherences. Another set of these four heteronuclear correlation experiments designed in the same lab is based on the recoupled polarization transfer (REPT) (195). Weak heteronuclear dipolar coupling constants can be extracted by means of the spinning sideband analysis extracted from the indirectly detected domain, and structural information can be obtained. Comparison of the REPT performance with that of HETCOR, MAS-J-HMQC is shown in L-tyrosine, which is abundant in nature. Applications to spectral editing are also shown. All these sequences work best at high-spinning speeds and are applicable to natural abundance samples encountered in most of the pharmaceutical applications. Saalwachter and Schnell (196) further built on the REDOR scheme, making it more sensitive to extract the $^{1}H-^{13}C$ couplings for up to medium range distances (REREDOR).

The same authors also introduced $^{1}H-^{13}C$ and $^{1}H-^{15}N$ correlation experiments with highly improved sensitivity by inverse detection on protons, applicable also to natural abundance ^{15}N samples (197,198). Ishii et al. (199,200) described the merits of inverse versus direct detection of $^{1}H-^{13}C$ correlation. Applications to unlabeled polymers are shown (Fig. 31). Indirect detection was also applied to peptides and proteins (201–204) and small molecules (205).

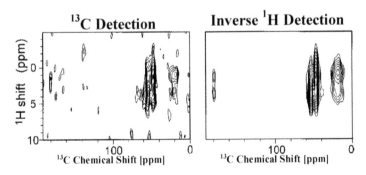

Figure 31 The standard carbon (*left*) and inverse proton dectection (*right*) of 2D $^{13}C-^{1}H$ heteronuclear correlation spectra of poly(methyl methacrylate). Note the sensitivity enhancement of the inverse-detected spectrum. *Source*: Modified from Ref. 200.

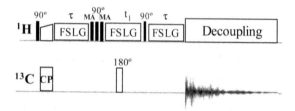

Figure 32 The MAS-J-HMQC (magic angle spinning, J-coupling based heteronuclear multiple quantum correlation) through bond correlation experiment. The homonuclear frequency shifted Lee-Goldburg (FSLG) decoupling is used during the J_{CH} coupling evolution period τ, as well as during proton t_1 evolution to decouple protons from each other while preserving the J_{CH} coupling during period τ. MA stands for pulses of magic angle (54.7°) flip angle. The carbon 180° pulse refocuses carbon chemical shift.

Building on the previously proved concept of HETCOR by J-coupling cross-polarization (206), Lesage and Emsley (207,208) introduced through-bond MAS-J-HSQC (J-coupling-based heteronuclear single quantum correlation) and MAS-J-HMQC (J-coupling-based heteronuclear multiple quantum correlation; Fig. 32) experiments, which correlate proton chemical shift by the J-coupling mechanism. Both sequences afford similar proton resolution on unlabeled samples, but MAS-J-HMQC offers slightly better sensitivity and less artifacts. For ^{13}C-labeled samples, MAS-J-HSQC provides a resolution advantage. The same lab showed applications of these techniques to the complete ^{13}C and ^{15}N resonance assignments of naturally abundant peptides in the absence of any other solution chemical shift or structural information (Fig. 33) (209). Based on their through-bond nature, these experiments assign the short and long range correlations very reliably without any artifacts, because no inter-molecular correlations are present in the spectra. The authors also devised a variant of the MAS-J-HSQC suitable for detection of sample mobilities. In essence, it is the WISE experiment coupled to the through bond, J-coupling-based ^{1}H–^{13}C polarization transfer, named J-WISE (210). The advantage of this experiment over WISE stems from the fact that J-WISE can distinguish bonded and nonbonded interactions. It was applied by the authors to localize bound water in the onion cell architecture. Total through-bond correlation spectroscopy (TOBSY) for solids was also devised (211–213). Alonso and Massiot (214) used different variants of through-bond heteronoclear correlation based on solution INEPT experiment to characterize mesostructured materials.

Another type of J-coupling-based, through-bond correlation is the solid-state analog of the well-known solution experiment 2D INADEQUATE (Fig. 34) (215). It is typically applied to carbons to trace the through-bond carbon–carbon connectivities. It is a very powerful technique for carbon assignments. It is based on the excitation and evolution of DQ carbon coherences and as such requires the presence of two connected ^{13}C nuclei. In unlabeled samples, on

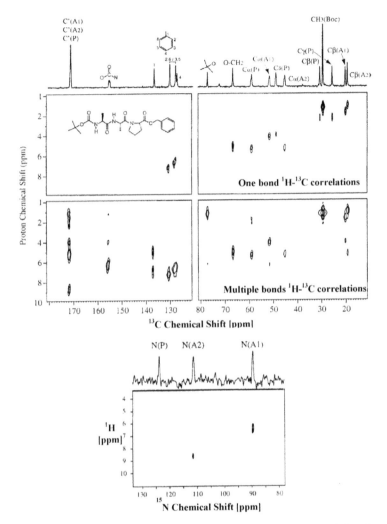

Figure 33 Through bond 2D MAS-J-HMQC spectra of the natural abundance tripeptide Boc–Ala–Ala–Pro–O–Bzl. The $^1H-^{13}C$ one bond and multiple bond correlation spectra were recorded with evolution times of 1.6 and 16 ms, respectively. The $^1H-^{15}N$ correlation spectrum was recorded with evolution time of 4.32 ms. *Source*: Modified from Ref. 209.

average, only one out of 10,000 carbons conforms with this scenario. Spin enrichment is, therefore, preferred, but the authors also show an INADEQUATE application to naturally abundant L-alanine. This experiment was also applied to observe N—H—N hydrogen bond directly, through $^{15}N-^{15}N$ J–coupling mediated by the H-bond (216). The refocused variant of this experiment was applied to carbon assignment of disordered solids, such as wood and

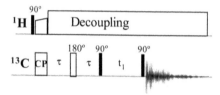

Figure 34 The 2D INADEQUATE for tracing carbon-carbon connectivities by through-bond $^1J_{CC}$-coupling based correlation. The carbon double quantum coherence is formed after the first carbon 90° pulse, evolved during t_1, and detected as an antiphase coherence after the second carbon 90° pulse.

cellulose (217). From the experimental T_{2C} values, which were in the order of 10 times longer than the apparent T_{2C}^* based on the inverse of line widths, authors conclude that the contribution to the line width must be through dispersion of isotropic chemical shifts. Moreover, authors noted that the distribution of chemical shift was highly correlated (i.e., if one carbon chemical shift changes due to the local change in the structure of the amorphous compound, the other bonded carbon reacts to the structural change in very well-determined way) and a much improved high resolution can be obtained in the sheared projection of the 2D spectra (218,219). The origin of the disorder (its type) can be characterized by reproducing the experimental spectra using simulations (220). Kono et al. (221–226) applied INADEQUATE and MAS-J-HMQC to study the polymorphism of cellulose triacetate and other biopolymers. Kaji and Schmidt-Rohr (227) devised a filtered INADEQUATE approach, where the observed magnetization is dephased by the difference of the recoupled anisotropic chemical shifts. It was applied to estimate trans/gauche ratios in double ^{13}C-labeled polymers.

Spiess group developed a correlation sequence based on DQ excitation with the use of back-to-back pulses (BABA). Initially, it was developed as a heteronuclear correlation sequence (228,229). In the indirect domain of such 2D spectrum, each proton resonance forms spinning sideband pattern with each sideband exhibiting narrow lines. Later, this sequence was generalized for any DQ excitation and typically applied to homonuclear correlations between abundant nuclei (230). Brown et al. (231) showed the application of BABA to investigate hydrogen-bonding structure of bilirubin. Analysis of the proton DQ spinning sideband patterns allowed for quantitative determination of proton–proton distances and the geometry. The shortest distances (the largest dipolar couplings) were obtained to a very high degree of accuracy (>0.01 nm).

Countless other recent multidimensional solid-state experiments can be found in the literature, especially in the field of structural characterization of proteins, which has received lots of attention recently. It is beyond the scope of this review to cover all the sequences applicable to the SSNMR characterization of proteins, since, in general, the techniques require samples with isotope enrichment and, as such, are of limited interest to pharmaceutical field. Typically, different types of recoupling of the rare nuclei (^{13}C or ^{15}N) anisotropic inter-actions are applied to reintroduce the structural information that are otherwise lost by the averaging effect of fast MAS (128,173,180,195,232–258). However, some of these recoupling schemes are applied to the abundant proton nuclei only and as such could be applied to pharmaceuticals. Dusold and Sebald (241) published a review of the different recoupling schemes. Some selected references to correlation experiments requiring isotope enrich-ment are given (181,212,259–298).

Chemical Shift Assignments

The knowledge of the site-specific chemical shift assignments of peaks to the different nuclei present in the molecule is not absolutely essential in order to extract the pharmaceutically relevant information, such as to differentiate crystal forms. However, the known assignments provide additional valuable infor-mation, so that one can relate the chemical shift differences between polymorphs to conformational changes at particular molecular sites. Additional structural insights can be gained by comparing the experimental assignments to calculated shifts. Lesage et al. (209) recently demonstrated that spectral assignments can be made by relying solely on the SSNMR data from the natural abundance samples. All carbon, nitrogens, and proton peaks of a tripeptide sample were com-pletely assigned in the solid state. However, for small molecules with a limited number of resonances and good spectral dispersion, most of the assignments can be done by direct comparison of the liquids and solid-state spectra. The SSNMR resonances usually fall within several parts per million from solution shifts. Unlike the solution, no solvent is present to affect the chemical shifts of solids, and the dynamics of molecules in the solid state is typically much slower so that only one kinetically frozen conformer (polymorph) typically exists, struc-ture of which may substantially differ from that in solution. There is a solid state-specific contribution to chemical shifts that does not exist in solution—the effect of crystal packing (the intermolecular contribution). The solution-solid state chemi-cal shift differences based on the different average molecular conformations typi-cally do not exceed 10 ppm. A smaller difference, typically not exceeding 3 ppm, comes from the packing of molecules in the crystal lattice (299).

The assignments of the well-resolved resonances can be inferred by comparing the solid-state spectra with their assigned solution counterparts. The remaining peaks can usually be assigned from the application of the spectra-editing techniques discussed previously. Due to the relative ease of implemen-tation and wide range of applicability, the short CP and dipolar dephasing editing are preferred. If assignment ambiguities still exist, such as when methyl-ene carbons resonate close to methines, more complex dipolar or J-coupling-based editing techniques should be applied. Finally, application of the various multidimensional correlation techniques discussed earlier should be able to remove any prevailing assignment ambiguities. Further check can be provided by the comparison of experimental shifts to empirical or ab initio quantum mechanical calculations of chemical shifts, which is likely to become an important tool for resonance assignments. Specifically, the ab-initio type calculations are likely to prove useful in predicting conformations of polymorphs by comparing the 3D molecular structure-dependent simulated chemical shifts with experimental data.

Molecular Structure from Chemical Shifts and Molecular Modeling

A knowledge of the experimentally determined principal values of the chemical shift interaction tensor, or at least the isotropic value, enables one to extract a wealth of structural information about the sample. This is true, especially when coupled to calculations. The speed of modern computers allows for relatively accurate predictions of chemical shielding molecules with multiple heavy atoms, such as carbons, oxygens, and nitrogens. By reproducing the experimentally determined chemical shift tensor values by means of ab initio or empirical calculations, the 3D structure of the compound can be determined. It should be noted that due to the presence of the crystal lattice, intermolecular interactions are considered to obtain accurate results. Some software packages allow for incorporation of the crystal lattice effect. In principle, chemical shift variations between the different molecules per asymmetric unit and different crystal forms can be predicted in this way. The importance of this modeling approach is likely to grow as the calculation methods improve.

Ando et al. (300) examined the ^{13}C chemical shifts of aromatic carbons as a probe for estimating the conformation of aromatic groups in the solid state. The chemical shifts of the biphenyl group show systematic correlations with the inter-ring torsion angles, when the substituent shielding effects are removed. The correlations can be used to estimate the conformation of biphenyl groups incorporated into larger molecules. The ab initio GIAO-CHF calculations agreed well with experimental results.

Maciejewska et al. (301) studied structure of 4-(4-methoxyphenyl)hexahydro-1H,3H-pyrido[1,2-c]pyrimidine by ^{13}C CPMAS and ab initio GIAO-CPHF chemical shielding calculations. The studied compound showed two molecules per asymmetric unit. To assign carbons specifically to the two molecules, the shielding constants were calculated using the X-ray geometry. Good fits with $R = -0.9988$ (molecule A) and $R = -0.9994$ (molecule B) were obtained between the experimental and calculated values (Fig. 35).

Buchanan et al. (302) compared SSNMR and X-ray to study the stereochemistry of 4-acetyl-, formyl-, and carboxy-benzo-9-crown3 ether. The differences between chemical shifts were rationalized by LORG calculations as a function of the aryl-O-CH$_3$ torsion angle. A complementary role of the two techniques was illustrated.

Dega-Szafran et al. (303) studied some N-(ω-carboxyalkyl)morpholine hydrohalides by spectroscopic techniques and PM3 calculations. Contrary to the solution, ^{13}C CPMAS spectra showed nonequivalence of the ring carbon

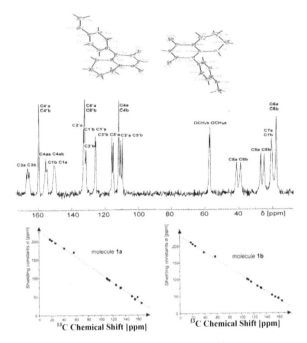

Figure 35 Chemical shift assignments of ^{13}C CPMAS spectrum (*middle*) of a compound shown at *top*. This compound contains two molecules per asymmetric unit. The correlation between the calculated shielding constants and the experimental chemical shifts for both asymmetric molecules is shown at *bottom*. *Abbreviation*: CPMAS, cross polarized magic angle spinning. *Source*: Modified from Ref. 301.

atoms, most likely due to the interactions of the chloride ion with ring hydrogen atoms. Good linear relationship ($r = 0.997$) was found between the experimental and calculated shifts (using GIAO-CPHF) based on both the X-ray and the semi-empirical PM3 geometries. This was used to confirm the chemical shift assignments originally based on comparison to solution shifts.

Harper et al. (304) studied the relative stereochemistry and molecular conformation using SSNMR tensors. This new SSNMR methodology involved a comparison of experimental ^{13}C tensor principal values (measured by using the FIREMAT experiment) with ab initio computed values for all possible stereoisomers. An accurate computation of the tensor values was required. About 3–15% and 1–2% error of the tensor values spaning for sp^3 and sp^2 carbons, respectively, was achieved. The method was tested on terrain, chosen as relatively rigid molecule with limited conformational diversity that would otherwise influence the calculation of the tensor values. Spectra-editing experiments and comparisons with calculated tensor value were used for carbon assignments.

Cyranski et al. (305) studied structure of 1-(2-hydroxy-4-bromophenyl)-4-methyl-4-imidazolin-2-ones. Comparison of solution and solid chemical shifts allowed for the identification of rigid and flexible structure fragments. Solution to solid differences were found significant for aromatic carbons, which are subject to the largest changes of the environment during reorientation of the imidazolin-2-one moiety. Due to the insufficient ^{1}H resolution in the solid state, the hydrogen bonds were also studied by FTIR spectroscopy. GIAO-HF/6-31 + G* chemical shift calculations were used.

Olejniczak et al. (306) studied 2-(3′,4′-dihydroxyphenyl)-7-β-D-Glucopyr-anos-1-O-yl-8-hydroxychroman-4-one using 2D NMR and DFT calculations. In order to correlate the chemical tensor elements to the molecular structure, the authors chose 2D PASS experiment to extract the tensor parameters by fitting the spinning sideband patterns of the F1 domain projections (Fig. 36), a very good match between the experimental and simulated data was achieved. B3PW91 functional DFT and 6-311G** basis set GIAO calculations were used. Because the calculations were done in the vacuum (no intermolecular inter-actions were considered), the experimental versus calculated chemical shift values showed linear relationship with considerable scatter (R^2 between 0.8 and 0.96 for the various tensor components).

Rajeswaran et al. (307) demonstrated the feasibility of an integrated approach to determine 3D structure of medium-sized organic molecules without using single crystals on N-(p-tolyl)-dodecylsulfonamide. The approach combined PXRD data, Monte Carlo, and ab initio computations, and SSNMR data. GIAO with DFT method using B3LYP/MIDI was used to predict the shielding constants that were converted to chemical shifts relative to tetramethyl-silane using linear equation developed over a diverse set of 37 organic com-pounds with RMS error of 3.6 ppm. A monomer geometry extracted based on PXRD data using PowderSolve software was used for the chemical shielding cal-culations. Good matches between the calculated and observed data was achieved. To confirm the right geometry, the computations were repeated for several com-binations of torsion angles. The authors propose the following work flow for crystal structure determination of polycrystalline samples: (*i*) collect PXRD and assess the crystallinity, (*ii*) obtain SSNMR and assess the presence of solvates, multiple phases, and number of molecules per asymmetric unit, (*iii*) confirm chemical composition using elemental analysis, (*iv*) measure sample density and compare with the X-ray calculation, (*v*) perform PXRD structural computations, and (*vi*) validate the structure using computed SSNMR shifts or select the most likely PXRD structure.

Aimi et al. (308) studied conformation of aromatic imide compounds con-taining diphenyl ether and benzophenone and diphenyl sulfide, diphenyl sulfone, and diphenylmethane (309) moieties. Dihydral angles were estimated using ^{13}C CPMAS approach combined with RHF/6-31G(d) ab initio chemical shift calculations. Chemical shift assignments were based on solution studies and interrupted decoupling solid-state spectral editing. TOSS mode was used to minimize spinning sidebands.

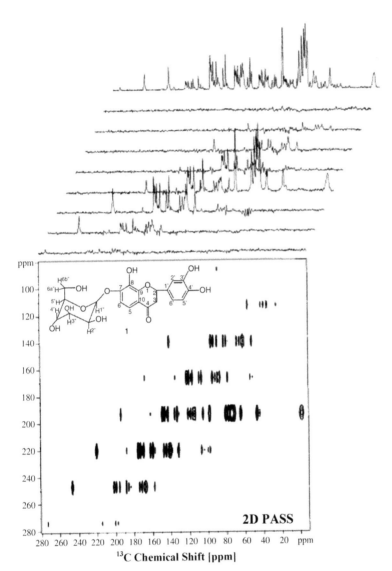

Figure 36 The 2D PASS experiment on flavanone. *Source*: Modified from Ref. 306.

Claramunt et al. (310) studied 1-benzoylazoles (imidazole, pyrazole, indole, benzimidazole, and carbazole). To understand the 9-benzoylcarbazole's unexpected ^{13}C CPMAS spectrum with additional splitting, GIAO/RFH/6-311G** ab initio calculations were performed. The splitting was proposed to be due to the conglomerate structure of the two enantiomers.

Marek et al. (311) studied ^{15}N naturally abundant and substituted purine derivatives. Geometric parameters obtained by using RHF/6-31G** were used to calculate chemical shielding (GIAO and IGLO). The predicted shielding was used to assign the experimental spectra to determine the parameters of the chemical shift anisotropy (CSA) tensor. The experimental CSA values were obtained by fitting the ^{15}N CPMAS spinning sidebands. Differences between solution and solid shifts were as large as 73 ppm.

Goward et al. (312) combined advanced SSNMR experiments and DFT-based chemical shift calculations to elucidate the supramolecular structure of benzoxazine oligomers. About 30 kHz spinning ^{1}H MAS spectra were compared with DFT-based calculation to assist with assignments. Good agreement was achieved. N–H protons were found to be spatially separated based on 2D ^{1}H–^{1}H double quantum (DQ) spectra of dimers and tetramers, but not trimers (Fig. 37), consistent with the proposed conformational model. The ^{15}N–^{1}H distances were also measured on ^{15}N-labeled sample using pulse-field gradient-enhanced REPT-HDOR (313) experiment.

Helluy and Sebald studied structure and dynamic properties of solid L-tyrosine-ethylester. Variable temperature spectra on the ^{13}C natural abundance sample showed considerable dynamics attributed to the π-flips of the phenyl ring. Zero quantum ^{13}C recoupling experiments on fully labeled sample afforded distance constraints that faithfully reproduced the molecular structure defined by short-range interactions but failed to uniquely characterize the complete conformation defined by longer range interactions.

Zolek et al. (314) recorded ^{13}C CPMAS spectra of a series of 11 coumarin samples and, in combination with GIAO-CHF calculations, characterized their conformations. Single resonance was found for each carbon of each compound. Assignments were done based on combination of liquid spectra and interrupted decoupling editing and were confirmed by the ab initio calculations. The calculations were performed for isolated molecules on their equilibrium geometries.

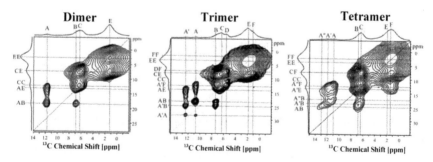

Figure 37 The 2D ^{1}H DQ spectra of benzoxazine oligomers. The *double letter* marked peaks in the double quantum domain describe connectivities between the peaks in the SQ domain. *Source*: Modified from Ref. 312.

When available, structures from single-crystal data were used. A general relationship between calculated shielding constants and experimental chemical shifts was developed for the coumarin series.

Mao et al. (146) studied biosolids-derived biocolloidal organic matter by TOSS-CPMAS, 2D HETCOR, and SUPER techniques. Two 2D HETCOR spectra were compared, one with Lee-Goldburg CP and one with standard Hartmann-Hahn CP. The LGCP showed correlations of protons and carbons separated by three or less bonds, whereas the HHCP showed correlations on 6 Å scale. Chemical shifts anisotropies were measured with SUPER technique to help to distinguish carbons from esters, amides, and carboxylates (144).

Harper et al. (315) compared the measured and computed chemical shift tensor principal values for all diastereomers of ambuic acid to predict the relative stereochemistry. No crystal structure was available for this compound. The tensor principal values were measured using the FIREMAT method. Eight model structures were prepared for the computation, differing in relative stereochemistry, hydrogen bonding, and conformations. Only tensors of carbons directly affected by the structural changes were considered. A dimeric structure was found to fit the experimental data best. This method of predicting a relative stereochemistry is of general applicability and especially useful for characterizing materials for which it is difficult to grow a good quality single crystal.

Olsen et al. (293,316,317) used a through-bond $^{13}C-^{13}C$ correlations to refine dynamic regions in the crystal structures of naturally abundant vitamin D_3. A scalar coupling-driven UC2QF COSY (double quantum correlation spectroscopy) developed by authors was used to completely assign the 54 carbon peaks of vitamin-D_3 (27 carbons per molecule and two molecules per asymmetric unit; Fig. 38). This required two weeks of acquisition on 500 MHz instrument. The ab-initio calculated chemical shifts using B3LYP functional and 6-311 + G(2d) basis were compared with the assigned shifts. A good agreement was observed.

Sun and Oldfield (318) showed that the structural constraints can be obtained based on SSNMR carbon chemical shifts. This was demonstrated experimentally by assigning a series of tryptophan-containing peptides with known X-ray structures, and comparing the experimental shifts with quantum chemical calculations. Nonprotonated aromatic carbons correlated well with the side-chain torsion angles.

Mehta et al. (319) studied the structure of alanine dipeptide in solution and the solid state. The ^{13}C chemical shift of alanine dipeptide in polycrystalline, lyophilized, and solvated state was compared. The structure was determined using DQNMR. The same secondary structure was observed with lyophilized form of H-bonding (H_2O) and non-H-bonding ($CHCl_3$) solvents.

Giavani et al. (320) devised experimental strategies for collecting ^{14}N MAS spectra of some amino acids exhibiting quadrupole coupling constant in the order of 1.2 MHz. At spinning frequency of 6 kHz, the spectra consisting of approximately 600 spinning sidebands (spectral width in access of 1 MHz) were fitted to extract the quadrupolar parameters. For some amino acids two different

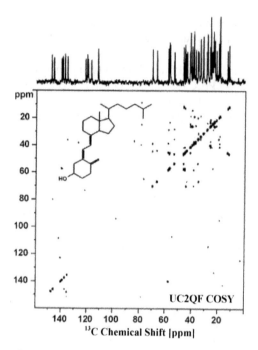

Figure 38 The 2D uniform-sign cross-peak double-quantum correlation spectroscopy (UC2QF COSY) on natural abundance level vitamin D_3. *Source*: Modified from Ref. 316.

^{14}N spinning manifolds were observed in accord with two sites per asymmetric unit predicted from X-ray structures.

Quantification Issues

The NMR has a great advantage over other techniques when it comes to quantification. Because NMR signal is directly proportional to the number of spins present in the sample, no external reference needs to be applied to calibrate the system response. In this sense, NMR is an absolute technique. However, in most pharmaceutical applications, only relative concentrations of mixtures of two or more polymorphs need to be determined even in complex dosage forms. Two cases should be distinguished: (*i*) direct polarization using one pulse-acquire sequence, and (*ii*) CPMAS-type of experiments. Direct polarization experiments require no special precautions, but for the sensitivity reasons, they typically apply only to such high abundance nuclei as ^{1}H, ^{19}F, or ^{31}P. Due to the CP step, the CPMAS signal intensities are not necessarily quantitative and some precautions discussed next have to be exercised in order to quantify accurately. In general, the basic requirement for successful quantification by SSNMR is that the NMR signals corresponding to the two or more components of the sample have to be discriminated according to one or more of their NMR parameters. An existence of differences in the chemical shifts (isotropic and anisotropic) of the components is the obvious differentiation. A good peak resolution simplifies the quantification substantially. A chemometric approach to peak deconvolution can be applied if partial peak overlap is present. Even in the case of totally overlapping resonances, differences in other NMR parameters, such as the dipolar coupling, J-coupling, or relaxation properties, could frequently be used to selectively filter in or out the signals from single components and the sample could be still quantified successfully. In particular, relaxation-based filters can turn out to be instrumental for identifying and quantifying the presence of an amorphous phase. Many pharmaceutical applications involve the quantification of an amorphous phase in the presence of a crystalline phase. Even though, amorphous spectra could show differences in chemical shifts when compared with their crystalline counterparts, it may still be difficult to precisely quantify small amounts of the amorphous phase. The amorphous peaks present in low concentration can be lost in the baseline due to their broad line character. Because the mobility of amorphous and crystalline phases might differ substantially, the mobility-induced relaxation differences allow for the relaxation filters to be employed (321,322). Another type of a "filter" can be employed specifically to discriminate amorphous phases. It is based on the fact that the glass transition temperature (T_g) of amorphous compounds is likely to be lower than the melting temperature of the crystalline phase. If CPMAS detection scheme is used, heating the sample above its T_g results in observation of only the crystalline phase. The signal from the amorphous phase is lost, because this phase reaches a liquidlike mobility, and therefore the dipolar coupling, which the CPMAS experiment relies on, is averaged out to zero. This may, however, be an invasive technique not suitable for all applications.

The easiest and usually the most accurate quantification of polymorphs is achieved when a sensitive nucleus, such as fluorine or phosphorus, is present in the molecule, and its spectrum shows good chemical shift differences between polymorphs. Direct polarization experiments of these nuclei are sensitive, and therefore the polymorphs can be quantified directly from the integrated peak ratios. However, attention must be paid to let the magnetization return fully to equilibrium between the consecutive scans. Good practice for an unknown sample is to run first a T_1 relaxation experiment to set the proper recycle delay. The limit of quantification when using sensitive nuclei can be well below 1% of one form in the presence of other forms. In the case of well-resolved peaks and fast relaxation, LOD of 0.01% is not unthinkable. As an added advantage, the spectra of these nuclei are usually very simple to analyze, because only a very limited number of resonances are typically present.

When no abundant nuclei (other than protons) are present in the sample, carbon-based SSNMR quantification is often the only remaining option. One might still choose to use a direct polarization acquisition. However, due to the long carbon T_{1C} relaxation times of typical crystalline organics, very long acquisition times will result. In spite of this, many literature examples of quantification using carbon direct polarization exist and are reviewed next. An alternative way to partially circumvent this issue of long experimental time is to calibrate the T_{1C} values on pure standards of the different forms in the mixture to be quantified. The intensities can then be corrected for incomplete relaxation without waiting for the full intensity equilibration (dictated by the slowest relaxing component) between successive scans. One has to make sure that the sameness of carbon relaxation times of the standards and unknowns is justified.

If the acquisition times for carbon direct polarization are prohibitive (rather typical case for pharmaceuticals), depending on the availability of pure form of standards, the analyst can choose to either construct a calibration graph or map the kinetics of CP. When pure forms are available, SSNMR spectra of mixtures with different known concentrations of polymorphs can be acquired, and a calibration curve can be constructed. To diminish the dependence on a precise setting of the experimental parameters, the calibration curve can be plotted as relative signal ratio of the two forms rather than as absolute intensities. The advantage of this approach is that the quantification of formulations is not hampered by the presence of excipients, assuming none or only very limited overlap of the excipient peaks with the peaks of interest. The calibration graph constructed with known API form mixtures can be directly applied for quantification of formulations. Logarithmic or other transformation of the resulting nonlinear calibration curves is desirable to optimize the fit. Because only a single spectrum has to be acquired for every unknown, the calibration curve method is usually faster than quantification based on mapping the CP kinetics, especially if large number of unknown samples have to be quantified. Unlike the CP kinetics approach, the only consideration in selecting the peaks for this analysis is that they originate from only one component at a time. Any overlap of peaks belonging to the same component (but well resolved from the other component) is tolerated, whereas, strictly, the peaks corresponding to single carbons are required using the CP kinetics approach. However, it should be noted that the calibration method automatically assumes the sameness of the samples (the standards and the unknowns) in terms of the NMR parameters, such as relaxation rates or CP dynamics. For well-defined crystalline samples this assumption may hold well, but it may be more troublesome when dealing with amorphous phases. It is well known that the mobility (and hence all the NMR parameters mentioned earlier) is a strong function of the relative humidity and the sample's thermal history. Even in the case of crystalline compounds, formulations processing as milling or compacting can dramatically change the relaxation and other NMR properties. Hence, the calibration method has to be applied judiciously.

CP Kinetics Approach

The CP is based on the pulse sequence depicted in Figure 12. The polarization transfer mechanism depends on the existence of strong dipolar coupling between the two cross-polarizing nuclei. According to Equations (12) and (13), the effective dipolar coupling, in turn, depends on the distance between the coupled nuclei and the orientation (with respect of the external magnetic field) of the vector connecting them and on the local dynamics of the molecular segment. The higher the frequency and the more isotropic character of the motions, the smaller the effective dipolar coupling is. Beside these internal sample-dependent factors, external conditions, such as temperature or MAS spinning speed, will also influence the effective dipolar coupling. The best cross-polarizing efficiencies are therefore achieved on rigid and static (nonspinning) samples. It also follows, that within the same molecule, different nuclei may cross-polarize in different rates, rendering the resulting peak intensities nonstochiometric. When a quick and only an approximate estimate of the polymorph content is sufficient, the intensity ratio of two corresponding carbon peaks, each belonging to the different form, can be used for quantification. Any additional knowledge of the system could prove useful in selecting the peaks, as the selected carbons should have comparable conformation and dynamics. For true polymorphs, this may be a reasonable assumption. The CP kinetics is then likely to be similar, reflecting the approximate quantitative ratio of the two forms.

In order to use CPMAS spectra for any accurate quantification purposes, one has to characterize the dynamics of the CP process. Examples of using CPMAS for quantification while ignoring the CP kinetics still abound in the literature. The dynamics of the CP was recently described in two excellent and exhaustive reviews describing cross-polarizing from abundant to rare nuclei (60) and CP between two abundant nuclei (59,323). An earlier detailed discussion of CP in terms of quantification was also described by Sullivan and Maciel (324) and by Harris (325). In the context of the quantification, only the abundant–rare CP kinetics is considered in this section, because in the case of the abundant–abundant system, it is assumed that the quantification can be achieved by direct polarization acquisition without the need for CP. The CP can be described by one of the two kinetic models: the simplified classical I-S model (Figs. 39 and 40) or the more rigorous I-I^*-S model (Fig. 41). The thermodynamic property of spin temperature is well suited to explain the CP process. In the classical model, the lattice is defined as having an infinite heat capacity and the abundant spins I have much larger heat capacity than the rare spins S. The spin temperature

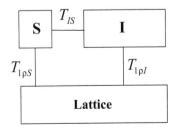

Figure 39 Classical I-S CP kinetics model.

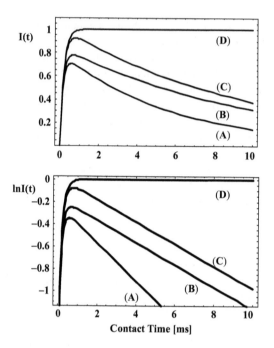

Figure 40 Calculated dependence of CPMAS intensity on the contact time using regular intensity scale (*top*) and logarithmic scale (*bottom*). Four different equations were used to generate these curves: (A) Equation 19, (B) Equation 20, (C) Equation 21, and (D) Equation 22. The parameters used to generate the curves were: $I_0 = 1$, $\varepsilon = 0.1$, $T_{IS} = 0.2$ ms, $T_{1\rho S} = 1$ ms and $T_{1\rho H} = 10$ ms. Note that for typical carbon CPMAS applications ($\varepsilon \sim 0.01$ and $T_{1\rho S} > 15$ ms) the curves (A), (B) and (C) would be almost exactly equivalent and the values of $\varepsilon = 0.1$ and $T_{1\rho S} = 1$ ms were chosen to illustrate the differences between Equations 19, 20, and 21. In the logarithmic representation, it is apparent that extrapolating the linear decaying part of the curve to zero contact time does not reproduce the correct intensity [$\ln I(0) = 0$]. When quantifying using the graphical extrapolation, a correction for the finite relaxation rates using the prefactors of Equations 19, 20, and 21 have to be applied. *Abbreviation*: CPMAS, cross polarization and magic angle spinning.

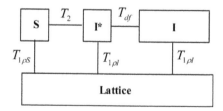

Figure 41 More rigorous I-I^*-S CP kinetics model. *Abbreviation*: CP, cross-polarization.

β is the so-called inverse spin temperature in the rotating frame, that is, the higher the β, the higher is the polarization difference between the energy levels, corresponding to stronger NMR signal. Just before CP, the abundant spin inverse temperature β_I^0 is high, whereas the rare spin β_S^0 is zero. The Hartman-Hahn match brings the two spin reservoirs into thermal contact, allowing the energy to flow from abundant to rare spins, building up the rare spin polarization with the rate proportional to the inverse of the CP time constant T_{IS}. This process continues until equilibrium is reached, when the spin temperatures of both reservoirs equalize. However, simultaneously with the CP process, the inverse spin temperatures of both I and S reservoirs are cooling down due to their contact with the lattice brought about by relaxation mechanism. The resulting rate of signal loss is proportional to the inverse of the rotating frame relaxation times $T_{1\rho S}$ and $T_{1\rho I}$. The following differential equations describe the overall CP process.

$$\frac{d\beta_I}{dt} = -\varepsilon\alpha^2 \frac{\beta_I - \beta_S}{T_{IS}} - \frac{\beta_I}{T_{1\rho I}}$$

$$\frac{d\beta_S}{dt} = \frac{\beta_I - \beta_S}{T_{IS}} - \frac{\beta_S}{T_{1\rho S}} \tag{18}$$

where ε is the spin population ratio ($\varepsilon = N_S/N_I$), and α is the Hartman-Hanh matching coefficient ($\alpha = \gamma_S B_{1S}/\gamma_I B_{1I} \approx 1$). It follows that for a rare–abundant spin system, ε is close to zero. Since the detected NMR CP signal $I(t)$ is proportional to the inverse temperatures, it is described by the general solution of Equation (18) as

$$I(t) = \frac{I_0}{a_+ - a_-} (e^{-(a_- t/T_{IS})} - e^{-(a_+ t/T_{IS})})$$

$$a_+ = c \cdot \left(1 \pm \sqrt{1 - \frac{d}{c^2}}\right)$$

$$d = \frac{T_{IS}}{T_{1\rho I}} \cdot \left(1 + \frac{T_{IS}}{T_{1\rho S}}\right) + \varepsilon\alpha^2 \frac{T_{IS}}{T_{1\rho S}} \tag{19}$$

$$c = \frac{1}{2} \cdot \left(1 + \varepsilon\alpha^2 + \frac{T_{IS}}{T_{1\rho S}} + \frac{T_{IS}}{T_{1\rho S}}\right)$$

This general equation is rather complicated and for the majority of practical applications involving CP to rare nuclei, it can be simplified using $\varepsilon\alpha^2 \approx 0$

$$I(t) = \frac{I_0}{1 + \frac{T_{IS}}{T_{1\rho S}} - \frac{T_{IS}}{T_{1\rho I}}} \cdot \left[e^{-(t/I_{1\rho I})} - e^{-t(1/T_{IS} + 1/T_{1\rho S})}\right] \tag{20}$$

Because the relaxation of rare nuclei is typically slow due to the lack of relaxation pathways, further approximation can be made $(T_{IS}/T_{1\rho S} \sim 0)$ leading to

$$I(t) = \frac{I_0}{1 - \frac{T_{IS}}{T_{1\rho I}}} \cdot \left(e^{-t/T_{1\rho I}} - e^{-t/T_{IS}}\right) \tag{21}$$

The typical double exponential character of the CP kinetics is clearly evident from the last two equations. According to Equation (21), the signal grows with the rate proportional to inverse of T_{IS} and decays with the rate proportional to inverse of $T_{1\rho I}$. It should be noted that there is a prefactor in both Equations (20) and (21). Instead of fitting the experimental curves to one of these equations, several literature studies suggested extracting the intensity I_0 by graphical means, by plotting the intensities in the logarithmic scale and extrapolating the linear portion of the curve to zero contact time. A care has to be taken, because I_0 scaled up by the factor of $(1 - T_{IS}/T_{1\rho I})$ is actually obtained in this way.

If, in addition, the relaxation of the abundant spins is very slow with respect to the CP time constant $(T_{IS}/T_{1\rho I} \sim 0)$, the Equation (21) further simplifies to

$$I(t) = I_0 \cdot (1 - e^{-(1/T_{IS})}) \tag{22}$$

The last equation predicts that, in the absence of any significant relaxation, the CP signal grows exponentially to reach the I_0 intensity for long enough cross-polarizing times. Many of the small organic molecule crystalline compounds fall into the category with relatively short $T_{1\rho I}$, and therefore either Equation (21) or even Equation (20) have to be used. It should be noted that the rotating frame relaxation times strongly depend on the strength of the B_1 irradiation field. In the slow motional regime, the stronger the irradiation, the longer the rotating frame relaxation is.

The asterisk in the I-I^*-S model (Figs. 41 and 42) denotes the physical proximity of the I^* spin (the abundant spin involved directly in the CP transfer) to the S spin. This model was designed to account for the oscillatory CP kinetics behavior found in many samples. It takes into consideration the fact that the spin temperature of all I spins might not be equivalent, that is, the spin diffusion between the abundant nuclei might be slow compared with the CP kinetics. The I^*S spins can be viewed as isolated spin pairs involved in cross-polarizing magnetization in the oscillatory manner, while the spin diffusion to the remaining surrounding abundant nuclei dampens these oscillations. For an isolated SI_n cluster, the cross-polarizing kinetics predicted by this model is

$$I(t) = I_0 \cdot e^{-(1/T_{1\rho I})}$$
$$\cdot \left[1 - \lambda \cdot e^{-(t/T_{df})} - (1 - \lambda) \cdot e^{-(3t/2T_{df})} \cdot e^{-(t^2/2T_2^2)}\right] \tag{23}$$

where $\lambda = 1/(n + 1)$, T_{df} is a spin diffusion constant, and the oscillatory behavior is replaced by constant T_2, characterizing the initial decay of the S magnetization. Unlike the classical model predicting a two-stage process (signal growth

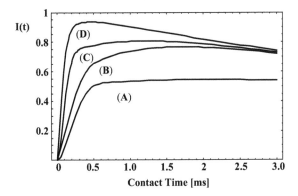

Figure 42 Calculated dependence of CPMAS intensity on the contact time using the *I-I*-S* CP kinetics model [Equation (23)]. The parameters used to generate the curves were: $I_0 = 1$ and $T_{1\rho H} = 10$ ms (all curves); (A) $\lambda = 0.5$, $T_2 = 0.2$ ms, $T_{df} = 5$ ms; (B) $\lambda = 0.5$, $T_2 = 0.2$ ms, $T_{df} = 1$ ms; (C) $\lambda = 0.3$, $T_2 = 0.1$ ms, $T_{df} = 1$ ms; (D) $\lambda = 0.2$, and $T_2 = 0.1$ ms, $T_{df} = 0.2$ ms. Note that the initial slope is controlled mostly by the parameter T_2, the maximum achieved after the initial raise by parameter λ and the double exponential character of the raising part of the curve by parameter T_{df}. *Abbreviations*: CPMAS, cross polarization and magic angle spinning; CP, cross-polarization.

followed by decay), the *I-I*-S* model predicts the CP kinetics to proceed in three stages: fast signal growth with possible oscillations, slow growth, and decay.

The *I-I*-S* model applies when the heteronuclear *I*-S* dipolar coupling is sufficiently strong as compared with the homonuclear *I–I* interaction. Samples exhibiting substantial degree of internal mobility comply with the classical model. In more rigid organic molecules, the *I-I*-S* model typically applies to CP kinetics of CH and CH_2 groups. Even for rigid compounds, the quarternary groups and fast rotating CH_3 groups usually conform with the classical model. To discriminate between the kinetic CP models, the early part of the CP kinetics curve must be accurately sampled to detect any oscillatory behavior or sudden intensity changes characteristic of the *I-I*-S* model. An example of fitting the CP kinetics of CAV11 molecule using Equation (23) is given in Figure 43. Carbonyl and methyl carbons were fitted using this model.

For accurate quantitative results, the CP kinetics should be mapped for every unknown sample. Clearly, this is a very time consuming proposition, especially for samples exhibiting long proton T_{1H} relaxation times or for formulations with low API loads. The alternative is to calibrate the two (or more) CP characteristic constants, $T_{1\rho H}$ and T_{CH}, on pure standards of all polymorphs present in the mixture and then use the known CP parameters to quantify the unknown mixtures. However, this scheme suffers the same insufficiencies discussed earlier in the context of the quantification by a calibration graph, namely the potential differences between relaxation and CP kinetic properties of the standards and the unknowns.

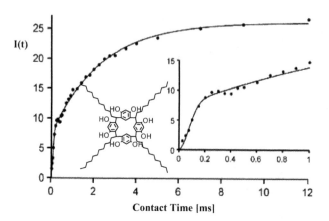

Figure 43   An example of fitting the CP kinetics of CAV11 molecule to Equation (23). *Abbreviation*: CP, Cross-polarization. *Source*: Modified from Ref. 496.

Quantification of Pharmaceutical Materials

When quantifying relative amounts of anhydrous carbamazepine and carbamazepine dihydrate, Suryanarayanan and Wiedmann combined the CPMAS approach together with T_1 filtering and an internal standard addition (326). The spectra of both anhydrate and dihydrate were practically identical, but the anhydrous carbamazepine showed very long T_{1H} relaxation. Setting the recycle delay to 10 seconds resulted in saturation of the anhydrous signal and acquisition of only the dihydrate. The authors used addition of known amount of glycine as an internal standard. By variable contact time CPMAS experiment, they determined T_{CH} and $T_{1\rho H}$ for both glycine and the dihydrate. Based on calibration mixture containing variable ratios of anhydrate and dihydrate with fixed amount of glycine (Fig. 44), they constructed calibration graph of peak area ratio (dihydrate/glycine) versus the mole fraction of dihydrate. This was done by choosing a single contact time (optimized for sensitivity) and correcting the intensities of both glycine and dihydrate peaks by T_{CH} and $T_{1\rho H}$ values.

Likar et al. (327) studied keto-enol equilibrium of trospectomycin sulfate bulk drug by ^{13}C CPMAS. Drying the bulk drug or formulation to low water levels dehydrated trospectomycin sulfate from the diol to the keto form. Variable contact time experiment was used to select the optimum contact time suitable for the quantification of the two forms. The quantification of the keto form was accomplished by monitoring the ratio of the carbonyl (keto form only) to the methyl group peaks (common for both forms).

Gao quantified three different delavirdine mesylate polymorphs and pseudopolymorphs by ^{13}C CPMAS (Fig. 45, top) (328). The CP dynamics was fully considered in the quantification as the data from the variable contact time experiment were fitted according to Equation (21) taking into account $T_{1\rho H}$ and T_{CH}

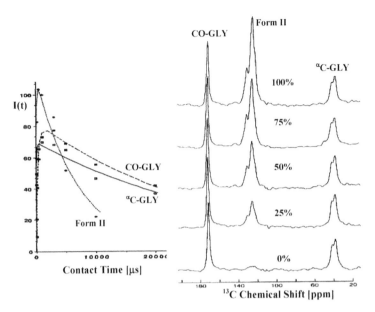

Figure 44 The CP kinetics of carbazepine dihydrate, Form-II, and that of α-carbon and carbonyl carbon of glycine (*left*). ^{13}C spectra of mixtures of anhydrate and dihydrate with constant content of glycine. Zero percent of dihydrate corresponds to 100% of anhydrate. *Abbreviation*: CP, cross-polarization. *Source*: Modified from Ref. 326.

(Fig. 45, bottom). An empirical detection limit of Form-VIII in Form-XI was found to be 2%. A discussion of the respective merits of using peak areas and peak intensities (heights) is provided. For forms with partially overlapped peaks, a difference spectrum (mixture—pure major form) was used to obtain the peak areas of the minor form. Sources of the potential experimental error were identified as (*i*) poor S/N, (*ii*) improper phase adjustments, and (*iii*) improper baseline correction. Peak heights were found to be less sensitive to the phase adjustments, but the quantification using peak heights were found to be less precise. In the follow-up study, Gao further characterized these three crystalline forms in addition to an amorphous form (329). The assignments of the carbon resonances were performed by combination of solution data, interrupted decoupling, and cross-polarization and polarization inversion (CPPI) experiments. Form-XII was found to consist of two molecules per asymmetric unit. Carbon shifts of methyl groups were found to be sensitive indicators of the intra-molecular geometries. T_{1H}, $T_{1\rho H}$, and T_{1C} relaxation times were also measured. A decrease in $T_{1\rho H}$ values correlated with increased disorder, which is rationalized by the dependence of the spin diffusion on the proximity of the neighboring protons. T_{1C} relaxation times of methyl groups (~1 second) were found to be approximately two orders of magnitude shorter than those of other carbons (~100 seconds), consistent with the fast rotation of the methyl groups. The rank order

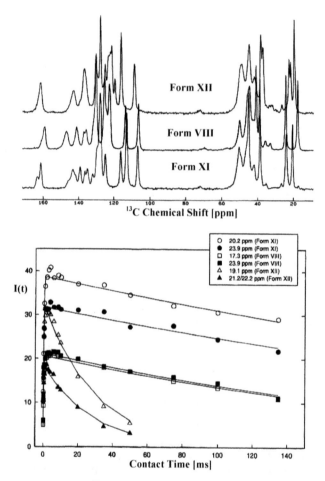

Figure 45 The ^{13}C CPMAS spectra of three forms of delavirdine mesylate (*top*). The CP kinetics of these three forms are also shown (*bottom*). *Abbreviations*: CPMAS, cross-polarization and magic angle spinning; CP, cross-polarization. *Source*: Modified from Ref. 328.

of relaxation times $T_{1\rho H}$ and T_{1C}, but not T_{1H} for nonmethyl groups was consistent. T_{1H} rank order was, however, consistent with that of methyl group T_{1C}, confirming the role of the methyl groups as T_{1H} relaxation sinks.

Gustafsson et al. (330) compared SSNMR and isothermal microcalorimetry to quantify the amorphous component (the degree of disorder) of lactose. Two principal least squares (PLS) models established from the NMR data, one for the whole range of concentrations 0–100% and the second one for 0–10%, accurately determined the amorphous component even when in mixture with acetylsalysilic acid

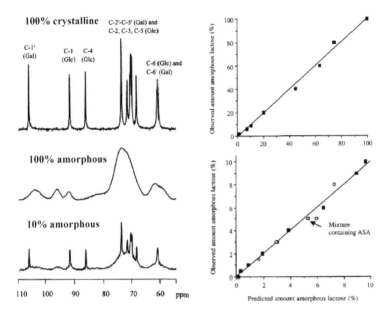

Figure 46 The ^{13}C CPMAS spectra of crystalline and amorphous lastose (*left*). PLS correlation plot for amorphous lactose (*right*), including a validation mixture containing acetylsalicylic acid (ASA). *Abbreviations*: CPMAS, cross polarization and magic angle spinning; PLS, principal least squares models. *Source*: Modified from Ref. 330.

(Fig. 46). The results from microcalorimetry and SSNMR were in agreement. Both techniques were able to detect lower than 0.5% amorphous component.

Lee et al. (331) applied ^{13}C CPMAS SSNMR to quantify impurities and excipients in "ecstasy" tablets. Chemical shift assignments were based on comparison with solution values in combination with spectral-editing experiments. Smaller than 2 ppm solution–solid chemical shift differences were observed. Various CP contact times were screened to find the optimal parameters affording quantitative data. Some of the excipients (lactose and sucrose) were found not to be fully relaxed even when 60-second recycle delay was applied.

Paris et al. (332,333) has used a line shape decomposition of ^{13}C CPMAS spectra to quantify the types of the glycosidic linkages in treated amorphous starchy substrates. The 2D WISE spectra were recorded to investigate the mobility. Variable contact time curves for all the deconvoluted peaks were found to fit the two proton reservoirs CP model [Equation (23)]. The extracted parameters of T_{df} and $T_{1\rho H}$ are correlated with the hydration level and to the local structure (domains larger than 1 nm are found in one of the structures). However, no attempt is shown to use the measured CP dynamics for the actual quantification of the types of the glycosidic linkages.

Kono et al. (334–336) used quantitative ^{13}C CPMAS to help in assignment of two native cellulose allomorphs (α and β). The assignment was achieved by using selectively labeled cellulose biosynthesized by two different routes. To determine the precise transition of the ^{13}C labeling to each position of cellulose, quantitative measurements were performed. A quantitative contact time was chosen based on variable contact time experiment.

Tozuka et al. (337) quantified two clarithromycin polymorphs by PXRD and SSNMR. A calibration curve method was used, plotting the carbonyl peak area ratio of one form with respect to the known content of this form.

Vickery et al. (338) quantified two crystalline polymorphs of roxifiban. A calibration graph method was also used. The detection limit was estimated to be 9% of Form-II in Form-I.

Apperley et al. (339) quantified mixtures of anhydrate and dihydrate of for-moterol fumarate by ^{13}C CPMAS. One molecule per asymmetric unit cell was found. The protonated carbons of one of the two phenyl rings were not observed at room temperature but gave an observable signal at $-50°$C that was consistent with the fast rotation of the phenyl ring at room temperature. Differences between some solution versus solid chemical shifts were justified by forming hydrogen bonding in the solid state. The experimental limit of detection in lactose was found to be 0.45% (15 hours acquisition). Dipolar dephasing experiment was chosen to minimize the intensities of lactose and its spinning sidebands. To facili-tate the quantification, T_{1H} was measured indirectly through CPMAS by proton saturation recovery, and T_{CH} and $T_{1\rho H}$ were estimated from variable contact time experiment. Similar CP dynamics allowed for using the intensity of the car-bonyl signal of the two forms quantitatively at one contact time. A correction was applied to account for different T_{1H} relaxation rates.

Fu et al. (340) devised a new LG-based CP approach affording nearly quan-titative intensities across the CPMAS spectra. During the contact time, proton magnetization is spin-locked by the LG sequence and the irradiation of the insen-sitive spins (^{15}N or ^{13}C) is modulated sinusoidally and with a constant rf amplitude. LG sequence lengthens the $T_{1\rho H}$ spin locked at the magic angle and the frequency modulation shortens the CP time for nonprotonated spins, while it has only limited effect on the protonated spins. Longer CP times can be used to afford a uniform CP signal for both protonated and non-protonated spins, making quantitative CP measurements possible. The approach is illustrated on lyophilized sample of ^{15}N-labeled L-histidine (Fig. 47). The two tautomeric forms of L-histidine (nonprotonated and protonated) are observed in the ^{15}N spec-trum. $T_{1\rho H}$ increased from 18 ms to 37 ms when spin locked using LG. The I-I^*-S model (Fig. 41) was used to evaluate the CP dynamics for protonated nitrogens, because the LG sequence attenuated the ^{1}H–^{1}H dipolar coupling, and therefore made the ^{1}H–^{15}N dipolar coupling stand out.

Byard et al. (341) quantified a presence of an amorphous component in micronized pharmaceutical development drug substance by SSNMR and PRXD. The limit of detection of PRXD for the amorphous component was

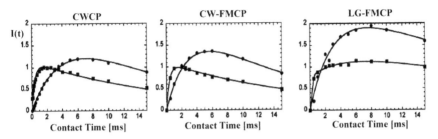

Figure 47 Differences between the standard continuous wave CWCP, continuous wave with frequency modulation CW-FMCP and Lee-Goldburg frequency-modulated LG-FMCP $^1H-^{15}N$ cross-polarization kinetics curves on L-Histidine. The *solid circles* and *squares* represent the nonprotonated and protonated forms of histidine, respectively. Based on direct polarization ^{15}N experiment, the ratio of nonprotonated to protonated intensities should be 1.94. Note that only LG-FMCP with contact time of 8 ms approximately reproduces this intensity ratio. *Source*: Modified from Ref. 340.

found to be >5% with an error of ±3%. NMR was used to quantify crystal defects at levels >3% with error ±2%. The defects arose due to the conformational differences attributed to the removal of ethanol and had only small effects on the crystal lattice and, hence, on PXRD spectra. Excellent linearity is demonstrated using standards. The stability study of the drug at stressed conditions showed formation of crystalline component. NMR is suggested to be used as a complementary technique alongside of PXRD.

Offerdahl et al. (69) quantified three crystalline and amorphous forms of anhydrous neotame by ^{13}C CPMAS. When in binary mixture, the detection limit for the crystalline forms was 1–2%. When mixed with amorphous phase, the detection limit of the amorphous phase was 5%. When quantifying using CPMAS, the authors rightfully stress the importance of recording the contact time dependence on each sample, as opposed to just standards. This is due to the possible differences in the CP and relaxation parameters of each new sample. The importance of long $T_{1\rho H}$ (relatively to T_{CH}) is also discussed. For example, for contact time of 2 ms and $T_{1\rho H}$ of the amorphous component of 11 ms, the signal has already relaxed by 25%. The differences between the $T_{1\rho H}$ of amorphous and crystalline components were used to partially filter out the crystalline one. Two intermittent acquisitions, one with standard contact time and the second one with long contact time, were subtracted by altering the phase of the receiver by 180° (Fig. 48). Even though a large penalty in S/N is incurred, the increased relative ratio of the amorphous component allows for better deconvolution.

Quantification of Natural Organic Matter, Wood, and Tar

This section on quantification of natural organic matter, wood, and tar is included due to the similarities of the NMR techniques used and the availability of only

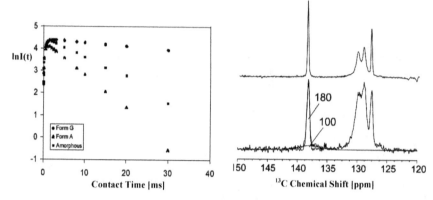

Figure 48 CP dynamics of different forms of neotame (*left*). ^{13}C CPMAS spectra of a mixture of small amounts of amorphous component (11%) in the crystalline form G. Standard CPMAS spectrum (*right, top*) and dual contact time (2 ms and 8 ms) spectrum (*right, bottom*). The deconvoluted peak areas (180 for crystalline and 100 for amorphous forms) have to be corrected by the CP dynamics to get the correct quantification. *Abbreviations*: CPMAS, cross polarization and magic angle spinning; CP, cross-polarization. *Source*: Modified from Ref. 69.

unlabeled materials. In many respects, the quantification of these materials is actually more challenging due to their multicomponent nature, where some of the components may show very fast relaxation and hence, do not cross-polarize efficiently.

Mao et al. (321) used direct polarization quantification corrected for incomplete relaxation. To avoid the time-consuming carbon relaxation experiment, they obtained the correction factor for relaxation in a clever way by comparing two simple 1D CPMAS experiments: The first one was a standard CPMAS, and the second one was CPMAS with T_1-filter placed before the carbon acquisition. The T_1-filter length was of the same duration as the recycle delay in the direct polarization experiment. These two CPMAS measurement allowed for correcting the direct polarization intensities.

Smernik and Oades (342) used spin counting to quantify natural organic matter. A detailed discussion of CP dynamics is provided. A number of materials including cellulose, pectin, lignin, and palmitic acid afforded quantitative spectra using both direct polarization and CPMAS. Some materials, including chitin, collagen, and HF-treated soil, gave better results with direct polarization. Both experiments performed poorly to quantify the charcoal, commercial humic acid, and whole soil. Signal losses in the CPMAS spectra were attributed to rapid relaxation rates ($T_{1\rho H}$) due to paramagnetic species and slow CP magnetization build-up rates. The same authors pointed out an existence of carbon background signal problems when quantifying soil organic matter (SOM) (343). The background signal was especially significant when using direct polarization acquisition. They attributed this background to the Kel-F material of the probe head. They suggested a pulse sequence, which suppresses the Kel-F signal by acquisition of spin echo while decoupling on proton, but not fluorine channels. Due to the strong $^{19}F-^{13}C$ coupling, the carbon signal originating from Kel-F is fully dephased before the start of the acquisition.

Direct polarization MAS technique was found to be more suitable as compared with CPMAS when quantifying tar pitches (344). Variable contact time experiments indicated large contact time-dependent signal differences for different carbons.

Smernik et al. (345) evaluated a performance of three different techniques to determine $T_{1\rho H}$: variable contact time (VCT), variable spin-lock (VSL) followed by CP and ^1H observed experiments. The authors discuss the consequences for quantification of charred and uncharred wood. VCT is shown to overestimate the $T_{1\rho H}$, mainly due to the presence of slowly cross-polarizing carbons. VSL was found to be better performing when analyzing the rapidly relaxing components. Directly proton-observed $T_{1\rho H}$ detected even faster relaxing components that were virtually undetectable by the indirect VCT and VSL techniques. The same group also studied impact of the remote protonation on CPMAS quantification of charred and uncharred wood (346). They used simultaneous fitting of the VCT and VSL curves to minimize the effect of slow and fast $T_{1\rho H}$ on the quantification. "Spin-accounting" term is proposed to account for all the wood carbon species by correcting the differential intensities by known CP parameters when quantifying at one preselected optimal CP time. It improved on the established spin-counting technique by correcting for the rapid $T_{1\rho H}$ relaxation and inefficient CP. Another editing technique called RESTORE (restoration of spectra via T_{CH} and $T_{1\rho}$ editing), generated three sub-spectra (well-represented CPMAS signal, under-represented signal due to rapid $T_{1\rho H}$ relaxation, and under-represented signal due to inefficient CP) based on the differences in the $T_{1\rho H}$ and T_{CH} (347,348). Acquisition of three CP spectra with different combinations of spin lock and contact time is required. RESTORE can identify the types of structures underestimated in CPMAS and the cause for their underestimation. Rapid $T_{1\rho H}$ mostly affected carbonyl and carbohydrate carbons, whereas inefficient CP mostly affected aromatic carbons. Smernik (349) also studied an effect of the magnetic field strength on the quantification (200 and 400 MHz). No improvement in spectral resolution was found at the higher field, but the spinning speed had to be doubled to keep the same spacing of the sidebands and the CP efficiency was reduced by 35–40%.

Similar analysis was applied by Bardet et al. (350) for quantitative characterization of an archeological wood. VCT CPMAS experiment was used to characterize the parameters of the CP and to choose an optimum contact time for further quantitative analysis of unknowns. The CPMAS intensities were then subjected to corrections to render quantitative results.

Mao et al. (351) studied suitability of five different ^{13}C SSNMR techniques to characterize humic acids of various origins and locations. The techniques, performance of which was compared at different magnetic field strengths, included direct polarization at 13 kHz spinning speed, a conventional CPMAS at 5 kHz, ramped CPMAS at 8 kHz, TOSS-CPMAS at 4.5 kHz, and direct polarization corrected by a factor obtained from TOSS-CPMAS-detected T_{1C} relaxation. The ramped CPMAS and TOSS-CPMAS gave consistently better results than those of the conventional CPMAS.

Direct polarization was also the preferred choice of Keeler and Maciel (352) for quantifying soil and organic soil fractions. An internal calibration sample of silicone rubber, placed in a capillary along the rotational axis of the spinner, was used and a correction due to different T_{1C} applied. Internal standard was also used for CPMAS characterization. Based on the measured values of the CPMAS parameters (T_{1H}, $T_{1\rho H}$, and T_{CH}), corrected intensities were used for the quantification. Paramagnetic components were identified to be largely responsible for the "invisible carbon" portion when CPMAS was used. The paramagnetic components shorten the $T_{1\rho H}$ of some proton, rendering the CP inefficient.

Van Lagen and de Jager (353) used steady state and "well-defined data processing" to improve CPMAS quantification of humic acids. By testing the reproducibility of the variable contact time experiments, the authors found the signal consistently decreased and leveled only after the first five experiments (total 2500 scans). Achieving the steady-state magnetization was found important for quantification. A well-defined processing procedure including spectra with large spectral widths (to improve baseline), linear prediction, and same phasing correction are recommended. Authors also recommend extrapolating the decaying part of the variable contact time experiment to zero time. Similar to a number of other studies, they seem to omit the necessary correction of the intensities I_0 extracted by this procedure by the relaxation factors of $1 - T_{IS}/T_{1\rho I}$ [Equation (21)]. Nonprotonated carbons are therefore overestimated (342).

Peuravuori et al. (354) assessed a practical accuracy for quantitative determination of natural organic matter (NOM) by carbon SSNMR. Different magnetic fields, MAS speeds, and applications of single and ramped amplitude CPMAS with TOSS was reviewed. Little difference between single and ramped CPMAS was found. When accompanied with faster spinning speeds, better performance was found at higher magnetic fields. A poor spectral repeatability performance and errors contributed by TOSS are discussed. The overall conclusion reached was that it is still difficult, if not impossible, to make SSNMR of NOM quantitative. Only qualitative analysis is recommended. On the contrary, Conte et al. (355) found CPMAS affording accurate representation of whole carbon content in NOM when compared with elemental analysis.

Khalaf et al. (356) compared sorption domains in molecular weight fractions of a soil humic acid. Integration of the ^{13}C CPMAS spectra of unfractioned and fractioned humic acids was used to show the relative carbon-type distribution. The ^{19}F MAS was used to characterize the absorptive uptake of hexafluorobenzene (HFB). At least three sorptive sites were observed, each showing different HFB mobility

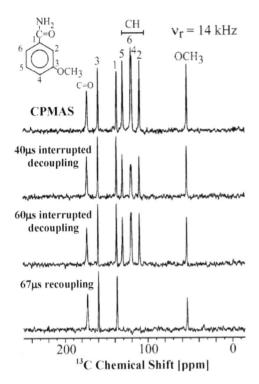

Figure 49 Failure of the standard-editing techniques using interrupted dipolar coupling coupled to high MAS spinning speeds (*middle two traces*). Interrupted decoupling with inclusion of carbon 180° pulse (causing recoupling of the ^1H–^{13}C dipolar coupling) properly dephases the CH carbons. *Source*: Modified from Ref. 357.

Mao and Schmidt-Rohr (357) devised accurate quantitation of aromaticity and nonprotonated aromatic carbon fraction in NOM. Direct polarization is used for the quantification. An improved dipolar dephasing sequence for high MAS spinning speeds is used for spectral editing (Fig. 49). The problem of overlap between aromatic and alkyl carbon resonances at 90–120 ppm is solved by ^{13}C chemical shift anisotropy filtering technique, selecting sp^3-hybridized carbons. Because the sp^3-hybridized carbons of O–C–O groups have smaller chemical shift anisotropy than the aromatic carbons, inclusion of the ^{13}C chemical shift anisotropy filter dephased the carbons with larger anisotropies. In addition nonprotonated O–C–O groups can be distinguished by adding dipolar dephasing step.

Polymorphism

One of the most important contributions of SSNMR in characterization of pharmaceutical solids is its exquisite sensitivity to differentiate the different crystalline and amorphous forms and, hence, detect the polymorphism. Polymorphs are typically differentiated based on their chemical shift differences (including the chemical shift anisotropy) but, in principle, differences in any other NMR property, such as dipolar coupling, J-coupling, or relaxation, can be used as well. A whole separate section describes the various applications of relaxation as selective filters or as the mobility probes. Chemical shifts are very sensitive to molecular conformations and crystal packing. In order for the polymorphs to be distinguished based on their chemical shifts, they must show differences in either the conformations or the crystal packing.

If a single crystal of suitable qualities can be grown, the single crystal X-ray diffraction provides a detailed description of the crystal structure, including molecular conformation and crystal packing. However, even the availability of single crystal data does not negate the need for additional studies. For example, the single crystal X-ray data contain only very limited information about the sample dynamics. SSNMR can fill the voids by probing molecular motions occurring at very different timescales. Also, pharmaceutical solid formulations are based on polycrystalline samples (not single crystals), and may actually contain mixtures of many crystal forms or other impurities. The industry "golden standard" for powder characterization remains to be the powder X-ray diffraction (PXRD). SSNMR and PXRD should be viewed as complementary techniques, as they take advantage of different phenomena to analyze polymorphism. PXRD is sensitive to differences in unit-cell dimensions, whereas SSNMR is sensitive to conformational changes. One can think of two purely hypothetical situations: (*i*) If two crystal forms differ exclusively in their unit-cell dimensions but the conformation of the molecule is preserved, the PXRD patterns are likely to be substantially different, whereas SSNMR might not register any changes. (*ii*) If the unit-cell dimensions are preserved while the conformation changes substantially, SSNMR is likely to pick up the differences, whereas the PXRD patterns will be identical. The real cases are always somewhere in between these two hypothetical situations, but based on the potentially biased author's personal experience with numerous crystal form studies by both SSNMR and PXRD techniques, at least several cases were encountered for which no differences in PXRD spectra were observed but showed characteristic SSNMR spectra. No case has yet been encountered for which the opposite would hold true.

The SSNMR has a number of important advantages over other spectroscopic techniques when it comes to determining polymorphism. One of the major advantages is its high degree of resolution making it a sensitive probe of conformation. Even minor changes of conformation or crystal packing between polymorphs produce different local environments and consequently different chemical shifts. SSNMR is usually more sensitive for polymorph detection than other techniques, including PXRD, IR, NIR, or Raman spectroscopies. In contrast to these more established spectroscopic techniques, SSNMR spectra are largely void of any dependence on physical sample properties as particle size, homogeneity, or residual solvent content. Therefore, pharmaceutical solids can be studied by NMR without a need for a special sample preparations. Note that if severe segregation of a multicomponent sample occurs, the quantification of the components might be effected due to their uneven excitation (Fig. 11). Samples viable for SSNMR analysis include a whole range of pharmaceutical formulations, such as tablets (typically crashed into powders; however, spinning of whole tablets has been also shown), lyophilized powders, capsules, suspensions, ointments, and others. SSNMR does not suffer from the preferred orientation phenomena, which often lead to an incorrect identification or quantification of polymorphs by PXRD. In addition, SSNMR is a nondestructive technique, which allows other analyses to be performed after the NMR spectrum is acquired. Another advantage of SSNMR is that the observed chemical shift differences between polymorphs can be related to particular molecular sites based on known assignments. The study of polymorphic transitions can be performed by variable temperature experiments. Probably, the most important advantage of SSNMR for pharmaceutical applications is its suitability for the analysis of complex formulations. Typically, excipients found in drug formulations resonate in limited spectral regions. In carbon spectra, most excipient peaks can be found between 60 and 100 ppm and as such do not completely overlap the active ingredient resonances, whereas fluorine spectra are always completely void from excipient contributions. Spectral regions containing only drug signals can often be found even in complex formulations.

A caution has to be exercised when evaluating the number of molecules per asymmetric unit. Usually the presence of multiple molecules per asymmetric unit is detected by SSNMR by comparing the number of expected resonances (based on the structure of the compound) and experimentally observed peaks. If more than the expected number of peaks is counted, presence of multiple molecules per asymmetric unit or a physical mixture is evident. However, even if the total count of the peaks does not exceed the expected number (due to the overlap of some peaks), the resolved areas of the spectrum can be used for the comparison, and the excess lines can still be detected. After the possibility of the presence of a physical mixture of polymorphs is excluded, at least one more factor has to be checked. The residual dipolar coupling to quadrupolar ^{14}N or ^{35,37}Cl nuclei, which is not completely averaged out by MAS, can cause carbon lines to split, typically exhibiting an asymmetric doublet pattern. Because this second-order effect is indirectly proportional to the magnetic field, it is the largest at lowest fields, and typically not important for magnetic field in excess of 11.7T (500 MHz proton frequency). This residual dipolar coupling is a well-known phenomenon and can even be used to extract information about the structure of the compound (157,358–374). It depends on the quadrupolar coupling parameters of the particular ^{14}N nucleus and on the orientation of the quadrupolar and dipolar tensors.

If more than the expected number of resonances are observed and the possibility of attributing it to the residual dipolar coupling to quadrupolar nuclei is eliminated, then the multiple lines should be interpreted as either the evidence for a mixture of several polymorphs or, conversely, for the presence of multiple asymmetric sites per unit. There are several ways to differentiate between these two instances. As a first approximation, when pure form of standards are absent, one can measure the relative peak intensities of the split resonances. If the peak ratios of all the resonances appearing as multiplets are approximately equal to one, multiple sites are likely to be present. This is, however, not a proof of multiple sites because, by coincidence, one could have an equal mixture of different polymorphs. In some cases, the build-up dynamics of CP, which is a dependence of CPMAS signal intensity on CP time, can be used to differentiate polymorphs from multiple sites (54). However, it is not straightforward to analyze the CP build-up intensity differences between carbons from the same molecular sites in terms of polymorphs versus multiple sites. Another way to distinguish mixtures from multiple sites is to measure proton T_{1H} or $T_{1\rho H}$ relaxation. Due to the efficient proton spin diffusion, the proton relaxation as detected indirectly on the split resonances will be either the same if multiple molecules per asymmetric unit are present or different if a mixture of forms is present. The difference of the proton relaxation values is the proof of a presence of mixture, but the opposite is not the proof of multiple molecules per asymmetric unit, because by coincidence, the components of the mixture may relax at the same rate. A positive proof can be gained by observation of cross-peaks between the split resonances in a through space correlation experiment. A positive proof for the existence of polymorphs can also be gained by varying the relative concentration of the two forms and observing the varying intensity ratios of corresponding carbons. This, in turn, requires sample manipulation, such as varying the temperature, humidity, or other conditions to change the ratio of polymorphs.

Stephenson et al. (375) studied polymorphic, isomorphic, and solvated forms of dirithromycin. Three different crystal forms out of the 12 known forms (two anhydrous forms, nine stoichiometric solvated forms, and one amorphous form) were studied. Carbon chemical shift assignment was achieved by comparison with solution shifts in combination with interrupted decoupling spectral editing technique. The resonances of the incorporated solvents were observed in the carbon spectra. More than one molecule per asymmetric unit was found in some of the forms. Steric hindrance of the solvent molecules was related to the *N,N*-dimethylamine peak coalescence in the variable temperature spectra. Carbon chemical shifts of the lactone carbon were related to the presence or absence of hydrogen bonding to this group. Combined spectroscopic approach (PXRD, IR, and SSNMR) was found to be advantageous to characterize crystalline forms in the absence of single crystal X-ray data.

McGeorge et al. (376) studied conformational analysis of three polymorphs of a disperse azobenzene dyestuff by ^{13}C and ^{15}N SSNMR in an attempt to understand the effect of polymorphism. Carbon chemical shifts indicated the possibility of forming an intramolecular hydrogen bond resulting in two conformations coexisting in the crystal structure of each polymorph and differing by internal rotation of a sidechain acetamido group. The conformer interchange can occur even in the solid state and can be explained by the open nature of the crystals as a result of steric effects of substituents.

Schmidt et al. (377) measured the temperature width of first-order single crystal to single crystal phase transition of three forms of 2-(2,4-dinitroben-zyl)-3-methylpyridine by ^{15}N (^{15}N enriched samples) and ^{13}C CPMAS. For molecular crystals, the classical first-order phase transition description is claimed to be too simplistic, because the different crystal phases can exhibit gradual onset of physical properties, such as heat capacity and volume over broad range of temperatures. This is due to fine microscopic details or nonequilibrium conditions. To shed light on the poorly understood contributions of the intermolecular potentials to the thermodynamic functions, the authors studied photoactive crystals by SSNMR. A comparison with solution, interrupted decoupling, REDOR, and TEDOR experiments were used for chemical shift assignments of all phases. Attention has been paid to proper temperature calibration. Temperature hysteresis of the phase transition of 7K was observed. The observations are rationalized in terms of two different phases having very different timescales. The first process constitutes a fast redistribution of the mole ratio of the coexisting phases, whereas the much slower second process involves macroscopic relaxation of the system. Existence of mesoscopic domains having different surface areas and different interfacial and strain energy contributions is therefore postulated.

Glasser and Shiftan (378) studied the solvatomorphism of (-)-scopolamine hydrobromide/hydrochloride salts. Regular CPMAS and spectral-editing experiments, such as dipolar dephasing, CPPI and T_{1C} inversion recovery, were used. The last experiment was used to filter out the fast relaxing CH_3 signals after 30 seconds delay. (T_{1C} of 50–60 seconds was observed on nonmethyl carbons.) The authors discuss in detail the carbon assignments, the observed line widths, and the differences between solution and solid carbon shifts. The otherwise commonly observed rapid topomerization was not observed. A π-flip of aromatics rings was observed in some of the solvatomorphs, and not in others. Its presence appeared to be correlated with the void-space above and below the ring. RHF GIAO-DFT chemical shielding calculations matched the observed chemical shift differences caused by γ-gauche effect. Correlations between torsion angles and the methylol and carbonyl chemical shifts were found, which could be utilized as diagnostic markers of the ester conformation. The authors conclude that SSNMR can be a powerful tool to unravel the contributions to weighted time-average structures observed in solution.

Apperley et al. (379) studied five polymorphs of sulfathiazole (Forms-I to -V) by ^{13}C CPMAS. Carbon spectra of all five forms were noticeably different, making SSNMR an excellent monitoring technique (Fig. 50). A combination of dipolar dephasing, comparison with solution shifts, and the broadening of the carbons close to ^{14}N was used to unambiguously assign nearly all the carbon resonances. Forms I–III and IV–V were found to have two and one molecule per asymmetric unit, respectively. The rotation of the aromatic group about the

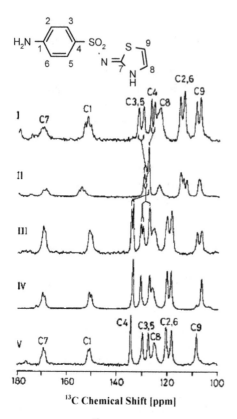

Figure 50 The ^{13}C CPMAS spectra of five polymorphs of sulfathiazole. *Abbreviation*: CPMAS, cross-polarization and magic angle spinning. *Source*: Modified from Ref. 379.

S–C bond was found to be low on the NMR timescale. The complex behavior of variable temperature spectra of Form-I was consistent with a strong anisotropic lattice expansion rather than the effect of the internal reorientation. Differences between chemical shifts of the different forms are discussed in detail and related to their intra- and intermolecular packing and H-bond patterns.

Dong et al. (380) described crystal structure and physical characterization of neotame methanol solvate. A single crystal X-ray structure is presented. The ^{13}C peaks were assigned based on the interrupted decoupling and previous studies with aspartame (381). Unlike the monohydrate, two crystallographically non-equivalent sites were observed for the methanol solvate. Padden et al. (381) compared the relative capabilities of carbon SSNMR and PXRD for analyzing mixtures of polymorphs of neotame. Altering the crystallization and drying conditions generated mixtures of solid forms. A systematic study was performed to observe conversion under vacuum of the monohydrate to a mixture of forms and then reconversion to the monohydrate upon exposure to moisture under ambient conditions. Up to seven forms were observed on the basis of the presence of distinct resonances for quaternary phenyl carbon. PXRD did not register any significant differences. SSNMR indicated a presence of many forms. The insensitivity of the PXRD to detect the different forms is explained by conformational changes of neotame not being accompanied by changes in the unit-cell parameters. The authors concluded that it is important that PXRD no longer is the exclusive technique for polymorph screening. Zell et al. (382,383) investigated polymorphism in aspartame and neotame. Aspartame was found to exist in three forms at room temperature, two of which had more than three molecules per asymmetric unit. Different 2D exchange experiments on ^{13}C-labeled aspartame based on spin diffusion and RFDR were run for assignment purposes. For crystal forms with multiple molecules per asymmetric unit, it was possible to assign the crystallographically inequivalent carbons. To obtain an optimal resolution on the ^{3}C-labeled aspartame, proton decoupling power combined with TPPM was found to be more critical than high-spinning speed to average out the ^{13}C$-^{13}$C and ^{13}C$-^{1}$H dipolar couplings. High-spinning speed MAS extending a period of several days was found to covert the neotame forms. PXRD was found to be insensitive to follow the interconversions. This was again rationalized as mostly the conformational changes, which SSNMR is sensitive to, are happening without significant changes in the unit-cell parameters. Dong et al. (384) studied seven neotame anhydrate polymorphs. Carbon SSNMR clearly distinguished between forms A, D, F, and G, the only forms produced in bulk. Form D was identified as having two molecules per asymmetric unit. Based on the different ^{13}C chemical shifts between the forms, a conformational polymorphism is suggested. Dong et al. (385) studied dehydration kinetics of neotame monohydrate. A carbon chemical shift of one of the aromatic neotame carbons was found to be characteristic of each crystal form and used to differentiate them. SSNMR data clearly indicated a partial dehydration of neotame monohydrate at 50°C to produce neotame anhydrate Form A.

Sack et al. (386) studied ^{15}N$-^{1}$H dipolar coupling and ^{15}N chemical shielding in two polymorphs of polyglycine by ^{1}H, ^{2}H, and ^{15}N SSNMR. Partial deuteration eliminated the largest intramolecular ^{1}H$-^{1}$H dipolar coupling, and it was possible to get structural information by orienting the dipolar vector in the molecular frame using the chemical shielding.

Lee et al. (387) studied polymorphism of phenylpyruvic acid. Partial assignments of the ^{13}C CPMAS is given for both Forms-I and -II. The Form-II was found to partially convert into Form-I upon the NMR analysis. The authors suspect that the high pressure produced by the spinning (4 kHz), and the long acquisition times (four hours) speeded up the form of transformation.

Chen et al. (388) characterized three hydrated, one water/methanol, and one amorphous forms of nedocromil by SSNMR and FTIR. The chemical shift assignments were accomplished based on combination of solution NMR, interrupted decoupling, and knowledge of crystal structures. More than one molecule per asymmetric unit was observed in some forms. FTIR and SSNMR are shown to be powerful complementary tools for probing the chemical environment. FTIR was found more useful for probing the interactions of water molecules, whereas SSNMR was superior in probing the interactions and comformations of the nedocromil ion.

Henck et al. (389) studied polymorphism of tedisamil dihydrochloride. The ^{13}C CPMAS spectra of the three enantiotropically related polymorphs were obtained at different temperatures: Form-II at 50°C (most stable form at ambient conditions), Form-I at 190°C, and Form-III at −60°C. The carbon spectra of the three forms were substantially different (Fig. 51). Even though single-crystal X-ray spectroscopy predicted only one molecule per asymmetric unit, splitting of most carbons was observed. This is explained by the reduced symmetry due to the disordered cyclopropyl groups. The authors noted increased resolution at higher temperatures and the significant temperature dependence of chemical shifts of some carbons, whereas others were largely unchanged.

Stephenson (390) presents a structure determination based on powder diffraction data obtained from conventional laboratory sources by application of Monte Carlo/simulated annealing methods. The ^{13}C CPMAS is used to confirm the predicted structure of two forms of acetohexamide.

Medek and Frydman (391) studied polymorphic forms of vitamin B_{12} recrystallized from various solvents by ^{13}C, ^{15}N, ^{31}P, and [59]Co SSNMR. The majority of the carbon resonances were assigned based on combination of solution assignments and solid-phase spectral-editing experiments (interrupted decoupling, short contact time CP, and INEPT-based experiment; Fig. 52). Variable temperature ^{13}C indicated reversible phase transformations. Some of

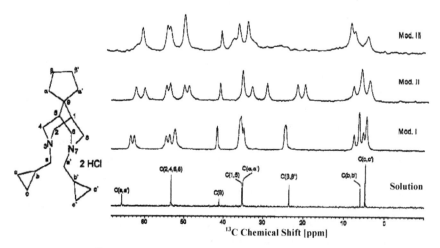

Figure 51 The [13]CPMAS solid-state and solution spectra of tedisamil dihydrochloride. *Abbreviation*: CPMAS, cross polarization and magic angle spinning. *Source*: Modified from Ref. 389.

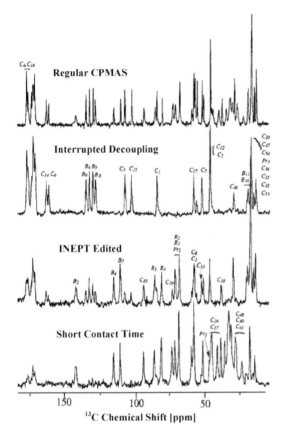

Figure 52 Spectral editing on vitamin B_{12}. *Source*: Modified from Ref. 391.

the nitrogens were not observed, presumably due to dipolar coupling to the ^{59}Co nucleus. The parameters of chemical shift and quadrupolar tensors were extracted by fitting the ^{59}Co static line shapes. The combined knowledge from this multi-nuclear approach allowed for structural predictions that correlated well with the known enzymatic role of B_{12}.

Variankaval et al. (392) characterized crystal forms of β-estradiol. SSNMR was used to monitor the changes in estradiol crystals as a function of temperature. Temperature-dependent 1D ^{1}H and 2D ^{1}H–^{13}C WISE experiments were utilized to follow the molecular mobility of estradiol during heating and cooling. The 2D WISE confirmed that the increased mobility at high temperatures observed in 1D ^{1}H MAS spectra was ensuing not from the steroidal skeleton but from the water in the sample. Differences between ^{13}C CPMAS spectra of each of the three forms (EA, EC, and ED) were observed. The broad nature of the ED peaks was attributed to improper crystallization. The presence of two molecules per asymmetric unit was found in forms EC and ED.

Wenslow et al. (393) studied polymorphism in an aza-steroid. Remarkably well-resolved ^{13}C CPMAS peaks of Form-I were observed, with some showing FWHM of 5 Hz, which was still limited only by the upper limit on acquisition time due to decoupling cycle constraints. In contrast to Form-I, Form-II showed multiple peaks per carbon, indicating two molecules per asymmetric unit.

Lee et al. (394) studied 3,4-methylenedioxyamphetamine hydrochloride and seven related compounds and their mixtures with lactose. Differences between solution- and solid-phase carbon spectra of the compounds are related to their crystal structures. The chemical shift differences are attributed to the variations in crystal packing. Lactose monohydrate was added to mimic the illicit production of "Ecstasy" tablets. An interaction between the 3,4-methylenedioxy-*N*-methylamphetamine hydrochloride and lactose monohydrate, as evidenced by chemical shift changes in several carbons, was found to be specific for this compound. It is attributed to lattice packing brought about by thorough dry-mixing with lactose at room temperature. It is concluded that lactose affects the crystal packing by reducing conformational rigidity, so that the molecule more closely resembles that in solution.

Crowley et al. (395) characterized mesomorphism of oleic acid and propranolol oleate. Spectral differences were observed upon γ-to-α phase transition. Carbon resonances were assigned by means of interrupted decoupling and solution data. Variable temperature ^{13}C CPMAS spectra and T_{1C} times were obtained to study molecular mobility and conformational disorder. T_{1C} relaxation for the α-phase indicated considerable conformational changes supporting the α-oleic acid classification as a conformationally disordered crystalline phase. Low values of T_{1C} times (fast relaxation) of propranolol oleate-I and -II suggested the absence of a rigid crystalline molecular lattice. The ^{13}C SSNMR is deemed as a powerful technique for pharmaceutical lipid characterization.

Harper and Grant (396) characterized carbon chemical shift tensors in polymorphs of verbenol. The verbenol sample was found to be a mixture of polymorphic materials containing two distinct lattice types. The major component contained three molecules per asymmetric unit, whereas the minor form contained four molecules per asymmetric unit. FIREMAT technique was used to correlate the isotropic with anisotropic carbon chemical shifts. The chemical shift tensorial parameters for each carbon were extracted from the anisotropic part of the spectrum. Only four different carbons showed the error bars of chemical shift calculations to be smaller than the predicted chemical shift differences upon varying the geometries. Verbenol has a rigid bicyclic structure. The experimental and ab initio computed chemical shift tensors demonstrated that most of the structural variations arose from differences in the dihedral angle of the hydroxyl hydrogen (H–O–C–C). The inability to differentiate the different molecules of the asymmetric unit by NMR (each of the three peaks for given carbon could not be specifically assigned to a given molecule in the asymmetric unit) prevented association of the calculated values to the specific molecule in the asymmetric unit.

Strohmeier et al. (397) investigated the polymorphs of dimethyl-3,6-dichloro-2,5-dihydroxyterephthalate. From previous X-ray study, the major structural difference between the white and yellow forms was the torsional angle between the ester group and the aromatic ring. The conjugation of the π-electronic system and the H-bond patterns are greatly affected. GIAO-DFT and HF quantum mechanical calculations (Gaussian 98) were coupled to experimentally derived carbon chemical shift tensorial values (extracted using FIREMAT technique) and to the carbon detected chlorine electric field gradient (EFG) values. The $^{35,37}Cl-^{13}C$ dipolar coupling parameters were extracted by fitting the variable magnetic field spectra (2.35, 4.7, and 9.4T) spectra (Fig. 53). Chlorine EFG parameters were taken from previous NQR study. Strong effect of the $^{35,37}Cl-^{13}C$ dipolar coupling was observed on the α- and β-carbon positions (with respect to the chlorine) at the lower fields, whereas only the α-carbon resonance was split at 9.4T. The conformational differences between the two polymorphs were found to have a great effect on the electronic properties of the molecules and, hence, on the chlorine EFG. Differences in the isotropic chemical shift values of up to 10.3 ppm were observed between the two forms. The conformational differences were found to be even better characterized by chemical shift tensor principal values (up to 14 ppm differences) than the isotropic chemical shifts alone. Good correlation between the experimental and calculated chemical shift tensorial values was obtained, especially when a stack of three molecules incorporating the intermolecular H-bonds, rather than just one molecule, was used for the calculation.

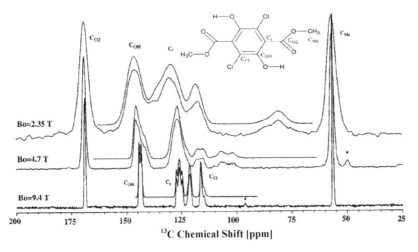

Figure 53 The ^{13}C CPMAS spectra at three different fields. Simulation of the residual $^{35,37}Cl-^{13}C$ dipolar coupling is included. *Abbreviation*: CPMAS, cross polarization and magic angle spinning. *Source*: Modified from Ref. 397.

Hanai et al. (398) compared vibrational and NMR spectroscopies to study cis-cinnamic acid polymorphs. Unlike the *cis*-58° crystal form, carbon NMR spectrum of the *cis*-68° crystal form showed splitting of several carbons, consistent with multiple molecules per asymmetric unit.

Fojud and Jurga (399) studied phase transitions in a series of *n*-alkylamonium chlorides by variable temperature ^{13}C CPMAS. The temperature of the transitions depended on the length of the alkyl chain and involved a change in the motional freedom of the methylene groups.

Yoshinari et al. (400) studied the moisture-induced polymorphic transition of mannitol and its morphological transformations. The ^{13}C CPMAS spectra of three polymorphs afforded distinctly different chemical shifts that could be used for the polymorphic characterization. The δ to β polymorphic transition was confirmed by SSNMR upon exposure of form δ to 97% RH. T_{1H} and T_{1C} relaxation was employed to estimate the molecular mobility characteristic of each structure. Both carbon and proton relaxation times were, on average, approximately twice longer in form δ. This was consistent with the δ-structure having rigidly linked H-bond network. The authors conclude that water acts as a molecular loosener by breaking the crystal's H-bonds to facilitate δ- to β-form conversion as a result of multi-nucleation.

Bauer et al. (401) studied the extraordinary example of ritonavir's conformational polymorphism. Only one crystal form was originally identified during development. Two years later, several lots failed the dissolution tests, because a new polymorph appeared. During the following several weeks, new Form-II began to appear in both the bulk drug and formulation areas. Due to the much lower solubility of Form-II, the formulation was unmanufacturable. The authors are not very specific about the SSNMR analysis, except for stating that ^{13}C CPMAS and ^{1}H MAS were used to distinguish the two forms. No spectra are shown.

Liang et al. (402) studied solvent-modulated polymorphism of sodium stearate crystals. Ternary crystalline dispersions of 10% sodium stearate in mixtures of water and propylene glycol were investigated. The conformation and mobility of the alkane chains in α- and β-forms were studied by ^{13}C CPMAS. The rigidity of the alkyl chains was compared for the freshly prepared and aged samples by measuring $T_{1\rho H}$ relaxation. Interestingly, longer $T_{1\rho H}$ values indicated more mobility. The authors quote a reference of Pursch et al. (403) in which a similar phenomena on alkyl chains attached to silica (chromatographic phases) is observed. A greater percentage of *trans*-chain conformation with higher rigidity was found in the α-form. Propylene glycol resonances appeared in the CPMAS spectra of both forms, indicating a strong association with the sodium stearate.

Giordano et al. (404) studied polymorphism of CHF1035 (Forms-I, -II, and -III). Comparison with solution NMR was used for carbon assignment purposes. All forms showed more than one molecule per asymmetric unit. Unlike in solution, distinct signals for each isobutyryl methyl carbon were observed, suggesting highly restricted motion in the solid state. Reversible changes were observed in variable temperature spectra of Form-I, suggesting some conformational changes.

Kimura et al. (405) characterized α, β, and γ polymorphs of a quinolinone derivative. Solution data were used to assign the ^{13}C CPMAS spectra of the three forms. Spectral differences were found between the three polymorphs, suggesting differences in the molecular conformation and crystal packing. Form-α showed two molecules per asymmetric unit.

Wiegerinck et al. (406) studied polymorphism of ^{13}C-labeled Org OD14 and Org 30659 drugs during drug processing. Different low API loading tablet formulations were examined by ^{13}C CPMAS. No form conversion was observed immediately after formulating and after seven days at 25°C/60%RH. Carbon SSNMR was found to be a powerful tool for the quantitative analysis of polymorphic pharmaceutical compounds in different stages of manufacturing of the drug product.

Lee et al. (407) characterized paclitaxel in its dihydrate, anhydrous, and amorphous forms. Significantly different ^{13}C CPMAS spectra of the compounds studied in terms of peak splitting and peak sharpness were observed.

Smith et al. (408) determined the crystal structure of anhydrous theophylline by PXRD and ^{13}C and ^{15}N SSNMR. SSNMR confirmed existence of only a single molecule in the crystallographic asymmetric unit. Dipolar dephasing was used to partially assign both nitrogen and carbon resonances. Nitrogen chemical shift supported only one type of H-bonds out of the several ab initio calculated models. This paper is a nice example how to use the synergies of PXRD and SSNMR for structure determination.

Wang et al. (409) characterized solubility, metastable zone width, and racemic mixture of the beta-blocker drug propranolol hydrochloride. The structural differences between the (R)- and (S)-enantiomers were investigated by PXDR, IR, and ^{13}C SSNMR. Carbon spectra of the (S)-enantiomer and the (R,S)-racemate were markedly different in both chemical shifts and the lineshapes, suggesting that the racemic mixture belongs to the racemic compound.

Zhang et al. (410) studied crystallization and transitions of sulfamerazine polymorphs. Tentative assignments of ^{13}C CPMAS spectra of Forms-I and -II were made based on solution data and interrupted decoupling. Multiple sites per asymmetric unit were observed in Form-I. Peak broadening of some carbons indicated significant molecular motion. The temperature induced from transformation is studied by variable temperature ^{13}C CPMAS.

Moynihan and O'Hare (411) studied monoclinic and orthorhombic forms of paracetamol. The ^{13}C CPMAS spectra of the monoclinic form and a spectrum of a mixture of forms prepared from a melt were obtained. A spectrum of the orthorhombic form was reconstructed as a difference of the two experimental spectra. Well-resolved differences in chemical shifts of both forms were observed. Asymmetric doublets were observed for carbons bound to nitrogen due to the residual ^{13}C–^{14}N dipole–dipole interaction not completely averaged out by MAS.

Garcia et al. (412) showed differences between polymorphism and desmotropy on examples of 3-phenyl- and 5-phenyl-1H-pyrazoles and 3-phenyl-1H-indazole. In addition to polymorphism, the tautomeric compounds can also present desmotropy. The term desmotropy refers to the crystallization of a compound in two different tautomers. The compounds were characterized by ^{15}N and ^{13}C CPMAS. The ^{15}N chemical shifts were found to be more sensitive to environmental effects, such as hydrogen bonds, than to tautomerism. The ^{13}C chemical shifts were found to be similar to the solution and constituted a precise tool to identify and determine the purity of the tautomers. In the two closely related compounds studied, the authors found a case of desmotropy and a case of concominant ("one pot") polymorphism.

Novoselsky and Glaser (413) studied conformational polymorphism of sertraline hydrochloride, an antidepressant drug. The carbon isotropic chemical shift values of Forms-I and -III were correlated with γ-gauche effect resulting from the respective antiperiplanar and synclinal torsion angles. RHF GIAO calculations of the magnetic shielding tensors were used to support the conclusions.

Zimmermann et al. (414) studied mesomorphism, isomerization, and dynamics of a series of pyramidic liquid crystals. Variable temperature and variable spinning speed carbon NMR was used to study the structure dynamics and mesomorphic properties of the crown and saddle mesophases of nonaalkanoyloxy tribenzocyclononene. Rotor synchronized carbon 2D exchange spectra were recorded, identifying the presence of dynamics by the off-diagonal cross-peaks (415). Chemical exchange (cross-peaks between different sidebands) and physical reorientation (cross-peaks linking sidebands of the same manifold) could be distinguished. The spectra of the crown mesophase exhibited dynamic features consistent with physical reorientation of planar threefold molecular jumps. In contrast, the saddle isomers did not show dynamic effects.

Potrzebowski et al. (416) devised a new method to distinguish enantiomers and racemates and to determine an enantiomeric purity. Chiral discrimination can be achieved through differences in $^1H-^1H$, $^1H-^{13}C$, and $^{13}C-^{13}C$ dipolar coupling, chemical shift anisotropies, and ^2H quadrupolar splitting. Typically, enantiomers and racemates crystallize in lattices belonging to different point groups and as such can show differences in the chemical shifts. The authors show that by using 1D ODESSA, differences between enantiomers and racemates can be recognized even when no differences in isotropic chemical shifts (magnetically equivalent nuclei) between the two exists. An advantage is taken of the fact that the molecular symmetries and packing of enantiomers and racemates are usually significantly different, giving rise to different angles and distances between the nuclei and hence the different dipolar coupling. By following the intensity of the entire ^{31}P spinning sideband set of cyclophosphamide, isophosphamide, and bromophosphamide as a function of the mixing time, the racemates and enantiomers of these compounds were distinguished. Good correlation with enantiomeric purity is shown.

Maurin et al. (417) studied polymorphism of roxifiban, which was found to exist in two polymorphic forms. No differences between the two forms were observed by DSC, IR, and Raman spectroscopy. Slight difference was detected by isothermal microcalorimetry. The PXRD and ^{13}C CPMAS SSNMR spectra of the two forms were substantially different. Carbon NMR observed lower symmetry in Form-I by splitting several peaks. In contrast, one molecule per asymmetric unit was observed for Form-II.

Wang et al. (418) studied polymorphic behavior of NK1 receptor antagonist by, among other techniques, ^{13}C CPMAS. Four anhydrous polymorphic forms I to IV were discovered. The compound can exist as Forms-I or -II at RT and as Forms-III and -IV at elevated temperatures. Polymorphic transitions were studied by variable temperature experiments. At 115°C, Form-I transforms into Form-III, spectrum of which shows less peaks, indicating less molecules in the asymmetric unit. At 80°C, Form-II transforms into Form-IV. Carbon spectrum of Form-IV differs from that of Form-III mainly in three peaks, shifted approximately by 0.5 ppm.

Pisklak et al. (419) combined ^{13}C CPMAS and molecular modeling to study three new derivatives of buspirone. Variable contact time experiments were recorded to optimize intensities of all carbons. More than one molecule per asymmetric unit was observed. GIAO-RHF ab initio calculations were performed to facilitate the resonance assignments. Pisklak et al. (419) studied coumarine anticoagulants, warfarin, and sintrom, by ^{13}C CPMAS and ab initio calculations (420). Two enol-ketal tautomers were found to exist for each. The solution–solid chemical shifts differences for the solution minor and major components were found to be of similar size, and no conclusions as far as the solid-state configuration and conformation could be reached. Good correlations between the experimental and GIAO-DFT calculated shifts were established.

Komber et al. (421) studied tautomerism of cyclo-*tris*(4-R-2,6-pyridylfor-mamidine). Dipolar dephasing was used to aid the assignment of ^{13}C CPMAS spectra. Three molecules of DMSO were found to be incorporated into the crystal lattice. Sharp signals from sp^2-hybridized pyridine and imino nitrogens and a broad signal from the amino nitrogens were observed in ^{15}N CPMAS spectra. The broadening of NH signals was caused by the residual dipolar coupling with the proton. Signal splitting of carbon and nitrogen resonances indicated a nonsymmetric structure. No signal coalescence was observed in the ^{13}C and ^{15}N CPMAS spectra, proving the absence of very low rate of proton exchange and therefore the existence of only one tautomer in the solid state.

Harper et al. (422) investigated solid-state polymorphism in 10-deacetyl baccatin III. Carbon chemical shift tensors were measured by using the FIREMAT technique for the DMSO solvate and the unsolvated forms. Both forms represented well-defined lattices, as indicated by narrow ^{13}C CPMAS peaks. The origin of the chemical shift differences between the two forms is studied by the aid of ab initio calculations. The carbon assignment was done by combination of spectra editing and comparison with solution and the calculated shifts. The near invariance of all nonaromatic signals in the variable temperature spectra suggested that only the benzoyl moiety was mobile. The authors discussed in detail whether the chemical shift differences between the forms were caused by lattice or conformation. In general, for molecules with moderate polarity, the lattice does not contribute significantly to chemical shifts. Conversely, in charged compounds and polar structures, the inclusion of lattice is necessary to explain the shifts. The incorporation of the DMSO molecule was found not to cause the shift differences. Only three out of 10 carbons showed evidence for the lattice contribution. The observed tensor differences appeared to arise primarily from conformational variations in ring substituents and the cyclohexyl ring.

Meejoo et al. (423) studied the structural disorder of β-polymorph of (E)-4-formylcinnamic acid. In the absence of the single crystal X-ray data, the structure of this form was determined from powder X-ray. The ^{13}C CPMAS was used to investigate structural disorder of the orientation of the formyl group. One molecule in the asymmetric unit was observed. The CHO group shows peak splitting. Because there is no NMR interaction that could give rise to such splitting, authors conclude that it must be due to the structural disorder of this group. The relative intensities of the two CHO peaks did not depend on the CP contact time, reflecting the relative populations of the two orientations. The peak that is shifted downfield is proposed to correspond to the orientation engaged in short intermolecular hydrogen bond, on the basis that intermolecular H-bonding to carbonyl groups is known to increase the ^{13}C isotropic chemical shift.

Iuliucci et al. (424) studied ring chain tautomerism of erythromycin A. The ^{13}C CPMAS resonances were assigned using spectral editing and chemical shift calculations with DFT B3LYP/D95** method. Previously acquired values of chemical shift anisotropy for all carbons are also shown (425). A chemical shift assignment using 2D correlation techniques was also previously performed (Fig. 54) (426). The calculated chemical shift tensors correlated well with the experimental shifts. Formation of cyclic hemiketal was observed upon heating or desolvation by desiccation. The principal values of the hemiketal carbon chemical shift tensor are reported.

Reutzel-Edens et al. (427) characterized the five crystal forms of LY334370, a 5HT1f agonist, three anhydrates, a dihydrate, and an acetic acid solvate. Single crystal data of only Form-I were available. The ^{13}C CPMAS spectra for these five forms showed the differences typically observed for polymorphs and solvates (Fig. 55). Splitting due to the residual dipolar coupling to quadrupolar ^{14}N and the distorted intensities of the carbons J-coupled to ^{19}F were observed. Single molecular conformation was detected in each crystal form. Some carbon resonances were found to show significant differences compared with the solution (up to 3 ppm), possibly due to the formation of H-bonds.

Zolek et al. (428) combined ^{1}H and ^{13}C SSNMR spectroscopy with GIAO-CPHF calculations to characterize chloramphenicol, thiamphenicol, and their pyrrole analogs. Unexpectedly, good resolution in the ^{1}H MAS (32 kHz spinning)

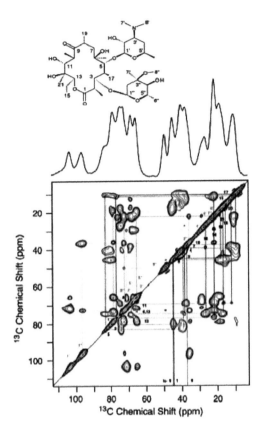

Figure 54 The 2D ^{13}C–^{13}C chemical shift correlation spectrum of [U-^{13}C]-labeled erythromycin. *Source*: Modified from Ref. 426.

spectrum was observed for chloramphenicol, resolving the signals of ortho- and *meta*-protons separated by 0.5 ppm. The observed differences of the carbon shifts between solution and solid were up to 6 ppm. Broadening or splitting of the carbon resonances of carbons linked to ^{14}N or to ^{35,37}Cl were observed as a result of the residual dipolar coupling to quadrupolar nuclei. Good correlation was observed between the GIAO-CPHF calculated and observed carbon chemical shifts. Some of the differences between the calculated and observed values are ascribed to H-bonding, because the calculations were performed on isolated molecules.

Mirmehrabi et al. characterized the enamine and nitronic acid tautomeric forms of ranitidine hydrochloride. The ^{13}C CPMAS identified a molecular disorder in Form-II, whereas Form-I was more ordered.

Vanhaecht et al. (429) studied cocrystallization phenomena in piperazine-based copolyamides by SSNMR and other techniques. The ^{1}H SSNMR and $T_{1\rho H}$ relaxation was used to investigate the influence of the composition on the

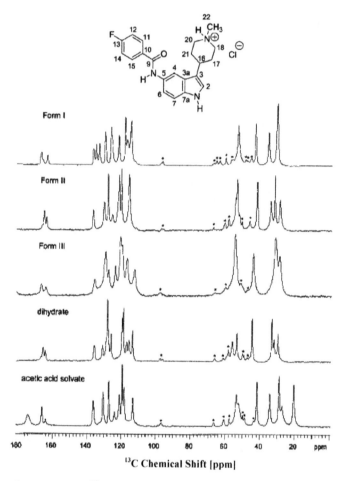

Figure 55   The ^{13}C CPMAS spectra of LY334370 hydrochloride crystal forms. *Abbreviation*: CPMAS, cross polarization and magic angle spinning. *Source*: Modified from Ref. 427.

percentage of the rigid phase of the copolyamides and delivered additional indications for cocrystallization.

Portieri et al. (430) studied the effects of polymorphic differences on three sulfanilamide forms observed by ^{13}C and ^{15}N SSNMR. The relaxation times were measured at different temperatures and analyzed in terms of mobilities. Nitrogen shielding parameters were obtained from static and MAS spectra of labeled samples and compared with calculations.

Kameda et al. (431) studied ^{13}C chemical shifts of the triclinic and monoclinic crystal forms of valinomycin. The 2D FIREMAT technique was used to characterize carbon chemical shift tensors (Fig. 56). The isotropic chemical shifts and the corresponding tensor components gave useful information about

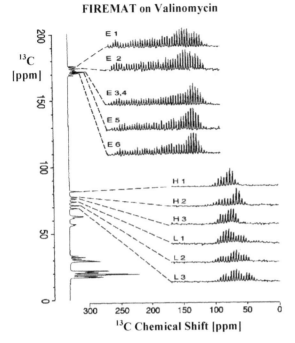

FIREMAT on Valinomycin

Figure 56 The 2D ^{13}C FIREMAT of triclinic valinomycin. *Abbreviation*: FIREMAT, five π-replicated magic angle turning. *Source*: Modified from Ref. 431.

the molecular structure of valinomycin in the solid state. An utility of the chemical shift tensors for the understanding of secondary structures of peptides was demonstrated.

Remenar et al. (432) determined the physical form of 2-tert-butyl-4-methoxy-phenol (BHA), ^{13}C labeled on the methoxy carbon, and wet granulated onto common pharmaceutical excipients. At 0.1% loading, BHA was shown to be amorphous and mobile (by mobility-induced loss of the ^{13}C CPMAS signal due to smaller average dipolar coupling) signal in the freshly prepared blends. At 0.5% loading, it was found amorphous on cellulose and crystalline on lactose, mannitol, calcium phosphate dehydrate, and croscarmellose sodium.

Sheth et al. (433) studied mechanochromism of piroxicam accompanied by intermolecular proton transfer. A proton transfer accompanied both solid-state disorder, and a change in color induced by mechanical stress, a phenomenon referred to by authors as mechanomorphism. Variable temperature and variable contact time ^{13}C CPMAS spectra of polymorphs I, II, amorphous and monohydrate were acquired (Fig. 57). Deprotonation of the acidic enol group provided more carbonyl-like character, and therefore a large downfield shift occurred. The protonated and deprotonated species could thus be quantified by integration of their respective peaks. Most of the amorphous piroxicam was found to consist

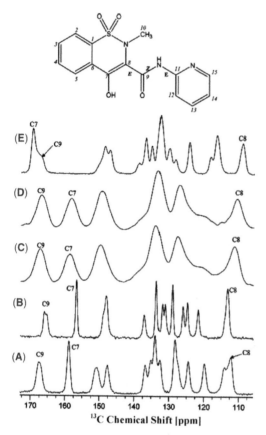

Figure 57   The ^{13}C CPMAS spectra of piroxicam polymorphs. (A)–(E) show Form-I, Form-II, Form-I cryoground for 60 minutes, Form-II cryoground for 60 minutes, and piroxicam monohydrate, respectively. *Source*: Modified from Ref. 433.

of neutral molecules. The charged species comprised of only about 8% of the amorphous phase. The ability to quantify the fractions of neutral and charged molecules in the amorphous phase highlighted the unique capability of SSNMR to quantify mixtures in the absence of standards.

Booy et al. (434) investigated the use of ^{13}C labeling to enhance the carbon sensitivity to study the effect of tabletting on the polymorphism of a steroidal drug at dosage levels of 0.5–2.5% active. The crystalline form could be readily analyzed in the tablets, and no change of form was observed. Labeling of neighboring carbons, which causes broadening due to the homonuclear dipolar coupling, should be avoided.

Schmidt (435) characterized the polymorphism of local anesthetic drugs of falicaine HCl and dyclonine HCl by ^{13}C and ^{15}N CPMAS. Differences between

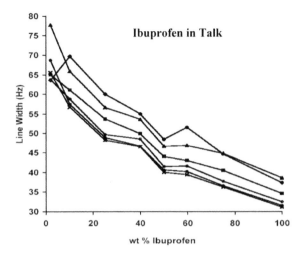

Figure 58 Line widths of ^{13}C CPMAS spectra of ibuprofen (IBP) when in physical mixture with the talk. Different IBP peaks correspond to different symbols. *Abbreviation*: CPMAS, cross polarization and magic angle spinning. *Source*: Modified from Ref. 437.

the polymorphs of both drugs were readily observed by SSNMR. Dipolar dephasing and comparison with solution help with assignments. An internal rotation of part of the drug molecule was confirmed by variable temperature study.

Suzuki and Kawasaki (436) evaluated forms of troglitazone by SSNMR. The SSNMR distinguished the hydrated and anhydrous forms better than PXRD. Diastereomers were found to coexist as a physical mixture. Unlike PXRD, SSNMR was able to characterize the crystal forms when formulated in tablets. A stable amorphous form was found in tablets.

Barich et al. (437) investigated SSNMR line widths of ibuprofen in drug formulations. The line widths were measured for different sample preparations, including crystallization from different solvents, melt-quenching, manual grinding, cryogrinding, compacting, and blending with excipients. Very interestingly, when in physical mixtures with most excipients, the line width decreased, affording better ^{13}C resolution. The exception was mixing with talc, for which the line width increase is attributed to the increased magnetic susceptibility anisotropy of talc (Fig. 58). The line width can thus provide information on physical characteristics of the sample.

Fluorine Applications

As a result of the structure-based rational drug dosing, many active pharmaceutical ingredients contain fluorine atoms. Fluorine groups may improve the potency or offset the infavorable pharmacological properties of the compound.

From the perspective of the SSNMR spectroscopist, the presence of fluorine atoms is very desirable since ^{19}F nuclei are almost as sensitive as ^{1}H, but do not face the same limitations as protons. Because the number of fluorine atoms per molecule is usually small, homonuclear ^{19}F–^{19}F couplings are relatively weak and can be removed by conventional MAS. However, a proton decoupling and/or a very fast MAS is desirable to cancel the strong ^{1}H–^{19}F dipolar interaction, sometimes necessitating the use of special NMR hardware. If CF$_3$ groups are present, a very fast MAS spinning should be applied to remove the homonuclear ^{19}F–^{19}F coupling, already scaled down considerably by the fast internal rotation of the CF$_3$ group. The proton decoupling is less important in this case, because the residual ^{1}H–^{19}F coupling is well removed just by the fast MAS. The fluorine line widths were found to be determined by mostly inhomogeneous contributions (438,439). The typical fluorine lines are approximately 1 ppm wide. Due to the large fluorine overall chemical shift range, this is typically sufficient to resolve the different polymorphs. The limit of detection based on sensitivity considerations is well under 1%, and in favorable cases LOD of 0.01% is not unusual. Fluorine NMR can be thought of as an ideal filter when applied to formulations. Because no typical excipient contains fluorine atoms, the only signal is due to the fluorinated API. Also the same detection limits apply when corrected for the dilution by excipients.

The presence of fluorine also enables measurements of fluorine–carbon distances by REDOR experiment without the need for isotope enrichment. Multidimensional homonuclear or heterunuclear experiments are also possible serving as an additional tool for structural investigations. It should be noted, however, that the presence of fluorine atoms complicates the observation of resonances in carbon spectra for carbons that are in physical proximity of the fluorines. It is very difficult to efficiently decouple fluorine and protons simultaneously. The ^{19}F–^{1}H dipolar coupling makes the proton decoupling of nearby carbons inefficient, because the ^{19}F–^{1}H coupling effectively increases the proton "chemical shift" range and the proton off-resonace affects then limits the ^{1}H–^{13}C decoupling efficiency (440).

Campbell et al. (441) studied two polymorphs of a piperidone-substituted benzopyran with a fluorinated sidechain by ^{13}C and ^{19}F SSNMR and other techniques. More than one molecule per asymmetric unit was observed for Form-II. A splitting of the carbons bound to ^{14}N due to the residual dipolar coupling was observed and calculated with literature values of the quadruplar tensor. An increased carbon resolution was achieved by the application of double (^{1}H and ^{19}F) decoupling. Low-temperature spectra did not show any extra information. Proton and fluorine relaxation parameters were obtained. The proton decoupled fluorine line widths were approximately 500 Hz.

Harris and Crowe (442) studied fluorinated diazadiphosphetidines to obtain information on structure and dynamics. Chemical shift anisotropy tensor were characterized by analysis of the spinning sideband manifolds. Proton and fluorine relaxation times were measured.

Brouwer et al. (443) studied supramolecular solids by carbon and fluorine NMR. The ^{19}F MREV-8 experiment, minimizing the homonuclear fluorine contribution, was used to characterize the shielding tensor of a static sample. The static CP spectra (no MREV-8 applied) were then simulated using fluorine inter-nuclear distances.

Fuchs and Scheler (444–447) and others used high-resolution ^{19}F SSNMR to quantitatively investigate structural changes like branching and cross-linking of irradiated polymers.

Holstein et al. (448) studied the semicrystallinity and polymorphism in poly(vinylidene fluoride) by ^{1}H–^{19}F double resonance SSNMR. Amorphous and crystalline regions provided distinct chemically shifted signals. Application of relaxation filters suppressed either the amorphous or crystalline parts. The polymorphism of the crystalline polymer was reflected in the observation of several fluorine resonances. Isolated spectra of the crystalline phase of different polymorphs could be obtained. Especially the immobile crystalline chains required application of high-power proton decoupling. Ando et al. (449,450) studied the crystalline and amorphous phases of the semicrystalline poly(vinylidene fluoride) using variants of [1]H–[19]F and [19]F–[1]H CPMAS experiments. An [1]H–[19]F CPMAS inversion recovery experiment, in which, after a short CP period, the phase of the fluorine spin-lock pulse is changed by 180° and the inverted fluorine signal decays and goes through zero, was used to observe details of the initial CP processes as the coherent magnetization oscillations (Fig. 59). The CP drain experiment was used to measure the contact time dependence of the residual proton magnetization after the proton background signal was removed by the double CP (Fig. 60). The spin lock served to remove any residual fluorine magnetization before the CP drain step. Significant differences were observed between the crystalline and amorphous phases in the CP and relaxation behavior and related motions in the amorphous domain. Su and Tzou (251) determined the relative content of the various domains of poly(vinylidene fluoride). The phase separation was analyzed by 2D ^{19}F exchange spectroscopy and chemical shift anisotropy was extracted by fitting the spinning sideband patterns. Wormald et al. further studied this polymer using combination of relaxation

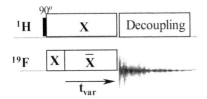

Figure 59   The [1]H–[19]F CPMAS sequence with variable contact inversion recovery to study details of the initial part of the CP process. *Abbreviations*: CPMAS, cross polarization and magic angle spinning; CP, cross-polarization.

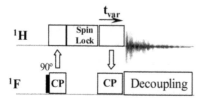

Figure 60   Double CPMAS (^{19}F–^{1}H and ^{1}H–^{19}F) experiment to study contact time dependence of proton magnetization after the removal of proton background. *Abbreviation*: CPMAS, cross polarization and magic angle spinning.

filters and two-dimensional RF-driven re-coupling (RFDR) and spin diffusion experiments (252).

Ando et al. (253) studied an amorphous perfluorated polymer by a combination of high-speed ^{19}F MAS, ^{19}F CRAMPS, and ^{19}F–^{13}C CPMAS experiments. The peak assignment was confirmed by GIAO shielding calculation. A single phase was confirmed based on relaxation measurements. The ^{19}F–^{13}C CP kinetics was followed by variable contact time studies and analyzed theoretically by the quantum mechanical master equation. C–F distances, CPMAS time constants, and parameters of homonuclear spin-diffusion among fluorines were determined.

Wenslow (454) studied crystalline and amorphous forms of a selective muscarinic M_3 receptor antagonist in both bulk and pharmaceutical dosage form samples (Fig. 61). Ambient and elevated temperature spectra were taken. A temporal line narrowing was observed at high temperatures, caused by the solvent escape, resulting in a high mobility of the drug. Observing the same effect on drug product, the conversion of crystalline to amorphous form was unambiguously confirmed in the low dosage tablets.

Solis et al. (455) studied the fluorine chemical shifts of organic molecules and analyzed them in terms of surface charge representation of the electrostatic embedding potential of crystalline inter-molecular effects. Different charge models (GRID and SCREED) and a cluster model were evaluated in order to include intermolecular interactions in the calculation of ^{19}F chemical shift tensors. The charge models allowed determination of the point charge distribution that mimicked the crystal field in the shielding calculations.

Glaser et al. (456) studied orientation of a peptide in lipid membranes based on fluorine dipolar coupling of the 4-CF_3-phenylglycine labels. A simple pulse experiment allowed for simultaneous measurement of both the anisotropic chemical shift and the homonuclear dipolar coupling of the CF_3 group in an oriented membrane sample, uniquely defining the effective tilt angle of the CF_3-axis with respect to the external field. The sign of the dipolar coupling was deduced from the anisotropic chemical shift of the oriented sample. The dipolar coupling provided an accurate measure of the orientation of the amino acid sidechain in the peptide backbone.

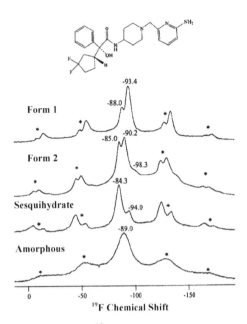

Figure 61 The ^{19}F MAS spectra (15 kHz spinning) of different forms of the compound shown at the *top*. No proton decoupling was applied. *Abbreviation*: MAS, magic angle spinning. *Source*: Modified from Ref. 454.

Park et al. (457) investigated crystalline structure of poly(vinylidene fluoride) hybrid composites. The long $T_{1\rho F}$ of the crystalline phase, which was significantly different from that of amorphous phase, was used to observe the crystalline domains selectively after spin-lock filter. The selective fluorine spectra were found to be very sensitive to the crystalline structures, and the relative populations of each morphological component could be extracted quantitatively.

Budarin et al. (458) studied surface energy and surface area of inorganic materials indirectly by measuring chemical shifts of adsorbed trifluoroacetic acid. The advantage of this technique over the standard surface area measurement by using N_2 stemmed from the fact that the "real-life" catalysis involves larger molecules that may not be able to access the entire surface. By NMR, the surface area can be measured specifically with fluorine analogs of the desired molecules.

Antonioli and Hodgkinson (440) studied $^{13}C–^{19}F$ interactions in carbon spectra of spinning solids. Carbon resonances of carbons close to fluorine nuclei typically showed significant line broadening. This is explained by the inefficiency of proton decoupling by the ^1H off-resonance effect caused by the $^1H–^{19}F$ dipolar coupling (Fig. 62). Sharper lines could be obtained by applications of more sophisticated decoupling schemes, such as SPINAL-64 (79). The lines were found to have an essentially inhomogeneous character ascribed

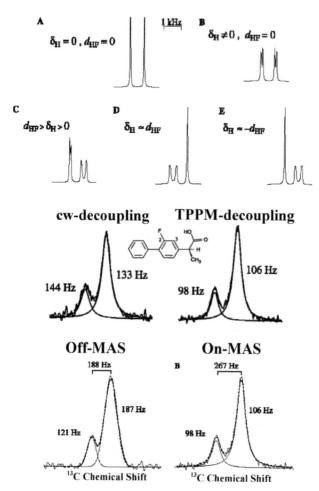

Figure 62 *Top*—simulation of CW-decoupled ^{13}C spectra in $^{13}C-^{19}F-^1H$ coupled system under MAS for different combinations of chemical shift anisotropy (δ_H) and $^{19}F-^1H$ dipolar coupling (d_{HF}). *Bottom*—experimental and simulated ^{13}C CPMAS spectra of flurbiprofen at different decoupling and MAS spinning conditions. *Abbreviations*: CW, continuous wake; MAS, magic angle spinning; CPMAS, cross polarization and magic angle spinning. *Source*: Modified from Ref. 440.

to a large anisotropy of bulk magnetic susceptibility. The line widths were also found to be very sensitive to the precise setting of magic angle. A misset of the magic angle resulted in asymmetric broadening of the $^1J_{CF}$ doublet components, degrading resolution. Another factor influencing the carbon line width was the effect of the finite lifetime of the ^{19}F states. If ^{19}F relaxes fast ($^1J_{CF} < 1/2\pi T_{2F}$; often true for rigid solids), the carbon $^1J_{CF}$ doublet collapses to a single line (it is "self-decoupled"). If the fluorine relaxation is slower, the carbon J-doublet is broadened. The dipolar interaction with fluorine also often makes the appearance of the J-doublet asymmetrical, due to the different intensity distribution in the spinning sideband manifolds from the two ^{19}F spin states.

Amorphous Phase Characterization

Gustafsson et al. (330) compared SSNMR and isothermal microcalorimetry in the assessment of the amorphous component of processed lactose. The PLS mathematical models established from NMR spectra accurately predicted results even when the lactose contained an additional component, acetylsalicylic acid. Both microcalorimetry and SSNMR could detect as low as 0.5% amorphous lactose. SSNMR has the advantage to provide additional structural information about the mixture. Since the authors used recycle delay of 2.5 seconds, the crystalline component was partially saturated due to its long T_{1H}. Moreover, this relaxation is likely to change from sample to sample, limiting the potential applications.

Utz (450–462) applied NMR to investigate structure and dynamics of amorphous solids. A global and local order that develops under plastic deformation is investigated by 2D DECODER. This technique correlates two different orientations of the sample to each other, while the sample undergoes flips or slow rotation. Comparison of the experimental results with molecular simulations indicated that thermal relaxation was active up to the entanglement length scale, even at temperatures far below the glass transition.

Cadars et al. (220) studied the disorder of amorphous compounds in chemical shift-correlated 2D experiments. This study is a follow-up on the previous publications from the same lab discussed earlier in the context of INADEQUATE experiment (Fig. 63) (217–219). A conditional probability of chemical shift distribution can be extracted from the experimental data. The tranverse dephasing times are important in determining the resolution of the distribution. The structural information contained in the distribution is important for the understanding of disorder.

Relaxation and Dynamics

NMR is an excellent technique to study dynamic processes, such as molecular motions, in solid samples. The sensitivity of NMR to motions arises from the orientational dependence (the anisotropy) of the nuclear spin interactions described in the introduction section. Types of motions and their frequencies can be inferred. Motions can be sampled by NMR directly, through coherent processes, or, indirectly, through incoherent processes, such as relaxation. The direct sampling of motions through changes in the nuclear spin interactions include monitoring 1D spectral line shapes or designing multidimensional experiments, such as chemical exchange experiments. The indirect sampling of motion stems

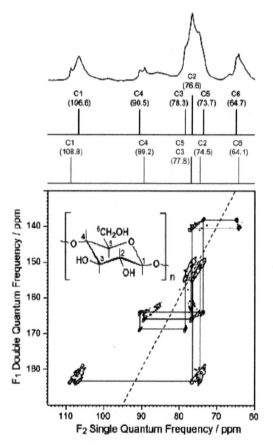

Figure 63 The 2D ^{13}C INADEQATE spectrum of 10% ^{13}C-labeled cellulose. Two different structural conformations were found. The chemical shift correlations are highlighted by the *lines*. The 1D stick spectra reproduce the chemical shifts of the two structural conformations. The regular 1D ^{13}C CPMAS spectrum of cellulose is shown at *top*. *Abbreviation*: CPMAS, cross polarization and magic angle spinning. *Source*: Modified from Ref. 219.

from the fact that the relaxation processes are driven by the random fluctuations in the nuclear spin interactions from molecular motions. Both the coherent and incoherent processes are sensitive to the motions in broad frequency range. As such, NMR can detect motions with frequencies ranging from near zero to 10^9 Hz.

Relaxation is a response of the system to perturbations, such as a series of rf pulses by driving the system back to its thermodynamic equilibrium. There are different types of NMR relaxation, including longitudinal spin-lattice (T_1), transversal spin–spin (T_2), longitudinal relaxation in the rotating frame ($T_{1\rho}$), and cross-relaxation (the NOE effect). Longitudinal relaxation corresponds to restoring the spin levels to their Boltzmann equilibrium distribution described by equation (3). The spin–spin relaxation induces the loss of the transversal coherence (the loss of NMR signal) due to gradual randomization of the phases of individual spins in the whole-spin ensemble. The longitudinal relaxation in the rotating frame represents the loss of coherence under the application of a continuous irradiation by a spin-locking pulse. If strong enough, the spin-locking pulse cancels the evolution due to coherent contribution of the various nuclear interactions (mainly dipolar coupling in the case of organic molecules), and the only loss of signal is due to the incoherent $T_{1\rho}$ relaxation. The actual values of the relaxation times depend on the type of the nuclear interaction causing the relaxation and on the motions present locally in the molecule. Typically, in the absence of paramagnetic species or quadrupolar nuclei, the dominant relaxation mechanism for organic molecules is the dipolar coupling. The motional frequency dependence of the relaxation process is expressed by so-called spectral density functions. They reflect the likelihood of the given nuclear interactions (e.g., dipolar coupling or combination of interactions) that molecular motion with a particular frequency causes the relaxation. This frequency dependence of the spectral density is very different for the T_1, T_2, and $T_{1\rho}$ types of relaxations. In general, T_2, $T_{1\rho}$, and T_1 are sensitive to very low frequency motions, mid-kHz (tens to hundreds kHz; on the order of the spin-locking field) and MHz (tens to hundreds MHz; on the order of the Larmor frequency) range, respectively. The motions are typically described by one or more correlation times, which are indirectly proportional to temperature. Figure 64 represents a schematic depiction of the temperature dependence of the T_1 and T_2 relaxation times for two different values of external magnetic field. The extreme narrowing regime is defined by the positive slope of the relaxation with increasing temperature and is also

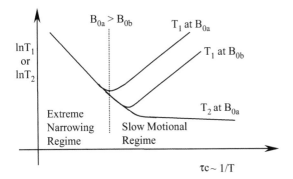

Figure 64 Schematic representation of temperature dependence of the T_1 and T_2 relaxation times. The *vertical dotted line* corresponds to the correlation time in the order of the inverse of the Larmor frequency at the field B_{0a}. This line divides the temperature dependence into two regions: the extreme narrowing regime with fast motions (with respect to the Larmor frequency) and the slow motion regime.

characterized by fast molecular motions (as compared with the Larmor frequency). Conversely, the slow motional regime is defined by the negative slope of the relaxation with increasing temperature.

Torchia and Szabo (463) presented a formalism for extracting spin-lattice relaxation in static and spinning polycrystalline solids in their important and highly quoted publication. General equations were developed for orientation-dependent frequencies and T_1. Various jump and diffusive models are considered. Explicit equations for dipolar and quarupolar values of T_1 are presented for two- and three-site jumps and the continuous diffusion. Torchia (464) also developed a CP-based sequence for determination of T_{1C}, which is routinely applied to measure carbon longitudinal relaxation.

Schaefer et al. (465) studied motion in glassy polymers of poly(methyl methacrylate), polycarbonate, poly(phenylene oxide), polystyrene, polysulfone, poly(ether sulfone), and poly(vinyl chloride) by ^{13}C NMR. T_{1C}, $T_{1\rho C}$, and T_{CH} have been measured for individual carbons. From adiabatic demagnetization in the rotating frame experiments, it was found that $T_{1\rho C}$ is dominated by spin–lattice rather than spin–spin processes, rendering the $T_{1\rho C}$ sensitive to the low- to mid-kHz frequency range of motions. T_{1C} contained information about motions in the Larmor frequency range. The $T_{CH}/T_{1\rho C}$ ratio was correlated to toughness and impact strength for all seven polymers.

Kitamaru et al. (466) used relaxation filters to detect crystalline-amorphous interphase of polyethylene. The T_{1C} relaxation times as long as 2750 seconds for the crystalline component were observed. In contrast, the T_{1C} relaxation times of the amorphous component were less than 1 second. Only little differences in T_{1C} were found. The crystalline components were eliminated by either short recycle delay (T_{1C} filter) or by T_{2C} filter (carbon direct polarization with interrupted proton decoupling). Further, the rubbery and interfacial amorphous phases could be distinguished based on their T_{2C} differences (2.4 vs. 0.044 ms). Two amorphous phases were found by the virtue of their different T_{1H} times. No proton spin diffusion between the two phases was observed, most likely due to a significant difference in mobility. The fully relaxed direct polarization spectra were quantified in terms of the different crystalline and amorphous components by means of line shape analysis.

Oksanen and Zografi (467) studied molecular mobility in mixtures of absorbed water and poly(vinylpyrrolidone) (PVP). At least two water populations with different timescales were identified as the water content decreased based on NMR diffusion measurements and T_{1H} (Fig. 65). Significant rotational and translational mobility of water existed even at very low water content. T_{1C} measurements of PVP showed that water, even at very low content, had a profound effect on the molecular mobility of PVP. The authors conclude, that more consideration should be given to the state of polymer and less to the extent to which water is bound or unbound.

Van den Dries et al. (468) studied the molecular mobility of water and carbohydrate protons in maltose samples as a function of water content and

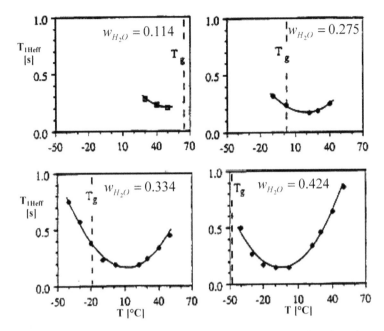

Figure 65 Effective T_{1Heff} of absorbed water in PVP as a function of temperature for divergent water contents. *Abbreviation*: PVP, polyvinylpyrrolidone. *Source*: Modified from Ref. 467.

temperature using ^1H NMR line shapes analysis and T_{2H} relaxation. Temperature-dependent mobile (mostly due to water, especially when below T_g of sugar) and immobile (sugar) proton fractions were observed. The mobility of the mobile water increased with water content and temperature. A small break in water mobility was observed at T_g, indicating that water molecules sensed the glass transition. The plasticizing effect of water is interpreted as disrupting the hydrogen bond network of the sugar molecules that is formed when the sugar is cooled under its T_g. Authors used the second moment analysis (as a measure of the strength of the dipolar coupling) of the broad immobile proton line shapes to learn about their mobility.

Yoshioka et al. (469) studied the effect of excipients on the molecular mobility of lyophilized formulations as measured by T_g and T_{mc} (NMR-based critical mobility temperature, defined as the temperature of appearance of the Lorentzian relaxation process due to liquid protons). The T_{mc} of dextran (PHEA), PVP, carboxymethylcellulose sodium salt (CMC-Na), hydroxypropylmethylcellulose (HPMC), and methylcellulose (MC) have been determined as a function of the plasticizing effect of the increasing water content. The formulations containing PHEA, MC, and HPMC exhibited comparable T_{mc} values and a significant increase in T_{2H} values of water peak (corresponding to higher water mobilities) at lower water levels compared with dextran and CMC-Na. They became micro-scopically liquidized (containing liquid protons as detected by NMR) at lower temperatures, suggesting higher molecular mobility of water. Higher mobility of water did not necessarily correspond to greater reduction in T_{mc}, because T_{2H} reflected the average of both bound and mobile water. The determined T_{mc} values were lower than the corresponding T_g values by 23–32°C, indicating that the formulations became microscopically liquidized below their T_g.

Yoshioka et al. (470) linked the molecular mobility of lyophilized protein formulations to the molecular mobility of polymer excipients by using ^{13}C NMR. In order to calculate the correlation time τ_c, T_{1C} was measured and related to τ_c assuming existence of a single correlation time, a slow motional regime, and either the dominance of the ^{1}H–^{13}C dipole–dipole interaction [Equation (24)] or chemical shift interaction [Equation (25)] for the relaxation.

$$1/T_{1C} = (1/10)\gamma_C^2\gamma_H^2 h^2 (2\pi)^{-2} r_{CH}^{-6}[1/(\omega_C - \omega_C)^2 + 3/\omega_C^2$$
$$+ 6/(\omega_C + \omega_C)^2]/\tau_c \tag{24}$$

The correlation time of dextran was found to follow the linear temperature dependence described by the Adam-Gibbs-Vogel equation (Arrhenius-like relationship) below T_g and the nonlinear temperature dependence by the Vogel-Tamman-Fulcher equation above T_g. The discontinuity was likely due to glass-rubber phase transition and was found to

$$1/T_{1C} = (6/40)\gamma_C^2 B_0^2 \delta_Z^2 (1 + \eta^2/3)(2/\omega_0^2\tau_c) \tag{25}$$

correspond to T_{mc} as determined by T_{2H}.

Zumbulyadis et al. (471) elucidated mixtures of polymorphs by ^{13}C CPMAS in combination with direct exponential curve resolution algorithm (DECRA). DECRA is a chemometric method for self-modeling mixture analysis. It employs a generalized rank annihilation method (GRAM), which requires two proportional data sets for the analysis, that was accomplished by splitting a single data set into two. The key to the analysis is the exponential dependence of the signal from the individual components on the physical property being measured (diffusion, T_1 or T_2 relaxations). Under optimal conditions, DECRA can quanti-tatively separate spectra of components whose relaxation differ only by 20%. It is, however, critical to keep the spectral artifacts originating from instrumental instabilities to a minimum. The authors show that pure-component carbon spectra of each of the two polymorphs can be unambiguously extracted from a spectrum of a mixture based on their different T_{1H} (Fig. 66). This technique should have a general application, because in majority of the cases, there are relaxation differences between polymorphs. This technique would not work to distinguish microphases with domains smaller than 30 nm, because spin diffusion is expected to equalize the proton relaxation. DECRA is relatively insensitive to poor S/N or spectral overlap.

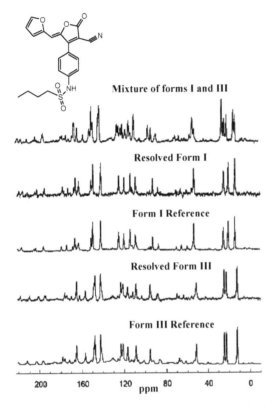

Figure 66 Example of resolving components of mixtures using the DECRA algorithm based on different relaxation properties of the components. The *top spectrum* was recorded with inversion recovery time of 0.2 s. *Abbreviation*: DECRA, direct exponential curve resolution algorithm. *Source*: Modified from Ref. 471.

Silva et al. (472) studied a multicomponent polymer blend by ^{1}H relaxation (carbon-detected T_{1H} and $T_{1\rho H}$) and spin diffusion. The 2D WISE sequence was also used as a T_{1H} filter. Some phase separation was detected based on the proton relaxation. This was confirmed by the WISE experiment. Merits of relaxation studies versus 2D WISE experiment to determine phase separation are discussed.

Aso et al. (473) explained the crystallization rate of amorphous nifedipine and phenobarbital from their molecular mobility was determined by carbon relaxation times, T_{1C} and $T_{1\rho C}$. T_{1C} of hydrogen-bound carbons are determined by molecular mobility within nanoseconds to microseconds timescale, and T_{1C} decreases with increasing molecular mobility in the slow motional regime. Molecular motions in the mid-kHz frequency range (as determined by $T_{1\rho C}$, which was sensitive to the fluctuations of the C–H vector; $T_{1\rho C}$ decreased if motions existed in the mid-kHz frequency range) became significant at temperatures between T_g-20 and T_g. The temperature dependence of the crystallization rate on T_g was found to be coincident with that of the relaxation rate calculated according to Adam-Gibbs-Vogel equation, indicating a correlation of the crystallization with molecular mobility.

Glaser et al. (474) studied eight-membered ring solid-state conformational conversion of nefopam methohalide salts via the atom flip mechanism. Editing experiments, such as interrupted decoupling, CPPI, and inversion recovery (20 second delay) were used for assignment purposes. Solid-state T_{1C} was measured and interpreted in terms of distance of the carbons from the closest protons. From variable temperature CPMAS spectra, the temperature-dependent occupancy factors for boat–boat and twisted chair–chair conformers are interpreted in terms of medium ring atom-flip facile interconversion between the two low energy conformations in crystals containing an appropriate size interaction void.

Kuwabara et al. (475) studied crystalline phase transitions of 1,20-eicosanediol. The ^{13}C CPMAS and ^{1}H CRAMPS revealed two magnetically different structures in the chain ends associated with hydrogen bonding in the M and M′ phases. The ^{13}C CPMAS line widths at intermediate temperatures showed broadening, which was explained by the authors as occurance of the molecular motion with the order of magnitude of the decoupling field (about 100 kHz). The differences in the chemical shifts as a function of temperature are explained in terms of the guache conformation effect. The T_{1C} relaxation was measured to compare the mobility of different carbon groups in each crystalline phase. The T_{1C} relaxation time as long as 2290 seconds was observed. Variable temperature ^{1}H CRAMPS data afforded detailed information about the hydroxyl groups.

Varner et al. (476) characterized molecular motions in the solid state by measuring T_{1C} at two different spinning angles. Because the C–H dipolar interaction has the same orientational dependence as C–D, the spectral densities responsible for the dipolar and the quadrupolar relaxation are the same. The orientational dependence of all three spectra densities [$J_0(0)$, $J_1(0)$, and $J_2(0)$] are thus determined.

Kakou-Yao et al. (477) showed hindered rotation of tert-butyl group in 2,2,5,5,-tetramethyl-3,4-haxandione monohydrazone by ^{13}C CPMAS. The broadening of carbon resonances of the tert-butyl group at lower temperatures was rationalized by higher activation energy of rotation, as compared with the free conformers as a result of packing constraints.

Rheingold et al. (478) studied superlattices and polymorphs of 2,6-di-tert-butylnaphthalene. Two polymorphs were identified which differed by a factor of 12 in the number of crystallographically independent tert-butyl group environments. The ^{13}C CPMAS showed only small and random chemical shift differences of no more than 1.6 ppm and only minor differences in line width, in spite of the averaging of 12 tert-butyl groups. Low-field variable temperature T_{1H} relaxation data were acquired. Polymorph E relaxation exhibited very complex behavior.

Beckmann et al. (479) studied relaxation properties of two polymorphs of 2,6-di-tert-butylnaphthalene. Single crystal X-ray structures of both forms A and E were obtained. One molecule per asymmetric unit was found by ^{13}C CPMAS. Methyl group and tert-butyl group re-orientations in solids are usually the only motions on the NMR timescale. All other motions, such as intramolecular vibrations, occur on a fast scale to effect NMR relaxation. Temperature-dependent T_{1H} relaxation was measured at two low Larmor frequency fields (8.5 and 22.5 MHz). In polymorph A, a very good fit of the temperature and Larmor frequency-dependent relaxation curves was achieved, predicting the three methyl groups to reorient at the same rate as the tert-butyl group. The polymorph A was found to be a textbook case of the effects of intramolecular reorientation on NMR relaxation rates. In contrast, the relaxation of form E was found to be significantly more complex. A unique fit could not be obtained since too many parameters were involved. The temperature- and Larmor frequency-dependent relaxation curves were very different and extremely sensitive to the differences between the two forms.

Vittadini et al. (480) studied mobility in cellulose by ^{1}H and ^{2}H peak intensities, relaxation times, and spectral line shapes as a function of moisture content. At higher moisture level, water molecules reoriented anisotropically, but the intensity of the mobile ^{1}H increased along with increases in deuterium T_1 and T_2 values, approaching the fast exchange regime (Fig. 67). At the moisture content below the monolayer value, the majority of the protons were immobile.

Perera et al. (481) characterized binary and ternary polymer blends by ^{1}H and ^{13}C NMR. T_{CH} CP constant and T_{1C} and $T_{1\rho C}$ relaxation times were interpreted in terms of changes in molecular motions. T_{CH} reflects the near static motions, whereas T_{1C} and $T_{1\rho C}$ reflect the MHz and mid-kHz regimes. T_{1H} and $T_{1\rho H}$ relaxation times were interpreted in terms of homogeneity of the polymer matrices. Heterogeneity of the blends on 1 to 6-nm length scale were confirmed

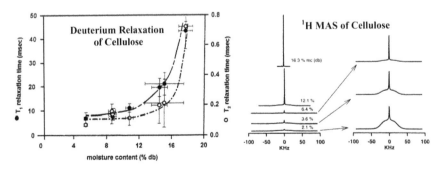

Figure 67 Deuterium T_1 and T_2 relaxation of cellulose as a function of moisture content (% dry basis; *left side*). ^{1}H MAS spectra of cellulose at different moisture contents (*right side*). *Abbreviation*: MAS, magic angle spinning. *Source*: Modified from Ref. 480.

based on calculation using equation (17), where the average literature values of diffusion coefficient of $D = 0.5$–0.8 nm^2/msec and $D = 0.05$ nm^2/msec were used for rigid and mobile polymers, respectively, and the time of diffusion was substituted with the two observed relaxation times.

Tang et al. (482) studied dynamics of α-D-galacturonic acid monohydrate by ^{13}C $T_{1\rho}$ and chemical shift anisotropy. The comparison of solution shifts allowed only for partial chemical shift assignment due to considerable spectral overlap. The ^{13}C T_1 relaxation was slow on a timescale of 100 seconds, indicating that ^{13}C T_1 is much longer than 100 seconds. $T_{1\rho}$ of all carbons showed a decreasing trend with temperature, implying that motion was on the low temperature side of the $T_{1\rho}$ minimum and the motional correlation time was longer than 10^{-5} seconds. The motional averaging of carbon chemical shift tensor as a function of temperature was observed. The tensor parameters were extracted based on fitting the 1 kHz MAS spinning sideband line shapes. Motions with frequency of the order of 10^4 Hz were detected.

Higgins et al. (483,484) studied the temperature dependence of T_{1H} and $T_{1\rho H}$ proton relaxation of bisphenol-A tetramethyl polycarbonate/polystyrene mixtures. Both pure components showed a maxima of T_{1H} around the respective T_g, indicating motions at frequencies inefficient for T_{1H} relaxation. The temperature dependence of the relaxation of the blend was predicted as an average of the single-component relaxations [Equation (26)].

$$1/T_{1H}^{\text{blend}} = \sum_a \frac{n_a}{T_{1H}^a} \tag{26}$$

The experimentally observed relaxation matched well with the predictions based on Equation (26). $T_{1\rho H}$ was found very sensitive to motions in temperature in the range 380–420 K. Some significant deviations were observed from the predicted $T_{1\rho H}$ values of the blend, most likely due to time-varying concentration fluctuations leading to strong deviations from homogeneous relaxation behavior.

Lopes et al. (485) developed a mathematical method of obtaining ^{13}C CPMAS subspectra of single components based on differences in relaxation. Application of this technique, which removes the need for obtaining the subspectra by trial-and-error approach, to cork cell walls is shown.

Tang et al. (486) introduced a proton relaxation-induced (PRICE; essentially a set of carbon detected, proton relaxation filters) technique, which used T_{1H}, T_{2H}, and $T_{1\rho H}$ proton relaxation properties to relate the dynamics of the plant cell walls to their domain structural details.

Nogueira and Tavares (487) studied behavior of alpha-methylstyrene-co-acrylonitrile. The variable contact time CP and $T_{1\rho H}$ relaxation were found to be the determinant factors to evaluate the dynamic molecular behavior and the homogeneity of the comonomer distribution. $T_{1\rho H}$ values were not homogeneous, reflecting a random sequencing of the comonomer insertion in the chains.

Lai et al. (488) studied poly(vinyl alcohol) gels by ^{13}C NMR. Information on the dynamics in the kHz range obtained from $T_{1\rho H}$ and $T_{1\rho C}$ data indicated a homogeneous arrangement of the amorphous or swollen polymeric chains.

Adriaensens et al. (489) studied polymeric mixtures by ^{13}C relaxation. No indication for phase separation was found as both the T_{1H} and $T_{1\rho H}$ relaxation behavior were indicative of homogeneous mixture on the nanometer length scale.

Brachais et al. (490) studied a heterogeneous character of polymeric mixtures interpreted based on interpenetrating networks. The rate of ^{1}H–^{13}C CP was composition-dependent, indicating that the chains of the polymeric components were interpenetrating at a very low spatial scale. This was also supported by T_{1H} and ^{13}C line width measurements as a function of temperature. The maxima in ^{13}C line widths of pure components were coincident with T_g. For the blend, ^{13}C line widths were sensitive to the composition, supporting the interpenetrating character.

Alamo et al. (491) measured T_{1C} of noncrystalline regions of semicrystalline polymers. Because the T_{1C} values of some carbons were found shorter than T_{1H} values, the relaxation showed multiexponential behavior due to the significant contribution of NOE effect. To eliminate this transient NOE contribution, weak proton saturation pulses were added during the relaxation delay. The authors used a modified CP-based saturation recovery sequence to filter out regions with long T_{1C}. It is based on alternating, in successive scans, the carbon magnetization in $+I_Z$ and $-I_Z$ directions immediately after the CP step and keeping the relaxation delay short.

Aso et al. (492) studied the effect of water on molecular mobility of sucrose and PVP in a colyophilized formulation. Due to its much higher T_g, PVP reduced the molecular mobility of sucrose as determined by T_{1C} and $T_{1\rho C}$. Contrary to the colyophilized dispersion, the decreasing T_{1C} with increasing relative humidity (RH) for pure PVP indicated that the MHz-order motions of PVP sidechains increased with RH. No RH dependence of T_{1C} (in 0–22% RH range) was found for both sucrose alone and the colyophilized dispersion. The presence of water apparently did not increase MHz-order motions. In the colyophilized dispersion, the formation of hydrogen bonds between PVP and sucrose prevented the increase in mobility with increasing RH. $T_{1\rho C}$ decreased with increasing RH for both sucrose and the sucrose-PVP dispersion, but the decrease was much larger for sucrose itself, suggesting that water molecules enhanced the mid-kHz-order motions by the plastisizing effect of water. PVP acted as an anti-plastisizing agent, which offsets any plastisizing effects of water.

Yoshioka et al. (493) studied the different motions in lyophilized protein formulations as determined by the temperature and RH dependence of T_{1H}, $T_{1\rho H}$, T_{1C}, and $T_{1\rho C}$ relaxation times. Unlike for the T_{1H} values, a minimum was reached in the temperature dependence of $T_{1\rho H}$ at higher humidities, above which the system was in the fast motional regime. Based on the temperature dependence of T_{1H}, $T_{1\rho H}$ of dextran protons in the lyophilized formulations, the dextran correlation time decreased from $\sim 10^{-6}$ seconds at $-100°C$ to 10^{-7}

seconds at 0°C. Multiple correlation times were found. Equations (27) and (28) were used to calculate each correlation time based on either T_{1H} or $T_{1\rho H}$. ω_0 and ω_1 are Larmor frequency and spin-locking field frequency, respectively. The factor A in Equation (28) was calculated from the observed temperature minimum of $T_{1\rho H}$. Above 0°C, the molecular motion of methylene groups was found to be too fast to be reflected in $T_{1\rho H}$, but it is still reflected in T_{1H}. Instead, motions of methine groups began to be reflected in $T_{1\rho H}$. Above the glass/rubber transition, both motions are greatly enhanced, substantially shortening the relaxation times.

$$1/T_{1H} = (9/8)\gamma_H^2\gamma_H^2 h^2 (2\pi)^{-2}[4\tau_c/15r_{HH}^6(1+\omega_0^2\tau_c^2)$$
$$+ 16\tau_c/15r_{HH}^6(1+4\omega_0^2\tau_c^2)] \tag{27}$$

$$1/T_{1\rho H} = A\tau_c/(1+4\omega_1^2\tau_c^2) \tag{28}$$

Sidhu et al. (494) studied polymorphism, conformation, and dynamics of the inclusion compound of 1,2-dichloroethane with tris(5-acetyl-3-thienyl) methane. The ^{13}C CPMAS spectra confirmed the phase purity of the slow cooling sample as being the triclinic form, whereas the sample prepared by slow evaporation resulted in a mixture of triclinic and monoclinic forms. Variable temperature solid-state ^{2}H spectra showed a dynamical conformational equilibrium of 1,2-dichloroethane consistent with either the *trans*-guache model with 120° jumps about the C—C bond or the twofold flip axis model. The ^{2}H T_1 relaxation rate measurements as a function of temperature supported the twofold flip axis model.

Prout (495) published a review titled *Molecules in Motion: A Study in the Synergy of X-ray Diffraction and SSNMR*. PXRD was found to give only an insufficient information on infrequent activated processes in the crystal or about some motional behavior related to phase changes. Multiple examples are shown, including pharmaceutically relevant compounds, such as phenoxypenicillin, for which a phenyl ring flip was detected only by SSNMR and not predicted by PXRD and deoxycholic acid clathrate with camphor, for which an order–disorder transition was detected by SSNMR.

Kuzmicz et al. (496) studied dimethylacetamide solvate of tetra(C-undecyl)calix[4]resorcinarene, a compound of interest for transport across membranes. Four resonances for each carbon were expected based on the crystal structure, but only two were observed, most likely due to the almost identical disposition of the molecules. CP kinetics was studied to provide information on the molecular mobility. The system was found to behave according to model I-I^*-S with different proton spin temperatures, described by Equation (23). Based on the fitted parameters for different carbons, the mobility of the compound was found greater than that of the crystallized solvent. The solvent molecules were found rigid, because they were trapped between layers of the compound.

Harris et al. (497) presented chemometric analysis of morphology and dynamics in irradiated cross-linked polyolefin by means of DECRA. The ^{13}C

DP and T_{1C}-filtered CPMAS experiments revealed peaks arising form the amorphous, crystalline, and interfacial components. The relative abundances of each phase were determined by subtracting DP spectra with 10-second recycle delay from those with 2000 second recycle delay. Corrections for incomplete relaxation, partial decay of crystalline component, and spinning sidebands were incorporated. The ^1H spin diffusion with ^{13}C detection was used to probe the thickness of the crystalline regions. The DECRA analysis was used to resolve the intermediate and amorphous phases.

Mao et al. (498) correlated poly(methylene)-rich amorphous aliphatic domains in humic substances. T_{1H} filter was applied using proton inversion recovery with CPMAS detection to filter out the components with short proton-relaxation times. The relaxation delay was chosen such that the signal of short T_{1H} components just passed the zero level. Only the long relaxing components are then detected. Direct polarization of ^{13}C MAS spectra were used for quantification.

Kitchin et al. (499) studied effects of polymorphism on functional group dynamics of α- and β-phases of L-glutamic acid by deuterium NMR. The contrasting dynamic properties of the α- and β-phases were investigated by ^2H line shapes, and ^2H T_1 relaxation times combined with simulations. The ^2H line shapes, were consistent with three-site jump 120° motions of the $-ND_3^+$ group. The variable temperatures of ^2H T_1 relaxation curves show minima at different temperatures for the two phases. The differences in the dynamics between the two polymorphs are rationalized in terms of different local environment in the crystal structures due to the different H-bonding geometries. An ability of ^2H to elucidate differences in functional group dynamics was demonstrated, even for cases when the structure differences between the polymorphs may be comparatively minor.

Leisen et al. (500) measured sorption isotherms by SSNMR. Instrumentation for automated recording of sorption isotherms at precisely controlled levels of temperature and RH was designed. The specially designed static probe was used to acquire the isotherm for a cotton fabric. Static proton DP spectra, T_{1H}, and T_{2H} times were recorded automatically as a function of temperature and RH. An analysis of time-domain data, Bloch decay, in terms of fast (rigid part) and slow (moisture exposed mobile part) ^1H T_2 relaxing components was made. An excess of the slowly relaxing component, when compared with gravimetric data, was attributed to plasticized segments of the formerly rigid cellulose matrix. Hysteresis similar to gravimetric sorption isotherm was observed. The NMR relaxation parameters were only dependent on the moisture content and independent of the exposure history. Thus, material properties, as defined by molecular mobilities, were found to be governed by moisture content alone.

Witkowski and Wawer (501) studied conformational dynamics in α-tocopherol esters. Differences in solution-solid carbon chemical shifts up to 6.9 ppm were observed. Two molecules per asymmetric unit were observed. By variable contact time studies, the CP time constant T_{CH} was found to be in the range of 0.55 to 1.22 ms for quarternary carbons, 0.12 ms for CH_2, and 0.32 to 0.37 ms for CH_3. The authors concluded that the α-tocopherol esters are flexible in two fragments of the molecule.

Lupulescu et al. (484) developed 2D relaxation-assisted separation (RAS) experiment using the different T_1 relaxation of powdered or disordered systems to get resolution in the indirect domain. Applications to ^{23}Na-, ^{35}Cl-, and ^{25}Mg- containing inorganic salts and ^{1}H of glycine sample were shown.

Harris et al. (502) devised a generally applicable numerical methodology to obtain more efficient relaxation filters to selectively retain or remove components based on relaxation times. The procedure uses linear combination of spectra with various recycle delays to obtain components that are both quantitative and pure. Exponential behavior is assumed. The intensity of NMR signal in DP experiments with a recycle delay t and a relaxation time T_1 is

$$I = 1 - e^{-t/T_1} \tag{29}$$

A Taylor expansion of this equation provides

$$I = t/T_1 - \frac{1}{2}(t/T_1)^2 + \frac{1}{2}(t/T_1)^3 - \frac{1}{24}(t/T_1)^4 + \cdots \tag{30}$$

In a system with two components with relaxation times of T_{1a} and T_{1b}, the selectivity of the component a can be expressed as I_a/I_b

$$I_a/I_b = \frac{T_{1b}}{T_{1a}} - t\frac{1}{2}\frac{T_{1b}}{T_{1a}}\left(\frac{1}{T_{1a}} - \frac{1}{T_{1b}}\right) + \cdots \tag{31}$$

The linear dependence of the selectivity on the T_{1b}/T_{1a} ratio prevents efficient and quantitative selection of the faster relaxing component in most systems. The maximum selectivity of the component a is when $t \ll T_{1a} \ll T_{1b}$ is T_{1b}/T_{1a}. A combination of multiple spectra can improve this selectivity. If, for example, one subtracts the spectrum acquired with twice the recycle delay from that with a single recycle delay, the first-order terms vanish:

$$2I_{(t)} - I_{(2t)} = \left(\frac{t}{T_{1a}}\right)^2 + \left(\frac{t}{T_{1b}}\right)^2 - \left(\frac{t}{T_{1a}}\right)^3 + \left(\frac{t}{T_{1b}}\right)^3 \cdots \tag{32}$$

and the maximum selectivity of the component a when $t \ll T_{1a} \ll T_{1b}$ is $(T_{1b}/T_{1a})^2$. Figure 68 shows the calculated signal intensity as a function of the relaxation time for different recycle delays and their linear combinations [Equation (32)]. Figure 69 shows the relative filtering efficiency of slow relaxing component with respect to the fast relaxing one. To selectively observe the faster relaxing component a, allowing for 3% of the undesired component b, the ratio of the relaxation times would have to be 1:115 for the standard DP experiment, but only 1:23 if a combination of two recycle delays are used, and even 1:13 if combination of three recycle delays are used. The limitation of this technique is reduction in S/N. Assuming the same experimental time, the two recycle delay methods will afford only 26% of the DP S/N.

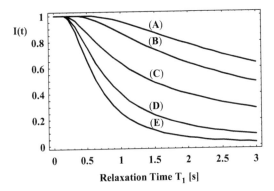

Figure 68 Calculated signal intensity as a function of the longitudinal relaxation time (T_1) for different recycle delays and linear combinations thereof. (A), (B) and (C) curves were calculated using Equation (29) with the recycle delays set to 3, 2, and 1 second, respectively. (D) curve is a linear combination of 2*(C) − (B) [Equation (32)], and (E) curve is a linear combination of 3*(C) − 2*(B) + (A). Note that the linear combinations of the recycle delays show faster cutoff and, as such, better filtering efficiency.

Similar procedures can be designed for filtering out the fast relaxing component with T_1 relaxation filter preceding acquisition. For example, a long relaxing component with $T_1 = 100$ seconds will lose 10% intensity when subjected to 10-second filtration which removes the fast relaxing components. It will

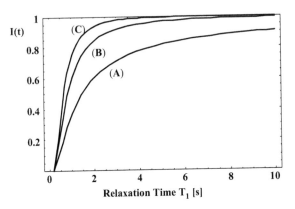

Figure 69 A relative filtering efficiency of a slow relaxing component ($T_1 > 0.2$ s) with respect to a fast relaxing component with $T_1 = 0.2$ seconds. (A) One relaxation delay of 1 second. (B) A linear combination of two relaxation delays, 1 and 2 seconds. (C) A linear combination of three relaxation delays 1, 2, and 3 seconds. Note that in all cases the fast relaxing component is observed quantitatively (RD > 5*T_1). The filtering efficiency of 1 corresponds to 100% suppression of the slow relaxing component while quantitatively retaining the fast relaxing one.

only lose 1% of its sensitivity when linear combination of two experiments with two filters (10 and 20 seconds) is used, and, at the same time, better filtration of the fast relaxing component is achieved. In addition, a combination of experiments can be designed shape modulation profiles, as for example, to filter out very fast and very slow relaxing components.

Lim et al. (503) studied dynamics in poly(ethylcrylate-co-sodium acrylate) ionomer ^{13}C CPMAS. $T_{1\rho C}$ relaxation times were studied as a function of temperature to address chain mobility, correlation times, and activation energies. Arhenius plots of correlation time versus inverse of temperature were constructed. Protonated carbons displayed much faster $T_{1\rho C}$ relaxation than the nonprotonated carbons. Molecular motions in the poly(ethylcrylate) homopolymer needed a higher activation energy compared with the ionomer.

Krushelnitsky et al. (504) studied hydration dependence of backbone and sidechain polylysine dynamics by naturally abundant carbon SSNMR. Temperature- and hydration level-dependent carbon relaxation experiments (T_{1C} and $T_{1\rho C}$) were performed. The relaxation data were simultaneously analyzed using the correlation function formalism and model-free approach. Unlike the backbone, the sidechains showed strong hydration response.

The dynamics of t-butyl group of 2-t-butyl-4-methylphenol was studied by low-frequency SSNMR proton relaxometry (505). From X-ray data, there were two inequivalent t-butyl groups in the asymmetric unit. Variable temperature and variable field proton T_2 relaxation data were fitted to two dynamics models, both of which were consistent with the X-ray data. Comparison of the t-butyl dynamics to other literature examples of a van der Waals organic solids is made. The synergies between X-ray and NMR were pointed out, where X-ray provided the boundary conditions for a set of dynamical models and NMR provided information on local angular anisotropies in the intra- and intermolecular potentials.

Masuda et al. (506) compared the molecular mobility in the glassy state between amorphous indomethacin and salicin based on T_{1C} relaxation times at temperatures below T_g. Salicin molecules exhibited multiple-exponential relaxation and were found to behave heterogeneously, showing fast local and slow mobilities. In contrast, the amorphous indomethacin was homogeneous and restricted, particularly in backbone carbons. The difference in mobilities of these two drugs were correlated with their different stabilities towards crystallization. At room temperature, 60% of amorphous indomethacin crystallizes after two months, whereas 100% of salicin crystallizes in 12 days.

Lubach et al. (507) investigated the effects of pharmaceutical processing on relaxation and its implications for formulation stability. Two crystalline lactose forms were subjected to pharmaceutical processing, including compaction, lyophilization, spray-drying, and cryogrinding. The compaction resulted in a decrease of the lactose monohydrate T_{1H} from 243 seconds to 79 seconds but did not produce any amorphous phase. Lyophilization and spray-drying produced amorphous lactose with T_{1H} of four seconds. Cryogrinding produced mixtures of

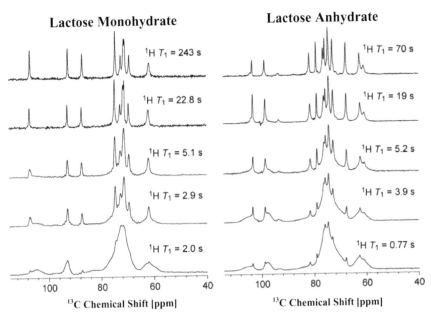

Figure 70 The ^{13}C CPMAS spectra and T_{1H} relaxation times of crystalline lactose forms cryoground (*from top*) for 0, 2, 10, 30, and 60 minutes, respectively. *Abbreviation*: CPMAS, cross polarization and magic angle spinning. *Source*: Modified from Ref. 507.

crystalline with amorphous lactose (Fig. 70). A short cryogrinding time (~2 minutes) reduced the T_{1H} by an order of magnitude without significant effect on the carbon spectrum. The content of the amorphous component increased with grinding time. After 60 minutes, a mostly amorphous sample resulted with some crystalline material left, showing T_{1H} of two seconds (shorter than the fully amorphous compound). The cryoground material showed some annealing at ambient conditions (Fig. 71). The authors concluded that the reduction of particle size, introduction of crystal defects, and formation of amorphous material all reduced the apparent T_{1H}. The reduction in relaxation rates suggest that processing can have significant negative impact on formulation stability. The authors propose replacing the lengthy and labor-intensive stability studies with solid-state relaxation measurements.

Stability

Markovic et al. (508) studied amorphous to crystalline drug transition in solid dispersion of potent cholesterol absorption inhibitor SCH 48461 with a polyethylene glycol. Carbon assignment was accomplished by comparison with solution. The broader than expected line width of the amorphous spectrum is explained by an incomplete averaging of the chemical shift tensor due to the presence of an

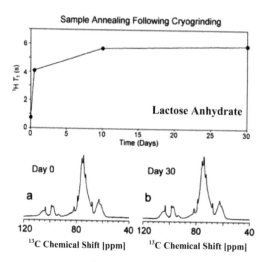

Figure 71 The ^{13}C CPMAS spectra and T_{1H} relaxation times upon annealing of anhydrous lactose cryoground for 60 minutes. *Source*: Modified from Ref. 507.

intermediate rate of molecular motion. To prove that the sharp peaks observed in the capsule formulation were due to crystalline drug (as opposed to liquid-like plastic or waxy solid in rapid motion), interrupted decoupling and cross-depolarization experiments were applied. The rigid, crystalline lattice was confirmed by selective dephasing of protonated carbons in these spectra. The amount of crystallinity was found to increase under different storage conditions ($25°C/60\%RH < 30°C/60\%RH < 40°C/80\%RH$).

Watanabe et al. (509) studied stability of amorphous indomethacin (IM) compounded with silica. Higher stability of amorphous IM compounded with SiO_2 was achieved by a cogrinding rather than a melt-quenching approach. The analysis of the ^{29}Si CPMAS line shapes indicated mechanochemical reaction between SiO_2 and IM by cogrinding. The ^{13}C CPMAS spectra after short cogrinding indicated a significant crystalline component, but fully amorphous line shapes were observed at longer cogrinding times. The observed decrease in stability of the short-ground samples were interpreted as due to a seeding effect of the remaining crystalline component.

Forster et al. (510) investigated physical stability of amorphous drug and drug/polymer melts using variable temperature SSNMR. Because the direct comparison of the absolute relaxation values between different compounds is not possible, the authors compared the temperature dependence of nifedipine and indomethacin $T_{1\rho H}$ instead. The relaxation data were found to reflect an overall "freeing up" of molecular mobility as the samples softened with increasing temperature. Based on the temperature-dependent relaxation behavior, the stability of the two drugs were predicted to be different, with the indomethacin samples showing a higher stability towards recrystallization.

Kalachandra et al. (511) examined a drug stability during film casting of five different drugs in ethylene vinyl acetate. This system promotes longer oral drug delivery. Based on the similarity of peaks between the pure drug and in the cast film, one of the drugs was found to be dispersed in the film as small crystallites rather than as a true solid solution. The phase separation most likely occurred during the drying process.

Domain Characterization

Spin diffusion properties of the abundant spins present an ideal tool to identify and characterize the different domains present in heterogeneous samples. By measuring the rate of spin diffusion, domain sizes can be estimated. Relaxation and dipolar coupling filters are typically employed to selectively observe the signal arising from a single phase.

Robyr studied local order in disordered solids by homonuclear (^{13}C and ^{15}N) and heteronuclear (^{2}H–^{13}C) techniques (512). Information on conformations of polymer chains and torsional angles along polymer backbone was obtained by these techniques.

Hu and Schmidt-Rohr (513) characterized crystallinity, domain sizes, and the highly mobile second amorphous phase of ultradrawn polyethylene fibres. The crystallinity was determined by the ^{1}H line shape deconvolution and by direct polarization carbon MAS with the knowledge of a correcting factor for incomplete relaxation obtained from the ^{13}C CPMAS T_1 filter measurement. The presence of a low-level, highly mobile, second amorphous phase was determined by ^{1}H NMR. Based on spin diffusion, domain sizes of the crystalline and amorphous regions were determined.

Colhoun et al. (514) studied orientation relaxation of polymers by 2D ^{13}C SSNMR. The 2D DECODER technique allowed for determination of the complete orientation distribution function for both amorphous and crystalline domains. The authors present a numerical method for extracting a biaxial molecular orientation distribution function of polystyrene from slow spinning speed MAS DECODER, based on the chemical shift anisotropy.

Asano and Takegoshi (515) studied free volume of amorphous polymers by ^{13}C line width experiments. A temperature dependence of carbon line width of six different amorphous polymers was examined around their glass-transition temperatures (T_g). The observed temperature dependence was explained by motional averaging over a dispersion of isotropic chemical shifts and an interference between local anisotropic motions and the dipolar decoupling. The latter was responsible for the line width versus temperature function to show maximum at one particular temperature. The derived fractional free volume at

T_g was much larger compared with that from rheological experiments. The difference was attributed to the differences in the amplitude of motions detected by both experiments. The rheological experiments observed an overall motion of polymers (cooperative arrangements of polymer segments), whereas the carbon line width was governed mainly by local fluctuations in dipolar interaction. The void space required for the local fluctuations is smaller than that for cooperative segmental motions.

Valtier et al. (516) used Goldman-Shen dipolar filter to remove magnetization from the rigid section of latex polymer and proton spin diffusion to quantify the sizes of inner core and shell domains.

Palmas et al. (517) studied curing and thermal ageing of elastomers by ^{13}C and ^1H NMR. A combination of high-resolution ^{13}C CPMAS, T_{2H} (proton-detected) and $T_{1\rho H}$ (carbon-detected) measurements and 2D WISE experiments provided identification of cross-linking, scission, and oxidation at low concentrations. A good spectrometer stability was required for the 2D experiments under slowly accelerated aging conditions.

Lin et al. (518) studied phase heterogeneity in the miscible blend of amorphous poly(benzyl methacrylate) with semicrystalline poly(ethylene oxide). Carbon-observed rotating frame proton relaxation ($T_{1\rho H}$) was measured to probe on possible molecular scales of heterogeneity. Depending on composition, one to three phases were observed.

Huster et al. (519) studied membrane protein topology by the rate of ^1H spin diffusion from the mobile lipids to the rigid polymer. Magnetization transfer in the mobile lipid was a slow and inefficient process. The slope of the spin diffusion build-up curves unambiguously determined the presence or absence of membrane-embedded protein domains. Quantitative analysis of the spin diffusion yielded an approximate insertion depth of the protein in the membrane. ^1H T_2 filter was used to select the mobile proton spins. The technique was also applied on unlabeled protein by detection of the lipid signals.

Brus et al. (520) studied mobility, structure, and domain sizes in polyimide-poly(dimethylsiloxane) networks of different composition by ^{13}C, ^1H, and ^{29}Si NMR. Large differences of chain mobilities were determined by 1D and 2D WISE. A correlation of the structure and morphology with mechanical and thermal properties is discussed.

Kretschmer et al. (521) quantified composition and domain sizes of industrial block copolymers using different ^1H SSNMR methods. Static measurements using a dipolar coupling filter step, destroying the magnetization of the rigid regions, to separate the mobile and rigid parts of the sample were used. A combination of high-spinning speed MAS (25 kHz) and high magnetic field (700 MHz) was used to get the structural information. Proton spin diffusion was used to determine the domain sizes.

Nozirov et al. (522) studied the rigid and mobile amorphous phases and the crystalline phase of poly[(R)-3-hydroxybutyric] acid. A spectral deconvolution of the ^{13}C MAS spectra showed different relative content of the various phases at different temperatures. A large difference in molecular mobility between the crystalline and amorphous regions was observed.

Murata et al. (523,524) studied a conformational stability of poly(L-alanine), poly(D-alanine), poly(L-isoleucine), polyglycine and poly(L-valine), and their blends using ^{13}CPMAS. The ^{13}C CPMAS and $T_{1\rho H}$ experiments were employed. The conformations of the polypeptides in their blends were found to be strongly influenced by the intermolecular hydrogen bonding interactions, which caused their miscibility at the molecular level. The $T_{1\rho H}$ relaxation times were measured and the domain sizes were estimated based on their values.

Jia et al. (525) presented an independent calibration of proton spin diffusion coefficients in amorphous polymers. The quantitative values of spin diffusion coefficients were obtained by using appropriately selected internal reference distances for spin diffusion within amorphous homopolymer. The 2D ^{1}H–^{13}C HETCOR with added spin diffusion evolution period was used due to the excellent resolution provided by carbon chemical shifts.

Qui and Ediger (526) studied the length scale of dynamic heterogeneity in supercooled D-sorbitol using multidimensional ^{13}C NMR. The experimental values of the dynamic heterogeneity length scale, obtained from 2D echo and 4D3CP experiments, were compared with various models.

Munson et al. (527) studied the stereo defect locations in the polylactide by carbon SSNMR. They synthesized a polymer with spin-labeled stereo defects. The carbonyl region of the spectrum was used to deconvolute the spectrum into amorphous and crystalline components. The locations of the stereodefect sites in either the crystalline or amorphous domains were determined by spectra subtraction of the corresponding unlabeled spectrum from the labeled spectrum. The surprising result was that the stereodefects were found to incorporate into both the crystalline and amorphous domains with almost equal probability. Only with increased defect concentration did the probability of the incorporation into the crystalline region decrease.

Werkhovan et al. (528) determine the domain sizes of a block copolymer by ^{1}H spin diffusion using WISE experiment. Two values of T_{1H} were obtained, consistent with two phases. No differences in mobilities were determined by WISE. By the spin diffusion measurements, the domain sizes were in the 20 nm range.

Akita et al. (529) studied the reversible order–disorder phase transition between γ- and α-phases of oleic acid. Discontinuous changes of ^{13}C CPMAS chemical shifts (as a function of temperature) were observed at the phase transition temperature. Variable temperature T_{1C} data showed the same tendency. The T_{1C} time became discontinuously short above the transition temperature.

Drug–Drug and Drug–Excipient Interactions

The sensitivity of the spin interactions, such as chemical shift or dipolar coupling, to the local environment can be exploited to characterize drug–drug or drug–excipient interactions. The sample mobility or relaxation can also be applied to characterize these interactions. In many cases, an observation of a proton transfer or a formation of H-bonds can point to the drug–drug or drug–excipient interactions. If applicable, multidimensional correlation experiments provide a positive proof of the presence of these interactions.

Rohrs et al. (530) studied the delavirdine mesylate tablet dissolution affected by a moisture-mediated solid-state interaction between drug and disintegrant. The ^{13}C CPMAS spectra of stressed tablets showed progressive conversion to the free base, which was quantified based on knowledge of the relaxation constants of both forms (328). The free base is known to be much less soluble in water. The authors postulated that the released methane sulfonic acid protonated the croscarmellose sodium, the cross-linked form of carboxymethylcellulose sodium used as disintegrant. Water content was found critical for the onset of this reaction and is assumed to have either a dissolution medium or a plasticizer role.

Puttipipatkhachorn et al. (531) studied the salicylic acid release from chitosan matrix dependent on the drug–polymer interaction and the physical state of the drug. An analysis of the ^{13}C CPMAS peaks revealed the drug–polymer interaction between salicylic acid and chitosan at an amino group, resulting in salicylate formation.

Watanabe et al. (532) studied the apparent equilibrium solubility of indomethacin compounded with silica by carbon SSNMR. The solubility of indomethacin was increased by cogrinding with silica. The change of short-range disorder was examined by carbon NMR. Variations in conformation, bond distortion, or nearest-neighbor distance upon a mechanical stress gives rise to peak broadening. The peak broadening was used as a quantitative measure of any changes in the molecular arrangement of the solid after prolonged griding. Combined PXRD and SSNMR data showed that the mechanical energy applied at early stages of grinding was used mainly to disrupt the long-range order, whereas the rest of the mechanical energy disrupted the short-range order. A high degree of correlation was found between the standardized peak width (SFWHM) and the solubility plotted on logarithmic scale. Therefore, carbon spectra could be used instead of the dissolution tests.

Interactions between the ibuprofen guest molecule and mesoporous silica host matrix was investigated by ^{1}H, ^{29}Si, and ^{13}C SSNMR (533). The DP ^{29}Si MAS experiments were used to quantitatively characterize the silica materials. Extremely high mobility of the ibuprofen molecules when the matrix was not functionalized were found. Very poor CPMAS efficiency, sharp ^{1}H spectra (both due to the reduction of dipolar coupling), and short T_{1C} (typical for liquids) data were all indicative of ibuprofen exhibiting liquidlike behavior. In contrast, when the silica matrix was functionalized by amino groups, the ability to acquire regular CPMAS spectra pointed to strong restrictions in mobility, suggesting interaction between the amino groups and the carboxylic groups. Chemical shift assignments were made by comparison with ibuprofen solution shifts.

Boland and Middleton (534) studied interactions between a drug and a membrane protein target by $^1H-^{19}F$ CPMAS. When the fluorinated ligand was not bound to its membrane target, no CP signal could be observed due to the averaging of the dipolar coupling by fast tumbling of the solution. However, when the ligand bound its target, CPMAS signal was observed. The binding could in turn be characterized based on the variable contact time CPMAS curves.

Kawada and Marchessault (535) studied crystalline complexes of amylose with lauric acid, *n*-butanol, *n*-pentanol, dimethyl sulfoxide, and thymol by ^{13}C CPMAS. The complexes formed single helices (V-polymorph) stable up to 90°C. Upon humidification (two days at 100% RH), a partial transformation to a double helix (B-polymorph) was observed. The authors concluded that the complexing agents were trapped in a matrix with crystal surfaces and dislocations accommodating the complexing. Amylose complexes were found to be present as partially ordered hydrophobically bonded structures. Smits et al. (536) studied interactions of crystalline amylose and crystalline and amorphous amylopectin with plasticizers, such as glycerol or ethylene glycol, in the absence of water by ^{13}C DP and CPMAS experiments. The appearance of the plasticizer in the CPMAS spectra suggested strong enough interaction to immobilize the plasticizer molecules. It was suggested that the amylopectin/glycerol interaction involved chemical exchange mechanism, in which the glycerol was continuously moving through the mixture and interacting alternatively with different sites. Heat treatment caused further immobilization of both glycerol and ethylene glycol.

Kozerski et al. (537) studied genistein (isoflavone) complexes with amines (morpholine and piperazine) by ^{13}C CPMAS. Carbon chemical shift assignments were accomplished by comparison with solution and by short contact time CPMAS. The extent of proton transfer between the free acid and complex was followed by carbon chemical shifts in the solid state. Most of the carbons exhibited rather small differences between the free acid and the complex carbon chemical shifts, typically smaller than ± 2 ppm. However, large shifts of 6.8 and 12.6 ppm were observed on one particular carbon for the two complexes, reflecting full proton transfer. This carbon was identified as being both donor and acceptor in the hydrogen bond network.

Zech et al. (538) characterized protein–ligand interactions by 1D ^{13}C spectra and 2D $^{13}C-^{13}C$ correlation experiments using ^{13}C-labeled protein. Differential chemical shifts (free protein versus protein with bound ligand) are used to distinguish between direct protein–ligand contacts and small conformational changes of the protein induced by ligand binding. Dipolar-assisted rotational resonance (DARR) (539) experiment was used to record the 2D $^{13}C-^{13}C$ correlation spectra. DARR uses rotational resonance modified with $^{13}C-^1H$ dipolar interaction (^1H rf field fulfils rotary-resonance condition).

Lubach et al. (540) studied pharmaceutical solids in polymer matrices. Two systems were studied: bupivacaine (free base and HCl salt) incorporated into tristearin and encapsulated using a solid protein matrix and ^{13}C-labeled asparagine

1:250 w/w in PVP and poly(vinyl alcohol). Different polymorphic forms of bupivacaine were generated upon heating or recrystallization. No polymorphic changes of bupivacaine were observed upon formulating in tristearin, but the tristearin itself was partially crystalized. The estimated limit of detection was 1:400. No ^{13}C CPMAS spectrum was observed for the protein matrix, indicating that the high mobility prevented efficient CP. Amorphous asparagine was identified when lyophilized in the polymer matrices. The polymer peaks broadened as well.

Schachter et al. (541) studied drug-poly(ethylene oxide) (PEO) solid solutions by ^{13}C CPMAS and variable temperature ^{1}H MAS. Ketoprofen was chosen as a model drug with poor water solubility. The hydrogen bonds between the carboxylic group of ketoprofen and the ether oxygen of PEO were the likely source of the high miscibility and formation of molecular dispersion.

Gjerde et al. (542) studied the anti-psychotic drug chlorpromazine (CPZ) interaction with phosphatidylserines by ^{13}C and ^{31}P SSNMR. ^{31}P spectra indicated the effect of acyl chain unsaturation on the anisotropic motion of the charged serine head group, implying restricted motions by intermolecular rather than intramolecular effects.

Apperley et al. (543) studied interaction of sildenafil citrate (Viagra) with water vapor. Unlike SSNMR, PXRD data showed no significant differences between samples equilibrated at different RH. The overall pattern of the SSNMR spectra as a function of RH remains unchanged, strongly suggesting only small changes of unit cell as water is absorbed or removed from the crystal lattice. Since only one set of ^{13}C CPMAS peaks is observed, water was said to move rapidly inside the lattice. The major changes were observed in the citrate anion and the drug's propyl group (Fig. 72). The shift patterns were consistent with formation of weak hydrogen bond. The site of protonation was studied by ^{15}N CPMAS. The lack of observing the characteristic ^{14}N–^{15}N residual dipolar splitting of the two neighboring nitrogens in the equilibrium state at high humidity (88% RH) was interpreted as a signature of the water dynamics-induced ^{14}N-enhanced relaxation. (The splitting is observed for anhydrous samples.) Comparison with solution- and solid-state dipolar dephasing editing was used to assign nitrogen resonances.

Bogdanova et al. (544) studied experimentally and by modeling the interactions of PVP with ibuprofen (IBP) and naproxen (NAP). A spontaneous conversion of PVP/IBP physical mixtures into a stable glass-like form (solid dispersion) was observed after storage. The enantiomer was found to react more strongly than the racemate. NAP did not interact with PVP. This was related to the differences in the drug crystal structures.

β-Cyclodextrin Complexes

A special class of drug–excipient interactions is the interaction of API with inclusion compounds, such as β-cyclodextrin (βCD). It is well known that the

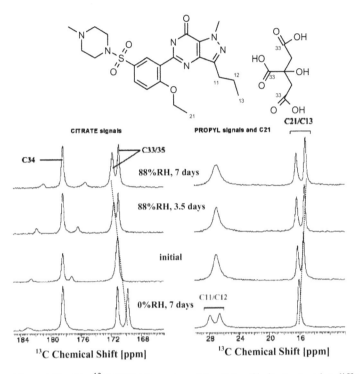

Figure 72 The ^{13}C CPMAS spectra of the sildenafil citrate stored at different conditions. *Abbreviation*: CPMAS, cross-polarization magic angle spinning. *Source*: Modified from Ref. 543.

structure of βCD and its polymers give rise to remarkable ability to form inclusion complexes with organic molecules, especially aromatic compounds, through host–guest interactions.

Schneider et al. (545) wrote a review on NMR studies of cyclodextrins and their complexes. It concentrates mostly on solution NMR studies, but a brief review of SSNMR of cyclodextrins is also included.

Kitchin and Halstead (546) studied methyl group dynamics in aspirin and aspirin βCD complexes by ^{2}H SSNMR. Variable temperature ^{2}H spectra and variable temperature ^{2}H inversion recovery experiments were recorded, and the activation energies of the aspirin-d$_3$ methyl group rotation and its correlation times were calculated. The authors found the aspirin-d$_3$ methyl group ^{2}H electric field gradient (EFG) tensor unaffected by the inclusion in βCD. They found the activation energy and the spin-lattice relaxation anisotropy smaller than for pure aspirin-d$_3$, suggesting that the rotation of the methyl group in the complex is less hindered than in the aspirin-d$_3$.

Crini et al. (547) studied βCD polymers cross-linked with epichlorohydrin. The ^{13}C MAS and CPMAS spectra were recorded as a function of temperature to overcome the spectral overlap of the polymerized epichlorohydrin with βCD peaks. The MAS spectra showed only the mobile components.

Garbow et al. (548) studied structure, dynamics, and stability of βCD inclusion complexes of aspartame and neotame. This was, in part, a solution-phase study showing that the inclusion complex formation accounts for the observed stability enhancement in solution. Studies of inclusion complexes in lyophilized solids provided an additional insight. The ^{13}C CPMAS line widths of the inclusion complex were broad, suggesting more amorphous environment of the sweetener, which was consistent with the formation of a true inclusion complex. Single $T_{1\rho H}$ relaxation time of the sweetener was observed for the inclusion compound, which was much shorter than for the pure sweetener and consistent with a presence of the single phase.

Saalwächter (549) studied chain dynamics and order of γCD inclusion compound with poly(dimethylsiloxane) by ^{1}H and ^{13}C SSNMR. Even at 30 kHz MAS spinning, the ^{1}H resonances of the individual protons were not resolved, indicating essentially rigid dipolar coupling environment. In contrast, signals from OH/H_2O hydrogen bonding region were sharp due to the dynamics-induced dipolar decoupling. This was confirmed by 1D DQ-filtered ^{1}H MAS. Assignment of the coupling partners was done by 2D SQ/DQ correlation. REPT-HDOR experiment was used to quantitatively determine ^{1}H–^{13}C dipolar coupling constant of the methyl groups. Two complimentary ways of extracting the coupling constant were employed: following of the REPT-HDOR intensity build-up and its sideband analysis. The dipolar couplings obtained from the REPT-HDOR technique were found to be influenced less by the remote couplings in the abundant proton spin bath than dipolar couplings obtained from BABA analysis of spinning sidebands. The order parameter for the methyl group, which is related to the ratio of the experimentally observed and static (rigid limit) dipolar couplings, was found to be 0.72. All methyl groups were found to be in the same rotation state. The definite validation of the rotation model was not possible from the NMR data alone. This work represents the first quantitative SSNMR investigation of local segmental fluctuations and order without the need for isotopic enrichment.

Wulff et al. (550) studied indomethacin complexes with cyclodextrins dispersed in PEG 6000 carrier by ^{13}C CPMAS. Complexation was identified for βCD and γCD but not for αCD. The distribution of chemical shift for the βCD complex was attributed to the possible formation of different types of complexes. Different parts of the indomethacin molecule could bind to βCD and, thereby, give several types of complexes. The change in chemical shifts of the γCD in the dispersion indicated a bond between γCD and the drug. A crystallinity of the dispersion was measured by ^{1}H SSNMR as a ratio between the broad peak of the amorphous PEG and the total area of the peak (broad amorphous plus sharp crystalline). The dispersions were found to be less crystalline compared with the pure polymer carrier.

Lu et al. (551) studied molecular motions in the supramolecular complexes between poly(ε-caprolactone)$-$PEO$-$poly(ε-caprolactone) and αCD and γCD. Guest$-$host magnetization exchange was observed by T_{1H}, $T_{1\rho H}$, and 2D heteronuclear correlation experiments. The CP dynamics of the inclusion complex was studied. The identical T_{1H} relaxation of the components of the inclusion complex pointed to the homogeneity of the domains on the length scale of 334Å. Spin diffusion was studied with 2D heteronuclear correlation experiment, using BLEW-12/BB-12 homonuclear decoupling during ^{1}H evolution and WIM-24 during CP to quench the ^{1}H$-^1$H spin diffusion. T_{1C} was used to study high-frequency motions (MHz) and line shape analysis of 2D WIM/WISE was used to investigate low-frequency motions (kHz). Temperature dependence of $T_{1\rho C}$ was also measured, but no $T_{1\rho C}$ minimum was observed, making the interpretation for kHz range dynamics difficult.

Yannakopoulou et al. (552) studied complex of (Z)-tetradec-7-en-1-al with βCD. Two different guest environments were detected by splitting of carbonyl and methyl groups. The guest molecules were found to be tightly trapped in the βCD torus and, therefore, resist oxidation. Two βCD molecules were found to form head-to-head dimmers that enclose one guest molecule disordered over two sites and packed in channels.

Cunha-Silva and Teixeira-Dias (553) studied inclusion of nonionic surfactant in βCD by ^{13}C CPMAS. Carbon multiplicity was reduced for the inclusion compounds studied, pointing to appreciable symmetrization of the βCD macrocycle. Two groups of spectral line shapes were observed when carbon spectra were recorded as a function of humidity. The spectra of sample equilibrated below 20% RH were different than those above 20% RH. Water seemed to contribute in a relevant way to the βCD macrocycle symmetrization.

Braga et al. (554) studied an inclusion complex of βCD with p-hydroxy-benzaldahyde. The free βCD exhibited multiple resonances for each type of carbon, which correlated with different torsion angles of the sugar linkages. The multiplicity was reduced to single peaks upon complexation, suggesting a symmetry increase upon inclusion. The same lab studied inclusion compound of S-IBP in βCD by ^{13}C CPMAS (555). Tight encapsulation was found in which the βCD adopted a more symmetrical conformation in the inclusion compound with each glucose unit in a similar environment. The same lab studied encapsulation of sodium nimesulide and its precursors in βCD (556). ^{13}C CPMAS confirmed that no chemical modification of the guests occurred upon formation of the inclusion complexes. The same lab also studied the humidity effect on 2-phenoxyethanol βCD inclusion compound (557). At RH higher than 20%, decreased multiplicities of the carbon resonances and diminished dispersion of chemical shifts was found, pointing to an improved symmetrization of the βCD macrocycle in the channel structure of this inclusion compound. Below 20% RH, the carbon spectra showed broad lines, characteristic of amorphous phase.

Lai et al. (558) studied imazalil βCD inclusion complex by ^{13}C CPMAS. Authors also observed a convergence of the multiple resonances of the carbons of the glucose units to the single peak between βCD and the complex. The imazalil peaks were broadened, following encagement into the hydrophobic cavity. Some of them split into multiple signals, indicative of a pronounced structural rearrangement of the imidazole and aryl rings. Carbon-detected rotating frame proton relaxation $T_{1\rho H}$ revealed the sample to be homogeneous. No change of mobility of the sugar rings in the βCD skeleton was detected by comparison of the free βCD $T_{1\rho H}$ with those of the inclusion compound. However, $T_{1\rho H}$ of imazalil decreased remarkably between free and complexed imazalil.

Excipient Characterization

Excipients can be characterized by SSNMR in a similar way a drug substance. The same suite of techniques is available and similar types of problems can be answered. Sugars appear as components of many excipient blends. Many examples of characterization of carbohydrates can be found in the literature (52,559,560).

Tanner et al. (561) studied crystalline forms of α- and β-chitin. Remarkably narrow carbon lines of a variety of chitin (polysaccharide similar to cellulose) samples were observed. Unlike previous reports, only a single molecule per asymmetric unit cell was observed. Spectra of at least three different forms of β-chitin were identified, differing in the hydration states. Spectra of the poorly ordered squid pen β-chitin were interpreted by using the spectra of the highly crystalline samples. The α-chitin spectra were found to be broad and hard to interpret. T_{1C} filter was used to filter out fast relaxing peaks, proving a mixture character of some samples.

Kunze et al. (562) characterized cellulose and cellulose ethers. Qualitative and quantitative analysis is shown, including quantitative characterization of the homogeneity and structural selectivity of substitution. Spectra of the cellulose polymorphs, cellulose-I (I_α and I_β), cellulose-II, Na-cellulose-I, NA-cellulose-II, and amorphous are shown, as are the interconversions between the cellulose forms upon the addition of NaOH. Mixtures of the forms are quantified without taking the CP dynamics into account. A degree of substitution could be quantified by choosing a resolved pair of carbon lines and a particular contact time, assuring equal intensity response.

Lefort et al. (563) assessed SSNMR and DSC methods for quantifying the amorphous content in solid dosage forms of ball-milled trehalose (Fig. 73). Samples with different amorphous fractions were prepared as physical mixtures of pure amorphous (obtained by mechanical milling) and crystalline components. After "calibration" on the physical mixtures, both methods were applied to study the trehalose amorphization during milling. NMR spectra for various lengths of milling could be reproduced as linear combinations of the pure crystalline and amorphous components. However, a different response was seen using DSC. The

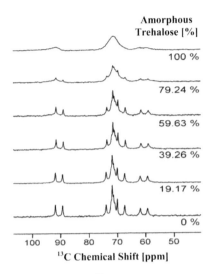

Figure 73 The ^{13}C CPMAS spectra of trehalose with different amorphous and crystalline contents. *Abbreviation*: CPMAS, cross polarization magic angle spinning. *Source*: Modified from Ref. 563.

DSC peaks spread over a very broad range of temperatures, and their enthalpies were smaller than expected. This is attributed to differences on a nanometer scale or larger, which do not affect the NMR spectra. NMR can, therefore, remain a successful technique for estimating the amorphous content, even if DSC might fail.

Atomic Distance Measurements

A determination of a three-dimensional structure of molecules by SSNMR relies on the knowledge of distances between atoms. The $1/r^3$ dependence of dipolar coupling [Equation (13)] can be used to directly extract the distances with high precision. Because the dipolar coupling is averaged out by fast MAS, it has to be reintroduced by any of the numerous recoupling schemes. The various recoupling techniques can restore either the heteronuclear or homonuclear dipolar coupling and the pairwise distances between atoms can be extracted. An isotope enrichment is typically required limiting the range of applications to pharmaceutical materials. A selective labeling scheme can be used to measure distances between the labeled nuclei. Procedures to measure multiple distances simultaneously in uniformly labeled samples were also devised. Until very recently, measurement of distances between protons were not successful, mainly due to the poor available resolution in proton spectra. With the advancement of modern homonuclear decoupling techniques, this becomes less of an issue. This opens up the possibility of solving crystal structures of small molecules based solely on SSNMR without the need for isotope enrichment (564).

Tong and Schaefer (565) characterized the interface of heterogeneous blends of fluorinated and selective carbon-labeled polymers by measuring $^{13}C-^{19}F$ distances using REDOR.

Fyfe et al. (566,567) compared $^{19}F-^{29}Si$ distances obtained from REDOR and TEDOR techniques on static and MAS spun samples of clathrasil octadecasil. The distances, which were obtained from fitting of the oscillatory CP behavior, agreed well with those known from X-ray, and the distance accuracy of 0.004 nm was achieved. Fyfe applied the methodology to study $^{19}F-^{29}Si$ distances in other compounds. Bertani et al. used spin counting θ-REDOR approach to study octadecasil (568).

Middleton et al. (569) determined the molecular conformations of four polymorphic forms of cimetidine from ^{13}C SSNMR distance and angle measurements. The doubly ^{13}C-enriched cimetidine was examined by ^{13}C CPMAS. The carbon spectrum of one of the four forms was consistent with two molecules per asymmetric unit. Splitting of the carbon resonances due to the residual dipolar coupling was evident. The distance between the two labeled carbons was determined from analysis of the rotational resonance magnetization exchange curves resulting from the reintroduction of the weak $^{13}C-^{13}C$ dipolar coupling. Eight-fold dilution by nonlabeled cimetidine was used to eliminate the inter-molecular dipolar coupling. Difference spectroscopy was used to remove the natural abundance background. The carbon distances of two forms for which single crystal structures were available were consistent with distances obtained from the single crystal structure. The DQ heteronuclear local field (2Q-HLF) experiment was used to determine the relative orientation of the $^{13}C-^{1}H$ bonds at the two labeled carbons. The experimental distances and the angles were used to constrain the molecular conformation calculations. This approach demonstrated that structures of polycrystalline compounds could be investigated in the absence of diffraction techniques.

Gilchrist et al. (438,439) measured inter-fluorine distances (from 0.5 to 1.2 nm) in organic compounds by ^{19}F recoupling methods. The RFDR method was used and the exchange curves were simulated by Monte Carlo–type algorithm (Fig. 74). An accuracy of 0.1 nm for short distances and 0.2 nm for long distances was achieved. The model compounds were selected based on four criteria: (*i*) resolved resonances, (*ii*) known (rigid) distances, (*iii*) short relaxation times, and (*iv*) dilution by matrix to prevent intermolecular dipolar coupling. The observed fluorine peaks were approximately 1 ppm wide, dominated by inhomogeneous contributions and not limited by proton decoupling (>100 kHz).

Using modified CPMG sequence, Salgado et al. (570) studied membrane-bound structure of a protein derived from angular and distance constraints from homonuclear ^{19}F dipolar coupling. A good agreement with the known structure was achieved. The chemical shift anisotropy of the macroscopically oriented sample enabled to determine the orientation of the peptide in the membrane bilayer.

Middleton et al. (571) used conformational analysis by SSNMR to aid restrained structure determination form PXRD. A molecular conformation is

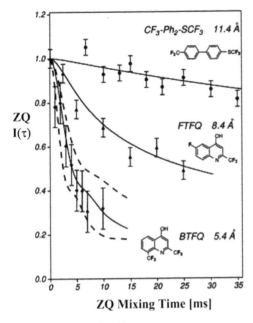

Figure 74 The $^{19}F-^{19}F$ distance measurements using RFDR experiment. Experimental and simulated RFDR exchange curves are shown. *Abbreviation*: RFDR, radiofrequency-driven re-coupling. *Source*: Modified from Ref. 438.

first determined approximately from a set of interatomic distances measured by REDOR. The conformation is then optimized against PXRD data using simulated annealing protocol. The $^{15}N-^{13}C$ REDOR distances were obtained on ^{15}N-labeled (and naturally abundant carbon) cimetidine. CPMAS with selective CH/CH_2 transfer was used to reduce spectral overlap (117). Restrained molecular dynamics was applied to obtain the molecular conformation from the SSNMR data, followed by PXRD optimization. With the combined SSNMR/PRXD approach, the problem of solving structures is transformed into a rigid body search.

Mehta et al. (572) applied REDOR sequences to measure carbon–fluorine and deuterium–fluorine distances on lyophilized bacterial whole cells (selectively ^{13}C-labeled), complexed with 4-fluorobiphenyl derivative of chloroeremomycin. The position of the drug was determined relative to the peptidoglycan stem and, therefore, did not require an absolute measurement of the REDOR full echo.

Grage et al. (573) applied $^2H\{^{19}F\}$ REDOR to study distances in biological solids using double resonance probes without proton decoupling. The lack of proton decoupling was found only to influence the accuracy of the obtained distances. The maximum distances, thus have to be smaller than 0.8 nm. Composite pulses were used on the fluorine channel, which greatly reduced the fluorine off-resonance effects.

Elena and Emsley (564) applied $^1H-^1H$ spin diffusion CRAMPS experiment to characterize structure and the crystallographic unit-cell parameters of naturally abundant small molecules. The spin diffusion was recorded using high-resolution 2D $^1H-^1H$ correlation experiment with eDUMBO-1 (102) homonuclear decoupling in both t_1 evolution and t_2 detection periods. The eDUMBO-1 afforded high enough resolution that a kinetic rate matrix model could be used to characterize the spin diffusion build-up curves between the single proton sites (Fig. 75). The comparison between the experimental data and simulations on

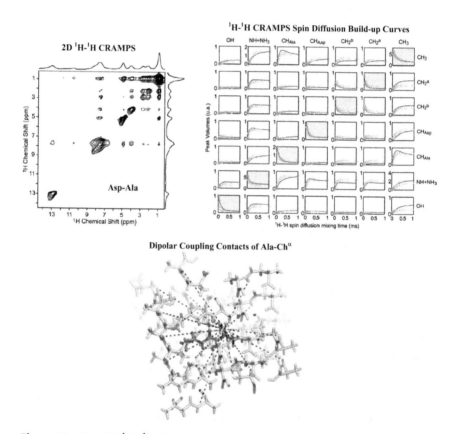

Figure 75 The 2D $^1H-^1H$ CRAMPS correlation experiment with variable time spin diffusion mixing period using eDUMBO-1 homonuclear decoupling (*top left*). The spin diffusion build-up curves together with their best fits (*top right*). The intra- and intermolecular dipolar coupling network of a single alanine CH^α (*bottom*). *Abbreviation*: CRAMPS, combined rotation and multiple-phase spectroscopy. *Source*: Modified from Ref. 564.

model dipeptide depended strongly on the crystal structure parameters, such as unit cell and the orientation of the molecule within it.

H-Bonds and Proton Transfer

The knowledge of H-bonds and proton transfer offer an important information about structure of the compound. Single-crystal X-ray diffraction is considered to provide the ultimate knowledge of H-bonds. However, because the positions of protons are inferred from X-ray data only indirectly, much remains to be learnt even when the single crystal structure is available. SSNMR is an excellent tool to study H-bonds and proton transfer. There are several different ways to gain knowledge on proton transfer by SSNMR. Chemical shielding is very sensitive to local distribution of electron densities, and the presence of H-bond is well reflected by sizeable chemical shifts. The calibrated chemical shift scales can be applied to precisely estimate the bond lengths of unknown samples. The dipolar coupling interaction provides another way to directly measure distances between nitrogens and protons with high precision. A picometer precision has been achieved (574). A definite confirmation of the presence of H-bonds can be obtained by observation of cross-peaks in correlation experiments based on J-coupling mechanism. The values of the H-bond J-coupling correlates well with the H-bond strength.

Limbach published a chapter on "Studies of Elementary Steps of Multiple Proton and Deuteron Transfers in Liquids, Crystals and Organic Glasses" (575). The focus of this study was put on NH...N proton transfer because (i) it is slower than OH...O, and (ii) ^{15}N is spin $\frac{1}{2}$ that can be followed easily by NMR. One of the major conclusions was that the proton transfer systems in the crystalline and amorphous states are subject to static perturbations of the reaction energy profiles of the proton motion.

McGregor et al. (76) studied the fast proton transfer in a series of substituted tetra-azaannulenes. The number of observed ^{13}C CPMAS peaks for all compounds except one was in accord with their symmetry as determined by crystal structures. The exception was the tetra-azaannulene, for which less than the expected number of lines were observed as a consequence of fast proton transfer accompanied with rapid molecular rotation. On cooling, a peak broadening was observed for this sample, as a result of the interference of the proton transfer with proton decoupling frequency of 60 kHz. The evidence from carbon spectra is in full agreement with previously reported ^{15}N data.

Hickman et al. (576) studied protonic conduction in imidazole by ^{15}N SSNMR. The 2D ^{15}N exchange study demonstrated that the conduction mechanism did not involve the reorientation of the imidazole ring.

Wu et al. (577) studied the intermolecular H-bonding effects on the amide ^{17}O electric field gradient and chemical shift tensor of benzamide. The ^{17}O MAS and stationary spectra were fitted to extract the ^{17}O quadrupolar and chemical shift parameters. Quantum mechanical calculation provided a way to relate these tensors to molecular structure.

Song and McDermott (578) studied proton transfer dynamics and N–H bond lengthening in N–H...N model systems by variable temperature ^{15}N CPMAS and 2D exchange experiments. Based on the nitrogen line shapes, short N...N distances and fast proton transfer through the N–H...N bridge was observed.

Zhao et al. (579) devised a new method for quantification of H-bonds by measurements of NH bond lengths using 2D ^{1}H–^{15}N dipolar-chemical shift correlation (with recoupling of the H–N dipolar couplings) combined with fast MAS. From the experimental ^{1}H–^{15}N dipolar coupling, the HN lengths were calculated with accuracy of less than 0.01 nm.

Kiss et al. (580) studied H-bonded networks of methoxy-substituted α-phenylcinnamic acids by spectroscopic and computational methods. It was possible to detect the aromatic C–H...O H-bonds responsible for the long-range ordering. An absence of olefinic H-bonds was confirmed by ^{13}C CPMAS spectra. Molecular modeling provided an additional proof for existence (or absence) of H-bonds.

Huelsekopf and Ludwig (581) correlated structural information, NMR, and IR techniques of N-methylacetamide. Nonadditive contribution of H-bonds was found to play a significant role. Calculated predictions were compared with experimental data and reasonable agreement was obtained.

Henry et al. (582) studied solid-state acid–base and tautomeric equilibria of lyophilized L-histidine and determined pH and pK values. A separate observation of all three acid–base pairs in the successive deprotonation of the carboxylic group was possible. The calculated solid-state pK values based on the ^{13}C CPMAS observed acid/base ratios were found similar to solution. Noninvasive in situ measurement of solid-state acidity was therefore possible. Differences in the ^{1}H CRAMPS spectra were also obsereved.

Foces-Foces et al. (583) studied H-bond structure and dynamics of pyrazole-4-carboxylic acid chains using ^{15}N CPMAS and other techniques. At low temperatures, NMR showed a proton disorder-order transition with localized protons. The proton transfer equilibrium detected by nitrogen NMR was found to be a sensitive tool for studying molecular structure that could not easily be detected by X-ray methods.

Totz et al. (584,585) studied the mechanism of proton conductivity in quasi-one-dimensional H-bonded crystals by 1D and 2D exchange ^{2}H NMR. The slow chemical exchange processes between H-bridges were studied over a wide temperature range. The ionic conductivity was realized by jumps of the protons between the different H-bonds. The proton mobility measured from the 2D experiments was related to the measured conductivity.

Takeda and Tsuzumitami (586) noted the slowing down of the proton transfer in H-bonds of benzoic acid crystals. Transfer rates were investigated by a temperature dependence of T_{1H} and T_{1D}. The results indicated that H and D pairs transfer in a concerted manner within the timescale of NMR (10^{-6} to 10^{-10} seconds).

Lorente et al. (587) studied proton and nitrogen chemical shielding (by line shape simulation), dipolar coupling (providing distances), and H-bond geometry correlations in a series of H-bonded acid–base complexes of collidine with carboxylic acid. The isotropic ^{15}N chemical shifts were found to be characteristic of N–H distances and could be used for evaluating H-bond geometries and solid-state acidities.

Goward et al. (588) investigated NH H-bond in solid benzoxazine dimer by ^{1}H–^{15}N correlations under fast MAS. The proton was found to be shared between the nitrogen and oxygen atoms. Different designs of REDOR-type of recoupling, including the sensitive inverse detection on proton channel, of the heteronuclear ^{1}H–^{15}N dipolar interactions were used. The spinning sidebands yielded the values of dipolar coupling and hence the N...H distances. Relatively long distances could be measured.

Ono et al. (589) studied H-bonding of amino acids and peptides by variable temperature ^{2}H static NMR. Parameters of the quadrupolar tensor were extracted by simulations using the ^{2}H dynamic NMR theory and correlated with N...O H-bond length.

Colsenet et al. (590) devised a straightforward detection of deprotonated conformers of malonic acid. By using slow spinning ^{13}C CPMAS spectra of samples lyophilized from solutions of different pH, different signals were observed for protonated and deprotonated species as a result of dramatic slow-down of the inter- and intramolecular proton exchanges. A planar intermolecularly H-bonded monocation was observed for the first time.

Brown et al. (216,591) determined the H-bond strengths by a quantitative measurement of homonuclear J-couplings. The existence of the through H-bond correlation peak in ^{15}N INADEQUATE confirmed the first ever directly detected solid-state H-bonding. The 2D MAS echo technique is used to accurately determine the through H-bond J_{NN} coupling strength, even in the cases in which no peak splitting is actually observed and, hence, the strength of the H-bond can be determined.

Lesage et al. (592) observed a H-agostic bond in a heterogeneous catalyst using 2D SSNMR. The authors used a variant of the SS-APT through bond J-coupling based technique, where t_1 evolution was incorporated to make it an 2D technique. By correlating the carbon chemical shifts with $^{1}J_{CH}$ coupling, the direct evidence for the H-agostic bond was provided.

Ilczyszyn and Ilczyszyn (593,594) studied betain-sulfamic acid (2:1) crystal by ^{13}C CPMAS and other techniques. It was found for the first time that the Raman and shielding tensors are strongly correlated. Very short and symmetric H-bonds were formed by acidic protons and betaine carboxylate groups.

Pawsey et al. (595) studied H-bonding interactions in self-assembled monolayers by ^{1}H fast MAS SSNMR. The 2D HETCOR and 2D ^{1}H double quantum correlations were extensively used.

Emmler et al. (596) devised a simple method for the characterization of OHO hydrogen bonds by ^{1}H rotor-synchronized echo SSNMR, which afforded higher resolution than the standard fast MAS spectra and the proton chemical shifts could be faithfully extracted. Using empirical valence bond order model, the experimental ^{1}H shifts were correlated with the H-bond geometries for a broad range of substances. For strong H-bonds, the deviations from the correlation curve are explained in terms of proton tautomerism. By applying the proton echo technique, it is now possible to routinely determine the position of the proton in a H-bond.

Leppert et al. (597) identified NH...N H-bonds in double-stranded RNA by 2D ^{15}N dipolar-chemical shift correlation spectra. The presence of H-bond was reflected in the cross-peaks between the donor and acceptor.

Schilf et al. (598) used ^{15}N and ^{13}C NMR to study intramolecular H-bonds in eight Schiff bases. The nitrogen chemical shifts of OH and NH tautomers were found especially useful to characterize the tautomerism. Four compounds were found to exist as OH tautomers, whereas the remaining four existed as NH tautomers with different stages of H transfer. This was confirmed by ^{13}C CPMAS. The position of protons in intramolecular H-bonds were found to be strongly determined by substituents in aromatic ring. In other studies of Schiff bases from the same lab, six out of eight compounds were found to form intramolecular H-bonds (599–604). The ^{15}N shifts and $^1J_{NH}$ were used to estimate the tautomeric equilibrium.

Li et al. (605) studied acid–base characteristics of bromophenol blue-citrate buffer systems in the lyophilized amorphous state. Raman and ^{13}C CPMAS SSNMR spectroscopies were used to monitor the ratio of ionized to unionized citric acid under various conditions as a function of pH. Downfield solid shifts were observed with increasing solution pH especially for the carboxylic carbons.

Chekmenev et al. (606–608) correlated nitrogen chemical shielding in labeled cyclic tripeptides (determined from MAS and stationary spectra) with X-ray structure. ^{1}H–^{15}N and ^{13}C–^{15}N dipolar couplings were used to accurately orient the chemical shielding tensor in the molecular frame (within 1°). The isotropic shifts were found dependent on the H-bonding and local conformation.

Diez-Pena et al. (609) studied the effect of H-bonding on the collapse of PMAA hydrogels by advanced ^{1}H SSNMR techniques. The ^{1}H MAS and 2D BABA-based correlation experiments are shown. Proton distances are obtained from the fitting of BABA spinning sidebands. Dependence on pH was also studied. Different populations of three COOH species (free mobile acid and two H-bonded dimers with different stabilities) were found to depend on pH and temperature.

Kawashima (610) studied coals soaked in pyridine by ^{15}N NMR. Two states of pyridine were observed: one with somewhat limited mobility and weak H-bonded pyridine and a second with even less mobility and strong H-bonded pyridine. The interaction between pyridine and coal was not strong enough to be observed by CPMAS.

Benhabbour et al. (611) studied imidazole-based proton-conducting composite materials to provide an insight into the mechanism of proton transfer. Application of variable temperature ^{1}H high-speed MAS and DQ filtered ^{1}H NMR using BABA sequence identified mobile protons. Comparison of macroscopic and microscopic mobilities are presented.

Gobetto et al. (612) studied H-bonded supramolecular adducts using ^{1}H MAS and ^{15}N CPMAS. Proton and nitrogen chemical shifts allowed a distinction between the intermolecular H-bonds of the N^+–H. . .O$^-$ (with proton transfer) and N. . .H–O (no proton transfer). Correlations between the chemical shifts and N–O distances were found. The localization of H-bonds by DFT calculations was in reasonable agreement with experimental data. The authors predict this approach to be especially fruitful for studies of partially disordered systems that cannot be studied by X-ray methods.

Schulz-Dobrick et al. (574) determined the geometry of H-bonds in solids with picometer accuracy by combination of NMR and quantum chemical calculations. The ^{1}H–^{15}N dipolar recoupling techniques (REREDOR- and HDOR-based) can determine H–N distances up to 250 pm with accuracy of ± 1 pm and ± 5 pm for short (100 pm) and long distances, respectively. The structures were computationally optimized based on the H–N distance constraints and checked for agreement between the experimental and predicted chemical shifts. With a precision of 1 pm, the vibrational averaging had to be considered, which reduced the measured dipolar coupling and, thus, the apparent distances derived from it.

Li et al. (613) studied acid–base interactions in complexes of heterocyclic bases and dicarboxylic acids. Combination of ^{15}N CPMAS and ab initio predictions showed the sensitivity of nitrogen chemical shifts to serve as a reliable probe for the degree of proton transfer in realistic systems (as opposed to models), ranging from no proton transfer (free base) through partial proton transfer (neutral crystal complexes or cocrystals) to full proton transfer (salts; Fig. 76).

Food and Cosmetic Applications

This short section on SSNMR applications in food and cosmetic sciences is included based on similarities of the problems studied and the similar limitations on obtaining isotopically labeled materials. A distinction between crystalline and amorphous phases and domain characterizations are among the common SSNMR applications to study the heterogeneous samples typically found in food and cosmetics materials. Because water plays a significant role in the physical and chemical properties of these samples, the distinction between free and bound water, mobility, and relaxation studies are frequent.

Waver and Zielinska (614) studied flavonoids by ^{13}C CPMAS to characterize their conformations. Carbon resonances were assigned based on solution shifts and solid spectral-editing techniques. Two molecules per asymmetric unit were observed in one of the compounds. An orientation of the hydroxyl

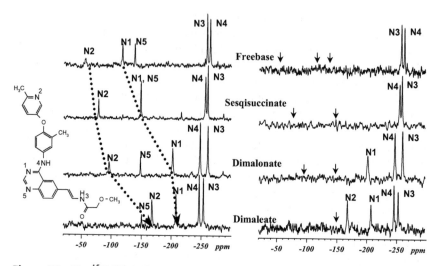

Figure 76 The ^{15}N CPMAS spectra of CP-724,714 free base and salts (*left*). The progressive proton transfer (from free base to dimaleate) is evident from the chemical shifts of nitrogens N1 and N2. Short-contact time CPMAS (protonated nitrogens only; *right*). *Abbreviation*: CPMAS, cross polarization magic angle spinning. *Source*: Modified from Ref. 613.

groups and the effects of hydrogen bonding were evaluated from the solution–solid differences of the carbon shifts.

Utility of SSNMR to study lyotropic materials formed from food grade monoglyceride mixtures was demonstrated (615). Carbon chemical shift anisotropy, as determined by MAS spinning sideband analysis, was used to distinguish the phase after addition of glucose guest molecule. Deuterium DQ spectroscopy was used to differentiate water in isotropic and anisotropic environments.

Table 1 Nuclear Spin Hamiltonians

Interaction	λ	C^λ	H_λ
Zeeman	Z	$-\gamma$	$-\gamma\hat{I} \cdot 1 \cdot \boldsymbol{B}_0$
Chemical shift	CS	γ	$\gamma\hat{I} \cdot A^{CS} \cdot \boldsymbol{B}_0$
J-coupling	J	—	$\hat{I} \cdot A^J \cdot \hat{S}$
Dipolar	D	$\frac{\mu_0}{2\pi}\gamma_1\gamma_2\hbar$	$\frac{\mu_0}{2\pi}\gamma_1\gamma_2\hbar\,\hat{I}_1 \cdot A^D \cdot \hat{I}_2$
Quadrupolar	Q	$\frac{eQ}{2I(2I-1)\hbar}$	$\frac{eQ}{2I(2I-1)\hbar}\hat{I}A^Q\hat{I}$
rf irradiation	rf	$-\gamma$	$-\gamma B_1(t)\hat{I}$

Table 2 Pharmaceutically Relevant NMR Nuclei

Isotope	Spin	Natural abundance (%)	Larmor frequency at 11.75T (MHz)
^1H	1/2	99.99	500.0
^{13}C	1/2	1.1	125.7
^{15}N	1/2	0.37	50.7
^{19}F	1/2	100.0	470.3
^{31}P	1/2	100.0	202.3
^2H	1	0.02[a]	76.7
^{14}N	1	99.6	36.1
^{35}Cl	3/2	75.8	49.0
^{37}Cl	3/2	24.2	40.8
^{79}Br	3/2	50.7	125.2
^{17}O	-5/2	0.04[a]	67.8
^{23}Na	3/2	100.0	132.2
^{25}Mg	-5/2	10.0	30.6
^{39}K	3/2	93.3	23.3

[a]Isotope enrichment is necessary for NMR studies.
Abbreviation: NMR, nuclear magnetic resonance.

Emulsifier phases of cosmetic emulsions were studied by ^1H spin diffusion and ^{31}P CPMAS (616). The ^{31}P CPMAS was used to suppress the dissolved ingredients. Even though the centrifugal forces were found to destroy the emulsion, leading to changes in chemical shift, new peaks not present in the direct polarization experiment were observed. The ^1H spin diffusion was used to identify bound and free water. A related previous paper (617) applied dynamic ^{31}P methods to the interfacially bound emulsifier to obtain details about the molecular dynamics at the interface. The ^{31}P T_1 and T_2 times indicated a restricted motion, independent of the droplet size in the emulsions.

Calucci et al. (618) studied hydration effects on gluten dynamics by proton and carbon relaxation. A resolution of the carbon spectra was affected by the heterogeneous character of the gluten samples. Hence, detailed interpretations of the spectra were not possible. Comparison between DP and CPMAS experiments were used to qualitatively assess the differences in the local chemical environment and in mobility. The signals enhanced in CPMAS, when compared with DP, were considered to originate from rigid domains. Further discrimination was achieved on the basis of different T_{1C} times, using different DP recycle delays, and on the basis of different T_{2H}, using delay between the proton 90° pulse and the CP transfer. The latter experiment was carried out with 40 µs delay, giving origin to a signal of only the carbons with strong enough dipolar coupling to give CP but coupled to protons with spin–spin relaxation longer than tens of microseconds. Hydration was found to cause an increased mobility of all protein moieties. Line shape of static ^1H spectra was fitted by the sum of Gaussian and exponential functions and analyzed based on the premise that Gaussian and exponential functions can be associated with ^1H nuclei in very rigid and relatively mobile environments, respectively. No liquidlike water was observed as no peaks with few Hertz line widths were seen. Partial assignment of the ^1H MAS spectra was made. T_{1H} with both direct proton and indirect carbon detection, and $T_{1\rho H}$ with indirect carbon detection, were used to assess the rigidity of the various sample components and the domain sizes. In dry gluten, the proteins were found not to mix with residual starch on the 200Å scale by T_{1H}, whereas $T_{1\rho H}$ showed heterogeneities within protein on $10-20$ Å scale. Hydration was found to mix the different protein fragments and the protein with starch.

REFERENCES

1. Slichter CP. Principles of Nuclear Magnetic Resonance. 3rd ed. New York: Springer-Verlag, 1990.
2. Abragam A. The Principles of Nuclear Magnetism. Oxford: Oxford University Press, 1985.
3. Mehring M. High Resolution NMR in Solids. 2nd ed. Berlin: Springer, 1983.
4. Schmidt-Rohr K, Spiess HW. Multidimensional Solid-State NMR and Polymers. New York: Academic Press, 1994.
5. Haeberlen U. Advances in Magnetic Resonance. In: Waugh JS, ed. New York: Academic Press, 1976.
6. Harris RK, Mann BE. NMR and the Periodic Table. New York: Academic Press, 1970.
7. Maciel GE. Science 1984; 226:282.
8. Oldfield E, Kirkpatrick RJ. Science 1985; 227:1537.
9. Turner GL, Kirkpatrick RJ, Risbud SH, Oldfield E. Am Ceram Soc Bull 1987; 66:656–663.
10. Andrew ER, Bradbury A, Eades RG. Nature 1958; 182:1659.
11. Lowe IJ. Phys Rev Lett 1959; 2:285.
12. Maricq MM, Waugh JS. J Chem Phys 1979; 70:3300–3316.
13. Hartmann SR, Hahn E. Phys Rev 1962; 128:2042.
14. Pines A, Gibby MG, Waugh JS. JCP 1973; 59:569.
15. Llor A, Virlet J. Chem Phys Lett 1988; 152:248–253.
16. Chmelka BF, Mueller KT, Pines A, Stebbins J, Wu Y, Zwanziger JW. Nature 1989; 339:42–43.
17. Mueller KT, Sun BQ, Chingas GC, Zwanziger JW, Terao T, Pines A. J Magn Reson 1990; 86:470–487.
18. Frydman L, Harwood JS. J Am Chem Soc 1995; 117:5367–5368.
19. Medek A, Harwood JS, Frydman L. J Am Chem Soc 1995; 117:12779–12787.
20. Pratum TK, Klein MP. J Magn Reson (1969–1992) 1989; 81:350–370.
21. Rajamohanan PR, Thangaraj A, Suryavanshi PM. Ganapathy S. Magn Reson 1991; 272–279.
22. Hill EA, Yesinowski JP. J Chem Phys 1997; 107:346–354.

23. Jeschke G, Jansen M. Angewandte Chemie, International Edition 1998; 37: 1282–1283.
24. Wi S, Frydman L. J Chem Phys 2000; 112:3248–3261.
25. Geffroy PM, Mabilat F, Bessada C, Coutures JP, Massiot D. Mater Sci Forum 2000; 325–326, 319–324.
26. Wi S, Frydman L. J Am Chem Soc 2001; 123:10354–10361.
27. Jakobsen HJ, Bildsoe H, Skibsted J, Giavani T. NATO Sci Ser, II: Mathematics, Physics and Chemistry 2002; 76:43–55.
28. Bachiller PR, Ahn S, Warren WS. J Magn Reson Ser A 1996; 122:94–99.
29. Ahn S, Lee S, Warren WS. Mol Phys 1998; 95:769–785.
30. Warren WS. Science (Washington, D.C.) 1998; 280:398–399.
31. Ahn S, Lisitza N, Warren WS. J Magn Reson 1998; 133:266–272.
32. Ernst RR, Bodenhausen G, Wokaun A. Principles of Nuclear Magnetic Resonance in One and Two Dimensions. Oxford: Clarendon, 1987.
33. Gerstein BC, Chow C, Pembleton RG, Wilson RC. J Phys Chem 1977; 81:595–599.
34. Taylor RE, Pembleton RG, Ryan LM, Gerstein BC. J Chem Phys 1979; 71:4541.
35. Bronnimann CE, Hawkins BL, Zhang M, Maciel GE. Analytical Chemistry 1988; 60:1743–1760.
36. Hafner S, Spiess HW. Solid State Nucl Magn Reson 1997; 8:17–24.
37. Hohwy M, Bower PV, Jakobsen HJ, Nielsen NC. Chem Phys Lett 1997; 273:297–303.
38. Lesage A, Duma L, Sakellariou D, Emsley L. J Am Chem Soc 2001; 123:5747–5752.
39. Brus J, Petrickova H, Dybal J. Monatshefte Fur Chemie 2002; 133:1587–1612.
40. Bielecki A, Kolbert AC, Levitt MH. Chem Phys 1989; 155:341–346.
41. Vinogradov E, Madhu PK, Vega S. Chem Phys Lett 2002; 354:193–202.
42. Vinogradov E, Madhu PK, Vega S. Chem Phys Lett 1999; 314:443–450.
43. Bosman L, Madhu PK, Vega S, Vinogradov E. J Magn Reson 2004; 169:39–48.
44. Madhu PK, Zhao X, Levitt MH. Chem Phys Lett 2001; 346:142–148.
45. Vinogradov E, Madhu PK, Vega S. Top Curr Chem 2004; 246:33–90.
46. Brittain HG, Bogdanowich SJ, Bugay DE, DeVincentis J, Lewen G, Newman AW. Pharm Res 1991; 8:963–973.
47. Bugay DE. Pharm Res 1993; 10:317–327.
48. Holzgrabe U, Diehl BWK, Wawer I. J Pharm Biomed Anal 1998; 17:557–616.
49. Burn SR, Pfeiffer RR, Stowell JC. Solid-State Chemistry of Drugs. 2nd ed. SSCI, Inc. 1999.
50. Duer Ann Rep NMR Spectrosc 2001; 43:1–58.
51. Smith ME. Nucl Magn Reson 2000; 29:251–315.
52. Potrzebowski MJ. In: Atta-ur-Rahman, ed. New Advances in Analytical Chemistry 2000, P1. 359–404.
53. Ye C, Ding S, McDowell CA. Annu Rep NMR Spectrosc 2000; 42:59–113.
54. Stephenson GA, Forbes RA, Reutzel-Edens SM. Advanced Drug Delivery Rev 2001; 48:67–90.
55. Bugay DE. Advanced Drug Delivery Rev 2001; 48:43–65.
56. Harris RK. Spec Publ-R Soc Chem 2001; 262:3–16.
57. Shapiro MJ, Goundarides JS. Biotechnol Bioeng 2001; 71:130–148.
58. Bryce D, Bernard GM, Gee M, Lumsden MD, Eichele K, Wasylishen RE. Canadian J Anal Sci Spectrosc 2001; 46:46–82.

59. Ando S, Harris RK, Scheler U. Encyclopedia Nucl Magn Reson 2002; 9:531–550.
60. Kolodziejski W, Klinowski J. Chem Rev 2002; 102:613–628.
61. Dybowski C, Bai S, Bramer SV. Anal Chem 2002; 74:2713–2718.
62. Auger M. J Anal Sci Spectrosc 2002; 47:184–189.
63. Reutzel-Edens SM, Bush JK. Am Pharm Rev 2002; 5:112–115.
64. Schmidt-Rohr K, Spiess HW. Annu Rep NMR Spectrosc 2002; 48:1–29.
65. Brown SP, Emsley L. Handbook of Spectroscopy 2003; 1:269–326.
66. Potrzebowski MJ. Eur J Org Chem 2003; 8:1367–1376.
67. Medek A. Pharmaceutical Applications of SSNMR. Vol. 5. Academic Press, Elsevier Science, 2003.
68. Tishmack Patrick A, Bugay David E, Byrn Stephen R. J Pharm Sci 2003; 92: 441–474.
69. Offerdahl TJ, Munson EJ, Offerdahl TJ, Salsbury JS, Dong Z, Grant DJW, Schroeder SA, Prakash I, Gorman EM, Barich DH, Munson EJ. In press, 109–112 .
70. Watts A. Nature Reviews Drug Discovery 2005; 4:555–568.
71. Aliev AE, Law RV. Nucl Magn Reson 2001; 30:214–310.
72. Aliev AE, Law RV. Nucl Magn Reson 2002; 31:225–288.
73. Aliev AE, Law RV. Nucl Magn Reson 2003; 32:238–291.
74. Aliev AE, Law RV. Nucl Magn Reson 2004; 33:233–305.
75. Aliev AE, Law RV. Nucl Magn Reson 2005; 34:253–327.
76. McGregor AC, Lukes PJ, Osman JR, Crayston JA. J Chem Soc, Perkin Trans 2 1995; 809–813.
77. Bennett AE, Rienstra CM, Auger M, Lakshmi KV. J Chem Phys 1995; 103: 6951–6958.
78. Gan Z, Ernst RR. Solid State Nucl Magn Reson 1997; 8:153–159.
79. Fung BM, Khitrin AK, Ermolaev K. J Magn Reson 2000; 142:97–101.
80. Ernst M, Zimmermann H, Meier BH. Chem Phys Lett 2000; 317:581–588.
81. Detken A, Hardy HE, Ernest M, Meier BH. Chem Phys Lett 2002; 356:298–304.
82. Khitrin AK, Fung BM, McGeorge G. Encyclopedia Nucl Magn Reson 2002; 9:91–98.
83. Ernst M, Samoson A, Meier BH. J Magn Reson 2003; 163:332–339.
84. Ashida J, Asakura T. J Magn Reson 2003; 165:180–183.
85. Ernst M. J Magn Reson 2003; 162:1–34.
86. Ernst M, Meier MA, Tuherm T, Samoson A, Meier BH. J Am Chem Soc 2004; 126:4764–4765.
87. Brauniger T, Wormald P, Hodgkinson P. Monatshefte fur Chemie 2002; 133: 1549–1554.
88. Paepe GD, Hodgkinson P, Emsley L. Chem Phys Lett 2003; 376:259–267.
89. Paepe GD, Giraud N, Lesage A, Hodgkinson P, Boeckmann A, Emsley L. J Am Chem Soc 2003; 125:13938–13939.
90. Paepe GD, Lesage A, Emsley L. J Chem Phys 2003; 119:4833–4841.
91. Waugh JS, Huber LM, Haeberlen U. Phys Rev Lett 1968; 20:180.
92. Burum DP, Rhim WK. J Chem Phys 1979; 71:944–956.
93. Burum DP, Linder M, Ernst RR. J Magn Reson 1981; 44:173–188.
94. Cory DG. J Magn Reson 1991; 94:526–534.
95. Demco DE, Hafner S, Spiess HW. J Magn Reson Ser A 1995; 116:36–45.
96. Nanz D, Ernst M, Hong M, Ziegeweid MA, Schmidt-Rohr K. J Magn Reson Ser A 1995; 113:169–176.

97. Hafner S, Spiess HW. J Magn Reson Ser A 1996; 121:160–166.
98. Hafner S, Spiess HW. Concepts Magn Reson 1998; 10:99–129.
99. Sakellariou D, Hodgkinson P, Lesage A, Emsley L. Chem Phys Lett 2000; 319:253–260.
100. Yamauchi K, Kuroki S, Ando I. J Mol Struct 2002; 602–603:9–16.
101. Igumenova TI, A EM. J Magn Reson 2003; 164:270–285.
102. Elena B, de Paepe G, Emsley L. Chem Phys Lett 2004; 398:532–538.
103. Lesage A, Sakellariou D, Hediger S, Elena B, Charmont P, Steuernagel S, Emsley L. J Magn Reson 2003; 163:105–113.
104. Lee M, Goldburg WI. Phys Rev Lett 1965; 20:180.
105. Rhim WK, Elleman DD, Vaughan RW. J Chem Phys 1973; 59:3740–3749.
106. Rhim WK, Elleman DD, Schreiber LB, Vaughan RW. J Chem Phys 1974; 60: 4595–4604.
107. Takegoshi K, McDowell CA. Chem Phys Lett 1985; 116:100–104.
108. Hohwy M, Nielsen NC. J Chem Phys 1997; 106:7571–7586.
109. Eden M, Levitt MH. J Chem Phys 1999; 111:1511.
110. Carravetta M, Eden M, Zhao X, Brinkmann A, Levitt MH. Chem Phys Lett 2000; 321:205–215.
111. Dixon WT. J Chem Phys 1982; 77:1800.
112. Dixon WT, Schaefer J, Sefcik MD, Stejskal EO, McKay RA. J Magn Reson (1969–1992) 1982; 49:341–345.
113. Opella SJ, Frey MG. J Am Chem Soc 1979; 101:5854–5856.
114. Wu X, Zilm KW. J Magn Reson Ser A 1993; 102:205–213.
115. Wu X, Zilm KW. J Magn Reson Ser A 1993, 104, 119–122.
116. Wu X, Burns ST, Zilm KW. J Magn Reson Ser A 1994; 111:29–36.
117. Burns ST, Wu X, Zilm KW. J Magn Reson 2000; 143:352–359.
118. Rossi P, Subramanian R, Harbison GS. J Magn Reson 1999; 141:159–163.
119. Kumashiro KK, Niemczura WP, Kim MS, Sandberg LB. J Biomol NMR 2000; 18:139–144.
120. DeVita E, Frydman L. J Magn Reson 2001; 148:327–337.
121. Burum DP, Bielecki A. J Magn Reson 1991; 95:184–190.
122. Sethi NK. J Magn Reson 1991; 94:352–361.
123. Sangill R, Rastrup-Andersen N, Bildsoe H, Jakobsen HJ, Nielsen NC. J Magn Reson Ser A 1994; 107:67–78.
124. Hu JZ, Harper JK, Taylor C, Pugmire RJ, Grant DM. J Magn Reson 2000; 142:326–330.
125. Hu JZ, Taylor CMV, Pugmire RJ, Grant DM. J Magn Reson 2001; 152:7–13.
126. Schmidt-Rohr K, Mao J-D. J Am Chem Soc 2002; 124:13938–13948.
127. Mao J-D, Schmidt-Rohr K. J Magn Reson 2003; 162:217–227.
128. Schmidt-Rohr K, Mao J-D. Chem Phys Lett 2002; 359:403–411.
129. Caldarelli S, Ziarelli F. J Am Chem Soc 2000; 122:12015–12016.
130. Lesage A, Steuernagel S, Emsley L. J Am Chem Soc 1998; 120:7095–7100.
131. Sakellariou D, Lesage A, Emsley L. J Magn Reson 2001; 19:40–47.
132. Duma L, Hediger S, Brutscher B, Boeckmann A, Emsley L. J Am Chem Soc 2003; 125:11816–11817.
133. Duma L, Hediger S, Lesage A, Emsley L. J Magn Reson 2003; 164:187–195.
134. Carravetta M, Zhao X, Bockmann A, Levitt MH. Chem Phys Lett 2003; 376:515–523.

135. Khitrin AK, Canlet C, Fung BM. Solid-State NMR Spectrosc 2001; 19:63–72.
136. Schmidt-Rohr K, Clauss J, Spiess HW. Macromolecules 1992; 25:3273–3277.
137. Gerardy-Montouillout V, Malveau C, Tekely P, Olender Z, Luz Z. J Magn Reson A 1996; 123(1):7–15.
138. Deazevedo ER, Hu W-G, Bonagamba TJ, Schmidt-Rohr K. J Am Chem Soc 1999; 121:8411–8412.
139. Deazevedo ER. J Chem Phys 2000; 112:8988–9001.
140. Reichert D, Baonagamba TJ, Schmidt-Rohr K. J Magn Reson 2001; 151:129–135.
141. Rossum BJv, Forster H, Groot HJMd. J Magn Reson 1997; 124:516–519.
142. Yao XL, Schmidt-Rohr K, Hong M. J Magn Reson 2001; 149:139–143.
143. Schmidt-Rohr K, Mao J-D. J Magn Reson 2002; 157:210–217.
144. Liu S-F, Mao J-D, Schmidt-Rohr K. J Magn Reson 2002; 155:15–28.
145. Mao J-D, Xing B, Schmidt-Rohr K. Environ Sci Technol 2001; 35:1928–1934.
146. Mao J-D, Hundal JS, Schmidt-Rohr K, Thompson ML. Environ Sci Technol 2003; 37:1751–1757.
147. Sozzani P, Simonutti R, Bracco S, Comotti A. Polym Prepr (Am Chem Soc, Div Polym Chem) 2003; 44:297–298.
148. DeLacroix SF, Titman JJ, Haemeyer A, Spiess HW. J Magn Reson 1992; 97:435–443.
149. Smith J, MacNamara E, Raftery D, Borchardt T, Byrn S. J Am Chem Soc 1998; 120:11710–11713.
150. Harper JK, McGeorge G, Grant DM. J Am Chem Soc 1999; 121:6488–6496.
151. Barich DH, Orendt AM, Pugmire RJ, Grant DM. J Phys Chem A 2000; 104:8290–8295.
152. Barich DH, Facelli JC, Hu JZ, Alderman DW, Wang W, Pugmire RJ, Grant DM. Magn Reson Chem 2001; 39:115–121.
153. Barich DH, Pugmire RJ, Grant DM, Iuliucci RJ. J Phys Chem A 2001; 105:6780–6784.
154. Strohmeier M, Orendt AM, Alderman DW, Grant DM. J Am Chem Soc 2001; 123:1713–1722.
155. Grant DM. Encyclopedia Nucl Magn Reson 2002; 9:73–90.
156. Barich DH, Hu JZ, Pugmire RJ, Grant DM. J Phys Chem A 2002;106:6477–6482.
157. Strohmeier M, Alderman DW, Grant DM. J Magn Reson 2002; 155:263–277.
158. Clawson JS, Strohmeier M, Stueber D, Orendt AM, Barich DH, Asay B, Hiskey MA, Pugmire RJ, Grant DM. J Phys Chem A 2002;106:6352–6357.
159. Ma Z, Barich DH, Solum MS, Pugmire RJ. J Agric Food Chem 2004; 2:215–221.
160. Strohmeier M, Grant DM. J Magn Reson 2004; 168:296–306.
161. Elena B, Hediger S, Emsley L. J Magn Reson 2003; 160:40–46.
162. Shao L, Crockford C, Geen H, Grasso G, Titmann JJ. J Magn Reson 2004; 167:75–86.
163. Gullion T, Schaeffer J. J Magn Reson 1989; 81:196–200.
164. Guillon T, Schaefer J. Adv Magn Reson 1989; 13:57–83.
165. Gullion T, McKay RA, Schmidt A. J Magn Reson (1969–1992) 1991; 94:362–369.
166. Gullion T, Schaefer J. J Magn Reson (1969–1992) 1991; 92:439–442.
167. Li Y, Evans JNS. J Chem Phys 1994; 101:10211–10216.
168. Mueller KT. J Magn Reson Ser A 1995; 113:81–93.
169. Tong G, Pan Y, Dong H, Pryor R, Wilson GE, Shaefer J. Biochemistry 1997; 36:9859–9866.

170. Tong G, Schaefer J. Macromolecules 1997; 30:7522–7528.
171. Mehta AK, Hirsh DJ, Oyler N, Drobny GP, Schaefer J. J Magn Reson 2000; 145:156–158.
172. Jaroniec, Younge, Tienstra, Herzfeld, Hriffin. J Magn Reson 2000; 146:132–139 .
173. Nishimura K, Fu R, Cross TA. J Magn Reson 2001; 152:227–233.
174. Kim SJ, Cegelski L, Studelska DR, O'connor RD, Mehta AK, Schaefer J. Biochemistry 2002; 41:6967–6977.
175. Bennett AE, Rienstra CM, Lansbury PT, Griffin RG. J Chem Phys 1996; 105:10289–10299.
176. Chan JC. Chem Phys Lett 2001; 335:289–297.
177. Vogt, Mattingly, Gibson, Mueller. J Magn Reson 2000; 147:26–35.
178. Petkova AT, Tycko R. J Magn Reson 2002; 155:293–299.
179. Gullion T, Pennington CH. Chem Phys Lett 1998; 290:88–93.
180. Jaroniec CP, Tounge BA, Herzfeld J, Griffin RG. J Amer Chem Soc 2001; 123:3507–3519.
181. Jaroniec CP, Filip C, Griffin RG. J Am Chem Soc 2002; 124:10728–10742.
182. Vogt FG, Gibson JM, Mattingly SM, Mueller KT. J Phys Chem B 2003; 107:1272–1283.
183. Grey CP, Veeman WS. Chem Phys Lett 1992; 192:379–385.
184. Blumenfeld AL, Coster DJ, Fripiat JJ. Chem Phys Lett 1994; 231:491–498.
185. Yap AT-W, Forster H, Elliot SR. Phys Rev Lett 1995; 75:3946–3949.
186. Grey CP, Vega AJ. J Am Chem Soc 1995; 117:8232–8242.
187. Kao H-M, Grey CP. J Phys Chem 1996; 100:5105–5117.
188. Deng F, Yue Y, Ye C. Solid State Nucl Magn Reson 1998; 10:151–160.
189. Ba Y, Kao H-M, Grey CP, Chopin L, Gullion T. J Magn Reson 1998; 133:104–114.
190. Hughes E, Gullion T, Goldbourt A, Vega S, Vega AJ. J Magn Reson 2002; 156:230–241.
191. Sack I, Balzs YS, Rahimipour S, Vega S. J Am Chem Soc 2000; 122:12263–12269.
192. Fu R. Chem Phys Lett 2003; 376:62–67.
193. Sinha N, Schmidt-Rohr K, Hong M. J Magn Reson 2004; 168:358–365.
194. Saalwachter K, Spiess HW. J Chem Phys 2001; 114:5707–5728.
195. Saalwachter K, Graf R, Spiess HW. J Magn Reson 2001; 148:398–418.
196. Saalwachter K, Schnell I. Solid State Nucl Magn Reson 2002; 22:154–187.
197. Schnell J, Langer B, Sontjens SHM, Gendereen MHPV, Sijbesma RP, Spiess HW. J Magn Reson 2001; 150:57–70.
198. Schnell I, Saalwachter K. J Am Chem Soc 2002; 124:10938–10939.
199. Ishii Y, Tycko R. J Magn Reson 2000; 142:199–204.
200. Ishii Y, Yesinowski JP, Tycko R. J Am Chem Soc 2001; 123:2921–2922.
201. Hong M, Yamaguchi S. J Magn Reson 2001; 150:43–48.
202. Schmidt-Rohr K, Saalwachter K, Liu SF, Hong M. J Am Chem Soc 2001; 123:7168–7169.
203. Reif B, Griffin RG. J Magn Reson 2003; 160:78–83.
204. Paulson EK, Morcombe CR, Gaponenko V, Dancheck B, Byrd RA, Zilm KW. J Am Chem Soc 2003; 125:15831–15836.
205. Khitrin AK, Fung BM. J Magn Reson 2001; 152:185–188.
206. Verhoeven RA, Verel R, Meier BH. Chem Phys Lett 1997; 266:465–472.
207. Lesage A, Sakellariou D, Steuernagel S, Emsley L. J Am Chem Soc 1998; 120:13194–13201.

208. Lesage A, Emsley L. J Magn Reson 2001; 148:449–454.
209. Lesage A, Charmont P, Steuernagel S, Emsley L. J Am Chem Soc 2000; 122:9739–9744.
210. Hediger S, Lasage A, Emsley L. Macromolecules 2002; 35:5078–5084.
211. Hardy EH, Detken A, Meier BH. J Magn Reson 2003; 165:208–218.
212. Hardy ED, Verel R, Meier BH. J Magn Reson 2001; 148:459–464.
213. Baldus M, Iuliucci RJ, Meier BH. J Am Chem Soc 1997; 119:1121–1124.
214. Alonso B, Massiot D. J Magn Reson 2003; 163:347–352.
215. Lesage A, Auger C, Caldarelli S, Emsley L. J Am Chem Soc 1997; 119:7867–7868.
216. Brown SP, Perez-Torralba M, Sanz D, Claramunt RM, Emsley L. J Am Chem Soc 2002; 124:1152–1153.
217. Lesage A, Bardet M, Emsley L. J Am Chem Soc 1999; 121:10987–10993.
218. Fayon F, Saout GL, Emsley L, Massiot D. Chem Commun 2002; 16:1702–1703.
219. Sakellariou D, Brown SP, Lesage A, Hediger S, Bardet M, Meriles CA, Pines A, Emsley L. J Am Chem Soc 2003; 125:4376–4380.
220. Cadars S, Lesage A, Emsley L. J Am Chem Soc 2005; 127:4466–4476.
221. Murata K, Kono H, Katoh E, Kuroki S, Ando I. Polymer 2003; 44:4021–4027.
222. Kono H, Erata T, Takai M. Macromolecules 2003; 36:5131–5138.
223. Kono H, Numata Y. Polymer 2004; 45:4541–4547.
224. Kono H. Biopolymers 2004; 75:255–263.
225. Kono H, Numata Y, Erata T, Takai M. Polymer 2004; 45:2843–2852.
226. Kono H, Numata Y, Erata T, Takai M. Macromolecules 2004; 37:5310–5316.
227. Kaji H, Schmidt-Rhor K. Macromolecules 2002; 35:7993–8004.
228. Sommer W, Gottwald J, Demco DE, Spiess HW. J Magn Reson Ser A 1995; 113:131–134.
229. Schnell J, Lupulescu A, Hafner S, Demco DE, Spiess HW. J Magn Reson 1998; 133:61–69.
230. Feike M, Demco DE, Graf R, Gottwald J, Hafner S, Spiess HW. J Magn Reson 1996; 122:214–221.
231. Brown SP, Zhu XX, Saalwaechter K, Spiess HW. J Am Chem Soc 2001; 123:4275–4285.
232. Tycko R, Dabbagh G. Chem Phys Lett 1990;173:461–465.
233. Lee YK, Kurur ND, Helmle M, Johannessen OG, Nielsen NC, Levitt MH. Chem Phys Lett 1995; 242:304–309.
234. Karlsson T, Helmle M, Kurur ND, Levitt MH. Chem Phys Lett 1995; 247:534–540.
235. Ernst M, Bush S, Kolbert AC, Pines A. J Chem Phys 1996; 105:3387–3397.
236. Gregory DM, Wolfe GM, Jarvie TP, Sheils JC, Drobny GP. Mol Phys 1996; 89:1835–1849.
237. Gross JD, Warshawski DE, Griffin RG. J Am Chem Soc 1997; 119:796–802.
238. Baldus M. Solid State Nucl Magn Reson 1998; 11:157–168.
239. Gross JD, Costa PR, Griffin RG. J Chem Phys 1998; 108:7286–7293.
240. Hohwy M, Jakobsen HJ, Eden M, Levitt MH, Nielsen NC. J Chem Phys 1998; 108:2686–2694.
241. Dusold S, Sebald A. Annu Rep NMR Spectrosc 2000; 41:184–264.
242. Antzutkin NO, Levitt MH. J Magn Reson 2000; 147:147–151.
243. Takegoshi K, Nakamura S, Terao T. Chem Phys Lett 2001; 344:631–637.
244. Raya J, Bianco A, Furrer J, Briand J-P, Piotto M, Elbayed K. J Magn Reson 2002; 157:43–51.

245. Verel R, Ernst M, Meier BH. J Magn Reson 2001; 150:81–99.
246. Fujiwara T, Khandelwal P, Akutsu H. J Magn Reson 2000; 145:73–83.
247. Ishii Y. J Magn Reson 2001; 114:8473–8483.
248. Goobes G, Vega S. J Magn Reson 2002; 154:236–251.
249. Kristiansen PE, Mitchell DJ, Evans JNS. J Magn Reson 2002; 157:253–266.
250. Williamson PTF, Verhoeven A, Ernst M, Meier BH. J Am Chem Soc 2003; 125:2718–2722.
251. Takegoshi K, Nakamura S, Terao T. J Chem Phys 2003; 118:2325–2341.
252. Thieme K, Schnell I. J Am Chem Soc 2003; 125:12100–12101.
253. Karlsson T, Popham JM, Long JR, Oyler N, Drobny GP. J Am Chem Soc 2003; 125:7394–7407.
254. Bjerring M, Nielsen NC. Chem Phys Lett 2003; 382:671–678.
255. Dvinskikh SV, Zimmermann H, Maliniak A, Sandstrom D. J Magn Reson 2004; 168:194–201.
256. Matsuki Y, Akutsu H, Fujiwara T. Magn Reson Chem 2004; 42:291–300.
257. Verhoeven A, Williamson PTF, Zimmermann H, Ernst M, Meier BH. J Magn Reson 2004; 168:314–326.
258. Dvinskikh SV, Zimmermann H, Maliniak A, Sandstroem D. J Chem Phys 2005; 122:044512/044511–044512/044512.
259. Ramamoorthy A, Gierasch LM, Opella SJ. J Magn Reson Ser B 1995; 109:112–116.
260. Peersen OB, Groesbeek M, Aimoto S, Smith SO. J Am Chem Soc 1995; 117:7228–7237.
261. Fujiwara T, Sugase K, Kamosho M, Ono A, Akutsu H. J Am Chem Soc 1995; 117:11351–11352.
262. Gu Z, Ridenour CF, Bronnimann CE, Iwashita T, Mcdermott A. J Am Chem Soc 1996; 118:822–829.
263. Feng X, Eden M, Brinkmann A, Luthman H, Eriksson L, Graslund A, Antzutkin ON, Levitt MH. J Am Chem Soc 1997; 119:12006–12007.
264. Verdegem PJE, Helmle M, Lugtenburg J, Groot HJMD. J Am Chem Soc 1997; 119:169–174.
265. Hong M, G Griffin, R. J Am Chem Soc 1998; 120:7113–7114.
266. Terao T. J Mol Struct 1998; 441:283–294.
267. Middleton DA, Robins R, Feng X, Levitt MH, Spiers ID, Schwalbe CH, Reid DG, Watts A. FEBS Lett 1997; 410:269–274.
268. Ishii Y, Hirao K, Terao T, Terauchi T, Oba M, Nishiyama K, Kainosho M. Solid State Nucl Magn Reson 1998; 11:169–175.
269. Tian F, Song Z, Cross TA. J Magn Reson 1998; 135:227–231.
270. Wilhelm M, Feng H, Tracht U, Spiess HW. J Magn Reson 1998; 134:255–260.
271. Williamson PTF, Grobner G, Spooner PJR, Miller KW, Watts A. Biochemistry 1998; 37:10854–10859.
272. Nishimura K, Naito A, Tuzi S, Saito H, Hashimoto C, Aida M. J Phys Chem B 1998; 102:7476–7483.
273. Baldus M, Petkova AT, Herzfeld J, Griffin RG. Mol Phys 1998; 95:1197–1207.
274. Verel R, Beek JDv, Meier BH. J Magn Reson 1999; 140:300–303.
275. Verdegem PJE, Bovee-Geurts PHM, Grip WJD, Lugtenburg J, Groot HJMD. Biochemistry 1999; 38:11316–11324.
276. Hong M. J Magn Reson 1999; 136:86–91.

277. Rossum B-Jv, Groot CPd, Ladizhansky V, Vega S, Groot HJMd. J Am Chem Soc 2000; 122:3465–3472.
278. Robyr P, Meier BH. Chem Phys Letts 2000; 327:319–324.
279. Mcdermott A, Polenova T, Bockmann A, Zilm KW, Paulsen EK, Martin RW, Montelione GT. J Biomol NMR 2000; 16:209–219.
280. Rienstra CM, Hohwy M, Hong M, Griffin RG. J Am Chem Soc 2000; 122:10979–10990.
281. Ishii Y, Tycko R. J Am Chem Soc 2000; 122:1443–1455.
282. Heindrichs ASD, Geen H, Titman JJ. J Magn Reson 2000; 147:26–35.
283. Huster D, Yamaguchi S, Hong M. J Am Chem Soc 2000; 122:11320–11327.
284. Yang J, Parkanzky PD, Khunte BA, Canlas CG, Yang R, Gabrys CM, Weliky DP. J Mol Graphics 19.
285. Carravetta M, EdeÇn M, Johannessen OG, Luthman H, Verdegem PJE, Lugtenburg J, Sebald A, Levitt MH. J Am Chem Soc 2001; 123:10628–10638.
286. Chan JCC, Brunklaus G. Chem Phys Lett 2001; 349:104–112.
287. Alia, Matysik J, Soede-huijbregts C, Baldus M, Raap J, Lugtenburg J, Gast P, Gorkom HJV, Hoff AJ, Groot HJMD. J Am Chem Soc 2001; 123:4803–4809.
288. Heindrichs, Geen, Giordani, Titman Chem Phys Lett 2001; 335:89–96.
289. Lange A, Luca S, Baldus M. J Am Chem Soc 2002; 124:9704–9705.
290. Asakura T, Yao J, Yamane T, Umemura K, Ulrich AS. J Am Chem Soc 2002; 124:8794–8795.
291. Vosegaard T, Nielsen NC. J Biomol NMR 2002; 22:225–247.
292. Rienstra CM, Hohwy M, Mueller LJ, Jaroniec CP, Reif B, Griffin RG. J Am Chem Soc 2002; 124:11908–11922.
293. Mueller LJ, Elliott DW, Kim K-C, Reed CA, Boyd PDW. J Am Chem Soc 2002; 124:9360–9361.
294. Wei Y, Lee D-K, McDermott AE, Ramamoorthy A. J Magn Reson 2002; 158:23–35.
295. Boer ID, Bosman L, Raap J, Oschkinat H, Groot HJMD. J Magn Reson 2002; 157:286–291.
296. Ladizhansky V, Griffin RG. J Am Chem Soc 2004; 126:948–958.
297. Katoh E, Takegoshi K, Terao T. J Am Chem Soc 2004; 126:3653–3657.
298. Carravetta M, Zhao X, Johannessen Ole G, Lai Wai C, Verhoeven Michiel A, Bovee-Geurts Petra HM, Verdegem Peter JE, Kiihne S, Luthman H, de Groot Huub JM, et al. J Am Chem Soc 2004; 126:3948–3953.
299. Padden BE. In: Graduate School. University of Minnesota, 2000:221.
300. Ando S, Hironaka T, Kurosu H, Ando I. Magn Reson Chem 2000; 38:241–250.
301. Maciejewska D, Herold F, Wolska I. J Mol Struct 2000; 553:73–77.
302. Buchanan GW, Astegar MF, G PAY. J Mol Struct 2001; 561:43–54.
303. Dega-Szafran Z, Gaszczyk I, Maciejewska D, Szafran M, Tykarska E, Wawer I. J Mol Struct 2001; 560:261–273.
304. Harper JK, Mulgrew AE, Li JY, Barich DH, Strobel GA, Grant DM. J Am Chem Soc 2001; 123:9837–9842.
305. Cyranski MK, Wawer I, Zielinska A, Mrozek A, Koleva V, Lozanova C. J Phys Org Chem 2001; 14:323–327.
306. Olejniczak S, Ganicz K, Tomczykowa M, Gudej J, Portzebowski MJ. J Chem Soc, Perkin Trans 2 2002:1059–1065.

307. Rajeswaren M, Blanton TN, Zumbulyadis N, Giensen DJ, Conesa-Moratilla C, Misture ST, Stephens PW, Huq A. J Am Chem Soc 2002; 124:14450–14459.
308. Aimi K, Fujiwara T, Ando S. J Mol Struct 2002; 602–603:405–416.
309. Aimi K, Fujiwara T, Ando S. J Mol Struct 2002; 602–603:417–428.
310. Claramunt RM, Cornago P, Sanz D, Foces-Foces C, Alkorta I, Elguerdo J. J Mol Struct 2002; 605:199–212.
311. Marek R, Brus J, Tousek J, Kovacs L, Hockova K. Magn Reson Chem 2002; 40:353–360.
312. Goward GR, Sebastiani D, Schnell I, Spiess HW, Kim H-D, Ishida H. J Am Chem Soc 2003; 125:5792–5800.
313. Helluy X, Sebald A. J Phys Chem B 2003; 107:3290–3296.
314. Zolek T, Paradowska K, Wawer I. Solid State Nucl Magn Reson 2003; 23:77–87.
315. Harper JK, Barich DH, Hu JZ, Strobel GA, Grant DM. J Org Chem 2003; 68:4609–4614.
316. Olsen RA, Struppe J, Elliott DW, Thomas RJ, Mueller LJ. J Am Chem Soc 2003; 125:11784–11785.
317. Mueller LJ, Elliott DW, Leskowitz GM, Struppe J, Olsen RA, Kim K-C, Reed CA. J Magn Reson 2004; 168:327–335.
318. Sun H, Oldfield E. J Am Chem Soc 2004; 126:4726–4734.
319. Mehta MA, Fry EA, Eddy MT, Dedeo MT, Anagnost AE, Long JR. J Phys Chem B 2004; 108:2777–2780.
320. Giavani T, Bildsoe H, Skibsted J, Jakobsen HJ. J Magn Reson 2004; 166:262–272.
321. Mao JD, Hu WG, Schmidt-Rohr K, Davies G, Ghabbour EA, Xing B. Soil Sci Soc Am J 2000; 64:873–884.
322. Hu W-G, Mao J, Schmidt-Rohr K, Xing B. Spec Publ-R Soc Chem 1999; 247:63–68.
323. Hazendonk P, K Harris, R, Galli G, Pizzanelli S. Physical Chemistry and Chemical Physics 2002; 4:507–513.
324. Sullivan MJ, Maclel GE. Anal Chem 1982; 54:1615–1623.
325. Harris RK. Analyst 1985; 110:649–655.
326. Suryanarayanan R, Wiedmann TS. Pharm Res 1990; 7:184–187.
327. Likar MD, Taylor RJ, Fagerness PE, Hiyama Y, Robins RH. Pharm Res 1993; 10:75–79.
328. Gao P. Pharm Res 1996; 13:1095–1104.
329. Gao P. Pharm Res 1998; 15:1425–1433.
330. Gustafsson C, Lennholm H, Iversen T, Nystrom C. Intern J Pharm 1998; 174:243–252.
331. Lee GSH, Craig DC, Kannangara GSK, Dawson M, Conn C, Robertson J, Wilson MA. J Forensic Sci 1999; 44:761–771.
332. Paris M, Bizot H, Emery J, Buzare JY, Buleon A. Int J Biol Macromol 2001; 29:127–136.
333. Paris M, Bizot H, Emery J, Buzare JY, Buleon A. Int J Biol Macromol 2001; 29:137–143.
334. Kono H, Yunoki S, Shikano T, Fujiwara M, Erata T, Takai M. J Am Chem Soc 2002; 124:7506–7511.
335. Kono H, Erata T, Takai M. J Am Chem Soc 2002; 124:7512–7518.
336. Kono H, Erata T, Takai M. Macromolecules 2003; 36:3589–3592.

337. Tozuka Y, Ito A, Seki H, Oguchi T, Yamamoto K. Chem Pharm Bull 2002; 50:1128–1130.
338. Vickery RD, Nemeth GA, Maurin MB. J Pharm Biomed Anal 2002; 30:125–129.
339. Apperley DC, Harris RK, Larsson T, Malmstrom T. J Pharm Sci 2003; 92:2487–2494.
340. Fu R, Hu J, Cross TA. J Magn Reson 2004; 168:8–17.
341. Byard SJ, Jackson SL, Smail A, Bauer M, Apperley DC. J Pharm Sci 2005; 94:1321–1335.
342. Smernik RJ, Oades JM. Geoderma 2000; 96:101–129.
343. Smernik RJ, Oades JM. Solid State Nucl Magn Reson 2001; 20:74–84.
344. Prauchner MJ, Pasa VMD, Menezes SMCd. J Wood Chem Technol 2001; 21:371–385.
345. Smernik RJ, Baldock JA, Oades JM, Whittaker AK. Solid State Nucl Magn Reson 2002; 22:50–70.
346. Smernik RJ, Baldock JA, Oades JM. Solid State Nucl Magn Reson 2002; 22:71–82.
347. Smernik RJ, Oades JM. Eur J Soil Sci 2003; 54:103–116.
348. Smernik RJ, Oliver IW, Merrington G. J Environ Qual 2003; 32:1523–1533.
349. Smernik RJ. Geoderma 2005; 125:249–271.
350. Bardet M, Foray MF, Tran Q-K. Anal Chem 2002; 74:4386–4390.
351. Mao J-D, Hu WG, Schmidt-Rohr K, Ding G, Davies G, Ghabbour EA, Xing B. Int J Environ Anal Chem 2002; 82:183–196.
352. Keeler C, Maciel GE. Anal Chem 2003; 75:2421–2432.
353. Van Lagen B, de Jager PA. Fresenius Eniron Bull 2003; 12:1211–1217.
354. Peuravuori J, Ingman P, Pihlaja K. Talanta 2003; 59:177–189.
355. Conte P, Piccolo A, Lagen Bv, Buurman P, Hemminga MA. Solid State Nucl Magn Reson 2002; 21:158–170.
356. Khalaf M, Kohl SD, Klumpp E, Rice JA, Tombacz E. Environ Sci Technol 2003; 37:2855–2860.
357. Mao JD, Schmidt-Rohr K. Environ Sci Technol 2004; 38:2680–2684.
358. Hexem JG, Frey MH, Opella SJ. J Am Chem Soc 1981; 103:224–226.
359. Hexem JG, Frey MH, Opella SJ. J Chem Phys 1982; 77:3847–3856.
360. Olivieri AC, Frydman L, Diaz LE. J Magn Reson (1969–1992) 1987; 75:50–62.
361. Olivieri AC, Fyrdman L, Grasselli M, Diaz LE. Magn Reson Chem 1988; 26:281–286.
362. Olivieri AC. J Magn Reson (1969–1992) 1989; 82:342–346.
363. Olivieri AC. J Magn Reson (1969–1992) 1989; 81:201–205.
364. Olivieri AC. J Chem Soc, Perkin Trans 2: Phys Org Chem (1972–1999) 1990; 85–89.
365. Frydman L, Olivieri AC, Diaz LE, Frydman B, Schmidt A, Vega S. Mol Phys 1990; 70:563–579.
366. Grasselli M, Diaz LE, Olivieri AC. Spectrosc Lett 1991; 24:895–907.
367. Harris RK, Olivieri AC. Prog Nucl Magn Reson Spectrosc 1992; 24:435–456.
368. Olivieri AC. J Magn Reson, Ser A 1993; 101:313–316.
369. Alarcon SH, Olivieri AC, Jonsen P. J Chem Soc, Perkin Trans 2: Phys Org Chem (1972–1999) 1993; 1783–1786.
370. Olivieri AC, Elguero J, Sobrados I, Cabildo P, Claramunt RM. J Phys Chem 1994; 98:5207–5211.

371. Odgaard L, Bak M, Jakobsen HJ, Nielsen NC. J Magn Reson 2001; 148: 298-308.
372. Hughes CE, Pratima R, Karlsson T, Levitt MH. J Magn Reson 2002; 159:25-35.
373. Strohmeier M, Grant DM. J Am Chem Soc 2004; 126:966-977.
374. Takegoshi K, Yano T, Takeda K, Terao T. J Am Chem Soc 2001; 123:10786-10787.
375. Stephenson GA, Stowell JG, Toma PH, Dorman DE,Greene JR, Byrn SR. J Am Chem Soc 1993; 116:5766-5773.
376. McGeorge G, Harris RK, Chippendale AM, Bullock JF. J Chem Soc, Perkin Trans 2 1996; 2:1733-1738.
377. Schmidt A, Kababya S, Appel M, Khatib S, Botoshansky M, Eichen Y. J Am Chem Soc 1999; 121:11291-11299.
378. Glaser R, Shiftan D, Drouin M. J Org Chem 1999; 64:9217-9224.
379. Apperley DC, Fletton RA, Harris RK, Lancaster RW, Tavener S, Threlfall TL. J Pharm Sci 1999; 88:1275-1280.
380. Dong Z, VG Young, J, Padden BE, Schroeder SA, Prakash I, Munson EJ, Grant DJW. J Chem Crystallography 1999; 29:967-975.
381. Padden BE, Zell MT, Dong Z, Schroeder SA, Grant DJW, Munson E. Anal Chem 1999; 71:3325-3331.
382. Zell MT, Padden BE, Grant DJW, Chapeau MC, Prakash I, Munson E. J Am Chem Soc 1999; 121:1372-1378.
383. Zell MT, Padden BE, Grant DJW, Schroeder SA, Wachholder KL, Prakash I, Munson EJ. Tetrahedron 2000; 56:6603-6616.
384. Dong Z, Padden BE, Salsbury JS, Munson EJ, Shroeder SA, Prakash I, Grant DJW. Pharm Res 2002; 19:330-336.
385. Dong Z, Salsbury JS, Munson EJ, Shroeder SA, Zhou D, Prakash I, Vyazovkin S, Wight CA, Grant DJW. J Pharm Sci 2002; 91:1423-1431.
386. Sack I, Macholl S, Wehrmann F, Albrecht J, Limbach HH, Fillaux F, Baron MH, Bunkowsky G. Appl Magn Reson 1999; 17:413-431.
387. Lee H-H, Kimura K, Takai T, Senda H, Kuwae A, Hanai K. Spectrochim Acta, Part A 1999; 55:2877-2882.
388. Chen LR, Padden BE, Vippagunta SR, Munson E, Grant DJW. Pharm Res 2000; 17:619-624.
389. Henck JO, Finner E, Burger A. J Pharm Sci 2000; 89:1151-1159.
390. Stephenson GA. J Pharm Sci 2000; 89:958-966.
391. Medek A, Frydman L. J Am Chem Soc 2000; 122:684-691.
392. Variankaval NE, Jacob KI, Dinh SM. J Cryst Growth 2000; 217:320-331.
393. Wenslow RM, Baum MW, Ball RG, Mccauley JA. J Pharm Sci 2000; 89:1271-1285.
394. Lee GSH, Taylor RC, Dawson M, Kannangara GSK, Wilson MA. Solid State Nucl Magn Reson 2000; 16:225-237.
395. Crowley KJ, Forbes RT, York P, Apperley DC, Nyqvist H, Camber O. J Pharm Sci 2000; 89:1286-1295.
396. Harper JK, Grant DM. J Am Chem Soc 2000; 122:3708-3714.
397. Strohmeier M, Orendt AM, Alderman DW, Grant DM. J Am Chem Soc 2001; 123:1713-1722.
398. Hanai K, Kuwae A, Takai T, Senda H, Kunimoto K-K. Spectrochimica Acta Part A 2001; 57:513-519.
399. Fojud Z, Jurga S. Mol Phys Rep 2001; 33:172-174.

400. Yoshinari T, Forbes RT, York P, Kawashima Y. Int J Pharm 2002; 247:69–77.
401. Bauer J, Spanton S, Henry R, Quick J, Dziki W, Porter W, Morris J. Pharm Res 2001; 18:859–866.
402. Liang J, Ma Y, Chen B, Munson EJ, Davis HT, Binder D, Chang H-T, Abbas S, Hsu F-L. J Phys Chem B 2001; 105:9653–9662.
403. Pursch M, Sander LC, Albert K. Anal Chem 1996; 68:4107–4113.
404. Giordano F, Rossi A, Moyano JR, Gazzaniga A, Massarotti V, Bini M, Capsoni D, Peveri T, Redenti E, Carima L, et al. J Pharm Sci 2001; 90:1154–1163.
405. Kimura N, Fukui H, Takagaki H, Yonemochi E, Terada K. Chem Pharm Bull 2001; 49:1321–1325.
406. Wiegerinck P, Booy K-J, Kellenbach E, Lambregts D, Vromans H, Vadar J, Kaspersen F. Synthesis and Applications of Isotopically Labelled Compounds. Proceedings of the International Symposium, 7th, Dresden, Germany, June 18–22, 2000, 2001, Vol. 7, 185–188.
407. Lee JH, Gi U-S, Kim J-H, Kim Y, Kim S-H, Oh H, Min B. Bull Korean Chem Soc 2001; 22:925–928.
408. Smith EDL, Hammond RB, Jones MJ, Roberts KJ, John JBO, Price SL, Harris RK, Apperley DC, Cherryman JC, Docherty R. J Phys Chem B 2001; 105:5818–5826.
409. Wang X, Wang XJ, Ching CB. Chirality 2002; 14:318–324.
410. Zhang GGZ, Gu C, Zell MT, Burkhardt RT, Munson EJ, Grant DJW. J Pharm Sci 2002; 91:1089–1100.
411. Moynihan HA, O'Hare IP. Int J Pharm 2002; 247:179–185.
412. Garcia MA, Lopez C, Claramunt RM, Kenz A, Pierrot M, Elguero J. Helvetica Chimica Acta 2002; 85:2763–2776.
413. Novoselsky A, Glaser R. Magn Reson Chem 2002; 40:723–728.
414. Zimmermann H, Bader V, Poupko R, Wachtel EJ, Luz Z. J Am Chem Soc 2002; 124:15286–15300.
415. De Jong AF, Kentgens APM, Veeman WS. Chem Phys Lett 1984; 109:337–342.
416. Potrzebowski MJ, Tadeusiak E, Misiura K, Ciesielski W, Bujacz G, Tekely P. Chem—Eur J 2002; 8:5007–5011.
417. Maurin MB, Vickery RD, Rabel SR, Rowe SM, Everlof JG, nemeth GA, Campbell GC, Foris CM. J Pharm Sci 2002; 91:2599–2604.
418. Wang Y, Wenslow RM, McCauley JA, Crocker LS. Int J Pharm 2002; 243:147–159.
419. Pisklak M, Perlinski M, Kossakowski J, Wawer I. Acta Poloniae Pharmaceutica 2002; 59:461–465.
420. Pisklak M, Maciejewska D, Herold F, Wawer I. J Mol Struct 2003; 649:169–176.
421. Komber H, Limbach H-H, Bohme F, Kunert C. J Am Chem Soc 2002; 124:11955–11963.
422. Harper JK, Facelli JC, Barich DH, McGeorge G, Mulgrew AE, Grant DM. J Am Chem Soc 2002; 124:10589–10595.
423. Meejoo S, Kariuki BM, Kitchin SJ, Cheung EY, Albesa-Jové D, Harris KDM. Helvetica Chimica Acta 2003; 86:1467–1477.
424. Iuliucci RJ, Clawson J, Hu JZ, Solum MS, Barich D, Grant DM, Taylor CMV. Solid State Nucl Magn Reson 2003; 24:23–38.
425. Alderman DW, McGeorge G, Hu JZ, Pugmire RJ, M.Grant D. Mol Phys 1998; 95:1113–1126.
426. Rienstra CM, Hatcher ME, Mueller LJ, Sun B, Fesik SW, Griffin RG. J Am Chem Soc 1998; 120:10602–10612.

427. Reutzel-Edens SM, Kleemann RL, Lewellen PL, Borghese AL, Antoine LJ. J Pharm Sci 2003; 92:1196–1205.
428. Zolek T, Paradowska K, Krajewska D, Rozanski A, Wawer I. J Mol Struct 2003; 646:141–149.
429. Vanhaecht B, Devroede J, Willem R, Biesemans M, Goonewardena W, Rastogi S, Hoffmann S, Klein PG, Koning CE. J Polym Sci, Part A: Polym Chem 2003; 41:2082–2094.
430. Portieri A, Harris RK, Fletton RA, Lancaster RW, Threlfall TL. Magn Reson Chem 2004; 42:313–320.
431. Kameda T, McGeorge G, Orendt AM, Grant DM. J Biomol NMR 2004; 29:281–288.
432. Remenar JF, Wenslow R, Ostovic D, Peresypkin A. Pharm Res 2004; 21: 185–188.
433. Sheth AR, Lubach JW, Munson EJ, Muller FX, Grant DJW. J Am Chem Soc 2005; 6641–6651.
434. Booy K-J, Wiegerinck P, Vader J, Kaspersen F, Lambregts D, Vromans H, Kellenbach E. J Pharm Sci 2005; 94:458–463.
435. Schmidt AC. Eur J Pharm Sci 2005; 25:407–416.
436. Suzuki N, Kawasaki T. J Pharm Biomed Anal 2005; 37:177–181.
437. Barich DH, Davis JM, Schrieber LJ, Zell MT, Munson EJ. In press.
438. Gilchrist LM, Monde K, Tomita Y, Iwashita T, Nakanishi K, McDermott AE. J Magn Reson 2001; 152:1–6.
439. Monde K, Tomita Y, ML Gilchrist, J, Mcdermott AE, Nakanishi K. Isr J Chem 2000; 40:301–306.
440. Antoniloli G, Hodgkinson P. J Magn Reson 2004; 168:124–131.
441. Campbell SC, Harris RK, Hardy MJ, Lee DC, Busby DJ. J Chem Soc, Perkin Trans 2 1997; 2:1913–1918.
442. Harris RK, Crowe LA. J Chem Soc, Dalton Trans 1999; 24:4315–4323.
443. Brouwer EB, Challoner R, Harris RK. Solid State Nucl Magn Reson 2000; 18:37–52.
444. Fuchs B, Scheler U. Macromolecules 2000; 33:120–124.
445. Lappan U, Fuchs B, Geibler U, Scheler U, Lunkwitz K. Polymer 2002; 43:4325–4330.
446. Dargaville TR, George GA, Hill DJT, Scheler U, Whittaker AK. Macromolecules 2003; 36:7138–7142.
447. Gleason KK, Hill DJT, Lau KKS, Mohajerani S, Whittaker AK. Nucl Instrum Methods Phys Res, Sect B: Beam Interactions With Materials and Atoms 2001; 185:83–87.
448. Holstein P, Scheler U, Harris RK. Polymer 1998; 39:4937–4941.
449. Ando S, Harrison RK, Reinsberg SA. Magn Reson Chem 2002; 40:97–106.
450. Ando S, Harris RK, Holstein P, Reinsberg SA, Yamauchi K. Polymer 2001; 42:8137–8151.
451. Su T-W, Tzou D-LM. Polymer 2000; 41:7289–7293.
452. Wormald P, Apperley DC, Beaume F, Harris RK. Polymer 2003; 44:643–651.
453. Ando S, Harris RK, Hirschinger J, Reinsberg SA, Scheler U. Macromolecules 2001; 34:66–75.
454. Wenslow RM. Drug Dev Ind Pharm 2002; 28:555–561.
455. Solis D, Ferraro MB, Facelli JC. J Mol Struct 2002; 602–603:159–164.

456. Glaser RW, Sachse C, Durr UHN, Wadhwani P, Ulrich AS. J Magn Reson 2004; 168:153–163.
457. Park J-W, Seo Y-A, Kim I, Ha C-S, Aimi K, Ando S. Macromolecules 2004; 37:429–436.
458. Budarin VL, Clark JH, Tavener SJ. Chem Commun (Cambridge, United Kingdom) 2004; 5:524–525.
459. Utz M. Annu Techn Conf-Soc Plast Eng 2002; 2:2015–2019.
460. Utz M, Atallah AS, Robyr P, Widmann AH, Ernst RR, Suter UW. Macromolecules 1999; 32:6191–6205.
461. Utz M, Robyr P, Suter UW. Macromolecules 2000; 33:6808–6814.
462. Utz M, Eisenegger J, Suter UW, Ernst RR. J Magn Reson 1997; 128:217–227.
463. Torchia DA, Szabo A. J Magn Reson (1969–1992) 1982; 49:107–121.
464. Torchia DA. J Magn Reson 1978; 30:613–616.
465. Schaefer J, Stejskal EO, Buchdahl R. Macromolecules 1977; 10:384–405.
466. Kitamaru R, Horii F, Murayama K. Macromolecules 1986; 19:636–643.
467. Oksanen CA, Zografi G. Pharm Res 1993; 10:791–799.
468. Dries IJvd, Dusschoten Dv, Hemminga MA. J Phys Chem B 1998; 102:10483–10489.
469. Yoshioka S, Aso Y, Kojima S. Pharm Res 1999; 16:135–140.
470. Yoshioka S, Aso Y, Kojima S, Sakurai S, Fujiwara T, Akutsu H. Pharm Res 1999; 16:1621–1625.
471. Zumbulyadis N, Antalek B, Windig W, Scaringe RP, Lanzafame AM, Blanton T, Helber M. J Am Chem Soc 1999; 121:11554–11557.
472. Silva NMd, Tavares MIB, Stejskal EO. Macromolecules 2000; 33:115–119.
473. Aso Y, Yoshioka S, Kojima S. J Pharm Sci 2000; 90:798–806.
474. Glaser R, Novoselsky A, Shiftan D. J Org Chem 2000; 65:6345–6353.
475. Kuwabara K, Horii F, Ogawa Y. J Mol Struct 2000; 525:163–171.
476. Varner SJ, Vold RL, Hoatson GL. J Magn Reson 2000; 142:229–240.
477. Kakou-Yao R, Pizzala H, Pietri N, Aycard J-P. J Chem Crystallography 2000; 30:593–598.
478. Rheingold AL, Figueroa JS, Dybowski C, Beckmann PA. Chem Commun 2000; 651–652.
479. Beckmann PA, Burbank KS, Clemo KM, Slonaker EN, Averill K, Dybowski C, Figueroa JS, Glatfelter A, Koch S, Liable-Sands LM, Rheingold AL. J Chem Phys 2000; 113:1958–1965.
480. Vittadini E, Dickinson LC, Chinachoti P. Carbohydr Polym 2001; 46:49–57.
481. Perera MCS, Ishiaku US, Ishak ZAM. Eur Polym J 2001; 37:167–178.
482. Tang HR, Belton PS, Davies SC, Hughes DL. Carbohydr Res 2001; 330: 391–399.
483. Higgins JS, Hodgson AW, Law RV. Polym Mater Sci Eng 2001; 84:990–991.
484. Lupulescu A, Kotecha M, Frydman L. J Am Chem Soc 2003; 125:3376–3383.
485. Lopes MH, Sarychev A, Neto CP, Gil AM. Solid State Nucl Magn Reson 2000; 16:109–121.
486. Tang H-R, Wang Y-L, Belton PS. Solid State Nucl Magn Reson 2000; 15:239–248.
487. Nogueira RF, Tavares MIB. J Appl Polym Sci 2002; 138–143.
488. Lai S, Casu M, Saba G, Lai A, Husu I, Masci G, Crescenzi V. Solid State Nucl Magn Reson 2002; 21:187–196.

489. Adriaensens P, Carleer R, Storme L, Vanderzande D, Gelan J. Polymer 2002; 43:7003–7006.
490. Brachais L, Laupretre F, Caille J-R, Teyssie D, Boileau S. Polymer 2002; 43:1829–1833.
491. Alamo RG, Blanco JA, Carrileo I, Fu R. Polymer 2002; 43:1857–1865.
492. Aso Y, Yoshioka S, Zhang J, Zografi G. Chem Pharm Bull 2002; 50:822–826.
493. Yoshioka S, Aso Y, Kojima S. J Pharm Res 2002; 91:2203–2210.
494. Sidhu PS, Enright GD, Ripmeester JA. J Phys Chem B 2002; 106:8569–8581.
495. Prout K. Croatica Chemica Acta 2002; 75:817–833.
496. Kuzmicz R, Dobrzycki L, Wozniak K, Benevelli F, Klinowski J, Kolodziejski W. Physical Chemistry and Chemical Physics 2002; 4:2387–2391.
497. Harris DJ, Alam MK. Polymer 2002; 43:5147–5155.
498. Mao J-D, Hundal LS, Thompson ML, Schmidt-Rohr K. Environ Sci Technol 2002; 36:929–936.
499. Kitchin SJ, Ahn S, Harris KDM. J Phys Chem A 2002; 106:7228–7234.
500. Leisen J, Beckham HW, Benham M. Solid State Nucl Magn Reson 2002; 22:409–422.
501. Witkowski S, Wawer I. J Chem Soc, Perkin Trans 2 2002; 433–436.
502. Harris DJ, Azevedo ERd, Bonagambab TJ. J Magn Reson 2003; 162:67–73.
503. Lim AR, Kwark Y-J, Kim J-S. J Appl Phys 2003; 94:7351–7355.
504. Krushelnitsky A, Faizullin D, Reichert D. Biopolymers 2004; 73:1–15.
505. Beckmann PA, Paty C, Allocco E, Herd M, Kuranz C, Rheingold AL. J Chem Phys 2004; 120:5309–5314.
506. Masuda K, Tabata S, Sakata Y, Hayase T, Yonemochi E, Terada K. Pharm Res 2005; 22:797–805.
507. Lubach JW, Xu D, Segmuller B, Munson EJ. In Press 2005.
508. Markovich RJ, Evans CA, Coscolluela CB, Zibas SA, Rosen J. J Pharm Biomed Anal 1997; 16:661–673.
509. Watanabe T, Wakiyama N, Usui F, Ikeda M, Isobe T, Senna M. Int J Pharm 2001; 226:81–91.
510. Forster A, Apperley D, Hempenstall J, Lancaster R, Rades T. Pharmazie 2003; 58:761–762.
511. Kalachandra S, Lin DM, Stejskal EO, Prakki A, Offenbacher S. J Mater Sci: Mater Med 2005; 16:597–605.
512. Robyr P. Polym Mater Sci Eng 2000; 82:135.
513. Hu WG, Schmidt-Rohr K. Polymer 2000; 41:2979–2987.
514. Colhoun FL, Armstrong RC, Rutledge GC. Polym Mater Sci Eng 2000; 82:136–137.
515. Asano A, Takegoshi K. J Chem Phys 2001; 115:8665–8669.
516. Valtier M, Drujon X, Wilhelm M, Spiess HW. Macromolecules 2001; 202:1262–1272.
517. Palmas P, Campion LL, Bourgeoisat C, Martel L. Polymer 2001; 42:7675–7683.
518. Lin R-H, Woo EM, Chiang JC. Polymer 2001; 42:4289–4297.
519. Huster D, Yao X, Hong M. J Am Chem Soc 2002; 124:874–883.
520. Brus J, Dybal J, Sysel P, Hobzova R. Macromolecules 2002; 35:1253–1261.
521. Kretschmer A, Drake R, Neidhoefer M, Wilhelm M. Solid State Nucl Magn Reson 2002; 22:204–217.
522. Nozirov F. Solid State Nucl Magn Reson 2002; 21:197–203.

523. Murata K, Kuroki S, Kimura H, Ando I. Biopolymers 2002; 64:26–33.
524. Murata K, Kuroki S, Ando I. Polymer 2002; 43:6871–6878.
525. Jia X, Wolak J, Wang X, White JL. Macromolecules 2003; 36:712–718.
526. Qiu X, Ediger MD. J Phys Chem B 2003; 107:459–464.
527. Munson EJ, Carlson LK, Jorving JE, Zelt MT, Abbott J, Hillmyer MA. Polym Prepr 2003; 44:325–326.
528. Werkhoven TM, Mulder FM, Zune C, Jerome R, M.deGroot HJ. Macromol Chem Phys 2003; 204:46–51.
529. Akita C, Kawaguchi T, Kaneko F, Yamamoto H, Suzuki M. J Phys Chem B 2004; 108:4862–4868.
530. Rohrs BR, Thamann TJ, Gao P, Stelzer DJ, Berger MS, Chao RS. Pharm Res 1999; 16:1850–1856.
531. Puttipipatkhachorn S, Nunthanid J, Yamamoto K, Peck GE. J Controlled Release 75:143–153.
532. Watanabe T, Hasegawa S, Wakiyama N, Kusai A, Senna M. Int J Pharm 2002; 248:123–129.
533. Babonneau F, Camus L, Steunou N, Ramila A, Vallet-Regi M. Mater Res Soc Symp Proc 2003; 775:77–82.
534. Boland MP, Middleton DA. Magn Reson Chem 2004; 42:204–211.
535. Kawada J, Marchessault RH. Starch/Staerke 2004; 56:13–19.
536. Smits ALM, Kruiskamp PH, Soest JJGv, Vliegenthart JFG. Carbohydr Polym 2003; 53:409–416.
537. Kozerski L, Kamienski B, Kawecki R, Urbanczyk-Lipkowska Z, Bocian W, Bednarek E, Sitkowski J, Zakrzewska K, Nielsen KT, Hansen PE. Org Biomol Chem 2003; 1:3578–3585.
538. Zech SG, Olejniczak E, Hajduk P, Mack J, McDermott AE. J Am Chem Soc 2004; 126:13948–13953.
539. Takegoshi K, Nakamura S, Terao T. Chem Phys Lett 344:631–637.
540. Lubach JW, Padden BE, Winslow SL, Salsbury JS, Masters DB, Topp EM, Munson EJ. Anal Bioanal Chem 2004; 378:1504–1510.
541. Schachter DM, Xiong J, Tirol GC. Int J Pharm 2004; 281:89–101.
542. Gjerde A, Holmsen H, Nerdal W. Biochimica et biophysica acta 2004; 1682: 28–37.
543. Apperley DC, Basford PA, Dallman CI, Harris RK, Kinns M, Marshall PV, Swanson AG. J Pharm Sci 2005; 94(3):516–523.
544. Bogdanova S, Pajeva I, Nikolova P, Tsakovska I, Mueller B. Pharm Res 2005; 22:806–815.
545. Schneider H-J, Hacket F, Rudiger V. Chem Rev 1998; 98:1755–1785.
546. Kitchin SJ, Halstead TK. Appl Magn Reson 1999; 17:283–300.
547. Crini, Bourdonneau, Martel, Piotto, Morcellet, Richert, Vebrel, Torri, Morrin. J Appl Polym Sci 2000; 75:1288–1295.
548. Garbow JR, Likos JJ, Schroeder SAJ. Agric Food Chem 2001; 49:2053–2060.
549. Saalwachter K. Macromol Rapid Commun 2002; 23:286–291.
550. Wulff M, Aden M, Egenfeldt J. Bioconjugate Chem 2002; 13:240–248.
551. Lu J, Mirau PA, Shin ID, Nojima S, Tonelli AE. Macromol Chem Phys 2002; 203:71–79.
552. Yannakopoulou K, Ripmeester JA, Mavridis IM. J Chem Soc, Perkin Trans 2 2002; 1639–1644.

553. Cunha-Silva L, Teixeira-Dias JJC. J Phys Chem B 2002; 106:3323–3328.

554. Braga SS, Aree T, Imura K, Vertut P, Boal-Palheiros I, Saenger W, Teixeira-Dias JJC. J Inclusion Phenomena and Macrocyclic Chemistry 2002; 43:115–125.

555. Braga SS, Gonçalves IS, Herdtweck E, Teixeira-Dias JJC. New J Chem 2003; 27:597–601.

556. Braga SS, Ribeiro-Claro P, Pillinger M, Goncalves IS, Pereira F, Fenandes AC, Romao CC, Correia PB, Teixeira-Dias JJC. Org Biomol Chem 2003; 1:873–878.

557. Cunha-Silva L, Teixeira-Dias JJC. New J Chem 2004; 28:200–206.

558. Lai S, Locci E, Piras A, Porcedda S, Lai A, Marongiu B. Carbohydr Res 2003; 338:2227–2232.

559. Anulewicz R, Wawer I, Piekarska-Bartoszewicz B, Temeriusz A. Carbohydr Res 1996; 281:1–10.

560. Temeriusz A, Piekarska-Bartoszewicz B, Wawer I. Carbohydr Res 1997; 304:335–340.

561. Tanner SF, Chanzy H, Vincendon M, Roux JC, Gaill F. Macromolecules 1990; 23:3576–3583.

562. Kunze J, Ebert A, Fink HP. Cellul Chem Technol 2000; 34:21–34.

563. Lefort R, De Gusseme A, Willart JF, Danede F, Descamps M. Int J Pharm 2004; 280:209–219.

564. Elena B, Emsley L. J Am Chem Soc 2005; 127:9140–9146.

565. Tong G, Schaefer J. Macromolecules 1997; 30:7522–7528.

566. Fyfe CA, Lewis AR, Chezeau J-M. Canadian J Chem 1999; 77:1984–1993.

567. Fyfe CA, Skibsted J, Grodney H, Altenschildesche HMz. Chem Phys Lett 1997; 281:44–48.

568. Bertani P, Raya J, Hirschinger J. Solid State Nucl Magn Reson 2002; 22: 188–203.

569. Middleton DA, Duff CSL, Peng X, Reid DG, Saunders D. J Am Chem Soc 2000; 122:1161–1170.

570. Salgado J, Grage SL, Kondejewski LH, Hodges RS, Mcelhaneyh RN, Ulrich AS. J Biomol NMR 2001; 21:191–208.

571. Middleton DA, Peng X, Saunders D, Shankland K, David WIF, Markvardsen A. J Chem Commun (Cambridge, United Kingdom) 2002; 17:1976–1977.

572. Mehta AK, Cegelski L, O'Connor RD, Schaefer J. J Magn Reson 2003; 163:182–187.

573. Grage SL, Watts JA, Watts A. J Magn Reson 2004; 166:1–10.

574. Schulz-Dobrick M, Metzroth T, Spiess HW, Gauss J, Schnell I. Chem Phys Chem 2005; 6:315–327.

575. Limbach H-H. Intermolecular Forces. Berlin Heidelberg: Springer-Verlag, 1991:280–295.

576. Hickman BS, Mascal M, Titman JJ, Wood IG. J Am Chem Soc 1999; 121:11486–11490.

577. Wu G, Kazuhiko Y, Dong S, Grondey H. J Am Chem Soc 2000; 122: 4215–4216.

578. Song X-J, McDermott AE. Magn Reson Chem 2001; 39:S37–S43.

579. Zhao X, Sudmeier JL, Bachovchin WW, Levitt MH. J Am Chem Soc 2001; 123:11097–11098.

580. Kiss JT, Felfodi K, Hannus I, Palinko I. J Mol Struct 2001; 565–566:463–468.

581. Huelsekopf M, Ludwig R. Magn Reson Chem 2001; 39:S127–S134.

582. Henry B, Tekely P, Delpuech J-J. J Am Chem Soc 2002; 124:2025–2034.
583. Foces-Foces C, Echevarria A, Jagerovic N, Alkorta I, Elguero J, Langer U, Klein O, Minguet-Bonvehi M, Limbach H-H. J Am Chem Soc 2001; 123:7989–7906.
584. Totz J, Michel D, Ivanov YN, Sukhovsky AA, Aleksandrova IP, Petersson J. Magn Reson Chem 2001; 39:S50–S58.
585. Michel D, Totz J, Ivanov YN, Sukhovsky AA, Aleksandrova IP, Petersson J. Ferroelectrics 2002; 267:303–310.
586. Takeda S, Tsuzumitani A. Magn Reson Chem 2001; 39:S44–S49.
587. Lorente P, Shenderovich IG, Golubev NS, Denisov GS, Buntkowsky G, Limbach H-H. Magn Reson Chem 2001; 39:S18–S29.
588. Goward GR, Schnell I, Brown SP, Spiess HW, Kim H-D, Ishida H. Magn Reson Chem 2001; 39:S1–S17.
589. Ono S, Taguma T, Kuroki S, Ando I, Kimura H, Yamauchi K. J Mol Struct 2002; 602–603:49–58.
590. Colsenet R, Gardiennet C, Henry B, Tekely P. Angew Chem Int Ed 2002; 41:4743–4745.
591. Brown SP, Perez-Torralba M, Sanz D, Claramunt RM, Emsley L. Chem Commun (Cambridge, United Kingdom) 2002; 1852–1853.
592. Lesage A, Emsley L, Chabanas M, Coperet C, Basset J-M. Angew Chem Int Ed 2002; 41:4535–4538.
593. Ilczyszyn MM, Ilczyszyn M. J Raman Spectrosc 2003; 34:693–704.
594. Godzisz D, Ilczyszyn MM, Ilczyszyn M. J Mol Struct 2002; 606:123–137.
595. Pawsey S, Mccormick M, Paul SD, Graf R, Lee YS, Reven L, Spiess HW. J Am Chem Soc 2003; 125:4174–4184.
596. Emmler T, Gieschler S, Limbach HH, Buntkowsky G. J Mol Struct 2004; 700:29–38.
597. Leppert J, Urbinati CR, Haefner S, Ohlenschlaeger O, Swanson MS, Goerlach M, Ramachandran R. Nucleic Acids Res 2004; 32:1177–1183.
598. Schilf W, Kamienski B, Szady-Chelmieniecka A, Grech E. J Mol Struct 2004; 700:105–108.
599. Schilf W, Kamienski B, Kolodziej B, Grech E. J Mol Struct 2004; 708:33–38.
600. Rozwadowski Z, Schilf W, Kamienski B. Magn Reson Chem 2005; 43:573–577.
601. Schilf W, Kamienski B, Szady-Chelmieniecka A, Grech E. J Mol Struct 2005; 743:237–241.
602. Szady-Chelmieniecka A, Grech E, Rozwadowski Z, Dziembowska T, Schilf W, Kamienski B. J Mol Struct 2001; 565–566:125–128.
603. Schilf W, Kamienski B, Dziembowska T, Rozwadowski Z, Szady-Chelmieniecka A. J Mol Struct 2000; 552:33–37.
604. Schilf W, Kamienski B, Szady-Chelmieniecka A, Grech E. Solid State Nucl Magn Reson 2000; 18:97–105.
605. Li J, Chatterjee K, Medek A, Shalaev E, Zografi G. J Pharm Sci 2004; 93:697–712.
606. Chekmenev EY, Zhang Q, Waddell KW, Mashuta MS, Wittebort RJ. J Am Chem Soc 2004; 126:379–384.
607. Chekmenev EY, Xu RZ, Mashatu MS, Wittenbort RJ. J Am Chem Soc 2002; 124:11894–11899.
608. Waddell KW, Chekmenev EY, Wittebort RJ. J Am Chem Soc 2005; 127: 9030–9035.

609. Diez-Pena E, Quijada-Garrido I, Barrales-Rienda JM, Schnell I, Spiess HW. Macromol Chem Phys 2004; 205:430–437.

610. Kawashima H. Energy & Fuels 2005; 19:538–543.

611. Benhabbour SR, Chapman RP, Scharfenberger G, Meyer WH, Goward GR. Chem Mater 2005; 17:1605–1612.

612. Gobetto R, Nervi C, Valfre E, Chierotti MR, Braga D, Maini L, Grepioni F, Harris RK, Ghi PY. Chem Mater 2005; 17:1457–1466.

613. Li ZJ, Abramov Y, Bordner J, Leonard J, Medek A, Trask AV. In press 2005.

614. Wawer I, Zielinska A. Magn Reson Chem 2001; 39:374–380.

615. Hughes E, Frossard P, Sagalowicz L, Nouzille A, Raemy A, Watzke H. Spec Publ-R Soc Chem 2002; 286:144–150.

616. Plass J, Emeis D. Magnetic Resonance in Colloid and Interface Science 2002; 375–382.

617. Plass J, Emeis D, Blumich B. Journal of Surfactants and Detergents 2001; 4:379–384.

618. Calucci L, Forte C, Galleschi L, Geppi M, Ghiringhelli S. Int J Biol Macromol 2003; 32:179–189.

Milton Keynes UK
Ingram Content Group UK Ltd.
UKHW020005071024
449327UK00031B/2660

9 780367 390938